INTRODUCTION TO PROBABILITY AND STATISTICS

STATISTICS: Textbooks and Monographs

A Series Edited by

D. B. Owen, Founding Editor, 1972–1991

W. R. Schucany, Coordinating Editor
Department of Statistics
Southern Methodist University
Dallas, Texas

1. The Generalized Jackknife Statistic, *H. L. Gray and W. R. Schucany*
2. Multivariate Analysis, *Anant M. Kshirsagar*
3. Statistics and Society, *Walter T. Federer*
4. Multivariate Analysis: A Selected and Abstracted Bibliography, 1957–1972, *Kocherlakota Subrahmaniam and Kathleen Subrahmaniam*
5. Design of Experiments: A Realistic Approach, *Virgil L. Anderson and Robert A. McLean*
6. Statistical and Mathematical Aspects of Pollution Problems, *John W. Pratt*
7. Introduction to Probability and Statistics (in two parts), Part I: Probability; Part II: Statistics, *Narayan C. Giri*
8. Statistical Theory of the Analysis of Experimental Designs, *J. Ogawa*
9. Statistical Techniques in Simulation (in two parts), *Jack P. C. Kleijnen*
10. Data Quality Control and Editing, *Joseph I. Naus*
11. Cost of Living Index Numbers: Practice, Precision, and Theory, *Kali S. Banerjee*
12. Weighing Designs: For Chemistry, Medicine, Economics, Operations Research, Statistics, *Kali S. Banerjee*
13. The Search for Oil: Some Statistical Methods and Techniques, *edited by D. B. Owen*
14. Sample Size Choice: Charts for Experiments with Linear Models, *Robert E. Odeh and Martin Fox*
15. Statistical Methods for Engineers and Scientists, *Robert M. Bethea, Benjamin S. Duran, and Thomas L. Boullion*
16. Statistical Quality Control Methods, *Irving W. Burr*
17. On the History of Statistics and Probability, *edited by D. B. Owen*
18. Econometrics, *Peter Schmidt*

19. Sufficient Statistics: Selected Contributions, *Vasant S. Huzurbazar (edited by Anant M. Kshirsagar)*
20. Handbook of Statistical Distributions, *Jagdish K. Patel, C. H. Kapadia, and D. B. Owen*
21. Case Studies in Sample Design, *A. C. Rosander*
22. Pocket Book of Statistical Tables, *compiled by R. E. Odeh, D. B. Owen, Z. W. Birnbaum, and L. Fisher*
23. The Information in Contingency Tables, *D. V. Gokhale and Solomon Kullback*
24. Statistical Analysis of Reliability and Life-Testing Models: Theory and Methods, *Lee J. Bain*
25. Elementary Statistical Quality Control, *Irving W. Burr*
26. An Introduction to Probability and Statistics Using BASIC, *Richard A. Groeneveld*
27. Basic Applied Statistics, *B. L. Raktoe and J. J. Hubert*
28. A Primer in Probability, *Kathleen Subrahmaniam*
29. Random Processes: A First Look, *R. Syski*
30. Regression Methods: A Tool for Data Analysis, *Rudolf J. Freund and Paul D. Minton*
31. Randomization Tests, *Eugene S. Edgington*
32. Tables for Normal Tolerance Limits, Sampling Plans and Screening, *Robert E. Odeh and D. B. Owen*
33. Statistical Computing, *William J. Kennedy, Jr., and James E. Gentle*
34. Regression Analysis and Its Application: A Data-Oriented Approach, *Richard F. Gunst and Robert L. Mason*
35. Scientific Strategies to Save Your Life, *I. D. J. Bross*
36. Statistics in the Pharmaceutical Industry, *edited by C. Ralph Buncher and Jia-Yeong Tsay*
37. Sampling from a Finite Population, *J. Hajek*
38. Statistical Modeling Techniques, *S. S. Shapiro*
39. Statistical Theory and Inference in Research, *T. A. Bancroft and C.-P. Han*
40. Handbook of the Normal Distribution, *Jagdish K. Patel and Campbell B. Read*
41. Recent Advances in Regression Methods, *Hrishikesh D. Vinod and Aman Ullah*
42. Acceptance Sampling in Quality Control, *Edward G. Schilling*
43. The Randomized Clinical Trial and Therapeutic Decisions, *edited by Niels Tygstrup, John M Lachin, and Erik Juhl*
44. Regression Analysis of Survival Data in Cancer Chemotherapy, *Walter H. Carter, Jr., Galen L. Wampler, and Donald M. Stablein*
45. A Course in Linear Models, *Anant M. Kshirsagar*
46. Clinical Trials: Issues and Approaches, *edited by Stanley H. Shapiro and Thomas H. Louis*
47. Statistical Analysis of DNA Sequence Data, *edited by B. S. Weir*
48. Nonlinear Regression Modeling: A Unified Practical Approach, *David A. Ratkowsky*
49. Attribute Sampling Plans, Tables of Tests and Confidence Limits for Proportions, *Robert E. Odeh and D. B. Owen*
50. Experimental Design, Statistical Models, and Genetic Statistics, *edited by Klaus Hinkelmann*
51. Statistical Methods for Cancer Studies, *edited by Richard G. Cornell*
52. Practical Statistical Sampling for Auditors, *Arthur J. Wilburn*
53. Statistical Methods for Cancer Studies, *edited by Edward J. Wegman and James G. Smith*

54. Self-Organizing Methods in Modeling: GMDH Type Algorithms, *edited by Stanley J. Farlow*

55. Applied Factorial and Fractional Designs, *Robert A. McLean and Virgil L. Anderson*

56. Design of Experiments: Ranking and Selection, *edited by Thomas J. Santner and Ajit C. Tamhane*

57. Statistical Methods for Engineers and Scientists: Second Edition, Revised and Expanded, *Robert M. Bethea, Benjamin S. Duran, and Thomas L. Boullion*

58. Ensemble Modeling: Inference from Small-Scale Properties to Large-Scale Systems, *Alan E. Gelfand and Crayton C. Walker*

59. Computer Modeling for Business and Industry, *Bruce L. Bowerman and Richard T. O'Connell*

60. Bayesian Analysis of Linear Models, *Lyle D. Broemeling*

61. Methodological Issues for Health Care Surveys, *Brenda Cox and Steven Cohen*

62. Applied Regression Analysis and Experimental Design, *Richard J. Brook and Gregory C. Arnold*

63. Statpal: A Statistical Package for Microcomputers—PC-DOS Version for the IBM PC and Compatibles, *Bruce J. Chalmer and David G. Whitmore*

64. Statpal: A Statistical Package for Microcomputers—Apple Version for the II, II +, and IIe, *David G. Whitmore and Bruce J. Chalmer*

65. Nonparametric Statistical Inference: Second Edition, Revised and Expanded, *Jean Dickinson Gibbons*

66. Design and Analysis of Experiments, *Roger G. Petersen*

67. Statistical Methods for Pharmaceutical Research Planning, *Sten W. Bergman and John C. Gittins*

68. Goodness-of-Fit Techniques, *edited by Ralph B. D'Agostino and Michael A. Stephens*

69. Statistical Methods in Discrimination Litigation, *edited by D. H. Kaye and Mikel Aickin*

70. Truncated and Censored Samples from Normal Populations, *Helmut Schneider*

71. Robust Inference, *M. L. Tiku, W. Y. Tan, and N. Balakrishnan*

72. Statistical Image Processing and Graphics, *edited by Edward J. Wegman and Douglas J. DePriest*

73. Assignment Methods in Combinatorial Data Analysis, *Lawrence J. Hubert*

74. Econometrics and Structural Change, *Lyle D. Broemeling and Hiroki Tsurumi*

75. Multivariate Interpretation of Clinical Laboratory Data, *Adelin Albert and Eugene K. Harris*

76. Statistical Tools for Simulation Practitioners, *Jack P. C. Kleijnen*

77. Randomization Tests: Second Edition, *Eugene S. Edgington*

78. A Folio of Distributions: A Collection of Theoretical Quantile-Quantile Plots, *Edward B. Fowlkes*

79. Applied Categorical Data Analysis, *Daniel H. Freeman, Jr.*

80. Seemingly Unrelated Regression Equations Models: Estimation and Inference, *Virendra K. Srivastava and David E. A. Giles*

81. Response Surfaces: Designs and Analyses, *Andre I. Khuri and John A. Cornell*

82. Nonlinear Parameter Estimation: An Integrated System in BASIC, *John C. Nash and Mary Walker-Smith*

83. Cancer Modeling, *edited by James R. Thompson and Barry W. Brown*

84. Mixture Models: Inference and Applications to Clustering, *Geoffrey J. McLachlan and Kaye E. Basford*

85. Randomized Response: Theory and Techniques, *Arijit Chaudhuri and Rahul Mukerjee*

86. Biopharmaceutical Statistics for Drug Development, *edited by Karl E. Peace*

87. Parts per Million Values for Estimating Quality Levels, *Robert E. Odeh and D. B. Owen*

88. Lognormal Distributions: Theory and Applications, *edited by Edwin L. Crow and Kunio Shimizu*

89. Properties of Estimators for the Gamma Distribution, *K. O. Bowman and L. R. Shenton*

90. Spline Smoothing and Nonparametric Regression, *Randall L. Eubank*

91. Linear Least Squares Computations, *R. W. Farebrother*

92. Exploring Statistics, *Damaraju Raghavarao*

93. Applied Time Series Analysis for Business and Economic Forecasting, *Sufi M. Nazem*

94. Bayesian Analysis of Time Series and Dynamic Models, *edited by James C. Spall*

95. The Inverse Gaussian Distribution: Theory, Methodology, and Applications, *Raj S. Chhikara and J. Leroy Folks*

96. Parameter Estimation in Reliability and Life Span Models, *A. Clifford Cohen and Betty Jones Whitten*

97. Pooled Cross-Sectional and Time Series Data Analysis, *Terry E. Dielman*

98. Random Processes: A First Look, Second Edition, Revised and Expanded, *R. Syski*

99. Generalized Poisson Distributions: Properties and Applications, *P. C. Consul*

100. Nonlinear L_p-Norm Estimation, *Rene Gonin and Arthur H. Money*

101. Model Discrimination for Nonlinear Regression Models, *Dale S. Borowiak*

102. Applied Regression Analysis in Econometrics, *Howard E. Doran*

103. Continued Fractions in Statistical Applications, *K. O. Bowman and L. R. Shenton*

104. Statistical Methodology in the Pharmaceutical Sciences, *Donald A. Berry*

105. Experimental Design in Biotechnology, *Perry D. Haaland*

106. Statistical Issues in Drug Research and Development, *edited by Karl E. Peace*

107. Handbook of Nonlinear Regression Models, *David A. Ratkowsky*

108. Robust Regression: Analysis and Applications, *edited by Kenneth D. Lawrence and Jeffrey L. Arthur*

109. Statistical Design and Analysis of Industrial Experiments, *edited by Subir Ghosh*

110. U-Statistics: Theory and Practice, *A. J. Lee*

111. A Primer in Probability: Second Edition, Revised and Expanded, *Kathleen Subrahmaniam*

112. Data Quality Control: Theory and Pragmatics, *edited by Gunar E. Liepins and V. R. R. Uppuluri*

113. Engineering Quality by Design: Interpreting the Taguchi Approach, *Thomas B. Barker*

114. Survivorship Analysis for Clinical Studies, *Eugene K. Harris and Adelin Albert*

115. Statistical Analysis of Reliability and Life-Testing Models: Second Edition, *Lee J. Bain and Max Engelhardt*

116. Stochastic Models of Carcinogenesis, *Wai-Yuan Tan*

117. Statistics and Society: Data Collection and Interpretation: Second Edition, Revised and Expanded, *Walter T. Federer*

118. Handbook of Sequential Analysis, *B. K. Ghosh and P. K. Sen*

119. Truncated and Censored Samples: Theory and Applications, *A. Clifford Cohen*

120. Survey Sampling Principles, *E. K. Foreman*

121. Applied Engineering Statistics, *Robert M. Bethea and R. Russell Rhinehart*
122. Sample Size Choice: Charts for Experiments with Linear Models: Second Edition, *Robert E. Odeh and Martin Fox*
123. Handbook of the Logistic Distribution, *edited by N. Balakrishnan*
124. Fundamentals of Biostatistical Inference, *Chap T. Le*
125. Correspondence Analysis Handbook, *J.-P. Benzécri*
126. Quadratic Forms in Random Variables: Theory and Applications, *A. M. Mathai and Serge B. Provost*
127. Confidence Intervals on Variance Components, *Richard K. Burdick and Franklin A. Graybill*
128. Biopharmaceutical Sequential Statistical Applications, *edited by Karl E. Peace*
129. Item Response Theory: Parameter Estimation Techniques, *Frank B. Baker*
130. Survey Sampling: Theory and Methods, *Arijit Chaudhuri and Horst Stenger*
131. Nonparametric Statistical Inference: Third Edition, Revised and Expanded, *Jean Dickinson Gibbons and Subhabrata Chakraborti*
132. Bivariate Discrete Distribution, *Subrahmaniam Kocherlakota and Kathleen Kocherlakota*
133. Design and Analysis of Bioavailability and Bioequivalence Studies, *Shein-Chung Chow and Jen-pei Liu*
134. Multiple Comparisons, Selection, and Applications in Biometry, *edited by Fred M. Hoppe*
135. Cross-Over Experiments: Design, Analysis, and Application, *David A. Ratkowsky, Marc A. Evans, and J. Richard Alldredge*
136. Introduction to Probability and Statistics: Second Edition, Revised and Expanded, *Narayan C. Giri*

Additional Volumes in Preparation

INTRODUCTION TO PROBABILITY AND STATISTICS

Second Edition, Revised and Expanded

Narayan C. Giri

Department of Mathematics
University of Montreal
Montreal, Quebec, Canada

Marcel Dekker, Inc.   New York • Basel • Hong Kong

Library of Congress Cataloging-in-Publication Data

Giri, Narayan C.
    Introduction to probability and statistics / Narayan C. Giri. --
2nd ed., rev. and expanded.
        p.   cm. -- (Statistics, textbooks and monographs ; v. 136)
    Includes bibliographical references and index.
    ISBN 0-8247-9037-5 (alk. paper)
    1. probabilities. 2. Mathematical statistics. I. Title. II. Series.
QA273.G526   1993
519.5--dc20   93-12127
                                                                    CIP

The publisher offers discounts on this book when ordered in bulk quantities.
For more information, write to Special Sales/Professional Marketing at the
address below.

This book is printed on acid-free paper.

Marcel Dekker, Inc.
270 Madison Avenue, New York, New York 10016

Current printing (last digit):
10 9 8 7 6 5 4 3 2 1

PRINTED IN THE UNITED STATES OF AMERICA

To my mother
and the memories of
my father and mother-in-law

Preface to the Second Edition

This is a revised and expanded edition of the original publication *Introduction to Probability and Statistics: Part 1, Probability, Part II, Statistics*. Nearly eighteen years have passed since the first edition of the book was published. During that time, undergraduate courses have changed greatly. Many first-year graduate topics are now included in undergraduate courses. This new edition attempts to bring the original book up to date by thorough revision, rewriting, addition of four new chapters, and addition of new problems and materials in each chapter. In preparing this volume I have tried to incorporate various comments by reviewers of the original book and by colleagues who have used it as a textbook and made numerous corrections to the original version. The comments of my own students and my long experience in teaching the course have also been utilized in the revision. This book is designed for a two-semester course in statistics, but can also be used for a one-semester course by suitable choice of chapters.

Narayan C. Giri

Preface to the First Edition

PART I, PROBABILITY

This book is intended to provide an introductory understanding of the theory of probability to students at the undergraduate level. Careful consideration has been given, in writing this text, to the needs of students with a special interest in statistics. It is hoped that the book will match the contents of courses on this subject for beginning and advanced undergraduates in most United States and Canadian universities. The basic definitions and ideas of the subject are covered, and an effort has been made to familiarize the student with fundamental results but not much other acquaintance with advanced mathematics will be found necessary. A thorough knowledge of calculus is assumed on the part of the reader. Some important results of matrix algebra have been included in the Appendix to supplement the readers' knowledge in this area and to ensure a better understanding of the probabilistic treatments of problems involving more than a single variable.

Chapter 1 provides a little historical background of the theory of probability and discusses the importance of the various concepts in actual models in natural sciences from a logical point of view. General concepts, including both classical and axiomatic definitions and the usual basic theorems have been included in Chapter 2. Chapter 3 is actually a translation of some of the concepts introduced in the earlier chapters into the characteristics of actual models among natural phenomena. Further results and systematic study of these characteristics are included in the last two chapters.

In keeping with the nature of an introductory text, many examples and motivations relevant to specific topics have been included. Exercises are given at the end of each section to make the mathematical treatment of the preceding theory more comprehensible.

I feel that it would be appropriate to spread the material discussed in this book over two three-hour, one-semester courses, However, the first three chapters may be used in a three-hour, one-semester course in elementary probability theory; but Section 2.2 and parts of Chapter 3 should be avoided if such a course is intended particularly for students with a narrower mathematical background than juniors in mathematics.

If the reader finds this book useful, the credit is entirely due to my own teachers and colleagues, in particular Professors C. M. Stein, S. Karlin, and E. Parzen of Stanford University, Professor J. Kiefer of Cornell University, Professors H. K. Nandi and P. K. Bose of Calcutta University and Dr. A. K. Gayen of I. I. T. Kharagpur, India, under whose influence I have come to appreciate the statistics of the present century.

The preparation and revision of the manuscript of this book would not have been an easy task without the help of Professor A. R. Roy and Dr. S. K. Basu, who helped me by reading the entire manuscript with great care and diligence and offering valuable suggestions at various stages. The comments of Professor Y. Lepage on Chapters 4 and 5 have been extremely helpful. I wish to express my gratitude and special thanks to all of them.

I have presented the materials in parts in different courses in the Department of Mathematics, University of Montreal. The comments of the students have helped me to improve the presentation. I express my thanks to all of them.

My wife Nilima, daughter Nabanita, and son Nandan have been very helpful and patient during the preparation of the book. I gratefully acknowledge their assistance. The assistance of my parents and my elder brothers is also gratefully acknowledged.

I would like to express my sincere thanks to the National Research Council of Canada and the Ministry of Education, Government of Quebec, for financial assistance during the preparation of the manuscript. Finally, I would like to express my gratitude to the secretaries of the Department of Mathematics, University of Montreal, for an excellent job in typing the manuscript. Special thanks are due to the editors of Marcel Dekker, Inc., for putting the original manuscript into its final form.

PART II, STATISTICS

This book is intended to be a systematic presentation of the introductory theory of some topics of mathematical statistics. The discussions have been specially tailored to match, in general, the contents of regular and advanced undergraduate courses in most U. S. and Canadian universities. Thus a knowledge of undergraduate mathematics, particularly of calculus, has been assumed for appreciating the mathematical treatments in the book; some important results of the matrix algebra have been included in Appendix A to supplement the readers' knowledge in this area.

Chapter 1 provides a tour of the various types of statistical problems and then discusses the broad nature of statistical inference. Data reduction techniques taking advantage of sufficiency are also included in this chapter. Probabilistic tools for the study of order statistics have been developed in Chapter 2. Different methods of parametric fixed sample estimation and testing of hypotheses have been discussed in detail in Chapters 3 and 4, respectively. Chapter 5 is concerned with sequential procedures of hypotheses testing, while Chapter 6 deals with the nonparametric methods for this aspect of statistical inference. A systematic study of the general linear hypothesis and analysis of variance has been presented in Chapter 7.

To maintain the general nature of introductory textbooks, we have tried to include many examples and motivations relevant to specific topics. Many exercises are also given at the end of each chapter to make the mathematical treatments of the preceding theory more comprehensible.

We feel that it will be appropriate to spread the materials of the book over two three-hour, one-semester basic courses in statistical inference for undergraduates. The first semester course may take care of Chapters 1, 2, 3, and part of Chapter 4, while the remaining portions of the book may be covered in the second semester.

If the readers find this book useful, the credit is entirely due to my teachers and colleagues like Prof. C. M. Stein, Prof. S. Karlin, Prof. E. Parzen of Stanford University, Prof. J. Kiefer of Cornell University, and Prof. H. K. Nandi and Prof. P. K. Bose of Calcutta University, under whose influence I have come to appreciate the statistics of the present century.

The preparation and the revision of the manuscript would not have been an easy task without the help of Dr. S. K. Basu who helped me by reading the entire manuscript with great care and diligence and offering valuable suggestions at various stages. I would like to express my gratitude and thanks to him. I have presented the materials in the book in parts in different courses in the Department of Mathematics, University of Montreal. The comments of the students were very useful for improving the presentation. My wife Nilima, daughter Nabanita, and son Nandan have also been very helpful and patient during the preparation of this book. I gratefully acknowledge their assistance. The assistance of my mother and brothers is also gratefully acknowledged.

I would like to thank John Wiley and Sons, Inc., for their permission to reproduce Tables 1B, 4B, and 5B from their publications and the editors of

Biometrika for their permission to reproduce Table 6B. I would also like to express my sincere thanks to the National Research Council of Canada and the Ministry of Education, Government of Quebec, for financial assistance for the preparation of the manuscript. Finally, I would like to express my gratitude to the secretaries of the Department of Mathematics, University of Montreal, for an excellent job in typing the manuscript.

Narayan C. Giri

Contents

Preface to the Second Edition   v

Preface to the First Edition   vii

1. INTRODUCTION   1

     1.1   Stochastic Model of Natural Phenomena   1
     1.2   Scientific Methodology and Statistics   4
         Bibliography   8

2. GENERAL CONCEPTS OF PROBABILITY   9

     2.1   Classical Definition of Probability   9
     2.2   Axiomatic Definition of Probability   34
     2.3   Bayes' Theorem   52
         Bibliography   54

3. RANDOM VARIABLES, PROBABILITY DISTRIBUTIONS, AND CHARACTERISTIC FUNCTIONS   55

     3.1   Random Variables   55
     3.2   Discrete Univariate Random Variables   58

3.3	Continuous Univariate Random Variables	61
3.4	Multidimensional Random Variables	70
3.5	Marginal Distributions of a Bivariate Random Variable	75
3.6	Conditional Distributions of a Bivariate Random Variable	78
3.7	Independence of Random Variables	81
3.8	Functions of a Random Variable	87
3.9	Functions of Several Random Variables	91
3.10	Distribution of Product and Ratio of Two Random Variables	96
3.11	Mathematical Expectation	99
3.12	**Moments**	**106**
3.13	**Probability-Generating Functions**	**114**
3.14	Mathematical Expectation of a Function of a Bivariate Random Variable	115
3.15	Conditional Mathematical Expectation	129
3.16	Two Important Inequalities	131
	Bibliography	135

4. STOCHASTIC CONVERGENCE AND LIMIT THEOREMS — **136**

4.1	**Two Types of Convergence**	**138**
4.2	Limit Theorems	155
	Bibliography	167

5. CONCEPTS OF STATISTICS — 169

5.1	Inductive Inference	169
5.2	The Nature of a Statistical Problem	170
5.3	The Nature of Statistical Inference	174
5.4	Transitions in the History of Statistical Methodology	178
5.5	Data Reduction and Sufficiency	180
5.6	The Exponential Family of Distributions	187
	Bibliography	191

6. UNIVARIATE DISTRIBUTIONS — 192

6.1	Standard Univariate Distributions	193
6.2	Sampling Distribution	204
	Bibliography	210

7. MULTIVARIATE DISTRIBUTIONS 211

7.1 Properties of Multivariate Distributions 211
7.2 Bivariate Normal Distribution 215
7.3 Multivariate Normal Distribution 219
7.4 Elliptically Symmetric Distributions 229
7.5 Multinomial Distribution 235
7.6 Distribution of Quadratic Forms 237
7.7 Multivariate Normal Case: Distribution of Sample
Mean and Sample Covariance Matrices 240
Bibliography 245

8. ORDER STATISTICS AND RELATED DISTRIBUTIONS 246

8.1 Order Parameters 246
8.2 Order Statistics 247
8.3 Some Related Distributions 251

9. STATISTICAL INFERENCE: PARAMETRIC POINT
ESTIMATION 259

9.1 Criteria for Judging Estimators 260
9.2 Completeness and the Best Unbiased Estimator 268
9.3 Most Efficient Estimator and Consistent Estimator 275
9.4 Various Methods of Estimation 285
9.5 Estimation of Parameters in $N_p(\mu,\Sigma)$ 297
9.6 Estimation of Parameters in Multinomial Population 310
Bibliography 312

10. TESTING OF STATISTICAL HYPOTHESES 313

10.1 Basic Definitions 314
10.2 Tests of Simple H_0 Against Simple H_1 317
10.3 Tests of Simple H_0 Against Composite H_1 328
10.4 Tests of Composite H_0 Against Composite H_1
(One-Parameter Case) 331
10.5 Unbiased Tests 335
10.6 Tests of Composite H_0 Against Composite H_1
(Multiparameter Case): Likelihood-Ratio Tests 342
10.7 Test of Zero Correlation 355
10.8 Confidence Intervals 358
10.9 Tests and Confidence Interval for Mean in $N_p(\mu,\Sigma)$ 361
Bibliography 369

11. LARGE-SAMPLE METHODS   370

    11.1   Introduction   370
    11.2   Edgeworth Approximation   373
    11.3   Variance Stabilizing Transformations   374
    11.4   Tests of Goodness of Fit   375
    11.5   Test of Independence in a Contingency Table   377
           Bibliography   380

12. STATISTICAL DECISION THEORY   381

    12.1   Basic Concepts   381
    12.2   Admissible Decision Rules   383
    12.3   Bayes' Decision Rule   384
    12.4   Minimax Decision Rule   390
    12.5   Admissibility of Bayes' Rules   396
           Bibliography   400

13. SEQUENTIAL ANALYSIS   401

    13.1   Sequential Probability Ratio Tests   402
    13.2   Fundamental Relationship Between A, B and the
           Error Probabilities α, β 405
    13.3   Properties of the Stopping Rule N in the SPRT   407
    13.4   Operating Characteristic Function   412
    13.5   Average Sampling Number of the SPRT   415
           Bibliography   417

14. NONPARAMETRIC METHODS   418

    14.1   One-Sample Methods   418
    14.2   Two-Sample Methods   420
    14.3   Rank Test for the One-Way Classification   428
           Bibliography   431

15. GENERAL LINEAR HYPOTHESIS AND ANALYSIS OF
    VARIANCE   432

    15.1   Least Squares Estimates of β 433
    15.2   Maximum-Likelihood Estimates of β and σ^2 437
    15.3   Properties of the Maximum-Likelihood Estimators $\hat{\beta}$
           and $\hat{\sigma}^2$ 439
    15.4   Test of Hypotheses (Analysis of Variance)   445
    15.5   Multiple Comparison: The Scheffé Method   457
           Bibliography   463

Contents

16. SOME APPLICATIONS OF ANALYSIS OF VARIANCE 464

    16.1 Randomized Block Design 465
    16.2 Latin Square Design 467
    Bibliography 474

APPENDIX A. VECTORS AND MATRICES 475

    A.1 Vectors 475
    A.2 Matrices 484
    Bibliography 517

APPENDIX B. STATISTICAL TABLES 518

    Table 1B Binomial Distribution Function 518
    Table 2B Poisson Distribution Function 524
    Table 3B Unit Normal Distribution 526
    Table 4B χ^2 Distribution 527
    Table 5B Student's t Distribution 528
    Table 6B F Distribution 529

Index 531

1

Introduction

1.1 STOCHASTIC MODEL OF NATURAL PHENOMENA

During the middle of the seventeenth century, games of chance were a popular pastime in fashionable society in France. Members of the French nobility spent their spare hours on a great variety of such games, searching for a rational method of playing them to their advantage. It was late in the seventeenth century that a French nobleman, Chevalier de Méré, who was an enterprising gambler, noticed the advantage of betting on the occurrence of a double six in a game of chance consisting of 25 throws of a pair of dice. He amassed a considerable fortune playing on this bet, but failing to comprehend why it worked, he approached the great French mathematician Pascal in search of an explanation. Pascal's answer was as follows: Given that the dice are fair, there is a greater probability of obtaining double six at least once than not to obtain it at all in 25 throws of a pair of dice. After this, Pascal and other mathematicians of his time concentrated their interest on solving problems related to various games of chance known or referred to them. In doing this they treated each game separately and did not aim at devising any unified theory applicable to these games. It was only during the beginning of the eighteenth century that such a theory began to take shape as a result of the vigorous efforts of the Swiss mathematician James Bernoulli and the French mathematician de Moivre. Since then the potentiality of the theory has been well recognized, and its growth has been maintained through the efforts of front-ranking mathematicians the world over. Today the study of probability is proceeding at a phenomenal rate, and the theory has proven to be an indispensable tool for general scientific inference.

The formulation and growth of the theory of probability during the eighteenth and nineteenth centuries brought about a sharp and important change in the basic premises of scientific thinking. Scientific investigators during this period began to note a close analogy between the laws of uncertainty governing the outcome of games of chance and the laws of variation observed by them in apparently uncontrolled phenomena in their own fields of study. This led many of them to believe that a stochastic or probabilistic approach could explain the variability of observations in fields of scientific enquiry where such variations were unavoidable. Thus geneticists found a close parallel between the sex distribution of newborn babies and the distribution of "heads" and "tails" as outcomes of repeated tosses of a coin ("heads" corresponding, say, to male births and "tails" to female births). They then hypothesized that nature chose the sex of a newborn in pretty much the same way as she chose the sides of a tossed coin—that is, by chance—and thus that the laws governing the outcome of the tossing of a coin should be able to explain the laws governing the sex of a newborn. A probabilistic or stochastic model yielded a realistic explanation of a genetic phenomenon.

The use of a stochastic model in explaining scientific phenomena was new to the scientific thinking of those days, which had been dominated by rapid and significant progress in the physical sciences, where the rigor of absolute certainty in mathematical reasoning was the accepted ideal model. This was the so-called deterministic model of the universe. Because of the brilliance of their research in the physical sciences, Newton and his supporters and successors held the reins of scientific thinking of the day, and they firmly believed that nature acted strictly according to causality and was basically deterministic. That is, once all causal variables relevant to any phenomenon were identified and measured, the phenomenon could be predicted exactly by deduction from a deterministic model that was supposed to describe in exact terms the nature of dependence of the observed phenomenon on the causal factors affecting it. To these scientists, if any phenomenon gave rise to observations that were variable in character, this variability, rather than being an integral component of the phenomenon, was due primarily to the experimenter's inability to control some of the causal variables. They felt that in the study of natural phenomena in which such variation was apparently unavoidable, efforts should be directed toward isolating and controlling those causal variables that contributed to such variation, in order to reduce the variability and thus improve predictability on a deterministic basis. In Newton's time, scientific research was directed mainly toward discovering natural laws involving macro units, and a deterministic model of the phenomena they studied offered a workable basis for research. Thus in the study of physics, chemistry, and the engineering sciences during this time such models proved very effective and led to the explosion of scientific knowledge that marked the nineteenth century.

It was, however, apparent, even as early as in the eighteenth century, that no matter how strongly one believed in the deterministic model one could not use it beyond certain limits. A stochastic model was clearly needed as a realistic basis for explaining natural phenomena characterized by inherent variability. This was particularly true of the social and biological sciences, in which such variability was found to be more natural than was apparent in the physical sciences. However, much control one might exercise on the genetic processes involved in giving birth, the sex of the newborn was always unpredictable; but in a large group of births, the outcome was very much like the outcome of the flip of a coin. It was the growing complexity of the physical sciences (in microphysical studies, for example), and later in biological and social sciences (microbiological and behavioral studies), that finally demonstrated the inadequacy of deterministic-mathematical models to explain observed facts characterized by inherent variability, and led to the gradual replacement of such models by a stochastic approach. This replacement was due to the conscious realization that under a fixed set of circumstances (causal factors) apparently relevant to the study of a phenomenon, there may be more than one observed outcome; the phenomenon thus exhibits a random character. The situation is then comparable to an experiment with several outcomes, such as the tossing of a coin, for which the stochastic approach has long been accepted as appropriate; thus the adoption of a stochastic-probabilistic model for any such situation is inescapable. In fact, at present the realization has firmly taken root that any scientific investigation should take a stochastic-probabilistic approach. To offer a physical reason in support of this it would be enough to mention that even granting that nature is deterministic, it would very often be impossible to list and measure all the causal variables contributing to a phenomenon, and even if all these variables were known and measurable, the relationships among them may be so complicated that the resulting mathematical model would be dreadfully confusing to manipulate. Generally, an observed natural phenomenon will be the result of actions and interactions of a multiplicity of causal factors, some of which may not be known, or may be deliberately ignored, in the interest of constructing a simpler model that would explain the phenomenon nearly as well. It is those factors that have been eliminated that impart the stochastic character to the model.

A few illustrations will bring out how widely stochastic models are being used today in various branches of scientific investigation. In genetics, the growth of a population depends on the uncertainty element attached to the births and deaths of the units of the population. In actuarial study, the tabulation of policies is dependent on the chance of survival of an individual person over a length of time. In physics, the theoreticians have developed the concept of elementary particles moving, colliding, and sometimes splitting according to some laws of chance. And in the behavioral sciences, psychologists analyze apparently variable behavior patterns in particular species by treating these variations as random, obeying certain statistical laws.

In the discussion above we describe nature as a gambler, playing with the universe, and suggest why and how stochastic models play a fundamental role in current models of reality. Although this viewpoint is widely accepted at present, it took the scientific world rather a long time to adopt it.

1.2 SCIENTIFIC METHODOLOGY AND STATISTICS

Science is concerned primarily with studying natural phenomena and formulating general laws to explain them. The components of general scientific methodology can be classified as follows:

1. Formulation of a hypothesis
2. Experimentation and collection of relevant data with a view to verifying the hypothesis
3. Interpretation of the data so collected as evidence for or against the hypothesis

The object of formulating a hypothesis is to describe a particular natural phenomenon in a logical way. The hypothesis becomes a law if it can logically justify a natural phenomenon. However, before a scientist goes about formulating a hypothesis he sets up a model, which is an abstraction and simplification of the phenomenon it is supposed to describe. Certain basic aspects of the phenomenon are isolated as being of primary importance, and an analogy is drawn between these aspects and some logical structures about which the scientist has detailed or partial knowledge. This procedure gives a model the phenomenon. A *model*, whether deterministic or stochastic, has a mathematical structure involving a set of parameters capable of taking different values under different circumstances. Any particular phenomenon under study is a specific form of the general phenomenon described by the model and is characterized by a specific set of values of the parameters of the model. A *hypothesis* is a statement regarding the values of these parameters. The object of setting up a hypothesis is to verify how accurately it can explain a particular phenomenon within the framework of the adopted model. For example, the stochastic model describing the general phenomenon of the sex distribution of newborn babies was derived by analogy with the results of the tossing of a coin ("heads" and "tails") and the model involved in its stochastic description a parameter p defining the probability of obtaining, say, a "head" when the coin is tossed (which is equated with the probability of, say, a male birth). A hypothesis in this case may be $p = \frac{1}{2}$, which may explain the sex distribution of newborn babies in a particular community. For another community or the same community at a different time, a different value of p, and thus a different hypothesis, may be more appropriate.

It would be proper to consider the various descriptions of a natural phenomenon, corresponding to different sets of values of the parameters in its mathematical model, as the different "variants" of the phenomenon. For example, the different values of the parameter p (values between 0 and 1) in the stochastic model of the phenomenon of sex distribution of newborn babies define the different variants of the phenomenon. It must be emphasized that when we speak of studying a phenomenon we generally imply that we have in view a specific variant of it, characterized by a specific set of values of the parameters in its model description. Thus, in the example above, in studying the sex distribution of newborn babies in a community we are confronted with a specific value of p, namely the one for the community under study. The specific variant of a phenomenon under study is sometimes called its *true variant*, and a problem arises because we do not know the values of the parameters characterizing its true variant.

It may be noted that repetitions of the true variant of a phenomenon under study lead to the same or varying outcomes according as its mathematical model is deterministic or stochastic in character. In other words, for a phenomenon described by a deterministic model, repetition of its true variant gives rise to a fixed outcome, while for a phenomenon described by a stochastic model repetition leads to uncertain and thus random outcomes, there being one outcome out of a set of many such outcomes (the set being well defined for the true variant under study) for each repetition. Thus in the example above, characterized by a stochastic model, the sex of a newborn baby is random. From a statistical point of view, we say that the set of all possible outcomes of the repetitions of the true variant of a phenomenon under study describes the *target population* under study.

It is to be noted that the concept of probability is attached to experiments that yield a multiplicity of results when repeated. Take the case of tossing a coin. The outcome of a single toss is either a "head" or a "tail" and is thus unpredictable. However, if the coin is tossed n times and m heads ($m \le n$) are observed, the ratio m/n, which is called the *relative frequency* of heads in n tosses, demonstrates a remarkable constancy, in the sense that when the number of tosses n is increased indefinitely, the relative frequency m/n converges to a stable value, say p. This is described by saying that the probability of a "head" occurring when a coin is tossed is p. If the coin is fair, the value of p is $\frac{1}{2}$. Thus, starting from a situation that is entirely indeterminate and uncertain, namely, the outcome of a "head" or "tail" when a coin is tossed, it is possible to arrive at conclusions with considerable certainty regarding the relative frequency of "heads" for a large number of tosses. It is this stability of the relative frequency concept of probability that imparts a realistic character to stochastic models for explaining natural phenomena and for evaluating scientific laws in all branches of science involving uncontrollable variation of observed data.

The next step in scientific methodology is to conduct an experiment in order to collect appropriate data for checking the validity of a hypothesis. Here again laws of probability prove useful for formulating optimum design of the experiment. Take the case of testing the relative effectiveness of two fertilizers, A and B, for growing a crop, say wheat. If the same quantities of A and B are applied to two plots of equal size where wheat is grown, and if we observe the yields x_A and x_B of the plots, will it be right to say that A is more effective than B if x_A is larger than x_B? A little reflection will show that it will not be a valid conclusion, since there are a host of other factors (fertility of the soil, meteorological conditions, the variety of wheat, etc.), besides the fertilizers, that would influence the yield, and we may draw an erroneous conclusion if we ascribe the relatively higher value of x_A only to the superiority of fertilizer A, ignoring all other factors that influence the yield of wheat. This describes the general situation in any scientific experimentation. The observed result of an experiment depends on a multiplicity of factors, some of which are measurable while others are not, and some of which are controllable while others are not. In the example above, the soil fertility is neither a measurable nor a controllable factor. In a scientific inquiry we single out some of these factors as relatively more important or relevant to the phenomenon under study, and we are interested in studying their effects to the exclusion of all other factors. The cumulative effect of all the factors excluded from the field of study introduces some variation in the outcome of the experiment when it is repeated. In other words, the experiment becomes random in character, with more than one outcome. It is this random character of the experiment that justifies a stochastic model for the phenomenon being studied.

Since the observed results of an experiment to study a phenomenon become variable as a result of excluding some relatively unimportant factors influencing the phenomenon from the field of study, it becomes imperative for the experimenter to exert all possible care to guarantee that the variation in the results of the experiment is such that it does not disturb the validity of the study of the effects of the factors in which the experimenter is interested. If this is to be done effectively, the experimenter must design the experiment carefully. This difficulty has given rise to the subject of design of experiments, which rests on probabilistic considerations and is now a branch of statistics. In the example of the fertilizers, since a single pair of plots is inadequate to supply a basis of comparison of the two fertilizers, it is obvious that the experiment requires replication over a number of pairs of plots, called blocks of plots, each block consisting of two plots of fertilizers A and B to the two plots in each block. One must be careful that the assignment of fertilizers A and B to the two plots in each block is not of a systematic nature, since any systematic assignment may vitiate the validity of a comparison between the fertilizers by making plots assigned to A systematically more productive than those assigned to B because of any of the other factors affecting the yield. (Soil fertility could be one such factor.) This type of criticism can be avoided if the assignment of the fertilizers to the two plots in a block is done by means of a random mechanism, such as the tossing of a coin: that is, according to some probabilistic rule. The problem of choice of the probability set up for such assignment is the subject matter of a particular branch of statistics, namely, sampling theory. A probabilistic approach is thus essential for imparting validity to the conclusions to be derived on the basis of the data collected in scientific experiments.

The last but not the least important stage in scientific inference is the problem of evolving efficient methods for interpreting the data collected from an experiment: that is, how to draw conclusions regarding the hypothesis at hand. The process of inference involved is perforce of an inductive nature—from the sample of observed data to the population of all such conceivable data that can be obtained on the basis of the hypothesis when the experiment is repeated indefinitely, or from a particular case to a general conclusion. Any experimental science is concerned primarily with making such inductive inferences regarding the hypothetical infinite population (called the target population earlier, and described by the result of infinite repetition of the experiment at hand) on the basis of the information supplied by the observed data (called a sample from the target population, and obtained by performing the experiment only once). Herein lies the special effectiveness of the modern science of statistics, which has evolved and is in a continuous process of evolving very powerful and scientifically sound methods based on the theory of probability to meet the challenging needs of such inductive inference.

To sum up, the modern science of statistics, based on the theory of probability, offers effective help at every stage of scientific investigation, from the formulation of a model to the drawing of a valid inductive inference (regarding the hypothesis formulated in the background of the target population in view) on the basis of the information contained in the experiment. It should be true that the science of statistics pervades scientific thinking today and has become an indispensable tool in planning experiments for any scientific inquiry and in drawing valid conclusions on the basis of the data collected therefrom.

BIBLIOGRAPHY

Born, M. (1949). *Natural Philosophy of Cause and Chance*, Oxford University Press, New York.

Laplace, P. S. (1820). *Theorie analytique des probabilités*, Paris.

Von Mises, R. (1957). *Probability, Statistics and Truth*, 2nd ed., Macmillan, New York.

2
General Concepts of Probability

2.1 CLASSICAL DEFINITION OF PROBABILITY

The classical definition of probability depends on a number of concepts, such as elementary events, events, the total number of exhaustive, mutually exclusive, and equally likely cases, the number of cases favorable to an event, and so on. All these entities are the results of a random experiment.

2.1.1 Structure of a Random Experiment; Elementary Events

There is a common logical structure in regard to all random experiments; a random experiment $\mathcal{E}$ is performed under a known or *fixed* set $\mathcal{C}$ of conditions. The experiment $\mathcal{E}$ can be equated with the set $\mathcal{C}$ in the sense that every time the experiment $\mathcal{E}$ is performed we can say that the set $\mathcal{C}$ has occurred, and a repetition of $\mathcal{E}$ can be taken as a repetition of the set $\mathcal{C}$. Besides the fixed set $\mathcal{C}$ of conditions for a random experiment $\mathcal{E}$, there will be, in general, a set $\mathcal{C}'$ of other conditions, not included in $\mathcal{C}$, which will be completely outside our control and often not even recognizable, influencing the outcome of the experiment $\mathcal{E}$. For example, if $\mathcal{E}$ corresponds to the tossing of a die, the set $\mathcal{C}$ may consist of the following conditions: (1) that the die is cubic, and (2) that the die is fair (i.e., its material composition is homogeneous); and the set $\mathcal{C}'$ may consist of the following

9

conditions: (1) the initial position of the die, (2) the impulse imparted by the thrower, and (3) the friction of the surface on which the die is tossed. We can thus say that a random experiment $\mathcal{E}$ corresponds to two mutually exclusive sets $\mathcal{C}$ and $\mathcal{C}'$ of conditions, which can be expressed as

$$\mathcal{E} \sim (\mathcal{C}, \mathcal{C}')$$

of which $\mathcal{C}$ is known or fixed and $\mathcal{C}'$ is unknown.

Although the conditions in set $\mathcal{C}$ for an experiment $\mathcal{E}$ remain unaltered for repetitions of $\mathcal{E}$, the same may not be true for conditions in set $\mathcal{C}'$. In the die-throwing experiment, the initial position of the die, the impulse of the throw of the die, and the friction of the surface encountered by the die when tossed will differ for different tosses of the die.

When an experiment $\mathcal{E}$ is repeated, the different variants of the conditions in the set $\mathcal{C}'$ associated with $\mathcal{E}$ give rise to different outcomes for $\mathcal{E}$. It is not possible to predict the outcome of the experiment $\mathcal{E}$ for any particular time it is performed, since the variants of the conditions in the set $\mathcal{C}'$ will not be known for any particular occurrence of the set $\mathcal{C}$. In the die-throwing experiment, the outcome may be any of the six integers 1, 2, 3, 4, 5, 6 on one of its six faces, but whether a particular face will show cannot be predicted: that depends on the initial position of the die, the impulse of the throw, and the friction of the surface on which the die is thrown, which are unknown.

Definition 2.1.1.1: Events, sample space, and sample points   In the classical definition of probability the various outcomes of a random experiment $\mathcal{E}$ are referred to as *elementary contingent events* or simply as *elementary events* associated with $\mathcal{E}$. The totality of elementary events associated with a random experiment $\mathcal{E}$ is called the *sample space* generated by $\mathcal{E}$, and the elementary events themselves are referred to as *points* of the sample space, or simply as *sample points*.

Thus for the die-throwing experiment the sample space consists of the set of the numbers 1, 2, 3, 4, 5, and 6, and the sample points are the individual numbers 1, 2, 3, 4, 5, and 6. If the experiment consists of throwing two dice, the sample space consists of the 36 sample points (i,j), $i,j = 1,2,\ldots,6$, where i and j are the numbers shown by the first die and second die, respectively.

Besides the elementary events associated with a random experiment, it is possible to define a host of other events for the experiment. If the experiment consists of throwing a die, "the number on the face of the die is even" is an example of an event. In an experiment consisting of the throw of two dice, "the sum of the numbers on the faces of the two dice is less than or equal to 5" is another example of an event. Such events are called *contingent events* or *random events*. We shall call them simply *events*.

It will be noted that the event "the number on the face of the die is even" in the single-die-throwing experiment happens if we realize any of the three sample points 2, 4, and 6 for the experiment. The event does not happen if the sample point observed is any of 1, 3, and 5. Similarly, the event "the sum of the numbers on the faces of the two dice is less than or equal to 5" in the experiment consisting of the throw of two dice materializes if we realize any of the sample points (i,j), $i + j \leq 5$ [i.e., the sample points (1,1), (1,2), (1,3), (1,4), (2,1), (2,2), (2,3), (3,1), (3,2), (4,1)], whereas the event does not materialize if we realize any of the remaining sample points. Generally, an event in relation to a random experiment may or may not materialize, depending on the specific outcome of the experiment. The structure of an event A for an experiment is as follows: We can divide the sample space S of $\mathcal{E}$ into two mutually exclusive subsets S_A and $\overline{S}_A$, say, such that the event A occurs if we observe a sample point in S_A and the negative of the event A occurs (or the event A does not occur) if we observe a sample point in $\overline{S}_A$, when the experiment $\mathcal{E}$ is performed. (Thus $\overline{S}_A$ is the set of all points in S that are not in S_A.) We say that the event A corresponds to the subset S_A of the sample space S and express this relation as

$$A \sim S_A.$$

The set S_A completely defines the event A. Every subset of a sample space S generated by a random experiment $\mathcal{E}$ will define a contingent event for $\mathcal{E}$. If the sample space S consists of N sample points, the number of events defined is obviously 2^N. It may be that many of these events may be abstractions without any physical meaning.

EXERCISES

Write down the sample space and the subset of the sample space defining the specified events in the following cases.

1. A coin is tossed five times in succession, and the event is: (a) obtaining more heads than tails; (b) obtaining two heads; (c) obtaining at least one head; (d) obtaining at most one head; (e) the difference between numbers of heads and tails is unity.
2. Three dice are thrown, and the event is: (a) the sum of the points is 2; (b) the sum of the points is 10; (c) the sum of the points is at least 10; (d) the point 6 occurs at least once.

3. An urn contains five white, seven black, and 10 green balls. One ball is drawn from the urn. The event is: (a) the ball is white; (b) the ball is not green.

4. An urn contains three white and two black balls. Consider the following experiments: (1) two balls are drawn simultaneously from the urn; (2) two balls are drawn one after the other from the urn without replacement; and (3) two balls are drawn one after the other from the urn with replacement. For each of the experiments above the event is: (a) the balls are of the same color; (b) the balls are white; (c) the balls are of opposite color.

5. A class contains 10 students, four of them being Canadians, three Englishmen, two Americans, and one Italian. Three students are chosen without replacement. The event is obtaining at least two Canadians.

6. In Exercise 5, the students are chosen with replacement and the event is as in Exercise 5.

7. Two cards are drawn without replacement from a deck of well-shuffled cards. The event is: (a) both the cards are aces; (b) one of the cards is an ace.

8. Six balls are distributed at random among three boxes, and the event is: (a) the first box contains three balls; (b) one of the boxes contains two balls; (c) none of the boxes is empty.

9. Three different objects, 1, 2, and 3, are distributed at random on three different sites, marked 1, 2, 3. The event is: (a) none of objects occupies the place corresponding to its number; (b) at least two of the objects occupy places corresponding to their numbers.

10. A coin is tossed repeatedly until it shows a head. The event is that the head shows in less than 10 tosses.

11. Each of three boxes, identical in appearance, has two drawers. The first box contains a gold coin in each drawer; the second contains a silver coin in each drawer; and the third contains a gold coin in one drawer and a silver coin in the other. Consider the following experiments: (1) a box is chosen at random; (2) a box is chosen at random, one of its drawers is opened, and a gold coin found in it. For each of the experiments above, the event is that the box contains coins of differing metals.

12. Two urns, U_1 and U_2, contain, respectively, two white and three black balls, and three white and two black balls. One ball is transferred from U_1 to U_2 and thereafter (a) a ball is drawn from U_2, the event being that the ball is white; (b) two balls are drawn from U_2, the event being that one of them is white and the other black.

2.1.2 Equally Likely and Favorable Cases

In everyday language the words *probability* and *probable* are used with different shades of meaning. By saying "probably it will be sunny tomorrow," we mean that there are more convincing indications justifying sunny weather than cloudy tomorrow. On the other hand, in the statement "there is little probability in the statement he made," the word *probability* is used in the sense of credibility. We shall, however, use the word *probability* to signify the "degree of credence" we may place on the materialization of contingent events. That this is possible can be seen from the following examples.

Example 2.1.2.1 John walks along a street either with his eyes open or blindfolded. Either way, he might be injured, although not necessarily. How do the probabilities of his getting injured compare in the two cases? Anyone, no doubt, will contend that the probability of John's getting injured is "greater" when he walks blindfolded along the street than when he walks with his eyes open.

Example 2.1.2.2 An urn contains an equal number of white and black balls, which are similar in all respects except their color. The balls are well shaken up inside the urn and one ball is drawn thereafter. It may be either black or white. How do the probabilities of these two cases compare? One almost instinctively answers: "The probabilities are equal." If, on the other hand, the urn contained 15 white balls and three black balls, and one ball is drawn from it, one would say assertively that the probability of drawing a white ball is greater than that of drawing a black ball.

It would thus appear that probability is something which admits of comparisons in magnitude. The basic problem, however, is that given a set of contingent events we should be in a position to index them as to the magnitudes of their probabilities. The only way this can be done is to give a measure to the probability of a contingent event in terms of a finite real number, and to use the natural ordering of real numbers by their magnitudes for indexing the probabilities of a given set of contingent events.

In attempting to measure probabilities by finite real numbers, we encounter difficulties similar to those arising in other fields where measurements are involved. These difficulties cannot be avoided except by making some ideal assumptions or postulates. In measuring lengths we are first required to set up a criterion of equality of lengths. Similarly, in giving a numerical measure to probabilities, we are concerned initially with setting up a criterion of equiprobability of two contingent events for a random experiment; in other words, we should be able to say when two contingent events are equally probable or equally likely. Bernoulli enunciated the following criterion of equiprobability.

Postulate I: Criterion of equiprobability: equally likely events   Two contingent events for a random experiment $\mathcal{E}$ are considered *equally probable* or *equally likely* if, after taking into consideration all relevant evidence in the background of the experiment $\mathcal{E}$, one of them cannot be expected to materialize in preference to the other when the experiment $\mathcal{E}$ is performed.

Obviously, there is an element of vagueness in the criterion of equiprobability above, since its application depends on individual common sense and judgment, but unfortunately in the nature of things it is hardly possible to replace it with a better one.

Definition 2.1.2.1: Mutually exclusive events   Two events in relation to a random experiment $\mathcal{E}$ are said to be *mutually exclusive* if the occurrence of one of them precludes the occurrence of the other every time the experiment is performed; in other words, the two events cannot materialize simultaneously. Similarly, a set of events is said to be mutually exclusive if any two of them cannot materialize simultaneously.

Definition 2.1.2.2: Exhaustive events   A set of events in relation to a random experiment $\mathcal{E}$ is said to be *exhaustive* if one of them must necessarily materialize every time the experiment $\mathcal{E}$ is performed.

In Section 2.1.1 we identified events in relation to a random experiment with subsets of the sample space S generated by $\mathcal{E}$. It is quite clear from the definitions above that events are mutually exclusive if and only if any two of the subsets defining the events do not have any point (sample point) in common; and a set of events is exhaustive if and only if the subsets defining the events are such that every point (sample point) of the sample space is included in at least one of these subsets.

For measuring probabilities by numbers we are required to make the following postulate in addition to Postulate I.

Postulate II   For any random experiment $\mathcal{E}$, the elementary events (or the sample points) form a set of exhaustive, mutually exclusive, and equally likely events or cases, as they are commonly called. If n is the total number of elementary events for the experiment $\mathcal{E}$, the probability of any one of these events or cases is $1/n$.

When a die is thrown, the possible numbers of points 1, 2, 3, 4, 5, 6 are obviously exhaustive and mutually exclusive. They are also equally likely if the die is fair, and the probability of any one of these numbers is $\frac{1}{6}$.

Definition 2.1.2.3: Cases favorable to an event Let ε be a random experiment admitting of n elementary events $a_1, a_2, \ldots, a_n$. In other words, the sample space S generated by the experiment ε consists of the sample points $a_1, a_2, \ldots, a_n$. An event A in relation to the experiment ε is then defined by a subset S_A of the sample space S (cf. Section 2.1.1). Let S_A consist of the points $a_{i_1}, a_{i_2}, \ldots, a_{i_m}$, where $i_1, i_2, \ldots, i_m$ are different integers between 1 and n. The elementary events or sample points $a_{i_1}, a_{i_2}, \ldots, a_{i_m}$ are called *cases favorable to the event A*.

It is to be noted that $a_{i_1}, a_{i_2}, \ldots, a_{i_m}$ are the several mutually exclusive particular forms for the materialization of the event A, in the sense that if any of these sample points is realized by the experiment ε, then the event A materializes.

We are now in a position to give a formal definition of mathematical probability in the classical sense.

Definition 2.1.2.4: Classical definition of probability
 (a) Let ε be a random experiment admitting of n exhaustive, mutually exclusive, and equally likely cases. This will be referred to simply as "total number of equally likely cases" for the experiment ε.
 (b) Let A be an event in relation to the experiment ε.
 (c) Let m $(m \leq n)$ be the number of cases favorable to the event A. Then the *probability* of the event A is defined as the ratio m/n.

Representing the probability of the event A as $P(A)$, we write

$$P(A) = \frac{m}{n}.$$

It is obvious that for any event A,

 $0 \leq P(A) \leq 1;$

that is, the probability of A is a nonnegative number less than or equal to unity.

Example 2.1.2.3 An urn contains three white and two black balls. Two balls are drawn at random from the urn without replacement. What is the probability that the balls are of different colors?

To find the probability from first principles, we proceed as follows.

Step I. Count the total number of equally likely cases. Let us number the three white balls as W_1 W_2, W_3, and the two black balls as B_1, B_2. We write the elementary event composed of the first ball drawn being W_1 and the second ball W_2 as W_1W_2. Since the balls are drawn without replacement, the sample space would consist of the following 20 equally likely cases:

$$W_1W_2, \quad W_1W_3, \quad W_1B_1, \quad W_1B_2, \quad W_2W_3, \quad W_2B_1, \quad W_2B_2, \quad W_3B_1,$$
$$W_3B_2, \quad B_1B_2$$
$$W_2W_1, \quad W_3W_1, \quad B_1W_1, \quad B_2W_2, \quad W_3W_2, \quad B_1W_2, \quad B_2W_2, \quad B_1W_3,$$
$$B_2W_3, \quad B_2B_1.$$

Step II. Count the total number of cases favorable to the event. Since the event is that balls are of different colors, the event materializes for any of the following 12 sample points:

$$W_1B_1, \quad W_1B_2, \quad W_2B_1, \quad W_2B_2, \quad W_3B_1, \quad W_3B_2$$
$$B_1W_1, \quad B_2W_1, \quad B_1W_2, \quad B_2W_2, \quad B_1W_3, \quad B_2W_3.$$

Thus

$$\text{probability of the event} = \frac{\text{number of cases favorable to the event}}{\text{total number of equally likely cases}}$$

$$= \frac{12}{20} = \frac{3}{5}.$$

The enumeration of the total number of equally likely cases and of the number of cases favorable to an event is often simplified by the use of well-known formulas for the numbers of permutations and combinations. It should, however, be emphasized that it is always more instructive if the enumeration is done by writing down the sample space and counting the number of sample points in the subset of the sample space defining the event. It is advised that for evaluating the probability of any event, the reader, in the interest of clarity, should develop the habit of mentally figuring out the sample space and the subset of the space defining the event, even though it may be arduous to write down these sets explicitly.

Example 2.1.2.4 An urn contains n tickets numbered 1 to n, and m tickets are drawn from it. What is the probability that k $(k \le m \le n)$ of the tickets drawn have numbers previously specified?

In solving this example, writing down the sample space for counting the total number of equally likely cases and the sample points favorable to the event is cumbersome. Instead, we shall count the cases by applying formulas for the number of permutations. The total number of ways in which m tickets can be drawn out of n tickets is obviously

$$C_m^n = \frac{n(n-1)\cdots(n-m+1)}{1\cdot 2\cdots m},$$

and this gives the total number of equally likely cases for the problem. To count the number of cases favorable to the event of k of the m tickets drawn having numbers specified previously, we argue as follows. Since k of the m tickets drawn have specified numbers, $m - k$ of them have unspecified numbers. These $m - k$ tickets with unspecified numbers can arise from $n - k$ tickets, which represent the total number of tickets with unspecified numbers for the problem. This number is

$$C_{m-k}^{n-k} = \frac{(n-k)(n-k-1)\cdots(n-m+1)}{1\cdot 2\cdots(m-k)}$$

and is the number of cases favorable to the event. Thus the required probability is

$$P = \frac{C_{m-k}^{n-k}}{C_m^n} = \frac{m(m-1)\cdots(m-k+1)}{n(n-1)\cdots(n-k+1)}.$$

Example 2.1.2.5 Two dice are thrown n times in succession. What is the probability of obtaining double six at least once?

As there are 36 cases for every throw of two dice, and each case for any throw can be combined with each case for any other throw, the total number of equally likely cases for the problem is 36^n. To count the number of cases favorable to the event, we note that in a single throw of two dice, double six will not appear in 35 cases out of the total of 36 cases, and thus in n throws of the two dice double six will not appear in 35^n cases out of the total of 36^n cases. Thus the number of cases for which double six will appear at least once is $36^n - 35^n$. Thus the required probability is

$$P = \frac{36^n - 35^n}{36^n} = 1 - \left(\frac{35}{36}\right)^n.$$

We note that $p > \frac{1}{2}$ if we have

$$1 - \left(\frac{35}{36}\right)^n > \frac{1}{2}$$

or

$$n > \frac{\log 2}{\log 36 - \log 35} = 24.6.$$

This implies that in 25 throws there is greater probability of obtaining a double six at least once than not to obtain it at all. This produced Pascal's explanation of Chevalier de Méré's problem, referred to in Chapter 1.

EXERCISES

13. Three balls are drawn at random from an urn containing five white and four black balls. What is the probability that two of them are white and one is black?

14. What is the probability of obtaining (a) a total of 13 points when three dice are thrown; (b) two heads and three tails when five coins are tossed?

15. In poker, five cards are selected at random from a full deck of 52. What is the probability that a hand in a poker game contains a "pair," two cards of the same face value?

16. Eight cards are drawn at random from a full deck of 52 cards. What is the probability that they contain (a) no ace; (b) at least one ace; (c) exactly two aces?

17. What is the probability that in a bridge game a player and his partner have (a) all 13 cards of a specified suit; (b) all 13 cards of any suit?

18. Twenty tickets are numbered 1 to 20. What is the probability that four tickets taken in succession will bear numbers in an increasing or decreasing order?

19. Two urns U_1 and U_2 contain, respectively, five white and three black balls, and four white and four black balls. One ball is transferred from U_1 to U_2, and thereafter one ball is drawn from U_2. What is the probability that it is white?

20. Nine balls are distributed among four boxes. What is the probability that one of the boxes contain three balls?

21. What is the probability that a group of 50 persons will have a repeated birthday?

22. How many persons should there be in a group so that the probability of a repeated birthday in the group is greater than $\frac{1}{2}$?

2.1.3 Theorems of Total and Compound Probability

In Section 2.1.1 we evaluated probabilities of events by the direct application of the definition of probability, and this involved direct enumeration of the total number of equally likely cases and the number of cases favorable to an event. As problems grow in complexity the difficulty of direct enumeration of these cases also grows, and the computation of probabilities by direct application of the definition gets more involved. We shall now develop some theorems in probability with a view to avoiding these complications.

A theorem in probability aims at deducing an algebraic relationship between probabilities of various related events, such that given the probabilities of some of these events, the probabilities of some other events that can be explained by them in some manner can be evaluated. That this is possible can be seen from the following simple example: If the probability that an event A will happen is p, the probability that the event A will not happen, usually represented as the event $\overline{A}$, is $1 - p$, whatever the event A may be. When the probability of A is given, the probability of $\overline{A}$ can be deduced immediately.

Our general aim will be to develop a calculus of probability based on theorems on probability. The general logical scheme of the calculus of probability is as follows: Given the probabilities of a set $\mathcal{S}$ of events for a set of random experiment $\mathcal{E}_1, \mathcal{E}_2, \ldots$, and given another random experiment $\mathcal{E}^*$, which is related to $\mathcal{E}_1, \mathcal{E}_2, \ldots$ in a known manner, we should be able to make statements about the probabilities of specified events for $\mathcal{E}^*$, which can be explained in terms of the set $\mathcal{S}$ of events for $\mathcal{E}_1, \mathcal{E}_2, \ldots$ in a certain sense.

Two theorems are of fundamental importance in developing a calculus of probability: the theorem of total probability and the theorem of compound probability. The statements of these theorems involve some basic concepts, such as mutually exclusive events, compound events, conditional probability of an event given another event, and mutually independent events. Of these, we have already defined mutually exclusive events. The definitions of the remaining concepts appear below.

Throughout we denote an event A by the symbol (A) and express the probability of (A) as $P(A)$. Consider an event A in conjunction with another event B. We adopt the following notation:

1. $(A \cup B)$ will represent the event composed of the occurrence of A or B (or both).
2. (AB) will represent the event composed of the simultaneous occurrence of A and B. A standard notation for AB is $A \cap B$, but for convenience we will write AB instead of $A \cap B$.
3. $(A|B)$ will represent the event giving the occurrence of A when it is known that B has occurred, or the conditional occurrence of A given that B is certain. In short, $(A|B)$ is described as the conditional occurrence of A, given B.

In the same manner, $(A \cup B \cup C \cup \cdots)$ represents the occurrence of A or B or C or ..., and $(ABC \cdots)$ represents the simultaneous occurrence of A and B and C and ... . In relation to a given set of events $A_1, A_2, \ldots, A_n$, the event $(A_1 \cup A_2 \cup \cdots \cup A_n)$ is described as a *total event*, since it represents the occurrence of any one of the given events; and events such as $(A_1A_2),(A_1A_2A_3),...$ are described as *compound events*, since each of them represents the simultaneous occurrence of the events within parentheses.

It is to be noted that if $A,B,C,...$ are events in relation to a random experiment such that their occurrences are not conditioned by any other event for the experiment, and the experiment generates the sample space with N points, the total number of equally likely cases for the events $A,B,C,...$ is N, irrespective of the nature of the events. If $S_A,S_B,S_C,...$ are the subsets of S defining the events $A,B,C,...$ containing $m_A,m_B,m_C,...$ points, respectively, then the total number of cases favorable to the events $A,B,C,...$ are $m_A,m_B,m_C,...$, respectively. This holds even if we define $(C) = (A \cup B)$ or $(C) = (AB)$. By definition, $P(A) = m_A/N$, $P(B) = m_B/N$, $P(C) = m_C/N,...$. What is significant is the fact that the sample space remains unaltered for all unconditional events for a random experiment, and is identical with the space generated by the experiment. This is no longer true if we consider conditional events. Take, for example, $(C) = (A|B)$. From the definition of $(A|B)$ it is obvious that the possibility of occurrence of $(A|B)$ arises only after (B) has occurred, or in other words, if the random experiment gives rise to a sample point in S_B, the subset of the sample space S defining B. Thus, logically, the sample space for $(A|B)$ is S_B, and thus the total number of equally likely cases for $(A|B)$ is m_B. This is true irrespective of the nature of the events A and B. If m_{AB} is now the number of cases favorable to (AB), we have, by definition,

$$P(AB) = \frac{m_{AB}}{N} \quad \text{and} \quad P(A|B) = \frac{m_{AB}}{m_B}.$$

It is to be noted that since the compound event (AB) represents the event of joint occurrence of both A and B, and m_{AB} cases favorable to (AB) must be included in the m_A cases favorable to (A) and also in the m_B cases favorable to (B). In fact, the m_{AB} cases favorable to (AB) are exactly those cases that are included in both the m_A cases favorable to (A) and the m_B cases favorable to (B). In terms of set representation of events, this implies that the set S_{AB} defining the compound event (AB) consists of points that are common to both sets S_A and S_B defining (A) and (B), respectively. In the same manner, since the event $(A \cup B)$ represents the occurrence of (A) or (B) or (AB), the set $S_{A \cup B}$ defining the event $(A \cup B)$ consists of points in S_A that are not in S_B, points in S_B that are not in S_A, and points in S_{AB}. We can express this as

$$(A \cup B) = ((\overline{AB}) \cup (\overline{A}B) \cup (AB)),$$

where $(A\overline{B})$, $(\overline{A}B)$, and (AB) are obviously mutually exclusive.

It can be easily verified from the physical meanings of $(A \cup B)$ and (AB) that

1. $(A \cup B) = (B \cup A)$ and $(AB) = (BA)$; that is, the notations for total and compound events are commutative.
2. $((A \cup B)C) = (AC \cup BC)$.
3. $((AB)(AC)) = (ABC)$.

Theorem 2.1.3.1: Theorem of total probability: form I (mutually exclusive events) If $A_1, A_2, \ldots, A_n$ are n mutually exclusive events,

$$P(A_1 \cup A_2 \cup \cdots \cup A_n) = P(A_1) + P(A_2) + \cdots + P(A_n).$$

Proof. Let N be the total number of equally likely cases, out of which m_1 cases are favorable to the event A_1, m_2 cases are favorable to the event $A_2, \ldots, m_n$ cases are favorable to the event A_n. By definition,

$$P(A_i) = \frac{m_i}{N}, \qquad i = 1, 2, \ldots, n.$$

Since $A_1, A_2, \ldots, A_n$ are mutually exclusive, m_i cases favorable to A_i are different from m_j cases favorable to A_j, for any i and j, $i \neq j$, $1 \leq i$, $j \leq n$. This implies that the number of cases favorable to the event $(A_1 \cup A_2 \cup \cdots \cup A_n)$ is $m_1 + m_2 + \cdots + m_n$. Thus, by definition,

$$
\begin{aligned}
P(A_1 \cup A_2 \cup \cdots \cup A_n) &= \frac{m_1 + m_2 + \cdots + m_n}{N} \\
&= \frac{m_1}{N} + \frac{m_2}{N} + \cdots + \frac{m_n}{N} \\
&= P(A_1) + P(A_2) + \cdots + P(A_n). \qquad \text{Q.E.D.}
\end{aligned}
$$

Expressed in physical terms, Theorem 2.1.3.1 states that the probability of any one of a finite set of mutually exclusive events is the sum of the probabilities of these events. The theorem is sometimes referred to as the theorem of addition of probabilities; in other words, the theorem defines the law of addition of probabilities.

In applying the theorem it is to be noted that very often it is possible to define an event A in several mutually exclusive forms, $A_1, A_2, \ldots, A_n$, so that (A) occurs whenever any of $(A_1), (A_2), \ldots, (A_n)$ occurs, and conversely. In that case we express (A) as

$$(A) = (A_1 \cup A_2 \cup \cdots \cup A_n),$$

and then $P(A)$ is given by Theorem 2.1.3.1 as

$$P(A) = P(A_1) + P(A_2) + \cdots + P(A_n).$$

In practice it will be easier to compute the probabilities $P(A_i), i = 1,2,...,n$, and to obtain $P(A)$ as the sum of these probabilities, than to compute $P(A)$ directly.

Corollary 2.1.3.1.1 For any two events A_1 and A_2, not necessarily mutually exclusive,

$$P(A_1 \cup A_2) = P(A_1) + P(A_2) - P(A_1A_2).$$

Proof. If we consider the happening of the event A_1 in conjunction with that of the event A_2, we note that A_1 can materialize in two mutually exclusive and exhaustive forms: (1) "A_1 and A_2" [i.e., (A_1A_2)], and (2) "A_1 and $\overline{A}_2$" [i.e., $(A_1\overline{A}_2)$]. In other words, we can write

$$(A_1) = (A_1A_2 \cup A_1\overline{A}_2)$$

and since A_1A_2 and $A_1\overline{A}_2$ are mutually exclusive, we have, by Theorem 2.1.3.1,

$$P(A_1) = P(A_1A_2) + P(A_1\overline{A}_2).$$

Similarly,

$$P(A_2) = P(A_1A_2) + P(\overline{A}_1A_2).$$

Thus

$$P(A_1) + P(A_2) = P(A_1A_2) + [P(A_1\overline{A}_2) + P(\overline{A}_1A_2) + P(A_1A_2)].$$

Again, it is easy to verify that $(A_1\overline{A}_2)$, $(\overline{A}_1A_2)$, and (A_1A_2) are three mutually exclusive and exhaustive forms of the event $(A_1 \cup A_2)$. Thus we have, by Theorem 2.1.3.1,

$$P(A_1 \cup A_2) = P(A_1\overline{A}_2) + P(\overline{A}_1A_2) + P(A_1A_2).$$

Combining the results, we have

$$P(A_1 \cup A_2) = P(A_1) + P(A_2) - P(A_1A_2). \qquad \text{Q.E.D.}$$

Theorem 2.1.3.2: Theorem of total probability: form II For any finite set of events $A_1,A_2,...,A_n$,

$$P(A_1 \cup A_2 \cup \cdots \cup A_n) = \sum_{i=1}^{n} P(A_i) - \sum_{\substack{i=1 \\ i<j}}^{n} \sum_{j=1}^{n} P(A_iA_j)$$

$$+ \sum_{\substack{i=1 \\ i<j<k}}^{n} \sum_{j=1}^{n} \sum_{k=1}^{n} P(A_iA_jA_k)$$

$$- \cdots + (-1)^{n-1} P(A_1A_2 \cdots A_n).$$

Proof. The proof follows by induction, by using Corollary 2.1.3.1.1, which is a statement of the theorem for $n = 2$. Thus the theorem holds for $n = 2$. Suppose that the theorem holds for $n = m$, that is,

$$P(A_1 \cup \cdots \cup A_m) = \sum_{i=1}^{m} P(A_i) - \sum_{\substack{i=1 \\ i<j}}^{m} \sum_{j=1}^{m} P(A_i A_j)$$

$$+ \sum_{\substack{i=1 \\ i<j<k}}^{m} \sum_{j=1}^{m} \sum_{k=1}^{m} P(A_i A_j A_k)$$

$$- \cdots + (-1)^{m-1} P(A_1 A_2 \cdots A_m). \tag{2.1}$$

We are to show that the theorem holds for $n = m + 1$; we have

$$P(A_1 \cup \cdots \cup A_m \cup A_{m+1}) = P(A \cup A_{m+1}), \tag{2.2}$$

where we write $(A) = (A_1 \cup \cdots \cup A_m)$. Applying Corollary 2.1.3.1.1, we have

$$P(A \cup A_{m+1}) = P(A) + P(A_{m+1}) - P(AA_{m+1}). \tag{2.3}$$

To evaluate $P(AA_{m+1})$, we note that

$$P(AA_{m+1}) = P\{(A_1 \cup \cdots \cup A_m)A_{m+1}\}$$

$$= P(A_1 A_{m+1} \cup A_2 A_{m+1} \cup \cdots \cup A_m A_{m+1})$$

$$= P(B_1 \cup B_2 \cup \cdots B_m), \tag{2.4}$$

where $(B_i) = (A_i A_{m+1})$, $i = 1,2,\ldots,m$. Since the theorem holds for m events, we have

$$P(B_1 \cup \cdots \cup B_m) = \sum_{i=1}^{m} P(B_i) - \sum_{\substack{i=1 \\ i<j}}^{m} \sum_{j=1}^{m} P(B_i B_j)$$

$$+ \sum_{\substack{i=1 \\ i<j<k}}^{m} \sum_{j=1}^{m} \sum_{k=1}^{m} P(B_i B_j B_k) - \cdots . \tag{2.5}$$

Now

$$P(B_i) = P(A_i A_{m+1})$$

$$P(B_i B_j) = P\{(A_i A_{m+1})(A_j A_{m+1})\} = P(A_i A_j A_{m+1})$$

$$P(B_i B_j B_k) = P\{(A_i A_{m+1})(A_j A_{m+1})(A_k A_{m+1})\} = P(A_i A_j A_k A_{m+1}), \text{ etc.}$$

Substituting in Eq. (2.5), we have

$$P(B_1 \cup \cdots \cup B_m) = \sum_{i=1}^{m} P(A_i A_{m+1}) - \sum_{\substack{i=1 \\ i<j}}^{m} \sum_{j=1}^{m} P(A_i A_j A_{m+1})$$

$$+ \sum_{\substack{i=1 \\ i<j<k}}^{m} \sum_{j=1}^{m} \sum_{k=1}^{m} P(A_i A_j A_k A_{m+1}) - \cdots . \qquad (2.6)$$

From Eqs. (2.1)–(2.4) and (2.6), it follows that

$$P(A_1 \cup \cdots \cup A_m \cup A_{m+1})$$

$$= P(A_1 \cup \cdots \cup A_m) + P(A_{m+1})$$

$$- \left[\sum_{i=1}^{m} P(A_i A_{m+1}) - \sum_{\substack{i=1 \\ i<j}}^{m} \sum_{j=1}^{m} P(A_i A_j A_{m+1}) \right.$$

$$\left. + \sum_{\substack{i=1 \\ i<j<k}}^{m} \sum_{j=1}^{m} \sum_{k=1}^{m} P(A_i A_j A_k A_{m+1}) - \cdots \right]$$

$$= \sum_{i=1}^{m} P(A_i) - \sum_{\substack{i=1 \\ i<j}}^{m} \sum_{j=1}^{m} P(A_i A_j)$$

$$+ \sum_{\substack{i=1 \\ i<j<k}}^{m} \sum_{j=1}^{m} \sum_{k=1}^{m} P(A_i A_j A_k) - \cdots$$

$$+ P(A_{m+1}) - \left[\sum_{i=1}^{m} P(A_i A_{m+1}) - \sum_{\substack{i=1 \\ i<j}}^{m} \sum_{j=1}^{m} P(A_i A_j A_{m+1}) \right.$$

$$\left. + \sum_{\substack{i=1 \\ i<j<k}}^{m} \sum_{j=1}^{m} \sum_{k=1}^{m} P(A_i A_j A_k A_{m+1}) - \cdots \right]$$

$$= \sum_{i=1}^{m+1} P(A_i) - \sum_{\substack{i=1 \\ i<j}}^{m+1} \sum_{j=1}^{m+1} P(A_i A_j)$$

$$+ \sum_{\substack{i=1 \\ i<j<k}}^{m+1} \sum_{j=1}^{m+1} \sum_{k=1}^{m+1} P(A_i A_j A_k) - \cdots .$$

Thus the theorem holds for $n = m + 1$.   Q.E.D.

Theorem 2.1.3.3: Theorem of compound probability For any two events A and B,

$$P(AB) = P(A)P(B|A).$$

Proof. Let N be the total number of equally likely cases out of which m_A cases are favorable to (A) and m_{AB} cases are favorable to (AB). Then, by definition,

$$P(A) = \frac{m_A}{N} \quad \text{and} \quad P(AB) = \frac{m_{AB}}{N}.$$

We write

$$P(AB) = \frac{m_{AB}}{m_A}\frac{m_A}{N} = P(A)\frac{m_{AB}}{m_A}.$$

To give a probability meaning to the ratio m_{AB}/m_A we note that since (AB) represents the event of joint occurrence of (A) and (B), the m_{AB} cases favorable to (AB) must be included in the m_A cases favorable to (A). Thus $0 \le m_{AB}/m_A \le 1$. Moreover, we observe that, assuming the occurrence of (A), there are m_A equally likely cases for any event conditioned on the occurrence of (A), the remaining $N - m_A$ cases being impossible for such events; and out of these m_A equally likely cases, m_{AB} cases are favorable to $(B|A)$. Thus

$$\frac{m_{AB}}{m_A} = P(B|A).$$

Combining, we have

$$P(AB) = P(A)P(B|A). \quad \text{Q.E.D.}$$

Corollary 2.1.3.3.1 For any two events A and B,

$$P(AB) = P(B)P(A|B).$$

Proof. This follows from Theorem 2.1.3.3 since

$$P(AB) = P(BA). \quad \text{Q.E.D.}$$

Corollary 2.1.3.3.2 For any n events $A_1, A_2, \ldots, A_n$,

$$P(A_1 A_n \cdots A_n) = P(A_1)P(A_2|A_1)P(A_3|A_1 A_2) \cdots P(A_n|A_1 \cdots A_{n-1}).$$

Proof. This follows by induction from Theorem 2.1.3.3. By Theorem 2.1.3.3 we have

$$P(A_1 A_2) = P(A_1)P(A_2|A_1).$$

Again,

$$P(A_1A_2A_3) = P(A_1A_2)P(A_3|A_1A_2) = P(A_1)P(A_2|A_1)P(A_3|A_1A_2), \text{ etc.}$$

Q.E.D.

Corollary 2.1.3.3.2 is sometimes referred to as the *theorem of compound probability for a finite number of events*.

The theorem of compound probability, as established in Corollary 2.1.3.3.2, which aims to give the probability of a compound event $(A_1A_2 \cdots A_n)$ in terms of the probabilities of the individual events $(A_1),(A_2),...,(A_n)$, is too cumbersome from the standpoint of application, since the probabilities of the individual events are conditioned on the occurrence of other compound events. The theorem takes a very simple and realistic form in the case where the events are "independent." *Independence of events* is strictly a probabilistic concept and plays a dominant role in the theory of probability. We give a formal definition of independent events before we go on to inquire about its physical meaning.

Given any two events (A) and (B), we have defined

$$P(B|A) = \frac{m_{AB}}{m_A},$$

where m_{AB} and m_A are the numbers of cases favorable to (AB) and (A), respectively. It is to be noted that $P(B|A)$ is defined only when $m_A \neq 0$. If N is the total number of equally likely cases, we can rewrite $P(B|A)$ as

$$P(B|A) = \frac{m_{AB}/N}{m_A/N} = \frac{P(AB)}{P(A)}.$$

$P(B|A)$ is defined only if $P(A) \neq 0$. To distinguish $P(B|A)$ from $P(B)$, we shall refer to $P(B)$ as the *unconditional probability* of (B) and $P(B|A)$ as the *conditional probability* of (B) given (A).

Definition 2.1.3.1: Independence of a pair of events Two events A and B are said to be *independent* if the conditional probability of (B) given (A) is equal to the unconditional probability of (B); that is, $P(B|A) = P(B)$, provided that $P(B|A)$ is defined.

Thus for two independent events A and B, Theorem 2.1.3.3 gives us

$$P(AB) = P(A)P(B|A) = P(A)P(B);$$

that is, the probability of the compound event (AB) is the product of the probabilities of the events (A) and (B).

It can also be verified that

$$P(B|A) = P(B) \Longleftrightarrow P(A|B) = P(A).$$

This is so because

$$P(A)P(B|A) = P(AB) = P(B)P(A|B).$$

Thus the independence of (A) and (B) implies that the unconditional probability of either event is equal to its conditional probability given the other event; and (A) and (B) are independent if and only if $P(AB) = P(A)P(B)$.

The definition of independence of events can easily be extended to any finite number of events. For this we adopt the following notation. Given n events $A_1,A_2,...,A_n$, $P(A_1|A_2A_3 \cdots A_n)$ denotes the conditional probability of the event A_1 given that the events $A_2,A_3,...,A_n$ have occurred. We can represent

$$P(A_1|A_2A_3 \cdots A_n) = \frac{P(A_1A_2A_3 \cdots A_n)}{P(A_2A_3 \cdots A_n)}$$

if $P(A_2A_3 \cdots A_n) > 0$; $P(A_1|A_2A_3 \cdots A_n)$ is undefined if $P(A_2A_3 \cdots A_n) = 0$. Physically, $P(A_1|A_2A_3 \cdots A_n)$ means the fraction of occurrences of the events $A_2,A_3,...,A_n$ on which the event A_1 also occurs.

Definition 2.1.3.2: Independence of a number of events n events $A_1,A_2,...,A_n$ are said to be *independent* if the conditional probability of any one of these events, say A_j, given one or more of the remaining events, is equal to the unconditional probability of A_j. This can be expressed as follows: The events are independent if for any choice of k integers $i_1 < i_2 < \cdots < i_k$ from 1 to n (for which the following conditional probability is defined) and for any integer j from 1 to n not equal to $i_1,i_2,...,i_k$, we have

$$P(A_j|A_{i_1}A_{i_2} \cdots A_{i_k}) = P(A_j).$$

Theorem 2.1.3.4: Theorem of compound probability: independent events If $A_1,A_2,...,A_n$ are independent events,

$$P(A_1A_2 \cdots A_n) = P(A_1)P(A_2) \cdots P(A_n).$$

Proof. By Corollary 2.1.3.3.2 we have for any events $A_1,A_2,...,A_n$,

$$P(A_1A_2 \cdots A_n) = P(A_1)P(A_2|A_1) \cdots P(A_n|A_1A_2 \cdots A_{n-1}).$$

Since $A_1,A_2,...,A_n$ are independent, we have

$$P(A_2|A_1) = P(A_2),...,P(A_n|A_1A_2 \cdots A_{n-1}) = P(A_n).$$

Substituting these in the expression above, the theorem follows.

$$\text{Q.E.D.}$$

Theorem 2.1.3.4 is called the *theorem on multiplication of probabilities*.

Our aim was to develop a calculus of probabilities. The theorem on addition of probabilities (Theorem 2.1.3.1) and the theorem on multiplication of probabilities (Theorem 2.1.3.4) are the two important components of this calculus.

We have seen that the theorem of addition of probabilities holds only for mutually exclusive events, and that of multiplication of probabilities holds only for independent events. Mutual exclusiveness and independence of events are thus two essential concepts we need for developing the calculus of probability. Of these, the physical concept of mutual exclusiveness of events is quite clear and follows immediately from definition. Two events A and B are mutually exclusive if the happening of one of them precludes the happening of the other. Physically, this means that the subsets S_A and S_B of the sample space, defining the events A and B, respectively, do not have any point in common. Such a simple set-theoretical exposition does not seem to be possible for independent events, and this obscures a lucid physical interpretation of independence of events. All that can be said in the way of offering a physical interpretation of independence of events is that a set of events are independent if the probability of any one of them is not affected by supplementary knowledge concerning the materialization of any of the remaining events. But this amounts to giving a physical description of the fact that the conditional probabilities of any of the events given one or more of the remaining events is the same as its unconditional probability.

Pairwise Independence and Independence

We have seen that if two events A and B are independent,

$$P(A|B) = P(A) \Longleftrightarrow P(B|A) = P(B) \tag{2.7}$$

and that the events A and B are independent if and only if

$$P(AB) = P(A)P(B).$$

Three events A, B, C are independent by Definition 2.1.3.2 if

$$P(A|BC) = P(A|B) = P(A|C) = P(A)$$
$$P(B|AC) = P(B|A) = P(B|C) = P(B)$$
$$P(C|AB) = P(C|A) = P(C|B) = P(C). \tag{2.8}$$

All the conditions above are not independent. We shall show that the conditions above hold if any of the following sets of four conditions is satisfied:

I: $P(A|B) = P(A)$, $P(A|C) = P(A)$, $P(B|C) = P(B)$,
$\quad P(C|AB) = P(C)$

II: $P(A|B) = P(A)$, $P(A|C) = P(A)$, $P(C|B) = P(C)$,
$\quad P(B|AC) = P(B)$

III: $P(B|A) = P(B)$, $P(B|C) = P(B)$, $P(A|C) = P(A)$,
$\quad P(C|AB) = P(C)$

IV: $P(B|A) = P(B)$, $P(B|C) = P(B)$, $P(C|A) = P(C)$,
$\quad P(A|BC) = P(A)$

V: $P(C|A) = P(C)$, $P(C|B) = P(C)$, $P(A|B) = P(A)$,
$\quad P(B|AC) = P(B)$

VI: $P(C|A) = P(C)$, $P(C|B) = P(C)$, $P(B|A) = P(B)$,
$\quad P(A|BC) = P(A)$.

$$(2.9)$$

It is easy to check up that the four conditions in any of the sets above are mutually independent in the sense that none of them can be derived from the others. Moreover, the six sets of conditions are similar in structure and can be obtained by permuting the letters A, B, C.

Suppose that the conditions in set I hold. From the first three conditions it follows by Eq. (2.7) that

$$P(B|A) = P(B), \qquad P(C|A) = P(C), \qquad P(C|B) = P(C).$$

All the conditions in Eq. (2.8) will hold if we can show further that

$$P(B|AC) = P(B) \quad \text{and} \quad P(A|BC) = P(A).$$

For this, we notice by Theorem 2.1.3.4 that

$$P(ABC) = P(A)P(B|A)P(C|AB)$$
$$= P(A)P(B)P(C)$$

because $P(C|AB) = P(C)$ by hypothesis, and $P(B|A) = P(B)$, as proved. On the other hand, we have, again by Theorem 2.1.3.4,

$$P(ABC) = P(A)P(C|A)P(B|AC)$$
$$= P(A)P(C)P(B|AC)$$

because $P(C|A) = P(C)$, as proved. Comparing with the preceding expression, we have

$P(B|AC) = P(B)$.

Similarly, by using $P(ABC) = P(B)P(C|B)P(A|BC) = P(B)P(C)P(A|BC)$, we have

$P(A|BC) = P(A)$.

The proof that if the conditions in any other set are satisfied, then the conditions (2.9) for independence of events A, B, C hold, follows from symmetry and is left to the reader as an exercise.

It is easy to verify that events A, B, C are independent if and only if

$$P(AB) = P(A)P(B), \quad P(AC) = P(A)P(C),$$
$$P(BC) = P(B)P(C), \quad P(ABC) = P(A)P(B)P(C). \quad (2.10)$$

Conditions (2.10) imply conditions (2.8), and conversely. This is left as an exercise for the reader.

Thus conditions (2.8), any set of conditions in the group (2.9), and conditions (2.10) offer equivalent conditions for the independence of three events A, B, C. From conditions (2.10) it can further be inferred that for the independence of three events it is not enough that any two of them are independent. In fact, there may be cases where any pair of three events are mutually independent but the three events are not independent among themselves. These are the cases where we have

$$P(AB) = P(A)P(B), \quad P(AC) = P(A)P(C), \quad P(BC) = P(B)P(C)$$

but

$$P(ABC) \neq P(A)P(B)P(C).$$

Example 2.1.3.1: Pairwise independent events that are not independent An urn contains four tickets bearing numbers 1234, 2341, 3412, and 4123, and one ticket is drawn. Define the events A, B, C as follows:

(A) = first digit of the ticket drawn is 1 or 4

(B) = second digit of the ticket drawn is 2 or 4

(C) = third digit of the ticket drawn is 3 or 4.

Then the following probabilities can easily be computed.

$$P(A) = P(B) = P(C) = \tfrac{1}{2}$$
$$P(AB) = P(AC) = P(BC) = \tfrac{1}{4}$$
$$P(ABC) = \tfrac{1}{4}.$$

Obviously,

$$P(AB) = P(A)P(B), \qquad P(AC) = P(A)P(C), \qquad P(BC) = P(B)P(C),$$

and thus A, B, C are pairwise independent. But

$$P(ABC) \neq P(A)P(B)P(C),$$

and thus A, B, C are not independent.

The analysis of independent events for the case of three events, as done above, can easily be extended to any finite number of events. For the independence of n events $A_1, A_2, \ldots, A_n$, the following equivalent sets of conditions are necessary.

1. For any choice of k integers $i_1 < i_2 < \cdots < i_k$ such that

$$1 \leq i_p \leq n, \qquad p = 1, 2, \ldots, k \tag{2.11a}$$

and for any integer j such that

$$1 \leq j \leq n, \qquad j \neq i_1, i_2, \ldots, i_k, \tag{2.11b}$$

we have

$$P(A_j | A_{i_1} A_{i_2} \cdots A_{i_k}) = P(A_j).$$

2. For all combinations $1 \leq i < j < k < \cdots \leq n$, the multiplication rules

$$P(A_i A_j) = P(A_i)P(A_j)$$
$$P(A_i A_j A_k) = P(A_i)P(A_j)P(A_k)$$

$$\cdot$$
$$\cdot$$
$$\cdot$$

$$P(A_1 A_2 \cdots A_n) = P(A_1)P(A_2) \cdots P(A_n) \tag{2.12}$$

hold.

Conditions (2.11a) and (2.11b) together are a reproduction of the conditions of independence of n events in Definition 2.1.3.2. These conditions guarantee that the conditional probability of any single event dependent on the occurrence of one or more of the remaining events should be equal to its unconditional probability. All the conditions in the set are not independent, in the sense that some of them can be derived from the others. This was demonstrated earlier for the case of three events, which can be generalized for any finite number of events. Thus the set of conditions (2.11a) and (2.11b) does not yield a compact set of a minimum number of independent conditions for the independence of n events. Such a set is produced by the set of conditions (2.12). That none of conditions (2.12) can be derived from the remaining conditions can easily be verified; and that conditions (2.11a) and (2.11b), defining the independence of n events, follow from conditions (2.12) is also easily verifiable. This was demonstrated for the case of three events. Conditions (2.12) thus give us a minimum set of independent conditions for the independence of n events. Physically, conditions (2.12) imply that the multiplication rule of probabilities holds for any number of the n events.

The first line of conditions (2.12) yields $\binom{n}{2}$ equations, the second line yields $\binom{n}{3}$ equations, and so on. We thus have

$$\binom{n}{2} + \binom{n}{3} + \cdots + \binom{n}{n} = 2^n - n - 1$$

conditions in (2.12), which must be satisfied and which give the minimum number of independent conditions for the independence of n events. It should be noted that although the minimum number of independent conditions for the independence of n events is $2^n - n - 1$, these conditions can be written down in more than one form. Conditions (2.12) give only one form of these conditions. Other forms of these conditions were discussed for three events, where it was shown that the four conditions in any of the six sets in (2.9) assure the independence of three events.

EXERCISES

23. Urn U_1 contains five white and three black balls; urn U_2 contains four white and five black balls; and urn U_3 contains three white and seven black balls.
     (a) One ball is drawn from each urn. What is the probability that two of them are black and one white?
     (b) An urn is selected at random and four balls are drawn from it without replacement. What is the probability that there is at least one of each color among them?
     (c) One ball is drawn from U_1 and is placed in U_2; thereafter one ball is drawn from U_2 and placed in U_3. Subsequently, one ball is drawn from U_3. What is the probability that it is white?
24. In a bridge game what is the probability that:
     (a) Each player has exactly one ace?
     (b) A player and his partner have three aces between them?
     (c) A player and his partner have k ($k \le 13$) cards of a specified suit?
     (d) A hand has n_1 clubs, n_2 diamonds, n_3 hearts, and n_4 spades, where $n_1 + n_2 + n_3 + n_4 = 13$?

25. If four dice are tossed, what is the probability that:
    (a) The total of the face numbers is 15, 16, or 17?
    (b) The difference between the highest and the lowest face numbers is at least 3?
26. Two persons toss a coin 20 times each. What is the probability that they both obtain the same number of heads?
27. An urn A contains five balls numbered 1, four balls numbered 2, and seven balls numbered 3. Urn B_1 contains eight black and 10 white balls; urn B_2 contains seven black and eight white balls; and urn B_3 contains nine black and six white balls. One ball is drawn from A, and if the number on the ball is i, a ball is drawn from the urn B_i. What is the probability that it is white?
28. An urn contains N balls numbered $1,2,...,N$; k balls are drawn from it with replacement. Find the probability that every number $1,2,...,N$ occurs at least once.
29. Six men and six women are seated on a bench. What is the probability that men and women alternate? Does the result change if they are seated around a round table instead of a bench?
30. A box contains k balls numbered $1,2,...,k$. They are drawn one by one without replacement in succession. If X_i is the number of the ball drawn at the ith stage, we say that a match occurs if $X_i = i$, $i = 1,2,...,k$. Find the probability of (a) exactly r matches; (b) at least r matches.
31. A die is tossed repeatedly until all faces have appeared at least once. What is the probability that (a) the minimum number of tosses required is 12; (b) exactly 12 tosses are required?
32. A box contains N balls numbered $1,2,...,N$, and n balls are drawn from it with replacement. What is the probability that k different balls are drawn?
33. Using unions, intersections, and complements of sets, and the definitions of total and compound events in terms of sets, prove by applying the axiomatic definition of probability that for any three arbitrary events A, B, C:
    (a) $P(A \cup B \cup C) = P(A) + P(B) + P(C) - P(A \cap B) - P(B \cap C) - P(C \cap A) + P(A \cap B \cap C)$.
    (b) $P(A \cap B \cap C) = P(A)P(B|A)P(C|A \cap B)$.
34. Prove that if A and B are independent events, then:
    (a) $\overline{A}$ and $\overline{B}$ are independent events.
    (b) $\overline{A}$ and B are independent events.
35. Prove that for arbitrary events A, B, C, the following statements do not hold in general:
    (a) $P(A|C) + P(\overline{A}|C) = 1$.
    (b) $P(A|C) + P(A|\overline{C}) = 1$.

(c) $P(A \cup B|C) = P(A|C) + P(B|C)$.
(d) $P(A \cap B|C) = P(A|C)P(B|C)$.
Investigate conditions, if any, under which each statement is correct.

2.2 AXIOMATIC DEFINITION OF PROBABILITY

In the earlier sections of this chapter we have seen that the results of a random experiment is a sample space S whose points describe the possible outcomes of the experiment, and that events are described by subsets of S. In defining the probability of an event on the sample space S, we assumed that the sample points were equally likely and the probability of an event A was then defined as the ratio of the number of points in the subset S_A defining A (number of cases favorable to A) to the total number of points in S (total number of equally likely cases). That is, if S contains n points, the probability of a sample point is $1/n$, and the probability of an event A is obtained by adding the probabilities of the sample points of the subset S_A.

The definition of probability above works as long as S contains a finite number of points (however large that number may be), but becomes ambiguous when the number of points in S is infinitely large or uncountable. The classical definition of probability thus heavily rests on the assumption that the sample space consists of a finite number of points, and is to this extent restrictive. It is to be pointed out that the assumption that all sample points are equally likely for the definition of classical probability is rather stringent and is capable of generalization. Thus if a sample space S contains n points, say $s_1, s_2, \ldots, s_n$, and we attach numbers of measures $P_1, P_2, \ldots, P_n$, $P_i \geq 0$ for $i = 1, 2, \ldots, n$, and

$$\sum_{i=1}^{n} P_i = 1$$

to the points $s_1, s_2, \ldots, s_n$, respectively, such that p_i is the probability of the sample point s_i, $i = 1, 2, \ldots, n$, we can define the probability of an event A within the framework of the classical definition of probability as follows: Let S_A be the subset of S defining the event A, and let S_A consist of the points $s_{A_1}, s_{A_2}, \ldots, s_{A_m}$, where the s_{A_i} are some of the points $s_1, s_2, \ldots, s_n$ of the sample space S; the probability of A is defined as

$$P(A) = \sum_{i=1}^{m} P_{A_i}.$$

It can easily be verified that this extended definition of probability within the framework of the classical setup (based on the assumption that a sample space contains a finite number of points) does not disturb any of the theorems of probability we developed in the preceding sections. The basic steps in the proofs of these theorems remain unaltered.

Generally speaking, the framework for the definition of classical probability theory can be summarized as consisting of a set S (called the sample space) and a family F of subsets of S (called the elementary events) whose measures (probabilities) are prescribed in advance. Finally, a set of rules is postulated whereby measures (or equivalently probabilities) of subsets of S (events) can be computed. The accepted rules for classical probability theory are as follows:

1. S consists of a finite number of points (elementary events).
2. Denoting by μ the measure (probability), we have

$$\mu(S) = 1, \qquad \mu(\emptyset) = 0$$

where $\emptyset$ is the subset that contains no element of S (i.e., $\emptyset$ is the empty set). Note that S corresponds to the "sure" event and $\emptyset$ to the "impossible" event.

3. If $B_n \subset S, n = 1,2,3,\ldots,k$ are disjoint, then

$$\mu\left(\bigcup_{n=1}^{k} B_n\right) = \sum_{n=1}^{k} \mu(B_n)$$

(cf. the additive law of probability).

Any definition of probability that retains the above accepted rules of computation of probability measures is capable of retaining the algebra of probability developed in earlier sections. The axiomatic definition of probability is one such generalization of the classical definition. In the axiomatic definition of probability the sample space S is treated as an abstract set, and events in the sample space are the "measurable" subsets of S. For an intelligible exposition of the definition we need to know some fundamental results of the algebra of sets and the theory of measures. The following two sections are devoted to these topics.

2.2.1 The Algebra of Sets

Definition 2.2.1.1: Set   A *set*, in an abstract sense, is any well-defined collection of objects. These objects may be a collection of objects, such as numbers, people, dogs, or electric bulbs. An object belonging to a particular set is called an *element* of the set.

Sets are commonly represented by capital letters ($A,B,C,...$), and elements of sets by lowercase letters ($a,b,c,...$). If an element a belongs to a set A, it is represented as

$a \in A$.

If, on the other hand, an element does not belong to a set A, it is represented as

$a \notin A$.

To specify that certain objects belong to a set, we use braces $\{\cdot\}$. Thus the set $A = \{1,2,3\}$ consists of the numbers 1, 2, 3.

Sets can be described in two ways. (1) by enumerating or listing all elements in a set (the tabular form of the set), and (2) by describing or defining the elements of a set by a property or a set of properties that are satisfied by the elements of the set but by no element outside the set (the defining form of the set). An example of a tabular form of a set is $A = \{1,2,3\}$. The corresponding defining form of the set A is $A = \{a : a$ is a positive integer less than or equal to 3$\}$. All sets need not have both tabular and defining forms. For example, the set of real numbers between 0 and 1 does not have a tabular form. Its defining form is $A = \{a : a$ is a real number between 0 and 1$\}$.

Definition 2.2.1.2: Inclusion subset of a set   If each element of a set B is also an element of a set A, B is called a *subset* of A and is represented as $B \subset A$ (or $A \supset B$.)

The set $B = \{1,2,3\}$ is a subset of the set $A = \{3,2,1,5,9\}$. The set of all residents of Montreal is a subset of the set of all residents of Quebec. In particular, for any set A, $A \subset A$.

Definition 2.2.1.3: Equal sets   A set A is said to be *equal* to a set B if $A \subset B$ and $B \subset A$. This is expressed as $A = B$.

Thus the two sets $A = \{3,a,4,b\}$ and $B = \{a,b,4,3\}$ are equal. Note that if two sets are equal, they must consist of the same elements and only the arrangements of these elements may differ for the two sets. The definition of equality of sets implies that the arrangement of elements in a set will be of no consequence in our considerations of the set. Also, if an element occurs more than once in the representation of a set, we shall ignore its repetition in the representation, since it will be of no concern to us. Thus the sets $A = \{5,2,7,a,9\}$ and the set $B = \{2,5,2,a,9,a,7\}$ are equal. Thus in the representation of a set all elements are distinguishable and the same element does not occur more than once.

It is to be noted that with reference to a particular problem there will be a largest set containing all elements relevant to the problem, and any set considered for this problem will be a subset of this largest set. For example, all the elementary events (or outcomes) resulting from a random experiment form a set S, called the *sample space* for the experiment, and any event in relation to the random experiment is a subset of S. The existence of such a largest set is of fundamental importance for set representation. We shall call this largest set the *universal set*. For practical applications we shall be concerned with sets that are subsets of a universal set.

We denote by the symbol $\emptyset$ the null or empty set, which contains no elements. It is to be noted that $\emptyset \subset S$. Furthermore, any set A considered for the problem is a subset of S ($A \subset S$) and generates a set $\overline{A}$, called the *complement* of A, which consists of all elements of S that are not contained in A; that is, $\overline{A} = \{x \in S : x \notin A\}$.

There are a few basic operations on sets, analogous to the algebraic operations on real numbers, which are very useful in handling sets. The algebra of sets is based on the rules guiding these operations. In other words, just as the algebra of real numbers is confined to the study of operations on numbers and the consequences of these operations, the algebra of sets is concerned with studying some analogous operations on sets and the consequences of these operations. The basic operations on sets are those of union, difference, and intersection.

Definition 2.2.1.4: Union of sets The *union* of two sets A and B is defined as the set of all elements that belong to A or B or both A and B, and is represented as $A \cup B$. Symbolically, $A \cup B$ can be expressed as

$$A \cup B = \{x : x \in A \quad \text{or} \quad x \in B\}.$$

Definition 2.2.1.5: Difference of two sets The *difference* of two sets (read "A difference B") is defined as the set of elements that belong to A but not to B, and is denoted by

$$A - B.$$

Symbolically, $A - B$ can be expressed as

$$A - B = \{x : x \in A \quad \text{but} \quad x \notin B\}.$$

It is evident that

$$A - B = A \cap \overline{B}. \tag{2.13}$$

Definition 2.2.1.6: Intersection of sets   The *intersection* of two sets A and B is defined as the set of all elements that belong to both A and B, and is represented by

    $A \cap B$.

Symbolically, $A \cap B$ can be expressed as

    $A \cap B = \{x : x \in A \quad \text{and} \quad x \in B\}$.

For example, if $A = \{2,1,3,5,9\}$ and $B = \{3,9,13,4\}$, then

    $A \cup B = \{1,2,3,5,9,4,13\}$

    $A - B = \{1,2,5\}$

    $A \cap B = \{3,9\}$.

A handy pictorical way to represent sets, relationships among sets, and operations on sets is to use a *Venn diagram.* A Venn diagram represents a set by a geometrical configuration on a plane such that the points inside and on the boundary of the configuration are supposed to constitute the set. Various types of geometric configurations can be used for the purpose. The basic concept is that all the elements of a set are represented pictorially by all the points inside and on the boundary of the configuration representing the set. Many of the concepts already discussed are shown by a Venn diagram in Figure 1.

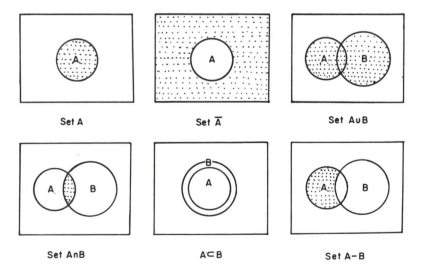

Figure 1.1   Venn diagram. The rectangle represents the universal set. The dotted regions represent the relevant sets.

For any set A we have, by definition,

$$A = A, \quad A \subset A.$$

Thus the equality and inclusion relations are *reflexive*. They are also *transitive*, as can be seen from the fact that

$$A = B, \quad B = C \quad \text{implies that} \quad A = C$$

$$A \subset B, \quad B \subset C \quad \text{implies that} \quad A \subset C.$$

Again it follows from definition that for any two sets A and B,

$$A \cup B = B \cup A, \quad A \cap B = B \cap A; \tag{2.14}$$

that is, the operations of union and intersection are *commutative*. But

$$A - B \neq B - A; \tag{2.15}$$

that is, the difference relationship is not commutative. Furthermore, it is easy to verify that for any three sets A, B, C,

$$A \cup (B \cup C) = (A \cup B) \cup C$$

$$A \cap (B \cap C) = (A \cap B) \cap C. \tag{2.16}$$

Thus the operations of union and intersection are *associative*. It can further be verified that the union and intersection operations obey the following distributive laws:

$$A \cap (B \cup C) = (A \cap B) \cup (A \cap C) \tag{2.17}$$

$$A \cup (B \cap C) = (A \cup B) \cap (A \cup C). \tag{2.18}$$

It would be instructive to verify the relations (2.15)–(2.18) by the use of Venn diagrams.

In view of Eqs. (2.16), it is customary to express the union and intersection of any number of sets without using parentheses. Thus we shall express

$$A_1 \cup A_2 \cup A_3 \cup \cdots \cup A_n = \bigcup_{i=1}^{n} A_i \quad \text{and}$$

$$A_1 \cap A_2 \cap A_3 \cap \cdots \cap A_n = \bigcup_{i=1}^{n} A_i,$$

where n can be infinite.

It is clear that union and intersection relations are related to the operation of taking complements by the following equations:

$$A \cap \overline{A} = \emptyset, \qquad A \cup \overline{A} = S$$
$$\overline{(A \cup B)} = \overline{A} \cap \overline{B}, \qquad \overline{(A \cap B)} = \overline{A} \cup \overline{B}, \tag{2.19}$$

where S is the universal set, A and B are any subsets of S, and $\emptyset$ is the null set. More generally.

$$\overline{\bigcup_{\alpha} A_{\alpha}} = \bigcap_{\alpha} \overline{A}_{\alpha}, \qquad \overline{\bigcap_{\alpha} A_{\alpha}} = \bigcup_{\alpha} \overline{A}_{\alpha}.$$

The four operations defined above (union, intersection, difference, and complementation) can be reduced to two in several different ways. For example, they can be expressed in terms of unions and complements. In fact, in view of Eqs. (2.13) and (2.16)–(2.19), the following duality is obvious.

Duality in Set Relations

There is a complete *duality* in set operations, in the sense that any true proposition about sets remains true if we interchange $\emptyset$ and S, $\cup$ and $\cap$, $\cap$ and $\cup$, $\subset$ and $\supset$, and leave $=$ and $-$ (complementation) unchanged all through. Thus any relation among a number of sets gives rise to a dual relation through the interchanges above.

The union of a countable number of sets $A_1, A_2, \ldots$ can be expressed as the union of a countable number of disjoint sets as follows:

$$\bigcup_{n=1}^{\infty} A_n = A_1 \cup (A_2 \cap \overline{A}_1) \cup (A_3 \cap \overline{A}_1 \cap \overline{A}_2) \cup \cdots . \tag{2.20}$$

Equations (2.14), (2.16), and (2.17) demonstrate a close relationship between the union and intersection of sets and the addition and multiplication of real numbers, the union of sets corresponding to addition of numbers and the intersection of sets to multiplication of numbers, since for real numbers a, b, c we have

$$a + b = b + a$$
$$a \cdot b = b \cdot a$$
$$a + (b + c) = (a + b) + c$$
$$a \cdot (b \cdot c) = (a \cdot b) \cdot c$$
$$a \cdot (b + c) = (a \cdot b) + (a \cdot c).$$

However, the second distributive law [Eq. (2.18)] does not have an analog in the domain of numbers, as the corresponding law would be

$$a + (b \cdot c) = (a + b) \cdot (a + c),$$

which is not true for all numbers a, b, c.

The algebra of sets is applicable to the representation of events with reference to a random experiment. If S is the sample space generated by a random experiment, then any event with reference to the experiment is a subset of S. In this sense the space S represents the universal set for the random experiment, and the terms *events* and *set* (a subset of S) can be used interchangeably. For example, if the random experiment consists of tossing two dice, the sample space consists of the 36 points (x,y), $x = 1,...,6$, $y = 1,...,6$. If E is the event that the sum of the face values of the two dice is 9, then E can be identified with the subset $\{(3,6),(4,5),(5,4),(6,3)\}$ of S; any event relevant to the experiment can similarly be identified with a subset of S by using its physical description. Henceforth we shall represent events as subsets of the underlying sample space.

If A and B are any two events with reference to a random experiment (subsets of the sample space generated by the random experiment), then $\overline{A}$, the complement of A, represents the event that A does not happen, $A \cup B$ the event that A or B (or both) happen, $A \cap B$ the event that A and B happen simultaneously, $A - B$ the event that A happens but B does not happen; if $A \subset B$, this implies that the materialization of the event A implies that of the event B, but the contrary may not be true. Set notation can be extended to represent the composition of any number of events.

Sequences of Sets, Upper and Lower Limits, and Monotone Sequences

Definition 2.2.1.7: Upper limit of a sequence of sets   The *upper limit* of a sequence of sets $\{A_n\}$ is the set of points that belong to A_n for infinitely many n, and is denoted as lim sup A_n. It can be easily verified that

$$\lim \sup A_n = \bigcap_{n=1}^{\infty} \bigcup_{m=n}^{\infty} A_m.$$

Definition 2.2.1.8: Lower limit of a sequence of sets   The *lower limit* of a sequence of sets $\{A_n\}$ is the set of points that belong to A_n for all but a finite number of n, and is denoted as lim inf A_n.

It can be verified that

$$\lim \inf A_n = \bigcup_{n=1}^{\infty} \bigcap_{m=n}^{\infty} A_m.$$

It is easy to show that

$$\lim \inf A_n \subset \lim \sup A_n. \tag{2.21}$$

Definition 2.2.1.9: Limit of a sequence of sets   If for a sequence of sets $\{A_n\}$, $\lim \sup A_n = \lim \inf A_n = A$, then A is called the *limit* of the sequence A_n, which is then said to converge to A, and is denoted by $\lim A_n = A$.

Definition 2.2.1.10: Monotone increasing and decreasing sequence of sets   A sequence of sets $\{A_n\}$ is *monotone increasing* if

$$A_1 \subset A_2 \subset A_3 \subset \cdots$$

and is *monotone decreasing if*

$$A_1 \supset A_2 \supset A_3 \supset \cdots.$$

Symbolically, we represent a monotone increasing sequence $\{A_n\}$ as $A_n \uparrow$ and a monotone decreasing sequence $\{A_n\}$ as $A_n \downarrow$.

Lemma 2.2.1.1   A monotone sequence of sets $\{A_n\}$, increasing or decreasing, converges, and the limit is $\bigcap_{n=1}^{\infty} A_n$ or $\bigcup_{n=1}^{\infty} A_n$ according as $\{A_n\}$ is decreasing or increasing.

Proof.   Let $\{A_n\}$ be decreasing. Then we have

$$\bigcap_{m=n}^{\infty} A_m = \bigcap_{m=1}^{\infty} A_m$$

and thus

$$\lim \inf A_n = \bigcap_{m=1}^{\infty} A_m.$$

We also have

$$\bigcup_{m=n}^{\infty} A_m = A_n$$

and thus

$$\lim \sup A_n = \bigcap_{n=1}^{\infty} A_n.$$

Thus

$$\lim A_n = \bigcap_{n=1}^{\infty} A_n.$$

The proof for increasing $\{A_n\}$ is similar.   Q.E.D.

Lemma 2.2.1.1 is an analog of the result for real numbers, which states that a monotone sequence of real numbers converges. The analogy is obvious if we write for a sequence of real numbers a_n,

$a_i \cup a_j$ = greater of the numbers a_i and a_j

and

$a_i \cap a_j$ = smaller of the numbers a_i and a_j.

Example 2.2.1.1
   (a)   If A and B are mutually exclusive, B cannot happen if A happens, and vice versa. Thus $A \cap B = \emptyset$ (the null set).
   (b)   With reference to the two events A and B, $A - B = A - A \cap B$ defines the event of the occurrence of A but not of both A and B.

Example 2.2.1.2   Let S be the sample space (universal set) consisting of all placements of four balls in five boxes. Then S consists of 4^5 points. For $i = 1,2,3,4,5$, define the events

E_i = ith box is empty

A_i^j = ith box contains j balls, $j = 1,2,3,4$.

The following relations can easily be verified:
   (a) $E_1 \cap E_2 \cap E_3 \cap E_4 = A_5^4.$
   (b) $A_i^j \cap A_i^{j'} = \emptyset$ for $j \neq j' = 1,2,3,4$ and $i = 1,2,3,4,5$.
   (c) $A_2^1 \cap A_3^1 \cap A_4^1 \cap A_5^1 \subset E_1.$
   (d) $A_1^j \cup A_2^j \cup A_3^j \cup A_4^j \cup A_5^j$ is the event that at least one of the boxes contains j balls, $j = 1,2,3,4$.

EXERCISES

36. Verify the following relations for any sets $A,B,C,\ldots$.
    (a) $\overline{A \cup B} = \overline{A} \cap \overline{B}$.
    (b) $A - B = A - (A \cap B) = A \cap \overline{B} = (A \cup B) - B$.
    (c) $(A \cap B) \cup (A - B) = A$.
    (d) $(A \cap B) \cap (A - B) = \emptyset$.
    (e) $(A \cup B) - (A \cap B) = (A - B) \cup (B - A)$.
    (f) $A \cap (B - C) = (A \cap B) - (A \cap C)$.
    (g) $(A \cup B) \cap C = C - C \cap (A \cup B)$.
    (h) $(A \cap B) \cap (B \cap C) = A \cap B \cap C$.
37. Let $A,B,C,\ldots$, be arbitrary sets. Verify the correctness or incorrectness of each of the following statements.
    (a) $(A \cap B) - C = A \cap B \cap C$.
    (b) $(A \cup B) - C = (A - C) \cup (B - C)$.
    (c) $\overline{(A \cup B)}C = (\overline{A} \cap C) \cup (\overline{B} \cap C)$.
    (d) $A \cup B = [A - (A \cap B)] \cup B$.
    (e) $(A \cap B) \cap (\overline{A} \cap C) = \emptyset$.
    (f) $A \cup B \cup C = A \cup (B - A \cap B) \cup (C - A \cap C)$.
    (g) $(A - B) \cup B = A$.
    (h) $A \cap \overline{B} \cap C \subset A \cup B$.
    (i) $(A - B) \cap (B - C) \subset A \cap B \cap C$.
38. Show that

    (a) $\left(\overline{\bigcup_{n=1}^{k} A_n} \right) = \bigcap_{n=1}^{k} \overline{A}_n, \quad \left(\overline{\bigcap_{n=1}^{k} A_n} \right) = \bigcup_{n=1}^{k} \overline{A}_n.$

    (b) $A \left(\bigcup_{n=1}^{k} B_n \right) = \bigcup_{n=1}^{k} (A \cap B_n).$

39. Let A,B,C be any three events. Using the operations of union, intersection, and complementation, give an expression for the following events.
    (a) None of the events occurs.
    (b) All three events occur.
    (c) At least one event occurs.
    (d) Only A occurs.
    (e) Only A and B occur.
    (f) Exactly one of the events occurs.
    (g) At most two of the events occur.
    (h) Exactly two of the events occur.
40. For a sequence of sets A_n, prove the following relation:

    $\lim \inf A_n \subset \lim \sup A_n.$

41. Let S consist of all integers from 1 to 50. Define the following subsets of S: E_1, all integers divisible by 5; E_2, all integers divisible by 7; E_3, all odd integers less than 36. Find each of the following sets.
    (a) $E_1 \cap E_2$.
    (b) $E_1 \cap E_2 \cap E_3$.
    (c) $E_1 \cap \bar{E}_2$.
    (d) $(E_1 \cap E_2) \cap \bar{E}_3$.
    (e) $(E_1 - E_2) \cap \bar{E}_3$.
    (f) $(E_1 - E_3) \cap (E_2 - E_3)$.
    (g) $(\bar{E}_1 \cap E_3) \cup (E_2 \cap E_3)$.
    (h) $E_1 \cap (E_2 \cup \bar{E}_3)$.
42. An urn contains four balls numbered 1, 2, 3, 4. They are drawn one after another without replacement. If A_i is the event that the ith ball appears at the ith draw, $i = 1,2,3,4$:
    (a) Show that $A_1 \cap A_2 \cap A_3 \subset A_4$; $A_1 \cap A_2 \cap \bar{A}_3 \subset \bar{A}_4$.
    (b) Find each of the following sets.

    $$\overline{A_1 \cup A_2 \cup A_3 \cup A_4}; \quad (A_1 \cap A_2) \cup (A_3 \cap A_4);$$
    $$(\overline{A_1 \cup A_2}) \cap (\overline{A_2 \cup A_3}).$$

43. Two dice are tossed. Define the following events.

    A_i: sum of the face values is i, $i = 2,3,\ldots,12$,

    B_1: difference of the face values is an odd integer,

    B_2: difference of the face values is an even integer or 0.

    Prove the following.
    (a) $A_i \subset B_1$ if i is an even integer.
    (b) $A_i \subset B_2$ if i is an odd integer.
    (c) $A_2 \cup A_4 \cup A_6 \cup A_8 \cup A_{10} \cup A_{12} = B_2$.
    (d) $A_1 \cup A_3 \cup A_5 \cup A_7 \cup A_9 \cup A_{11} = B_1$.

2.2.2 Set Function and Measure

Definition 2.2.2.1: Real-valued point function on a set A *real-valued point function f* defined on a set S is a rule that associates a real number with every point (or element) of S. The number associated with a particular point is called the *value* of the function for the point and is denoted by $f(x)$ for $x \in S$. The set S is called the *domain* of the function f, and the set of values $\{f(x) : x \in S\}$ is called the *range* of the function f.

It is to be noted that in the definition of a point function the same number can be associated with more than one point of S. If a different number is associated with each different point of S, the function is called *single valued*; otherwise, it is called *multiple valued*.

For example, S may be the set of residents of a city and f may be the point function that gives the heights of the residents, or the ages of the residents. As a second example, let $S = \{x : x$ is a real number between 0 and 1$\}$; then $f(x) = x^2$ is a single-valued function defined over the domain S with range equal to S.

Definition 2.2.2.2: Class of sets   A *class* of sets is a collection of sets.

Unlike sets, for which elements are points, the elements of a class are sets. Using script letters to represent classes, examples of classes are

$\mathcal{A} = \{\{a\}, \{b\}\}$

$\mathcal{B} = \{\{a\}, \{b\}, \{a,b\}\}$

$\mathcal{C} = \{\emptyset, \{a\}, \{b\}, \{a,b\}\}$

$\mathcal{D} = \{$all subsets of a set $S\}$,

where a, b are any specific objects and $\emptyset$ is the null set.

Definition 2.2.2.3: Closed class of sets   A class $\mathcal{F}$ of sets is said to be *closed* with respect to unions if $A \cup B$ is an element of $\mathcal{F}$ for all A and B belonging to $\mathcal{F}$. Similarly, a class $\mathcal{F}$ of sets is said to be closed with respect to intersections if $A \cap B$ is an element of $\mathcal{F}$ for all A and B belonging to $\mathcal{F}$. A class $\mathcal{F}$ is said to be closed if it is closed with respect to both unions and intersections.

It can easily be verified that of the examples of classes of sets given above, the class $\mathcal{A}$ is closed neither with respect to unions nor with respect to intersections, and thus is not closed; the class $\mathcal{B}$ is closed with respect to unions but not with respect to intersections (since $\emptyset = \{a\} \cap \{b\}$ is not an element of the class) and thus is not closed, whereas the classes $\mathcal{C}$ and $\mathcal{D}$ are closed.

A set function is usually defined over a closed class of sets.

Definition 2.2.2.4: Real-valued set function   Let $\mathcal{F}$ be a closed class of sets. A rule f that associates a real number with each set element of $\mathcal{F}$ is called a *real-valued set function* defined on $\mathcal{F}$. The number associated with a particular element A of $\mathcal{F}$ is called the *value* of the function f for the element $A \in \mathcal{F}$, and is denoted by $f(A)$. The class $\mathcal{F}$ is called the *domain* of the function f, and the set of values $\{f(A) : A \in \mathcal{F}\}$ is called the *range* of the function f.

For example, let $\mathfrak{F}$ consist of the families in a locality (a family is a set of persons residing in the locality). The total income of a family is a set function with $\mathfrak{F}$ as its domain and the set of all different total incomes as its range.

Additive Systems of Sets

The accepted rules for classical probability theory were summarized at the beginning of Section 2.2, S for finite sample space, where the events are subsets of S, in terms of the set representations $(A_1 \cup A_2)$ and $(A_1 \cap A_2)$. In order that the algebra of probability as developed in Section 2.1 remain meaningful, it is necessary that the class $\mathfrak{B}$ of sets B (events) for which probability is defined should satisfy rules 1 to 4, which can be rephrased as follows:

1. $\mathfrak{B}$ contains the null set $\emptyset$.
2. If $B \in \mathfrak{B}$, then $\overline{B} \in \mathfrak{B}$.
3. If $B_i \in \mathfrak{B}$, $i = 1,2,\ldots,n$, then $\cup_{i=1}^{n} B_i \in \mathfrak{B}$ for any finite n.
   Because of duality of set representation it follows that
4. $\cap_{j=1}^{n} B_i \in \mathfrak{B}$.

This implies that $\mathfrak{B}$ is a closed class of sets containing the null set. Such a set is called a finitely additive class of sets.

Definition 2.2.2.5: Finitely additive class of sets   A class $\mathfrak{B}$ of sets is called *finitely additive* if
   (a) $\emptyset \in \mathfrak{B}$.
   (b) If $B \in \mathfrak{B}$, then $\overline{B} \in \mathfrak{B}$.
   (c) If $B_i \in \mathfrak{B}$, $i = 1,2,\ldots,n$, then $\cup_{i=1}^{n} B_i \in \mathfrak{B}$ for any finite n.

To extend the definition of probability to a sample space S that is not necessarily finite and can be taken to be an abstract set, capable of retaining the algebra of probability, it is necessary that the class of sets (subsets of S) for which probability is defined is not only finitely additive but is completely additive.

Definition 2.2.2.6: Completely additive class of sets   A class $\mathfrak{B}$ of sets is called *completely additive* if
   (a) $\emptyset \in \mathfrak{B}$.
   (b) If $B \in \mathfrak{B}$, then $\overline{B} \in \mathfrak{B}$.
   (c) If $B_i \in \mathfrak{B}$, $i = 1,2,\ldots$, then $\cup_{i=1}^{\infty} B_i \in \mathfrak{B}$.
   A completely additive class of sets is also called a σ-*field*. It is evident that a σ-field is also closed under countable intersections.

It is evident that the intersection of any number (not necessarily countable) of σ-fields is again a σ-field. Moreover, given a family 𝔉 of sets, the intersection of all σ-fields containing 𝔉 is a uniquely defined minimal σ-field containing 𝔉. The existence of at least one σ-field containing a given family 𝔉 of sets (subsets of S) is trivial, since the class of all subsets of S is a σ-field. We thus have the following theorem:

Theorem 2.2.2.1. Any given family of subsets of a set S is contained in a unique minimal additive class $\mathcal{B}_0$ of subsets and in a unique minimal σ-field $\mathcal{B}$ of subsets.

For an example of a finitely additive class of sets, consider the family $\mathcal{F}_0$ of rectangles $a_i \le x_i < b_i$ ($i = 1,2,\dots,n$) in the Euclidean space E^n. $\mathcal{F}_0$ is not an additive class, but is contained in the minimal finitely additive class $\mathcal{B}_0$ consisting of all elementary figures and their complements. An *elementary figure* is the union of a finite number of such rectangles.

It is evident that the intersections of sets of a finitely additive (or completely additive) system with a fixed set of the system form a finitely additive (or completely additive) subsystem of the original system.

Definition 2.2.2.7: Finitely additive real-valued set function   Let $\mathcal{B}$ be a finitely additive class of sets (subsets of a set S). A real-valued set function μ, defined on $\mathcal{B}$, is called *finitely additive* if:

(a) $\mu(\emptyset) = 0$.
(b) If $B_i \in \mathcal{B}$, $i = 1,\dots,n$ are disjoint for any n, then

$$\mu\left(\bigcup_{i=1}^{n} B_i\right) = \sum_{i=1}^{n} \mu(B_i).$$

An additive set function need not always be real valued, but we shall be concerned with real-valued set functions. The condition that $\mu(\emptyset) = 0$ holds, in view of the additivity condition (b), if $\mu(B)$ is finite for at least one set $B \in \mathcal{B}$.

For a simple example of a finitely additive real-valued set function we may take $\mathcal{B}$ as the minimal finitely additive class containing the family of rectangles $a_i \le x_i < b_i$ ($i = 1,2,\dots,n$) in E^n, and $\mu(B)$ as the volume of B, where B is an elementary figure.

If the additivity property of a set function extends to countable systems of sets, the function is called completely additive. We thus have the following formal definition.

Definition 2.2.2.8: Completely additive real-valued set function   Let $\mathcal{B}$ be a completely additive class (σ-field) of sets (subsets of a set S). A real-valued set function defined on $\mathcal{B}$ is called *completely additive* if:
  (a) $\mu(\emptyset) = 0$.
  (b)   If $B_i \in \mathcal{B}$, $i = 1,2,\dots$ are disjoint, then

$$\mu\left(\bigcup_{i=1}^{\infty} B_i\right) = \sum_{i=1}^{\infty} \mu(B_i).$$

A completely additive set function is well defined even if $\mathcal{B}$ is finitely additive, but $\bigcup_{i=1}^{\infty} B_i$ belongs to $\mathcal{B}$ whenever B_i, $i = 1,2,\dots$ belongs to $\mathcal{B}$.

For an example of a completely additive set function we can take $\mu(B)$ to be the number of elements (finite or infinite) in B for all subsets B of a set S.

For an example of a finitely additive set function that is not completely additive, let S be any infinite set, and let

$$\mu(B) = \begin{cases} 0 & \text{if } B \text{ is a finite subset of } S \\ 1 & \text{if } B \text{ is an infinite subset of } S. \end{cases}$$

Now let B be a countable set of elements $(s_1, s_2, \dots)$ of S. Then $\mu(\{s_n\}) = 0$ for every n, and thus

$$\sum_{n=1}^{\infty} \mu(\{s_n\}) = 0.$$

But $\mu(B) = 1$. Thus μ is not completely additive. That μ is finitely additive is obvious.

Definition 2.2.2.9: Measure and measurable sets   A nonnegative, additive, real-valued set function defined on an additive class of sets is called a *measure*.

In other words, if $\mathcal{B}$ is an additive class of subsets of a set S (universal set), a real-valued set function μ, defined on $\mathcal{B}$, is called a measure if:
  (a) $\mu(B) \geq 0$ for every $B \in \mathcal{B}$.
  (b) $\mu(\emptyset) = 0$.
  (c)   If $B_i \in \mathcal{B}$ are disjoint such that $\cup_i B_i \in \mathcal{B}$, then

$$\mu\left(\bigcup_i B_i\right) = \sum_i \mu(B_i).$$

The measure μ is called *finitely additive* if the additivity property (c) is valid only for unions of finite numbers of disjoint sets; it is called *completely additive* if the additivity property extends also to unions of a countable infinity of disjoint sets. The sets B are called *measurable sets* (with respect to the measure μ). The measure μ is called *bounded* (or *finite*) if $\mu(S) < \infty$.

It follows from the definition that a measure defined over a finitely additive system $\mathcal{C}$ of sets is necessarily finitely additive and may be completely additive on $\mathcal{C}$ if

$$\mu\left(\bigcup_{i=1}^{\infty} C_i\right) = \sum_{i=1}^{\infty} \mu(C_i)$$

for $C_i \in \mathcal{C}$, $i = 1,2,\ldots$ disjoint, whenever

$$\bigcup_{i=1}^{\infty} C_i \in C.$$

A measure defined on a σ-field is always completely additive.

We are now ready to give an axiomatic set-theoretical definition of probability as a measure defined on a σ-field $\mathcal{B}$ of subsets of an abstract set S. The set S corresponds to the sample space underlying a random experiment, and the σ-field $\mathcal{B}$ corresponds to the class of all events (subsets of S) under consideration.

Definition 2.2.2.10: Axiomatic definition of probability   Let
  (a) S be any abstract set.
  (b) $\mathcal{B}$ be a σ-field of subsets of S.
  (c) μ be a measure defined on $\mathcal{B}$ such that $\mu(S) = 1$. Then μ is called a *probability measure* defined on $\mathcal{B}$. The triplet $(S,\mathcal{B},\mu)$ is called a *probability space* or *sample space*.

So far we have been designating the sample space generated by a random experiment consisting of all possible outcomes of the experiment by a single letter S. The triplet $(S,\mathcal{B},\mu)$ gives a more generalized expression for the sample space, in the sense that it not only signifies the space S of all possible outcomes of the underlying random experiment, but goes further by specifying the set $\mathcal{B}$ of events we are interested in and the probability measure μ appropriate for the situation under which the experiment is performed. Henceforth, whenever we make reference to a sample space by the single letter S, we shall have in mind the underlying σ-field of events and the probability measure under consideration.

In applications of the theory of probability, probabilities are attached to events that can be explained in terms of the outcomes of a random experiment. If the space S of these outcomes is finite, the σ-field $\mathscr{B}$ in the axiomatic definition (Definition 2.2.2.10) of probability coincides with the class of all subsets of S; in this case, Definition 2.2.2.10 conforms to the requirements of the classical definition of probability, as outlined in rules 1 to 4; a probability measure is attached to each point of S such that the sum of the probability measures of all points of S is 1, and then the probability measure of an event is the sum of the probability measures of all points of the subset of S defining the event.

If S is not finite, a simplified approach, such as that just given, is not available. Nevertheless, for cases in which S is infinite, our primary interest would usually be directed toward a well-defined family $\mathscr{F}$ of events (subsets of S) which need not form an additive system, and it is customary to take the σ-field $\mathscr{B}$ in the axiomatic definition of probability as the unique minimal σ-field containing $\mathscr{F}$. As an example, consider a target-shooting experiment for which contestants are required to fire shots from a specified distance at a flat circular surface with a 1-foot radius, the bull's-eye being at the center of this target. The ability of a shooter is determined by the distance between the center of the target and the position of his or her shot. In this case S consists of all points lying in the unit interval $[0,1]$ on the real line, and the family $\mathscr{F}$ of subsets of S in which we are primarily interested consists of all half-open intervals $[a,b]$, $0 < a < b < 1$, which is not an additive system. The σ-field relevant to this case is the unique completely additive class of elementary figures which are finite or countable unions of these intervals together with the interval $[0,1]$. Without any loss of generality we can take the σ-field to consist of all elementary figures that are finite unions of these intervals: by the Heine–Borel theorem, for every infinite covering of a closed subinterval of $[0,1]$, by open intervals there exists a finite subcovering.

Assuming that with reference to any random experiment, there is a well-defined family $\mathscr{F}$ of subsets of the underlying sample space S in which we are primarily interested, and that $\mathscr{B}$ in Definition 2.2.2.10 is the minimal σ-field of subsets of S containing $\mathscr{F}$, it becomes imperative for the general application of the axiomatic definition of probability that we know how to construct the actual minimal σ-field containing a given family of sets that are subsets of a given set. The details of this construction have been given by Hausdorff (1927, p. 85). In actual applications, however, the sample space S is usually a subset of E^k, the k-dimensional Euclidean space, and the family $\mathscr{F}$ of events of primary interest is the class of rectangles $[a_i,b_i]$, $i = 1,2,\ldots,k$, of S, yielding the unique σ-field $\mathscr{B}$ consisting of the elementary figures (countable unions of these rectangles); in this case the measure involved is called a *Lebesgue measure*. For further study the reader is referred to Loève (1955).

2.3 BAYES' THEOREM

We conclude this chapter by proving a major theorem due to Bayes. The theorem is simple, but the concepts involved are of great importance in the decision-theoretical approach to statistical inference.

Theorem 2.3.1: Bayes' theorem Let
(a) S be any sample space.
(b) $\{E_i, E_2, \ldots, E_n\}$ be a partition of S [i.e., $E_1, E_2, \ldots, E_n$ define a set of mutually exclusive and exhaustive events on S] such that $P(E_i) \neq 0$ for $i = 1, 2, \ldots, n$.
(c) $F \subset S$ be any event such that $P(F) \neq 0$. Then

$$P(E_k|F) = \frac{P(E_k)P(F|E_k)}{\sum_{i=1}^{n} P(E_i)P(F|E_i)}, \qquad k = 1, 2, \ldots, n \qquad (2.22)$$

where n may be countable or uncountable infinity. If n is uncountable, the summation in the denominator of Eq. (2.22) is replaced by the corresponding integral.

Proof. It is evident that $\{F \cap E_1, F \cap E_2, \ldots, F \cap E_n\}$ is a partition of the set F, and as such we have by Theorems 2.1.3.1 and 2.1.3.3 that

$$P(F) = \sum_{i=1}^{n} P(F \cap E_i) = \sum_{i=1}^{n} P(E_i)P(F|E_i).$$

Again by Theorem 2.1.3.3, we have for $k = 1, 2, \ldots, n$,

$$P(E_k|F) = \frac{P(E_k \cap F)}{P(F)} = \frac{P(E_k)P(F|E_k)}{P(F)}.$$

Combining the two expressions, the result follows.   Q.E.D.

In the statement of Bayes' theorem, the probabilities $P(F|E_k)$, $k = 1, 2, \ldots, n$ are called the *a priori probabilities* of the event F, and the probabilities $P(E_k|F)$, $k = 1, 2, \ldots, n$ are called the *a posteriori probabilities* of the events E_k. Bayes' theorem gives a rule for generating the system of a posteriori probabilities corresponding to a system of a priori probabilities. Bayes' theorem is sometimes referred to as *Bayes' rule*.

Example 2.3.1 Two urns U_1 and U_2 contain, respectively, three white and five black balls, and four white and six black balls. One ball is drawn from U_1 and placed in U_2. Then a ball is drawn from U_2 and turns out to be black. What is the probability that the ball drawn from U_1 was white?

Let E_1 be the event that the ball drawn from U_1 is white, and E_2 be the event that the ball chosen from U_1 is black. Let F be the event that the ball chosen from U_2 is black. Then $P(E_1|F)$ is the probability that the ball drawn from U_1 is white, given that the ball chosen from U_2 is black. To compute $P(E_1|F)$ we note that

$$P(E_1) = \frac{3}{8}, \qquad P(E_2) = \frac{5}{8}, \qquad P(F|E_1) = \frac{6}{11}, \qquad P(F|E_2) = \frac{7}{11}$$

and thus applying Bayes' rule, we have

$$P(E_1|F) = \frac{P(E_1)P(F|E_1)}{P(E_1)P(F|E_1) + P(E_2)P(F|E_2)}$$

$$= \frac{(3/8) \cdot (6/11)}{(3/8) \cdot (6/11) + (5/8) \cdot (7/11)}$$

$$= \frac{18}{53}.$$

EXERCISES

43. Three urns U_1, U_2, U_3 contain, respectively, two white and three black balls, three white and four black balls, and one white and two black balls.
    (a) Two balls, one from urn U_1 and the other from urn U_2, are drawn and placed in urn U_3; thereafter one ball is drawn from U_3 and is found to be black. What is the probability that the ball drawn from U_1 was white?
    (b) One ball is drawn from U_1 and placed in U_2; thereafter two balls are drawn from U_2 and placed in U_3. Then a ball is drawn from U_3 and is found to be white. What is the probability that the two balls drawn from U_2 are both white?
44. An urn contains five white and six black balls.
    (a) Three balls are drawn from the urn one after another without replacement, and it is found that one of them is white and the other two are black. What is the probability that the first ball drawn was white? What is the probability that the first ball drawn was white if the three balls were drawn with replacement?
    (b) Three balls are drawn from the urn and discarded. Then a fourth ball is drawn from it. What is the probability that it is white?

45. The families in a community are divided into four income groups G_1, G_2, G_3, G_4, consisting of N_1, N_2, N_3, N_4 families, respectively. Each group is composite in the sense that it consists of German, English, Canadian, and American families, their respective proportions for the group G_j being q_{j1}, q_{j2}, q_{j3}, q_{j4}, $\Sigma_{k=1}^4 q_{jk} = 1$, $j = 1,2,3,4$. A family is chosen at random and is found to be German. What is the probability that it belongs to the group G_j?

BIBLIOGRAPHY

Bernoulli, J. A. (1713). *Ars Conjectandi*, Basel.

Chung, K. L. (1968). *A Course in Probability Theory*, Harcourt Brace & World, New York.

Feller, W. (1957). *An Introduction to Probability Theory and its Applications*, Wiley, New York.

Hausdorff, F. (1927). *Mengenlehre*, De Gruyter, Berlin.

Loève, M. (1955). *Probability Theory*, Van Nostrand, New York.

Parzen, E. (1960). *Modern Probability Theory*, Wiley, New York.

Uspensky, J. V. (1937). *Introduction to Mathematical Probability*, McGraw-Hill, New York.

3

Random Variables, Probability Distributions, and Characteristic Functions

3.1 RANDOM VARIABLES

The outcome of a random experiment may be any one of a great variety of objects: heads and tails in tossing a coin; the integers 1, 2, 3, 4, 5, 6 when a die is tossed; red, white, and black balls when a ball is drawn from an urn containing balls of these colors. Thus the outcomes are not always real numbers. However, when the outcomes are not real numbers, it is possible to associate a real number with each outcome by means of a real single-valued function defined on the set Ω, where $(\Omega, \mathfrak{F}, P)$ is the sample space associated with a random experiment, the points ω of the set Ω representing the outcomes of the experiment; the elements of $\mathfrak{F}$ represent events connected with the random experiment, and $P(\cdot)$ is a probability function defined on $\mathfrak{F}$. We know from Chapter 2 that $\mathfrak{F}$ is a σ-field of subsets of Ω.

Example 3.1.1 Consider a simple random experiment of tossing a fair coin four times. The sample space Ω consists of the following 16 points:

$$\omega_1 = HHHH \quad \omega_5 = HHTH \quad \omega_9 = THHH \quad \omega_{13} = TTHH$$
$$\omega_2 = HHHT \quad \omega_6 = HHTT \quad \omega_{10} = THHT \quad \omega_{14} = TTHT$$
$$\omega_3 = HTHH \quad \omega_7 = HTTH \quad \omega_{11} = THTH \quad \omega_{15} = TTTH$$
$$\omega_4 = HTHT \quad \omega_8 = HTTT \quad \omega_{12} = THTT \quad \omega_{16} = TTTT.$$

55

The elements of $\mathcal{F}$ are all possible subsets of Ω (e.g., the subset corresponding to the event of getting two heads is $\{\omega_4,\omega_6,\omega_7,\omega_{10},\omega_{11},\omega_{13}\}$). For any element A of Ω,

$P(A) = \frac{1}{16} \times$ (number of points in A)

(e.g., $P\{$two coins show heads$\} = \frac{6}{16}$). We may define a random variable $X(\omega)$ on Ω in this case, representing the total number of heads in the four tosses:

$$X(\omega_1) = 4 \qquad X(\omega_5) = 3 \qquad X(\omega_9) = 3 \qquad X(\omega_{13}) = 2$$
$$X(\omega_2) = 3 \qquad X(\omega_6) = 2 \qquad X(\omega_{10}) = 2 \qquad X(\omega_{14}) = 1$$
$$X(\omega_3) = 3 \qquad X(\omega_7) = 2 \qquad X(\omega_{11}) = 2 \qquad X(\omega_{15}) = 1$$
$$X(\omega_4) = 2 \qquad X(\omega_8) = 1 \qquad X(\omega_{12}) = 1 \qquad X(\omega_{16}) = 0.$$

$X(\omega)$ takes five values: 0, 1, 2, 3, 4.

Example 3.1.2 Consider a random experiment of tossing a fair coin until a head appears for the first time. The sample space Ω consists of a countably infinite sequence of points

$$\omega_1 = H, \quad \omega_2 = TH, \quad \omega_3 = TTH, \quad \omega_4 = TTTH, \quad \ldots$$
$$\omega_j = TT\ldots(j-1) \text{ times}, H, \text{ etc.}$$

$\mathcal{F}$ consists of all possible subsets of Ω. For any subsets A of $\mathcal{F}$, $P(A)$ is equal to the sum of the probabilities of all points in A. It may be noted that

$$P(\omega_j) = (\tfrac{1}{2})^j, \qquad j = 1,2,\ldots,\infty.$$

In this case $X(\omega)$, representing the number of times a coin is tossed before the head appears for the first time, is a random variable; obviously, $X(\omega_j) = j - 1$.

Example 3.1.3 The birth of a baby is a random phenomenon. The sample space Ω is the set of all newborn babies. A function $X(\omega)$ that associates with each ω in Ω its height $X(\omega)$ is a random variable. The members of $\mathcal{F}$ are sets of the form $\{\omega : a < X(\omega) < b\}$ for some real numbers a, b.

Definition 3.1.1: Univariate random variable A function $X(\omega)$ defined on a probability space $(\Omega,\mathcal{F},P)$ is called a *univariate random variable* if it is a real single-valued function defined on Ω into R and for every Borel set B of the real line the set $\{\omega : X(\omega) \in B\}$ belongs to $\mathcal{F}$.

It is customary to drop ω in $X(\omega)$ and write $X(\omega)$ simply as X. This is mainly because, without any loss of generality, many manipulations involving random variables can be made without specific reference to the sample space Ω. By $P(X = a)$ (where a is a real number) we denote the probability of the event $\{\omega : X(\omega) = a\}$ and by $P(X \leq a)$ we denote the probability of the set $\{\omega : X(\omega) \leq a\}$.

An observed value of X is a value that has been observed in an actual performance of the experiment. We will denote it by the lowercase letter x. In general, we denote random variables by capital letters X, Y, Z,..., and their observed values by the corresponding lowercase letters x, y, z,.... Sometimes we will also use different subscripts with the same letter to denote different random variables; for example, $X_1, X_2,...,X_n$ will denote a sequence of n different random variables.

Definition 3.1.2: Distribution function of a univariate random variable Let X be a univariate random variable defined on a sample space $(\Omega, \mathfrak{F}, P)$. Then $F_X(x) = P(X \leq x)$ is called the *distribution function* of the random variable X.

A distribution function $F_X(x)$ satisfies the following properties:

1. $F_X(x)$ is nondecreasing in x; that is, $F_X(b) \geq F_X(a)$ if $b \geq a$, since

$$F_X(b) = P(X \leq b)$$
$$= P(X \leq a) + P(a < X \leq b)$$
$$= F_X(a) + P(a < X \leq b).$$

Since probability is a nonnegative function, $P(a < X \leq b) \geq 0$, and hence $F_X(b) \geq F_X(a)$.

2. It is obvious from the definition of $F_X(x)$ that

$$F_X(-\infty) = 0, \qquad F_X(\infty) = 1.$$

3. $F_X(x)$ is continuous at least to the right. To show this, let $x_1 > x_2 > \cdots > x$ be a decreasing sequence of arbitrary real numbers converging to x. Let B_k be the event given by

$$B_k = \{\omega : x < X(\omega) \leq x_k\}, \qquad k = 1, 2,... .$$

It is clear that $B_j \subset B_i$ if $i < j$ and x does not belong to any of B_i. Thus $\cap_{i=1}^{\infty} B_i$ is a null event, so that $P(\cap_{i=1}^{\infty} B_i) = 0$. Hence

$$\lim_{i \to \infty} P(B_i) = \lim_{i \to \infty} P\{x < X \leq x_i\}$$

$$= \lim_{i \to \infty} (F_X(x_i) - F_X(x))$$

$$= 0.$$

In other words,

$$\lim_{i \to \infty} F_X(x_i) = F_X(x).$$

Since $\{x_i\}$ is an arbitrary decreasing sequence converging to x, it follows that $F_X(x)$ is continuous to the right. If $F(x)$ is discontinuous to the left at $X = x$, then

$$\lim_{\epsilon \to 0} \{F_X(x) - F_X(x - \epsilon)\} = F_X(x) - F_X(x - 0)$$

$$= P(X = x)$$

is positive. $P(X = x)$ measures the jump of $F_X(x)$ at the jump point $X = x$. This is sometimes called the "saltus" of F_X at $X = x$. However, if $F_X(x)$ is continuous at $X = x$, then $P(X = x) = 0$.

It may be added here that every function $F_X(x)$ satisfying conditions 1 to 3 above can be regarded as the distribution function of some random variable X, a result that follows from a celebrated theorem of A. N. Kolmogorov (1933).

3.2 DISCRETE UNIVARIATE RANDOM VARIABLES

In most practical applications we deal with random variables that can by and large be divided into two classes: discrete and continuous.

Definition 3.2.1: Discrete univariate random variable   A univariate random variable X belongs to the *discrete* class if its distribution function $F_X(x)$ consists of a finite or a countably infinite number of jumps only.

In other words, a discrete random variable takes only a finite or at most a countably infinite number of values with positive probability. These values are called *jump points* of its distribution function, and the jumps are their corresponding probabilities. Let $x_1, x_2, \ldots, x_n, \ldots$ be the finite or infinite sequence of jump points with jumps $P(X = x_i) = p_i$, $i = 1, 2, \ldots, n$, of the distribution $F_X(x)$ of the discrete random variable X, and let $x_1 < x_2 < \cdots < x_n < \cdots$; then

$$F_X(x_i) = \sum_{j=1}^{i} p_j, \qquad \sum_{j=1}^{\infty} p_j = 1. \tag{3.1}$$

The function $P(X = x_i) = p_i$, $i = 1, 2, \ldots, n, \ldots$, is called the *probability mass function* of the discrete random variable X. If some constants appear in the description of a distribution function, they are called parameters of the distribution function. If $\theta_1,\ldots,\theta_k$ are the parameters of a distribution function $F_X(x)$, we shall sometimes denote this function by $F_X(x|\theta_1,\ldots,\theta_k)$.

We now list some important discrete random variables that are frequently encountered in probability and statistics. A random variable often derives its name from its probability mass function. For example, we call a random variable X *binomial* or *Poisson* if its probability mass function follows a binomial probability law or a Poisson probability law.

Example 3.2.1: Bernoulli random variable with parameter p A *Bernoulli* random variable X takes only two distinct values a and b, $a < b$, with probability given by the probability mass function

$$p_X(i|p) = P(X = i) = \begin{cases} 1 - p & \text{if } i = a \\ p & \text{if } i = b \\ 0 & \text{otherwise.} \end{cases}$$

The distribution function $F_X(x)$ is given by

$$F_X(i) = \begin{cases} 0 & \text{if } i < a \\ 1 - p & \text{if } a \le i < b \\ 1 & \text{if } i \ge b. \end{cases}$$

In many practical situations we assign to a the value 0 and to b the value 1. However, situations where the Bernoulli random variable takes values different from 0, 1 are not rare in the literature.

The following is an example of a Bernoulli random variable. Consider the random experiment of tossing a coin. The possible outcomes are heads (H) and tails (T), which we denote by 1 and 0, respectively. X is then a Bernoulli random variable with $P(X = 0) = 1 - p$ and $P(X = 1) = p$.

Example 3.2.2: Binomial random variable with parameters n, p, where n is a positive integer and $0 \le p \le 1$ A random variable X is binomial with parameters n, p if it takes values $0,1,\ldots,n$ such that the probability that X takes the value j, $j = 0,1,\ldots,n$ is given by

$$p_X(j|n,p) = p(X = j) = \begin{cases} \binom{n}{j}p^j(1 - p)^{n-j}, & j = 0,1,\ldots,n \\ 0, & \text{otherwise,} \end{cases}$$

which represents the probability mass function of X. We are using $\binom{n}{j}$ for the total number of combinations of n things taken j at a time. By the binomial theorem,

$$\sum_{j=0}^{n} \binom{n}{j} p^j(1 - p)^{n-j} = (1 - p + p)^n = 1.$$

The distribution function $F_X(x)$ is given by

$$F_X(k) = \begin{cases} 0 & \text{if } k < 0 \\ \sum_{j=0}^{k} \binom{n}{j} p^j (1-p)^{n-j} & \text{if } k = 0,1,\ldots,n-1 \\ 1 & \text{if } k \geq n. \end{cases}$$

Here $F_X(x)$ has $n + 1$ jumps and the jump at the point $X = i$ is $\binom{n}{i} p^i (1-p)^{n-i}$. An example of the binomial random variable with parameters n,p is the total number of heads in n independent tosses of a coin, in which the probability of a head turning up in a single toss is p.

Example 3.2.3: Geometric random variable with parameter p $(0 \leq p \leq 1)$ X takes all positive integral values with probability

$$p_X(j|p) = P(X = j) = \begin{cases} p(1-p)^{j-1} & \text{if } j = 1,2,\ldots,\infty \\ 0 & \text{otherwise.} \end{cases}$$

This represents the probability density function of a geometric random variable with parameter p, since for $0 \leq x \leq 1$, $(1-x)^{-1} = 1 + x + x^2 + x^3 + \cdots$, we get

$$\sum_{j=1}^{\infty} p(1-p)^{j-1} = p(1-1+p)^{-1} = 1.$$

The distribution function $F_X(x)$ is given by

$$F_X(k) = \begin{cases} 0 & \text{if } k \leq 0 \\ \sum_{j=1}^{k} p(1-p)^{j-1} & \text{if } k > 0 \end{cases}$$

$$F_X(\infty) = 1.$$

An example of a geometric random variable is the waiting time for heads to turn up for the first time in a sequence of independent tosses of a coin in which the probability of heads in each toss is p.

Example 3.2.4: Poisson random variable with parameter λ   This variable takes all positive integral values, including 0, with probability

$$p_X(k|\lambda) = P(X = k) = \begin{cases} \dfrac{e^{-\lambda}\lambda^k}{k!} & \text{if } k = 0,1,\ldots,\infty \\ 0 & \text{otherwise,} \end{cases}$$

where λ is a positive constant; $p_X(k|\lambda)$ is the probability mass function of a Poisson random variable with parameter λ, for since

$$e^x = 1 + x + \frac{x^2}{2!} + \frac{x^3}{3!} + \cdots,$$

we see that

$$\sum_{k=0}^{\infty} \frac{e^{-\lambda}\lambda^k}{k!} = e^{-\lambda} \cdot e^{\lambda} = 1.$$

The distribution function $F_X(x)$ is given by

$$F_X(k) = \begin{cases} 0 & \text{if } k < 0 \\ \sum_{j=0}^{k} \frac{e^{-\lambda}\lambda^k}{j!} & \text{for } k = 0, 1, \ldots, \infty \end{cases}$$

$$F_X(\infty) = 1.$$

The Poisson probability law is widely used to approximate various natural phenomena, such as suicide rate per year among women in a given city, demand for service in a big department store, or the major accident rate per day on a certain major highway. The mathematical justification for this lies in the fact that the Poisson probability law can be derived from the binomial probability law when n increases and p decreases in such a way that $np = \lambda$ is constant. In other words, the binomial probability law of rare events gives rise to the Poisson probability law. To see this mathematically, observe that

$$\lim_{n \to \infty} \binom{n}{k} p^k (1 - p)^{n-k} = \lim_{n \to \infty} \frac{n(n - 1) \cdots (n - k + 1)p^k(1 - p)^{n-k}}{k!}$$

$$= \lim_{n \to \infty} \left(1 - \frac{1}{n}\right)\left(1 - \frac{2}{n}\right) \cdots$$

$$\times \left(1 - \frac{k - 1}{n}\right)\lambda^k \frac{[1 - (\lambda/n)]^{n-k}}{k!}$$

$$= \frac{\lambda^k e^{-\lambda}}{k!}.$$

3.3 CONTINUOUS UNIVARIATE RANDOM VARIABLES

A univariate random variable X belongs to the continuous class if for every real number x its distribution function $F_X(x)$ is continuous (no jumps) and can be represented as

$$F_X(x) = \int_{-\infty}^{x} f_X(y) \, dy, \tag{3.2}$$

where $f_X(x)$ is continuous everywhere and $f_X(x) \geq 0$ for all x. It is obvious from the definition that

$$\int_{-\infty}^{\infty} f_X(x) = F_X(\infty) - F_X(-\infty) = 1.$$

The function $f_X(x)$ is known as the probability density function of the continuous random variable X.

We will now list some important continuous random variables that we will use frequently throughout the book.

Example 3.3.1: Normal random variable with parameters μ, σ   The probability density function $f_X(x)$ of this random variable is given by

$$f(x|\mu,\sigma^2) = \begin{cases} \dfrac{1}{\sqrt{2\pi}\sigma} \exp\left[-\dfrac{1}{2}\left(\dfrac{x-\mu}{\sigma}\right)^2 \right] & \text{if } -\infty < x < \infty \\ 0 & \text{otherwise,} \end{cases}$$

where $-\infty < \mu < \infty$ and $\sigma^2 > 0$.

The distribution function of the normal random variable X is given by

$$F_X(x) = \int_{-\infty}^{x} f_X(y)\, dy.$$

Clearly, $F_X(-\infty) = 0$ and

$$F_X(\infty) = \int_{-\infty}^{\infty} \frac{1}{\sqrt{2\pi}\sigma} \exp\left[-\frac{1}{2}\left(\frac{x-\mu}{\sigma}\right)^2 \right] dx$$

$$= \int_{-\infty}^{\infty} \frac{1}{\sqrt{2\pi}} e^{-(1/2)y^2}\, dy$$

$$= \int_{0}^{\infty} \frac{1}{\sqrt{2\pi}} e^{-(1/2)z} z^{(1/2)-1}\, dz$$

$$= \frac{1}{\sqrt{2\pi}} \frac{\Gamma(1/2)}{\sqrt{1/2}}$$

$$= 1.$$

It is known that

$$\frac{\Gamma(p)}{\alpha^p} = \int_{0}^{\infty} e^{-\alpha x} x^{p-1}\, dx,$$

where $\alpha > 0$, $p > 0$, and $\Gamma(p) = (p-1)\Gamma(p-1)$, with $\Gamma(\tfrac{1}{2}) = \sqrt{\pi}$ and $\Gamma(1) = 1$.

A normal random variable with $\sigma^2 = 1$ and $\mu = 0$ is called a *unit normal* or *standard normal random variable*, and its distribution function $F_X(x)$ is usually given by

$$P_X(x) = \int_{-\infty}^{x} \frac{1}{\sqrt{2\pi}} e^{-(y^2/2)} \, dy. \tag{3.3}$$

Exact evaluation of $P_X(x)$ in the closed form is not possible. However, numerically, it is a well-tabulated function. Table 3B in Appendix B presents values of $P_X(x)$ for different values of x.

The normal probability density function plays a dominant role in the theory of probability and statistics. It derives its importance mainly from the fact that there are numerous random physical phenomena that follow this probability law fairly well. Further, we will show later that the binomial probability law for large n approximately follows the normal law. From a mathematical viewpoint the normal random variable is important because its probability density function possesses various nice properties; for example, it is symmetric about μ and is bellshaped, with a peak at μ (see Figure 3.2). From the point of view of application, it is important because it is well tabulated and because the theory of statistical testing and estimation is well developed for this case.

Example 3.3.2: Gamma random variable $G(\lambda,r)$ with parameters λ, r
This random variable has probability density function with parameters $\lambda > 0$ and $r \geq 1$ (see Figure 3.3):

$$f(x|\lambda,r) = \begin{cases} \dfrac{\lambda}{\Gamma(r)} (\lambda x)^{r-1} e^{-\lambda x} & \text{if } x \geq 0 \\ 0 & \text{otherwise.} \end{cases}$$

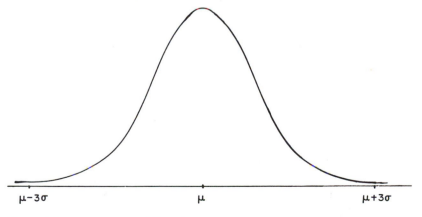

Figure 3.2 Normal probability density function.

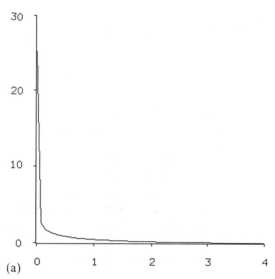

(a)

Figure 3.3 Gamma probability density function: (a) with $r = \lambda = \frac{1}{2}$; (b) with $r = \lambda = 2$; (c) with $r = \lambda = 10$.

The gamma probability density function with $r = 1$ is called the *exponential probability density function*, and with $\lambda = \frac{1}{2}$ it is known as the *chi-square* (χ^2_{2r}) *probability density function* with $2r$ degrees of freedom. Using properties of gamma integrals, we have

$$\int_{-\infty}^{\infty} f(x|\lambda,r)\, dx = \int_{0}^{\infty} \frac{\lambda^r}{\Gamma(r)} x^{r-1} e^{-\lambda x}\, dx$$

$$= \frac{\lambda^r}{\Gamma(r)} \frac{\Gamma(r)}{\lambda^r} = 1.$$

The distribution function

$$F_X(x) = \int_{0}^{x} \frac{\lambda^r}{\Gamma(r)} (\lambda y)^{r-1} e^{-\lambda y}\, dy$$

was well tabulated by Karl Pearson (1922). An example of the gamma random variable is the waiting time for the rth suicide to occur for the first time in a certain city when the suicide cases follow Poisson probability law with parameter λ.

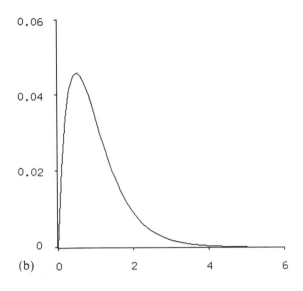

(b)

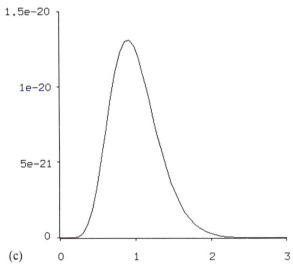

(c)

Example 3.3.3: Rectangular random variable The probability density function of the rectangular random variable x is given by

$$f(x|a,b) = \begin{cases} \dfrac{1}{b-a} & \text{if } a < x < b \\ 0 & \text{otherwise,} \end{cases}$$

where a, b are real.

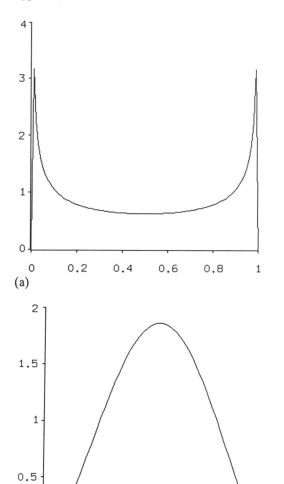

(a)

(b)

Figure 3.4 Beta probability density function: (a) with $\rho = q = \frac{1}{2}$; (b) with $b = q = 3$.

In this case

$$\int_a^b \frac{1}{b-a}\,dx = 1.$$

The distribution function $F_X(x)$ is given by

$$F_X(x) = \begin{cases} \dfrac{x-a}{b-a} & \text{if } a < x < b \\ 1 & \text{if } x \geq b \\ 0 & \text{if } x \leq a. \end{cases}$$

In the literature this function is sometimes called the *uniform probability density function*.

Example 3.3.4: Beta random variable, $B(p,q)$ This random variable has probability density function with parameters p, q (positive real):

$$f(x|p,q) = \begin{cases} \dfrac{\Gamma(p+q)}{\Gamma(p)\Gamma(q)}\, x^{q-1}(1-x)^{p-1} & \text{if } 0 < x < 1 \\ 0 & \text{otherwise}, \end{cases}$$

where p, q must both be greater than zero. By using the properties of beta integrals, we have (see Figure 3.4)

$$\int_0^1 \frac{\Gamma(p+q)}{\Gamma(p)\Gamma(q)}\, x^{q-1}(1-x)^{p-1}\,dx = \frac{\Gamma(p+q)}{\Gamma(p)\Gamma(q)}\frac{\Gamma(p)\Gamma(q)}{\Gamma(p+q)} = 1.$$

The distribution function

$$F_X(x) = \int_0^x \frac{\Gamma(p+q)}{\Gamma(p)\Gamma(q)}\, y^{q-1}(1-y)^{p-1}\,dy$$

has been tabulated by Karl Pearson (1934).

If X has continuous probability density function, the probability that X takes a single value b is 0, since

$$P(X = b) = \lim_{n\to\infty}\left[F_X\left(b + \frac{1}{n}\right) - F_X\left(b - \frac{1}{n}\right)\right] = 0.$$

EXERCISES

1. Let X be a continuous random variable with probability density function

$$f_X(x) = \begin{cases} e^{-x} & \text{for } x > 0 \\ 0 & \text{otherwise}. \end{cases}$$

Find, for any real a, $P\{X^3 < a\}$.

2. Let X be a uniformly distributed random variable over the interval $(0,1)$. Find (a) $P\{(1/X) < 2\}$; (b) $P\{(1/X^2) < 2\}$.

3. Two people toss an unbiased coin n times each. Find the probability that they will score the same number of heads.

4. Let $Z = \min(X,Y)$. Find $P(0 < Z \le 3)$ for each of the following cases:
   (a) X, Y are independent, identically distributed normal random variables with mean 0 and variance 4.
   (b) X, Y are independent, identically distributed binomial random variables with parameters $n = 10$, $p = 0.2$.
   (c) X, Y are independent, identically distributed uniform random variables over the interval $(0,2)$.

5. For an infinite number of independent tosses of an unbiased coin, find the probability density function of the random variable $U =$ number of tosses needed to get two heads together for the first time. Also find $P\{U \ge 4\}$.

6. Find the probability mass function of the random variable X, defined to be the sum of the two numbers showing in the toss of two dice.

7. The Scholastic Aptitude Test scores in mathematics in a certain year vary from 200 to 800. Assume that they are normally distributed with mean 450 and variance $(120)^2$.
   (a) What percentage of students score between 500 and 700?
   (b) What test score represents the 90th percentile?

8. For each of the following find the constant c so that $f_X(x)$ satisfies the condition of being a probability density function or a probability mass function of a random variable X.

   (a) $f_X(x) = \begin{cases} c(\frac{2}{3})^x, & x = 1,2,\dots,\infty \\ 0, & \text{otherwise.} \end{cases}$

   (b) $f_X(x) = \begin{cases} c \sin x, & 0 \le x \le \dfrac{\pi}{2} \\ 0, & \text{otherwise.} \end{cases}$

   (c) $f_X(x) = \begin{cases} ce^{-|x|}, & x \text{ real} \\ 0, & \text{otherwise.} \end{cases}$

9. Verify that each of the following $F_X(x)$ represents a distribution of a random variable X. Sketch the distribution function. Is the distribution function continuous? If so, find the corresponding probability density function.

   (a) $F_X(x) = \begin{cases} 0, & x < -2 \\ \frac{1}{2}, & -2 \le x < 0 \\ 1, & x \ge 0. \end{cases}$

(b) $F_X(x) = \begin{cases} 0, & x < 0 \\ 1 - \frac{1}{2}e^{x/3} - \frac{1}{3}e^{-[x/3]}, & \text{if } x \geq 0, \end{cases}$

where $[x/3]$ represents the greatest integer less than or equal to $x/3$.

(c) $F_X(x) = \begin{cases} \frac{1}{2}e^{-(x/50)^2}, & x \leq 0 \\ 1 - \frac{1}{2}e^{-(x/50)^2}, & x \geq 0. \end{cases}$

(d) $F_X(x) = \begin{cases} 0, & x < 0 \\ \sqrt{x}, & 0 \leq x \leq 1 \\ 1, & x > 1. \end{cases}$

(e) $F_X(x) = \begin{cases} \frac{1}{2}(x + \theta), & |x| < \theta \\ 0, & \text{otherwise.} \end{cases}$

10. Given that X is a normal random variable with mean 5 and variance 16, compute (a) $P\{|X - 5| \geq 3\}$; (b) $P\{|X - 4| < 2\}$; (c) $P\{X > 3\}$.

11. Let T represent the lifetime of a specific component of an electrical machine, and let $f_T(t)$, $F_T(t)$ be its probability density function and its distribution function, respectively. Let

$$g_T(t) = \frac{f_T(t)}{1 - F_T(t)}.$$

The function $g_T(t)$ is called the *failure rate* of the component at time t. The distribution function $F_T(t)$ is called an *increasing* or a *decreasing* failure-rate distribution if $g_T(t)$ increases with t or decreases with t, respectively. Classify each of the following distribution functions according to its failure rates:

(a) $F_T(t) = \begin{cases} 1 - e^{-t^\alpha}, & t > 0, \ \alpha > 0 \\ 0 & \text{otherwise.} \end{cases}$

(b) $F_T(t) = \begin{cases} \frac{1}{2}\left[1 - \exp\left(\frac{-t}{\alpha_1}\right)\right] + \frac{1}{2}\left[1 - \exp\left(\frac{-t}{\alpha_2}\right)\right], \\ \qquad \alpha_1 > 0, \ \alpha_2 > 0, \ t > 0 \\ 0, \qquad \text{otherwise.} \end{cases}$

12. A bomber carrying three bombs flies directly over a railroad track. If a bomb falls within 40 feet of the track, the track will be sufficiently

damaged to disturb the traffic. For a certain bomb site, the points of impact of a bomb have the probability density function

$$f_X(x) = \begin{cases} \dfrac{100 + x}{10,000} & \text{if } -100 \le x \le 0 \\[2mm] \dfrac{100 - x}{10,000} & \text{if } 0 < x \le 100 \\[2mm] 0 & \text{otherwise,} \end{cases}$$

where x represents the horizontal deviation in feet from the aiming point (the railroad track). Find the distribution function of X. If all bombs are used, what is the probability that the track will be damaged?

13. A filling station is supplied with gasoline once a week. If its weekly volume of sales (in thousand gallons) X has the following probability density function:

$$f_X(x) = \begin{cases} 5(1 - x)^4 & \text{if } 0 < x < 1 \\ 0 & \text{otherwise} \end{cases}$$

what must be the capacity of its tank in order that the probability that its supply will be exhausted in a given week shall be 0.01?

14. The length of time (in minutes) X that a certain lady speaks on the telephone is a random variable with probability density function

$$f_X(x) = \begin{cases} ce^{-x/5} & \text{if } x \ge 0 \\ 0 & \text{otherwise.} \end{cases}$$

(a) Find the value of the constant c.
(b) Find the probability that the duration of her conversation will be between 5 and 10 minutes.

3.4 MULTIDIMENSIONAL RANDOM VARIABLES

We defined a univariate random variable as a real-valued function of the outcome of a random experiment (Definition 3.1.1). However, in many cases, the outcomes of a random experiment cannot be fully expressed by a single (univariate) random variable, and we need to consider several such variables to give a clear expression of the outcome. Consider Example 3.1.3. The birth of a baby cannot be fully expressed by stating the height of the baby. The weight of the baby will definitely give more information that is not directly included in the height. In a similar fashion one can think of random experiments where we need to consider p simultaneously observed random variables $X_1, \ldots, X_p$ to express the outcomes in a meaningful way. We consider the vector $X = (X_1,\ldots,X_p)$ as a collection of p real single-valued functions, each defined on Ω into R, such that the p-tuple $(X_1,\ldots,X_p)$ assigns to every ω in Ω a point in Euclidean p-space. We will call $X = (X_1,\ldots,X_p)$ a p-variate random variable. We consider here the bivariate case ($p = 2$) in some detail, and will state the results for the general case. However, we discuss the general case in detail in Chapter 7.

Definition 3.4.1: p-Variate random variable A p-tuple $(X_1,\ldots,X_p)$ defined on $(\Omega,\mathfrak{F},P)$ is called a p-*variate random variable* if each X_i is a real single-valued function defined on Ω into R and if for reals $a_1,\ldots,a_p,\ b_1,\ldots,b_p$ in R, the set $\{\omega : a_i < X_i(\omega) < b_i\}$ belongs to $\mathfrak{F}$.

Thus a p-variate random variable is an ordered collection of p univariate random variables.

Definition 3.4.2: Distribution function of a p-variate random variable For every p-tuple of reals $(x_1,\ldots,x_p)$ the function $F_{X_1,\ldots,X_p}(x_1,\ldots,x_p) = P(X_1 \le x_1,\ldots,X_p \le x_p)$ is called the *distribution function* of the p-variate random variable $(X_1,\ldots,X_p)$.

Let us now consider the bivariate case in detail. As in the case of the distribution function $F_X(x)$ of a single random variable X, one can show the following for $F_{X_1,X_2}(x_1,x_2)$:

1. $F_{X_1,X_2}(x_1,x_2)$ is nondecreasing and continuous at least from the right in each variable.
2. $F_{X_1,X_2}(-\infty,x_2) = F_{X_1,X_2}(x_1,-\infty) = 0,$

$$F_{X_1,X_2}(+\infty,+\infty) = 1.$$

In addition to 1 and 2 above, $F_{X_1,X_2}(x_1,x_2)$ must also possess the following property:

3. For $x_{11} \le x_{12},\ x_{21} \le x_{22},$

$$F_{X_1,X_2}(x_{12},x_{22}) - F_{X_1,X_2}(x_{12},x_{21})$$
$$- F_{X_1,X_2}(x_{11},x_{22}) + F_{X_1,X_2}(x_{11},x_{21}) \ge 0.$$

The last condition is needed to assert that the probability of any rectangle in the plane is nonnegative. The fact that the first two conditions do not necessarily guarantee the last one can easily be demonstrated by the following distribution function:

$$F_{X_1,X_2}(x_1,x_2) = \begin{cases} 0 & \text{if } x_1 + x_2 < 0 \\ 1 & \text{if } x_1 + x_2 \ge 0. \end{cases}$$

$F_{X_1,X_2}(x_1,x_2)$ is nondecreasing and continuous to the right. Consider in each (x_1,x_2) a rectangle with three vertices above and one below the line $X_2 = -X_1$. Let the vertices be $(2,2)$, $(2,0)$, $(-1,2)$, and $(-1,0)$. Then

$$F_{X_1,X_2}(2,2) - F_{X_1,X_2}(2,0) - F_{X_1,X_2}(-1,2)$$
$$+ F_{X_1,X_2}(-1,0) = -1.$$

As in the case of the univariate random variable, we consider only discrete and continuous bivariate random variables. For convenience, we write a bivariate random variable as (X,Y).

Definition 3.4.3: Discrete bivariate random variable   A bivariate random variable (X,Y) belongs to the discrete class if for every pair of values (x_i,y_k) of (X,Y), its distribution function $F_{X,Y}(x,y)$ can be written as

$$F_{X,Y}(x,y) = \sum_{\substack{x_i \leq x \\ y_k \leq y}} p_{X,Y}(x_i,y_k), \tag{3.4}$$

where $p_{X,Y}(x_i,y_k) = P(X = x_i, Y = y_k)$ and $p_{X,Y}(x_i,y_k) > 0$ for a finite number or at most for a countably infinite number of points (x_i,y_k) for which

$$\sum_{x_i,y_k} p_{X,Y}(x_i,y_k) = 1.$$

We will call $p_{X,Y}(x_i,y_k)$ the jump of $F_{X,Y}(x,y)$ at the jump point (x_i,y_k); $p_{X,Y}(x_i,y_k)$ will be called a *bivariate probability mass function*.

Example 3.4.1   Consider a random experiment of n tossings of a coin that is capable of standing on edge. Let the probabilities of falling head or tail upward in a single toss be p_1 and p_2, respectively. Since standing on edge is a possible outcome when the coin is tossed, $p_1 + p_2$ need not be 1. Obviously, $p_1 + p_2 < 1$ and $1 - p_1 - p_2$ is the probability of the coin standing on edge in a single toss. Let X denote the total number of heads and Y denote the total number of tails in n tosses. One can show that

$$P_{X,Y}(x,y) = P(X = x, Y = y)$$

$$= \begin{cases} \dfrac{n!}{x! \, y! \, (n - x - y)!} \, p_1^x p_2^y (1 - p_1 - p_2)^{n-x-y} \\ \quad \text{if } x,y = 0,1,\ldots,n \quad \text{and} \quad 0 \leq x + y \leq n \\ 0 \quad \text{otherwise.} \end{cases}$$

$P_{X,Y}(x,y)$ represents the probability mass function of the bivariate random variable (X,Y). We have

$$\sum_{\substack{0 \le x \le n \\ 0 \le y \le n \\ 0 \le x+y \le n}} p_{X,Y}(x,y) = \sum_{x=0}^{n} \frac{n!}{x!\,(n-x)!}\, p_1^x$$

$$\times \sum_{y=0}^{n-x} \frac{(n-x)!}{y!\,(n-x-y)!}\, p_2^y (1 - p_1 - p_2)^{n-x-y}$$

$$= \sum_{x=0}^{n} \frac{n!}{x!\,(n-x)!}\, p_1^x (1 - p_1)^{n-x}$$

$$= 1.$$

Definition 3.4.4: Continuous bivariate random variable   A bivariate random variable (X,Y) belongs to the *continuous* class if $F_{X,Y}(x,y)$ is continuous in both x and y and there exists a nonnegative and continuous function $f_{X,Y}(x,y)$ in both variables such that for every pair of reals (x,y), $F_{X,Y}(x,y)$ can be written as

$$F_{X,Y}(x,y) = \int_{-\infty}^{x} \int_{-\infty}^{y} f_{X,Y}(x',y')\, dx'\, dy' \tag{3.5}$$

or

$$f_{X,Y}(x,y) = \frac{\partial^2}{\partial x\, \partial y} F_{X,Y}(x,y).$$

The function $f_{X,Y}(x,y)$ is known as the *joint probability density function* of X, Y. It is obvious that

$$\int_{-\infty}^{\infty} \int_{-\infty}^{\infty} f_{X,Y}(x,y)\, dx\, dy = F_{X,Y}(+\infty, +\infty) = 1.$$

Example 3.4.2: Bivariate normal random variable with parameters m_1, m_2, σ_1^2, σ_2^2, ρ; $-\infty < m_1 < \infty$, $-\infty < m_2 < \infty$, $\sigma_1^2 > 0$, $\sigma_2^2 > 0$, $-1 < \rho < 1$ The probability density function of this random variable is (see Figure 3.5)

$$f(x,y \mid m_1, m_2, \sigma_1^2, \sigma_2^2, \rho)$$

$$= \begin{cases} \dfrac{1}{2\pi\sigma_1\sigma_2(1-\rho^2)^{1/2}} \exp\left[-\dfrac{1}{2(1-\rho^2)} \left[\left(\dfrac{x-m_1}{\sigma_1} \right)^2 \right.\right. \\ \left.\left. \quad - 2\rho\left(\dfrac{x-m_1}{\sigma_1} \right)\left(\dfrac{y-m_2}{\sigma_2} \right) + \left(\dfrac{y-m_2}{\sigma_2} \right)^2 \right]\right] \\ \qquad \text{if } -\infty < x < \infty, \ -\infty < y < \infty \\ 0 \qquad \text{otherwise.} \end{cases}$$

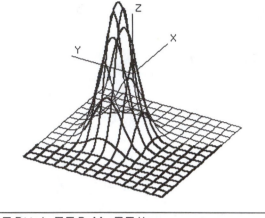

□□ Pitch □□ Roll □□ Yaw

(a)

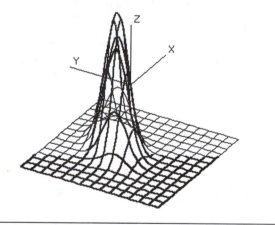

□□ Pitch □□ Roll □□ Yaw

(b)

Figure 3.5   Bivariate normal probability density function: (a) with $\rho = 0$; (b) with $\rho = \frac{1}{2}$.

Its distribution satisfies

$$F_{X,Y}(+\infty, +\infty) = \int_{-\infty}^{\infty} \frac{1}{\sqrt{2\pi}\,\sigma_2} \exp\left[-\frac{1}{2}\left(\frac{y - m_2}{\sigma_2}\right)^2\right]$$

$$\times \left(\int_{-\infty}^{\infty} \frac{1}{\sqrt{2\pi}\,\sigma_1(1 - \rho^2)^{1/2}} \exp\left\{-\frac{1}{2(1 - \rho^2)\sigma_1^2}\right.\right.$$

$$\left.\left.\times \left[x - m_1 - \rho\,\frac{\sigma_1}{\sigma_2}(y - m_2)\right]^2\right\} dx\right) dy$$

$$= \int_{-\infty}^{\infty} \frac{1}{\sqrt{2\pi}\,\sigma_2} \exp\left[-\frac{1}{2}\left(\frac{y - m_2}{\sigma_2}\right)^2\right] dy$$

$$\times \int_{-\infty}^{\infty} \frac{1}{\sqrt{2\pi}} e^{-(1/2)z^2}\, dz$$

$$= 1,$$

where

$$z = \frac{x - m_1 - \rho(\sigma_1/\sigma_2)(y - m_2)}{\sigma_1(1 - \rho^2)^{1/2}}.$$

3.5 MARGINAL DISTRIBUTIONS OF A BIVARIATE RANDOM VARIABLE

Given the joint probability mass or density function of a bivariate random variable (X,Y), we may also be interested in the probability mass or density function of either of the single random variables X and Y. We will call this the *marginal probability density function* of X or Y.

Let (X,Y) be a bivariate discrete random variable that takes values (x_i, y_k), $i = 1,\ldots,m$, $k = 1,\ldots,n$ with unit probability. In other words, if $P(X = x_i, Y = y_k) = p_{ik}$, then

$$\sum_{i=1}^{m} \sum_{j=1}^{n} p_{ij} = 1.$$

We note that the event $(X = x_i)$ materializes if any of n mutually exclusive events $(X = x_i, Y = y_1), (X = x_i, Y = y_2),\ldots,(X = x_i, Y = y_n)$ materializes. Thus we have

$$P(X = x_i) = \sum_{k=1}^{n} P(X = x_i, Y = y_k)$$

$$= \sum_{k=1}^{n} p_{ik}$$

$$= p_{i\cdot}. \tag{3.6}$$

Similarly, we have

$$P(Y = y_k) = \sum_{i=1}^{m} p_{ik} = p_{\cdot k}. \tag{3.7}$$

Then $p_X(x_i) = P(X = x_i)$ and $p_Y(y_k) = P(Y = y_k)$ are the marginal probability density functions of X and Y, respectively. We can represent these marginal probability functions in the following tabular form.

k \ i	1	2	$\cdots$	m	Total
1	p_{11}	p_{21}	$\cdots$	p_{m1}	$p_{\cdot 1}$
2	p_{12}	p_{22}	$\cdots$	p_{m2}	$p_{\cdot 2}$
.					
.					
.					
n	p_{1n}	p_{2n}	$\cdots$	p_{mn}	$p_{\cdot n}$
Total	$p_{1\cdot}$	$p_{2\cdot}$		$p_{m\cdot}$	1

In this table $p_{\cdot k}$ is the sum of the probabilities that Y takes the value y_k and X takes all possible values; $p_{i\cdot}$ is the sum of the probabilities that X takes the value x_i and Y takes all possible values.

Example 3.5.1 In Example 3.4.1, the marginal probability mass function of X is

$$
p_X(x) = \begin{cases}
\displaystyle\sum_{y=0}^{n-x} \frac{n!}{x!\,y!\,(n-x-y)!}\, p_1^x p_2^y (1 - p_1 - p_2)^{n-x-y} \\
\qquad \text{if } x = 0,\dots,n \\
\displaystyle\frac{n!}{x!\,(n-x)!}\, p_1^x \sum_{y=0}^{n-x} p_2^y (1 - p_1 - p_2)^{n-x-y}\, \frac{(n-x)!}{y!(n-x-y)!} \\
\qquad \text{if } x = 0,\dots,n \\
\displaystyle\frac{n!}{x!\,(n-x)!}\, p_1^x (1 - p_1)^{n-x} \\
\qquad \text{if } x = 0,\dots,n \\
0 \qquad \text{otherwise.}
\end{cases}
$$

Similarly,

$$
p_Y(y) = \begin{cases}
\displaystyle\frac{n!}{y!\,(n-y)!}\, p_2^y (1 - p_2)^{n-y} & \text{if } y = 0,\dots,n \\
0 & \text{otherwise.}
\end{cases}
$$

Given the distribution function $F_{X,Y}(x,y)$ of the continuous random variable (X,Y), the marginal distribution function $F_X(x)$ of X is given by

$$
F_X(x) = P(X \le x) = F_{X,Y}(x,\infty) = P(X \le x, Y \le \infty). \tag{3.8}
$$

Clearly, we have

$$
F_X(x) = \int_{-\infty}^{x} \int_{-\infty}^{\infty} f_{X,Y}(u,v)\, du\, dv
$$

$$
= \int_{-\infty}^{x} \left[\int_{-\infty}^{\infty} f_{X,Y}(u,v)\, dv \right] du
$$

$$
= \int_{-\infty}^{x} f_X(u)\, du,
$$

where

$$
f_X(u) = \int_{-\infty}^{\infty} f_{X,Y}(u,v)\, dv, \qquad f_X(x) = \frac{d}{dx} F_X(x)
$$

is called the *marginal probability density function* of X. Similarly,

$$
f_Y(y) = \frac{d}{dy} F_Y(y) = \int_{-\infty}^{\infty} f_{X,Y}(x,y)\, dx \tag{3.9}
$$

is called the marginal probability density function of Y.

Example 3.5.2 Bivariate normal random variable with parameters m_1, m_2, σ_1, σ_2, ρ The marginal density function of X is given by

$$
\begin{aligned}
f_X(x) &= \int_{-\infty}^{\infty} \frac{1}{2\pi \, \sigma_1\sigma_2(1 - \rho^2)^{1/2}} \exp\left\{ -\frac{1}{2(1 - \rho^2)} \left[\left(\frac{x - m_1}{\sigma_1} \right)^2 \right. \right. \\
&\quad \left. \left. + \left(\frac{y - m_2}{\sigma_2} \right)^2 - 2\rho\left(\frac{x - m_1}{\sigma_1} \right)\left(\frac{y - m_2}{\sigma_2} \right) \right] \right\} dy \\[2mm]
&= \frac{1}{\sqrt{2\pi}\, \sigma_1} \exp\left[-\frac{1}{2}\left(\frac{x - m_1}{\sigma_1} \right)^2 \right] \int_{-\infty}^{\infty} \frac{1}{\sqrt{2\pi}\, \sigma_2(1 - \rho^2)^{1/2}} \\[2mm]
&\quad \times \exp\left\{ -\frac{1}{2(1 - \rho^2)\sigma_2^2} \left[y - m_2 - \rho\frac{\sigma_2}{\sigma_1}(x - m_1) \right]^2 \right\} dy \\[2mm]
&= \begin{cases} \dfrac{1}{\sqrt{2\pi}\, \sigma_1} \exp\left[-\dfrac{1}{2}\left(\dfrac{x - m_1}{\sigma_1} \right)^2 \right] & \text{if } -\infty < x < \infty \\[2mm] 0 & \text{otherwise.} \end{cases}
\end{aligned}
$$

Thus the marginal probability density function of X is normal with parameters m_1 and σ_1^2. Similarly, one can show that the marginal probability density function of Y is normal with parameters m_2 and σ_2^2.

3.6 CONDITIONAL DISTRIBUTIONS OF A BIVARIATE RANDOM VARIABLE

Given two events A, B, we have defined (Chapter 2) the conditional probability of A given B. In this section we use this notion to define conditional probability mass and probability density functions.

Discrete Case

Let $p_{X,Y}(x_i,y_j) = p_{ij}$, $i = 1,...,m$, $j = 1,...,n$ be the probability mass function of a bivariate discrete random variable (X,Y). The marginal probability mass function of X is

$$
p_X(x_i) = p_{i\cdot} = \sum_{j=1}^{n} p_{ij}, \qquad i = 1,...,m
$$

and that of Y is

$$
p_Y(y_j) = p_{\cdot j} = \sum_{i=1}^{m} p_{ij}, \qquad j = 1,...,n.
$$

The expression

$$p_{Y|X}(y_j|x_i) = P(Y = y_j|X = x_i)$$

$$= \frac{p_{X,Y}(x_i,y_j)}{p_X(x_i)}$$

$$= \begin{cases} \dfrac{p_{ij}}{p_{i\cdot}} & \text{if } j = 1,\dots,n \\ 0 & \text{otherwise} \end{cases} \tag{3.10}$$

is called the *conditional probability mass function* of Y given $X = x_i$. It may be noted that $p_{Y|X}(y_j|x_i)$ is an honest probability mass function, that is,

$$p_{Y|X}(y_j|x_i) \geq 0$$

$$\sum_{j=1}^{n} p_{Y|X}(y_j|x_i) = 1.$$

Similarly, $p_{X|Y}(x_i|y_j) = (p_{ij})/(p_{\cdot j})$ is the conditional probability mass function of X given $Y = y_j$.

Example 3.6.1   Let $p_{X,Y}(x_i,y_j)$ be given by Example 3.4.1.

$$p_{X|Y}(x|y) = \frac{[n!/x!\,y!\,(n - x - y)!]p_1^x p_2^y(1 - p_1 - p_2)^{n-x-y}}{[n!/y!\,(n - y)!]p_2^y(1 - p_2)^{n-y}}$$

$$= \begin{cases} \dfrac{(n - y)!}{x!\,(n - x - y)!} \left(\dfrac{p_1}{1 - p_2}\right)^x \left(1 - \dfrac{p_1}{1 - p_2}\right)^{n-x-y} \\ \qquad \text{if } x = 0,1,\dots,n \\ 0 \qquad \text{otherwise.} \end{cases}$$

Continuous Case

Let the bivariate random variable (X,Y) have continuous distribution function $F_{X,Y}(x,y)$ in both x and y. We define

$$F_{X|Y}(x|y) = P(X \leq x|Y = y). \tag{3.11}$$

Then $F_{X|Y}(x|y)$ satisfies

$$F_{X|Y}(x,y) = \int_{-\infty}^{y} F_{X|Y}(x|y')\,dF_Y(y'). \tag{3.12}$$

This is because

$$F_{X,Y}(x,y) = P(X \le x, Y \le y)$$

$$= \int_{-\infty}^{y} P(X \le x, Y = y') \, dy'$$

$$= \int_{-\infty}^{y} P(X \le x|Y = y')f_Y(y') \, dy'$$

$$= \int_{0}^{y} F_{X|Y}(x|y') \, dF_Y(y'). \tag{3.13}$$

We now define the conditional probability density function of X given $Y = y$ as $f_{X|Y}(x|y)$ such that

$$F_{X|Y}(x|y) = \int_{-\infty}^{x} f_{X|Y}(x'|y) \, dx'. \tag{3.14}$$

It is easy to verify that

$$f_{X|Y}(x|y) = \frac{\partial}{\partial x} F_{X|Y}(x|y). \tag{3.15}$$

We would now like to show that

$$f_{X|Y}(x|y) = \frac{f_{X,Y}(x,y)}{f_Y(y)} \quad \text{for } f_Y(y) > 0. \tag{3.16}$$

To show this we have, from Eq. (3.13),

$$\frac{\partial}{\partial y} F_{X,Y}(x,y) = F_{X|Y}(x|y)f_Y(y) \tag{3.17a}$$

$$\frac{\partial^2}{\partial x \, \partial y} F_{X,Y}(x,y) = f_{X|Y}(x|y)f_Y(y). \tag{3.17b}$$

The left-hand side of Eq. (3.17b) is $f_{X,Y}(x,y)$, thereby yielding Eq. (3.16). Similarly,

$$f_{Y|X}(y|x) = \frac{f_{X,Y}(x,y)}{f_X(x)} \quad \text{for } f_X(x) > 0 \tag{3.18}$$

is the conditional probability density function of Y given that $X = x$. It should be emphasized that the conditional probability density functions described in Eqs. (3.1), (3.16), and (3.18) exist only if the denominators in these expressions are nonzero.

Example 3.6.2   Let (X,Y) be bivariate normal as given in Example 3.4.2. Here we have

$$f_{X|Y}(x|y) = \begin{cases} \dfrac{1}{[2\pi\sigma_1^2(1 - \rho^2)]^{1/2}} \exp\left\{\dfrac{1}{2\sigma_1^2(1 - \rho^2)}\right. \\ \left. \times \left[x - m_1 - \rho\dfrac{\sigma_1}{\sigma_2}(y - m_2)\right]^2\right\} \\ \qquad \text{if } -\infty < x < \infty \\ 0 \qquad \text{otherwise.} \end{cases}$$

Thus if (X,Y) is bivariate normal with parameters m_1, m_2, σ_1^2, σ_2^2, ρ, the conditional probability density function of X given that $Y = y$ is normal with parameters

$$m_1 + \rho\left(\frac{\sigma_1}{\sigma_2}\right)(y - m_2), \qquad \sigma_1^2(1 - \rho^2).$$

In deriving the probability density function from a distribution function F we have assumed that there exists some f such that $f = dF$. A function f such that $f = dF$ may not always exist. The necessary and sufficient condition on the distribution function F that guarantees the existence of the probability density function f is that F must be absolutely continuous. We then say that dF is the Radon–Nikodym derivative of F (cf. the Radon–Nikodym theorem in measure theory). We shall always assume the existence of a probability density function corresponding to a particular distribution function.

3.7   INDEPENDENCE OF RANDOM VARIABLES

In Chapter 2 we defined the independence of any two random events A, B by the condition $P(A \cap B) = P(A)P(B)$. We will use this notion to define the independence of two random variables X, Y. For any pair of real numbers (x,y), let $A = \{X : X \leq x\}$ and $B = \{Y : Y \leq y\}$; then we have

$$F_{X,Y}(x,y) = P(A \cap B)$$
$$F_X(x) = P(A)$$
$$F_Y(y) = P(B).$$

Definition 3.7.1: Independence of a pair of random variables Two random variables X, Y are *independent in distribution* if for every pair of reals (x,y),

$$F_{X,Y}(x,y) = F_X(x)F_Y(y).$$

In the case of discrete random variables (X,Y), $A = \{X = x\}$, $B = \{Y = y\}$ are random events. Hence the definition of independence in the discrete case is

$$p_{X,Y}(x,y) = p_X(x)p_Y(y).$$

In the case of continuous random variables (X,Y), the analogous definition in terms of the probability density function is

$$f_{X,Y}(x,y) = f_X(x)f_Y(y).$$

This follows from the fact that

$$f_{X,Y}(x,y) = \frac{\partial^2}{\partial x\, \partial y} F_{X,Y}(x,y)$$

$$= \frac{\partial^2}{\partial x\, \partial y} F_X(x)F_Y(y) \qquad \text{(by independence)}$$

$$= \frac{\partial}{\partial x} F_X(x) \frac{\partial}{\partial y} F_y(y)$$

$$= f_X(x)f_Y(y).$$

The concept of independence is better understood in terms of conditional probability mass or density functions. In the discrete case two random variables X, Y are independent if

$$p_{X|Y}(x|y) = p_X(x).$$

In the continuous case two random variables X, Y are independent if

$$f_{X|Y}(x|y) = f_X(x).$$

Thus two random variables are independent if their conditional distributions coincide with their marginal distributions.

The converse of the definitions of independence above is always true. It may also be understood that the definitions of independence in terms of the distribution function and in terms of the probability density function and the conditional probability density function are equivalent. If the random variables X, Y are not independent, they are called *dependent*.

Table 3.1 Probability Density Function of (X,Y)

x \ y	1	2	3	4	5	6	Marginal
1	$\frac{1}{36}$	$\frac{1}{36}$	$\frac{1}{36}$	$\frac{1}{36}$	$\frac{1}{36}$	$\frac{1}{36}$	$\frac{1}{6}$
2	$\frac{1}{36}$	$\frac{1}{36}$	$\frac{1}{36}$	$\frac{1}{36}$	$\frac{1}{36}$	$\frac{1}{36}$	$\frac{1}{6}$
3	$\frac{1}{36}$	$\frac{1}{36}$	$\frac{1}{36}$	$\frac{1}{36}$	$\frac{1}{36}$	$\frac{1}{36}$	$\frac{1}{6}$
4	$\frac{1}{36}$	$\frac{1}{36}$	$\frac{1}{36}$	$\frac{1}{36}$	$\frac{1}{36}$	$\frac{1}{36}$	$\frac{1}{6}$
5	$\frac{1}{36}$	$\frac{1}{36}$	$\frac{1}{36}$	$\frac{1}{36}$	$\frac{1}{36}$	$\frac{1}{36}$	$\frac{1}{6}$
6	$\frac{1}{36}$	$\frac{1}{36}$	$\frac{1}{36}$	$\frac{1}{36}$	$\frac{1}{36}$	$\frac{1}{36}$	$\frac{1}{6}$
Marginal	$\frac{1}{6}$	$\frac{1}{6}$	$\frac{1}{6}$	$\frac{1}{6}$	$\frac{1}{6}$	$\frac{1}{6}$	Total: 1

Example 3.7.1 Consider a random experiment of tossing an unbiased die twice. Let X denote the outcome of the first toss and Y denote the outcome of the second toss. The probability mass function of (X,Y) is given in Table 3.1. Here

$$\frac{1}{36} = p_{X,Y}(x,y) = \frac{1}{6} \cdot \frac{1}{6} = p_X(x)p_Y(y)$$

for $x,y = 1,\ldots,6$. Thus X, Y are independent. Since the tosses are independent, the outcome of the first toss does not depend on the outcome of the second toss. Hence, for $x,y = 1,\ldots,6$, $p_{X|Y}(x|y) = \frac{1}{6}$.

Example 3.7.2 Let (X,Y) be a bivariate discrete random variable; X takes values 0, 1, 2 and Y takes values 0, 1. Its probability mass function is given in Table 3.2. For $x = 0,1,2$, $y = 0,1$, $p_{X,Y}(x,y) \neq p_X(x)p_Y(y)$, and similarly, $p_{X|Y}(x|y) \neq p_X(x)$.

Table 3.2 Probability Density Function of (X,Y)

y \ x	0	1	2	Marginal
0	$\frac{1}{6}$	$\frac{1}{6}$	$\frac{1}{6}$	$\frac{1}{2}$
1	$\frac{1}{12}$	$\frac{1}{3}$	$\frac{1}{12}$	$\frac{1}{2}$
Marginal	$\frac{1}{4}$	$\frac{1}{2}$	$\frac{1}{4}$	1

Example 3.7.3   Let (X,Y) be a bivariate random variable with probability density function

$$f_{X,Y}(x,y) = \begin{cases} e^{-x-y} & \text{if } x > 0, \, y > 0 \\ 0 & \text{otherwise.} \end{cases}$$

The marginal probability density functions of X and Y are given by

$$f_X(x) = \begin{cases} e^{-x} & \text{if } x > 0 \\ 0 & \text{otherwise} \end{cases}$$

$$f_Y(y) = \begin{cases} e^{-y} & \text{if } y > 0 \\ 0 & \text{otherwise.} \end{cases}$$

Thus $f_{X,Y}(x,y) = f_X(x)f_Y(y)$. Hence X,Y are independent.

Example 3.7.4   Let (X,Y) be bivariate normal as given in Example 3.4.2. The marginal probability density functions of X, Y are normal with parameters m_1, σ_1^2 and m_2, σ_2^2, respectively. It is easy to see that if $\rho \neq 0$, then

$$f_{X,Y}(x,y) \neq f_X(x)f_Y(y)$$

and thus X, Y are not independent. However, if $\rho = 0$, then

$$f_{X,Y}(x,y) = f_X(x)f_Y(y)$$

and hence X, Y are independent. We have shown in Example 3.6.2 that the conditional probability density function of X given $Y = y$ is normal, with parameters $m_1 + \rho(\sigma_1/\sigma_2)(y - m_2)$, $\sigma_1^2(1 - \rho^2)$. If $\rho = 0$, this conditional probability density function reduces to the marginal probability density function of X.

     The notion of independence can be extended to an arbitrary finite or countably infinite sequence of random variables $X_1, \ldots, X_n, \ldots$.

Definition 3.7.2: Independence of a finite sequence of random variables   An arbitrary finite sequence of random variables $X_1, \ldots, X_n$ is said to be *independent* if for any n-tuple of reals $(x_1, \ldots, x_n)$,

$$F_{X_1, \ldots, X_n}(x_1, \ldots, x_n) = \prod_{i=1}^{n} F_{X_i}(x_i).$$

By the same argument, as in the case of bivariate random variables (X,Y), the sequence of discrete random variables $X_1, \ldots, X_n$ is said to be independent if

$$p_{X_1, \ldots, X_n}(x_1, \ldots, x_n) = \prod_{i=1}^{n} p_{X_i}(x_i),$$

and the sequence of continuous random variables $X_1,\ldots,X_n$ is said to be independent if

$$f_{X_1,\ldots,X_n}(x_1,\ldots,x_n) = \prod_{i=1}^{n} f_{X_i}(x_i).$$

In passing, we want to prove the following important property of an independent sequence of random variables: If the sequence of random variables $X_1,\ldots,X_n$ is independent, then every arbitrary subsequence of $k(2 \le k \le n)$ random variables $X_{l_1},\ldots,X_{l_k}$ is independent.

To show this, assume for simplicity that $l_1 = 1,\ldots,l_k = k$. We then have

$$F_{X_1,\ldots,X_k}(x_1\ldots,x_k) = F_{X_1,\ldots,X_n}(x_1,\ldots,x_k,\infty,\ldots,\infty)$$

$$= \prod_{i=1}^{k} F_{X_i}(x_i) \prod_{i=k+1}^{n} F_{X_i}(\infty)$$

$$= \prod_{i=1}^{k} F_{X_i}(x_i).$$

Definition 3.7.3: Independence of a sequence of random variables A countably infinite sequence of random variables $X_1,\ldots,X_n,\ldots$, is said to be *independent* if for $k = 2,\ldots,n,\ldots$, every arbitrary subsequence of k random variables $X_{l_1},\ldots,X_{l_k}$ is independent.

Definition 3.7.4: Identically distributed random variables A sequence of random variables $X_1,\ldots,X_n$ is said to be *identically distributed* if for any real number x,

$$F_{X_1}(x) = \cdots = F_{X_n}(x).$$

Example 3.7.5 Let $X_1,\ldots,X_n$ be n independent random variables with probability density functions

$$f_{X_i}(x_i) = \begin{cases} \dfrac{1}{\sqrt{2\pi}\,\sigma} \exp\left[-\dfrac{1}{2}\left(\dfrac{x_i - \mu}{\sigma}\right)^2 \right] & \text{if } -\infty < x_i < \infty \\ 0 & \text{otherwise.} \end{cases}$$

Then $X_1,\ldots,X_n$ are identically distributed.

EXERCISES

15. Find parameters m_1, m_2, σ_1, σ_2, and ρ such that the following probability density functions can be written in the form of a bivariate normal density function. In each case, find the marginal probability density

functions $f_X(x)$, $f_Y(y)$ and the conditional probability density function $f_{X|Y}(x|y)$.

(a)

$$f_{X,Y}(x,y) = \frac{1}{2\pi} \exp\left[-\frac{1}{2} (x^2 + y^2 + 6x - 10y + 34) \right].$$

(b)

$$f_{X,Y}(x,y) = \frac{1}{2\pi\sqrt{3}} \exp\left[-\frac{1}{2} (x^2 + \frac{1}{4} y^2 \right.$$
$$\left. -\frac{1}{2} xy - 2x + \frac{1}{2} y + 1) \right].$$

16. Suppose that the joint probability density function of the random variables X, Y is given by

(a)

$$f_{X,Y}(x,y) = \begin{cases} 1 & \text{if } 0 \le x \le 1, \quad 0 \le y \le 1 \\ 0 & \text{otherwise} \end{cases}$$

(b)

$$f_{X,Y}(x,y) = \begin{cases} \pi^{-1} & \text{if } x^2 + y^2 \le 1 \\ 0 & \text{otherwise} \end{cases}.$$

In each case, find the marginal probability density function $f_X(x)$, $f_Y(y)$ and the conditional probability density function $f_{X|Y}(x|y)$.

17. Let the joint probability density function of the random variables X, Y be given by

$$f_{X,Y}(x,y) = \begin{cases} \frac{3}{2}, & 0 \le x \le 1, \quad -(x-1)^2 \le y \le (x-1)^2 \\ 0 & \text{otherwise.} \end{cases}$$

Verify that X, Y are not independent.

18. Let X, Y be a bivariate discrete random variables with joint probability density function

$$p_{X,Y}(x,y) = \begin{cases} \dfrac{1}{n^2}, & x = 1,\dots,n, \quad y = 1,\dots,n \\ 0 & \text{otherwise.} \end{cases}$$

Are X and Y independent?

19. Prove that the joint distribution function $F_{X,Y}(x,y)$ of the random variables X, Y is continuous in (x,y) if and only if each $F_X(x)$, $F_Y(y)$ is continuous with respect to x and y, respectively.

20. Consider two consecutive tosses of a coin. The random variable X takes on the value 0 or 1 according to whether a head or a tail appears as a result of the first toss. The random variable Y takes the value 0 or 1 according to whether a head or a tail appears as a result of the second toss. Are X, Y independent?

21. A box contains five tickets marked with the numbers 1, 2, 3, 4, and 5. Three tickets are drawn at random from this box one by one with replacement. Let X, Y denote the maximum and the minimum scores on the tickets drawn. Find the joint probability density function of X and Y.

22. Let X, Y be two random variables having the following joint probability density functions:

(a)

$$f_{X,Y}(x,y) = \begin{cases} cx^3 e^{-x(y+1)} & \text{if } x > 0, \quad y > 0 \\ 0 & \text{otherwise.} \end{cases}$$

(b)

$$f_{X,Y}(x,y) = \begin{cases} c(1 + x + y)^{-n} & \text{if } x > 0, \quad y > 0, \quad n > 2 \\ 0 & \text{otherwise.} \end{cases}$$

Find (a) the constant c; (b) the marginal probability density function of X and Y; (c) the conditional probability density function of X given $Y = y$.

23. Two discrete random variables X, Y have the joint probability density function

$$p_{X,Y}(x,y) = \begin{cases} \dfrac{\lambda^x e^{-\lambda} p^y (1 - p)^{x-y}}{y! \, (x - y)!} & \text{if } y = 0, 1,\ldots,x, x = 0,1,2,\ldots \\ 0 & \text{otherwise,} \end{cases}$$

where λ, p are constants with $\lambda > 0$ and $0 < p < 1$.
(a) Find the marginal probability density functions of X and Y.
(b) Find the conditional probability density function of X given Y.

24. Show that if X, Y are independently distributed Poisson random variables with parameters λ, μ, respectively, the conditional density function of X given $X + Y = n$ is a binomial (n,p) with $p = \lambda/(\mu + \lambda)$.

3.8 FUNCTIONS OF A RANDOM VARIABLE

Let $Y = g(X)$ be a function of a continuous random variable X. We are interested in finding the distribution of Y. In this context we will always take g to be a Borel function; that is, for any real c, the set $\{x : g(x) \le c\}$

is a Borel set. An example of such function is any continuous function with a finite or at most a countably infinite number of discontinuities. The distribution function of Y is

$$F_Y(y) = P(Y \le y) = P\{X : g(X) \le y\}.$$

We will consider three cases:

Case 1

The function $g(X)$ is differentiable, $g'(X)$ is nonzero for all X, and $g(X)$ is either a monotonically increasing or a decreasing function of X. In this case Y is a continuous random variable with probability density function

$$
\begin{aligned}
f_Y(y) &= \frac{d}{dy} F_Y(y) \\
&= \begin{cases} f_X(g^{-1}(y)) \left| \dfrac{d}{dy} g^{-1}(y) \right| & \text{if } a < y < b \\ 0 & \text{otherwise,} \end{cases}
\end{aligned}
\tag{3.19}
$$

where a, b are, respectively, the lower and upper bounds of the range of Y, and $|\cdot|$ stands for the absolute value. This is because if g is a monotonically increasing function of X, then for a $a < y < b$,

$$P(Y \le y) = P(X \le g^{-1}(y)) = F_X(g^{-1}(y)),$$

so that

$$f_Y(y) = f_X(g^{-1}(y)) \frac{d}{dy} g^{-1}(y) \tag{3.20}$$

and if g is a monotonically decreasing function of X, then for $a < y < b$,

$$
\begin{aligned}
P(Y \le y) &= P(X \ge g^{-1}(y)) \\
&= 1 - P(X < g^{-1}(y)) \\
&= 1 - F_X(g^{-1}(y)),
\end{aligned}
$$

so that in either case

$$f_Y(y) = f_X(g^{-1}(y)) \left| \frac{d}{dy} g^{-1}(y) \right|. \tag{3.21}$$

It may be noted that

$$\frac{dg^{-1}(y)}{dy} = \left(\frac{dy}{dx} \right)^{-1} = |g'(x)|^{-1}. \tag{3.22}$$

In the case of discrete random variable X, the random variable $Y = g(X)$ is also a discrete random variable and has probability density function

$$P_Y(y) = P(g(X) = y)$$
$$= P(X = g^{-1}(y))$$
$$= P_X(g^{-1}(y)). \tag{3.23}$$

Example 3.8.1 Suppose that the continuous random variable X has probability density function $f_X(x)$. Let $Y = aX + b$, where $a \neq 0$, $-\infty < b < \infty$. The probability density function of Y is

$$f_Y(y) = \begin{cases} f_X\left(\dfrac{y-b}{a}\right)\left|\dfrac{1}{a}\right| & \text{if } -\infty < y < \infty \\ 0 & \text{otherwise.} \end{cases} \tag{3.24}$$

If X is discrete, then $Y = aX + b$ is a discrete random variable with probability density function

$$P_Y(y) = P_X\left(\frac{y-b}{a}\right). \tag{3.25}$$

Case 2

Suppose that $Y = g(X)$ is a differentiable function of the continuous random variable X. Let $g(X)$ be monotonically increasing for some values of X and be monotonically decreasing for the remaining values of X. The probability density function of the continuous random variable Y is obtained by expressing its distribution function in terms of the distribution of X and then differentiating it with respect to y. For illustration, consider $Y = X^2$; Y is monotonically increasing for positive values of X and monotonically decreasing for negative values of X.

$$F_Y(y) = \begin{cases} P(-\sqrt{y} \leq X \leq \sqrt{y}) & \text{if } y > 0 \\ 0 & \text{otherwise.} \end{cases}$$

Since $P(-\sqrt{y} \leq X \leq \sqrt{y}) = F_X(\sqrt{y}) - F_X(-\sqrt{y})$, we get

$$f_Y(y) = \begin{cases} \dfrac{f_X(\sqrt{y}) + f_X(-\sqrt{y})}{2\sqrt{y}} & \text{if } y > 0 \\ 0 & \text{otherwise.} \end{cases} \tag{3.26}$$

Example 3.8.2 Let X be a normal random variable with parameters 0, 1. Then $Y = X^2$ is monotonically increasing for positive values of X and is monotonically decreasing for negative values of X. Then the probability density function of Y is

$$f_Y(y) = \frac{1}{2} \left(\frac{1}{\sqrt{2\pi}} e^{-(1/2)y} + \frac{1}{\sqrt{2\pi}} e^{-(1/2)y} \right) y^{-1/2}$$

$$= \begin{cases} \dfrac{1/2}{\Gamma(1/2)} \left(\dfrac{1}{2} y \right)^{(1/2)-1} e^{-(1/2)y} & \text{if } y > 0 \\ 0 & \text{otherwise.} \end{cases}$$

Thus the square of a unit normal random variable is a chi-square random variable with one degree of freedom (see Example 3.3.2). It is now left to the reader to prove that if X is a normal random variable with parameters (μ, σ^2), then $Y = [(X - \mu)/\sigma]^2$ is a chi-square random variable with one degree of freedom.

Case 3

Let X be a continuous random variable and $Y = g(X)$ be a differentiable function of X with a nonzero derivative for all x except for a finite or at most for a countably infinite number of values of X. For a given value y of Y, let $x_1(y), \ldots, x_{k(y)}(y)$ be the values of x such that for $i = 1, \ldots, k(y)$, $g(x_i(y)) = y$, and $g'(x_i(y)) \neq 0$. Then

$$f_Y(y) = \begin{cases} \displaystyle\sum_{i=1}^{k(y)} f_X(x_i)|g'(x_i(y))|^{-1} & \text{if } k(y) > 0 \\ 0 & \text{otherwise.} \end{cases} \tag{3.27}$$

Because of notational difficulties we have written $f_X(x_i(y))$, $g'(x_i(y))$, but actually they should be expressed in terms of y. The proof of this assertion is left to the reader. As an illustration, let us take $Y = |X|$, where X $(-\infty < X < \infty)$ is a continuous random variable. For $y > 0$

$$F_Y(y) = P(-y \le X \le y) = F_X(y) - F_X(-y).$$

Hence the probability density function of Y is given by

$$f_Y(y) = \begin{cases} f_X(y) + f_X(-y) & \text{if } y > 0 \\ 0 & \text{otherwise.} \end{cases} \tag{3.28}$$

Example 3.8.3 Let X be a normal random variable with parameters μ, σ^2. Then $Y = |X|$ is a continuous random variable with probability density function

$$f_Y(y) = \begin{cases} \dfrac{2}{\sqrt{2\pi}\,\sigma} \exp\left[-\dfrac{1}{2\sigma^2}(y^2 + \mu^2) \right] \\ \quad \times \left[1 + \dfrac{(\mu y/\sigma)^2}{2!} + \dfrac{(\mu y/\sigma)^4}{4!} + \cdots \right] & \text{if } y > 0 \\ 0 & \text{otherwise.} \end{cases}$$

3.9 FUNCTIONS OF SEVERAL RANDOM VARIABLES

Let $X_1,\ldots,X_n$ be a sequence of n continuous random variables having a joint probability density function $f_{X_1,\ldots,X_n}(x_1,\ldots,x_n)$. Let

$$Y_i = g_i(X_1,\ldots,X_n), \qquad i = 1,\ldots,n \tag{3.29}$$

be a continuous one-to-one transformation of the random variables $X_1,\ldots,X_n$. Let us assume that the functions $g_1,\ldots,g_n$ have continuous partial derivatives with respect to $x_1,\ldots,x_n$. Let the inverse functions be denoted by

$$X_i = h_i(Y_1,\ldots,Y_n), \qquad i = 1,\ldots,n. \tag{3.30}$$

Denote by J the Jacobian of the inverse transformation. Then

$$J = \det \begin{pmatrix} \dfrac{\partial x_1}{\partial y_1} & \cdots & \dfrac{\partial x_1}{\partial y_n} \\ \vdots & \cdots & \vdots \\ \dfrac{\partial x_n}{\partial y_1} & \cdots & \dfrac{\partial x_n}{\partial y_n} \end{pmatrix},$$

where "det" denotes the determinant of the square matrix. We will assume here that there exists a region R of points $(x_1,\ldots,x_n)$ on which J is different from zero. Let S be the image of R under the transformations given in Eq. (3.29). Then we have

$$P((X_1,\ldots,X_n) \in R)$$

$$= \underbrace{\int \cdots \int_R f_{X_1,\ldots,X_n}(x_1,\ldots,x_n) \; dx_1,\ldots,dx_n}_{n \text{ integrals}}$$

$$= \int \cdots \int_S f_{X_1,\ldots,X_n}(h_1(y_1,\ldots,y_n),\ldots,h_n(y_1,\ldots,y_n))|J| \; dy_1,\ldots,dy_n$$

$$= P(Y_1,\ldots,Y_n) \in S). \tag{3.31}$$

Thus it follows that the joint probability density function of the random variables $Y_1,\ldots,Y_n$ is given by

$$f_{Y_1,\ldots,Y_n}(y_1,\ldots,y_n)$$
$$= \begin{cases} f_{X_1,\ldots,X_n}(h_1(y_1,\ldots,y_n),\ldots,h_n(y_1,\ldots,y_n))|J| & \text{if } (y_1,\ldots,y_n) \in S \\ 0 & \text{otherwise.} \end{cases}$$

For the change of variables in multiple integrals, the reader is referred to Apostol (1957, p. 253). We use the symbol $\{(y_1,\ldots,y_n) \in S\}$ to denote the set of all points $(y_1,\ldots,y_n)$ belonging to S.

To find the distribution of a sum of n continuous random variables $X_1,\ldots,X_n$ having a joint probability density function, we take

$$Y_1 = X_1 + \cdots + X_n$$

$$Y_i = X_i, \qquad i = 2,\ldots,n.$$

It is easy to see that $J = 1$. The inverse transformation is

$$X_1 = Y_1 \Sigma \sum_{i=2}^{n} Y_i$$

$$X_i = Y_i, \qquad i = 2,\ldots,n.$$

Hence

$$f_{Y_1}(y_1) = \underbrace{\int_{-\infty}^{\infty} \cdots \int_{-\infty}^{\infty}}_{(n-1)\ \text{integrals}} f_{X_1,\ldots,X_n}\left(y_1 - \sum_{i=2}^{n} y_i, y_2,\ldots,y_n\right) dy_2 \cdots dy_n.$$

Example 3.9.1: Sum of n independent normal random variables   Let $X_1,\ldots,X_n$ be independent and identically distributed normal random variables with parameters $\mu = 0$, $\sigma^2 = 1$. Let

$$Y_1 = \sum_{i=1}^{n} X_i.$$

Then the probability density function of the continuous random variable Y_1 is

$$f_{Y_1}(y_1) = \int_{-\infty}^{\infty} \cdots \int_{-\infty}^{\infty} \frac{1}{(2\pi)^{n/2}} \exp\left\{ -\frac{1}{2}\left[\sum_{i=2}^{n} x_i^2 + \left(y_1 - \sum_{i=2}^{n} x_i\right)^2 \right] \right\}$$

$$dx_2,\ldots,dx_n$$

$$= \int_{-\infty}^{\infty} \cdots \int_{-\infty}^{\infty} \frac{1}{(2\pi)^{(n-1)/2}\sqrt{2}}$$

$$\times \exp\left\{ -\frac{1}{2}\left[\sum_{i=2}^{n-1} x_i^2 + \frac{1}{2}\left(y_1 - \sum_{i=2}^{n-1} x_i\right)^2 \right] \right\} dx_2,\ldots,dx_{n-1}$$

$$= \int_{-\infty}^{\infty} \cdots \int_{\infty}^{\infty} \frac{1}{(2\pi)^{(n-1)/2}\sqrt{3}}$$

$$\times \exp\left\{ -\frac{1}{2}\left[\sum_{i=2}^{n-2} x_i^2 + \frac{1}{3}\left(y_1 - \sum_{i=2}^{n-2} x_i^2\right) \right] \right\} dx_2,\ldots,dx_{n-2}$$

$$\vdots$$

$$= \begin{cases} \dfrac{1}{\sqrt{2n\pi}} \exp\left(-\dfrac{1}{2n} y_1^2\right) & \text{if } -\infty < y_1 < \infty \\ 0 & \text{otherwise.} \end{cases} \tag{3.32}$$

Thus the sum of n independent, identically distributed unit normal random variables is a normal random variable with $\mu = 0$, $\sigma^2 = n$. It is left to the reader to prove that the sum of n independent, identically distributed normal random variables with parameters μ, σ^2 is a normal random variable with parameters $\mu = n\mu$, $\sigma^2 = n\sigma^2$.

Example 3.9.2: Sum of n independent gamma random variables Let $X_1,\ldots,X_n$ be independently distributed gamma random variables with parameters (λ, p_i), $i = 1,\ldots,n$. The joint probability density function of $X_1,\ldots,X_n$ is

$$f_{X_1,\ldots,X_n}(x_1,\ldots,x_n)$$

$$= \begin{cases} \dfrac{\lambda^{\sum_{i=1}^n p_i}}{\prod_{i=1}^n \Gamma(p_i)} \exp\left(-\lambda \sum_{i=1}^n x_i\right) \prod_{i=1}^n x_i^{p_i-1} & \text{if } x_i > 0,\ i = 1,\ldots,n \\ 0 & \text{otherwise.} \end{cases}$$

Let $Y_1 = \sum_{i=1}^n X_i$; the probability density function of Y_1 is

$$f_{Y_1}(y_1) = \int_0^\infty \cdots \int_0^\infty \frac{\lambda^{\sum_{i=1}^n p^i}}{\prod_{i=1}^n \Gamma(p_i)} \exp(-\lambda y_1)\left[\prod_2^n y_i^{p_i-1}\left(y_1 - \sum_{i=1}^n y_i\right)^{p_1-1}\right]$$

$$dy_2,\ldots,dy_n. \tag{3.33}$$

Noting that

$$\left(y_1 - \sum_{i=2}^n y_i\right)^{p_1-1} = \left(y_1 - \sum_{i=2}^{n-1} y_i\right)^{p_1-1}\left(1 - \frac{y_n}{y_1 - \sum_{i=2}^{n-1} y_i}\right)^{p_1-1}$$

and for $0 < x < \infty$, $z = y/x$,

$$\int_0^\infty y^{a-1}(x - y)^{b-1}\,dy = x^{a+b}\int_0^1 z^{a-1}(1 - z)^{b-1}\,dz$$

$$= x^{a+b}\frac{\Gamma(a)\Gamma(b)}{\Gamma(a + b)},$$

we get from Eq. (3.33),

$$f_{Y_1}(y_1) = \begin{cases} \dfrac{\lambda^{\sum_1^n p_i}}{\Gamma(\sum_{i=1}^n p_i)} e^{-\lambda y_1} y_1^{\sum_1^n p_i - 1} & \text{if } y_1 > 0 \\ 0 & \text{otherwise.} \end{cases} \tag{3.34}$$

Hence the sum of n independent gamma random variables with parameters λ, p_i, where $i = 1,\ldots,n$, is a gamma random variable with parameters λ, $\sum_{i=1}^n p_i$. Thus we note that the sum of n independent chi-square random variables with degrees of freedom p_i, $i = 1,\ldots,n$ is again a chi-square random variable with $\sum_{i=1}^n p_i$ degrees of freedom.

In the case of discrete random variables $X_1,\ldots,X_n$ with joint probability mass function $p_{X_1,\ldots,X_n}(x_1,\ldots,x_n)$, the probability mass function of

$$Y = \sum_{i=1}^{n} X_i$$

is given by

$$p_Y(y) = \sum_{x_1} \cdots \sum_{x_n} p_{X_1,\ldots,X_n}(x_1,\ldots,x_n), \tag{3.35}$$

where

$$\sum_{i=1}^{n} x_i = y.$$

Example 3.9.3: Sum of n independent binomial random variables Let $X_1,\ldots,X_n$ be n independent binomial random variables with parameters n_i, p. Let

$$Y = \sum_{i=1}^{n} X_i.$$

The joint probability mass function of $X_1,\ldots,X_n$ is given by

$$P_{X_1,\ldots,X_n}(x_1,\ldots,x_n) = \prod_{i=1}^{n} \binom{n_i}{x_i} p^{\sum_{i=1}^{n} x_i} (1 - p)^{\sum_{i=1}^{n} n_i - \sum_{i=1}^{n} x_i}$$

for $x_i = 0,1,\ldots,n_i$, $i = 1,\ldots,n$. From (3.35) we get

$$p_Y(y) = p^y(1 - p)^{\sum_{i=1}^{n} n_i - y} \sum_{x_1=0}^{n_1} \cdots \sum_{x_n=0}^{n_n} \prod_{i=1}^{n} \binom{n_i}{x_i}, \tag{3.36}$$

where

$$\sum_{1}^{n} x_i = y.$$

Let us now consider the coefficient of x^y in the following identity:

$$\prod_{i=1}^{n} (1 + x)^{n_i} = (1 + x)^{\sum_{i=1}^{n} n_i}.$$

The coefficient of x^y on the left-hand side is

$$\sum_{x_1} \cdots \sum_{x_n} \prod_{i=1}^{n} \binom{n_i}{x_i} \tag{3.37}$$

where

$$\sum_{i=1}^{n} x_i = y,$$

and that on the right-hand side is

$$\left(\begin{array}{c} \sum_{i=1}^{n} n_i \\ y \end{array} \right). \tag{3.38}$$

Since the terms (3.37) and (3.38) must be equal, we get from Eq. (3.36)

$$p_Y(y) = \begin{cases} \left(\begin{array}{c} \sum_{i=1}^{n} n_i \\ y \end{array} \right) p^y (1-p)^{\Sigma_{i=1}^n n_i - y} & \text{if } y = 0,1,\dots, \sum_{i=1}^{n} n_i \\ 0 & \text{otherwise.} \end{cases}$$

Example 3.9.4: Sum of n independent Poisson random variables Let $X_1,\dots,X_n$ be n independent identically distributed Poisson random variables with parameter λ. Their joint probability mass function is given by

$$P_{X_1,\dots,X_n}(x_1,\dots,x_n) = \frac{e^{-n\lambda}\lambda^{\Sigma_{i=1}^n x_i}}{\prod_{i=1}^{n} x_i!} \qquad \text{if } x_i = 0,1,\dots,\infty, \quad i = 1,\dots,n.$$

Let $Y = \Sigma_{i=1}^{n} X_i$. Then

$$P_Y(y) = e^{-n\lambda}\lambda^y \sum_{x_1} \cdots \sum_{x_n} \frac{1}{x_1! \cdots x_n!}, \tag{3.39}$$

with

$$\sum_{i=1}^{n} x_i = y.$$

It is well known that

$$(a_1 + a_2 + \cdots + a_n)^y = \sum_{x_1} \cdots \sum_{x_n} \frac{y!}{x_1! \cdots x_n!} a_1^{x_1} \cdots a_n^{x_n},$$

with

$$\sum_{1}^{n} x_i = y.$$

Taking, $a_i = 1/n$ for all i, we get

$$\sum_{x_1} \cdots \sum_{x_n} \frac{1}{x_1! \cdots x_n!} = \frac{n^y}{y!}, \tag{3.40}$$

with

$$\sum_1^n x_i = y.$$

Thus from Eqs. (3.39) and (3.40) we get

$$P_Y(y) = \begin{cases} \dfrac{e^{-n\lambda}(n\lambda)^y}{y!} & \text{if } y = 0,1,\ldots \\ 0 & \text{otherwise.} \end{cases}$$

3.10 DISTRIBUTION OF PRODUCT AND RATIO OF TWO RANDOM VARIABLES

Given the probability density function $f_{X,Y}(x,y)$ of the bivariate continuous random variables (X,Y), we are now interested in the probability density functions of $U = XY$ and $V = X/Y$. For $U = XY$, let us take

$$U = XY \qquad X = \frac{U}{Y}$$

$$\text{or}$$

$$Y = Y \qquad Y = Y.$$

The Jacobian of the inverse transformation from (U,Y) to (X,Y) is

$$\left(\det \begin{matrix} \dfrac{1}{Y} & \dfrac{-U}{Y^2} \\ 0 & 1 \end{matrix} \right) = \frac{1}{Y}.$$

Thus the joint probability density function of (U,Y) is

$$f_{U,Y}(u,y) = \begin{cases} f_{X,Y}\left(\dfrac{u}{y}, y\right) \dfrac{1}{|y|} & \text{if } -\infty < y < \infty \\ 0 & \text{otherwise.} \end{cases}$$

Hence the marginal probability density function of U is given by

$$f_U(u) = \int_{-\infty}^{\infty} f_{X,Y}\left(\frac{u}{y}, y\right) \frac{1}{|y|} \, dy. \tag{3.41}$$

For $V = X/Y$, let us take

$$V = \frac{X}{Y} \qquad X = VY$$

or

$$Y = Y \qquad Y = Y.$$

The Jacobian of the inverse transformation $(V,Y) \to (X,Y)$ is Y. The joint probability density function of (V,Y) is

$$f_{V,Y}(v,y) = f_{X,Y}(vy,y)|y|.$$

Thus the marginal probability density function of V is given by

$$f_V(v) = \int_{-\infty}^{\infty} f_{X,Y}(vy,y)|y| \, dy. \tag{3.42}$$

Example 3.10.1: Product of two independent beta random variables Let X, Y be independently distributed beta random variables with parameters (p_1,q_1), (p_2,q_2), respectively, such that $p_1 = p_2 + q_2$. The probability density function of $U = XY$ is [cf. Eq. (3.41)]

$$f_U(u) = \int_0^1 \frac{\Gamma(p_1 + q_1)\Gamma(p_2 + q_2)}{\Gamma(p_1)\Gamma(p_2)\Gamma(q_1)\Gamma(q_2)} u^{p_2-1}(x - u)^{q_2-1}(1 - x)^{q_1-1} \, dx$$

$$= \begin{cases} \dfrac{\Gamma(p_1 + q_1)}{\Gamma(p_2)\Gamma(q_1 + q_2)} u^{p_2-1}(1 - u)^{q_1+q_2-1} & \text{if } 0 < u < 1 \\ 0 & \text{otherwise.} \end{cases} \tag{3.43}$$

Example 3.10.2: F random variable Let X, Y be two independent chi-square random variables with degrees of freedom m, n, respectively. The probability density function of

$$F = \frac{X/m}{Y/n} = \frac{n}{m} \frac{X}{Y}$$

is then [cf. Eq. (3.42)]

$$f_F(f) = \int_0^{\infty} f_{X,Y}\left(\frac{m}{n} fy, y\right) \frac{m}{n} y \, dy$$

$$= \frac{m}{n} \frac{1}{2^{(m+n)/2}\Gamma(m/2)\Gamma(n/2)}$$

$$\times \int_0^{\infty} \exp\left[-\frac{1}{2} y\left(1 + \frac{m}{n} f\right)\right]\left(\frac{m}{n} f\right)^{(m/2)-1} y^{(m+n)/2-1}$$

$$= \begin{cases} \dfrac{m}{n} \dfrac{\Gamma[(m+n)/2]}{\Gamma(m/2)\Gamma(n/2)} \dfrac{[(m/n)f]^{(m/2)-1}}{[1 + (m/n)f]^{(m+n)/2}} & \text{if } f > 0 \\ 0 & \text{otherwise.} \end{cases} \tag{3.44}$$

EXERCISES

25. Show that if the random variable X has uniform distribution over the interval $(0,1)$, the random variable $-2 \log X$ has a chi-square distribution with 2 degrees of freedom.
26. Show that if the random variable X has probability density function

$$f_X(x) = \begin{cases} \dfrac{\sqrt{2}}{\pi} \left(\dfrac{x^2}{\sigma^3} \right) \exp\left(-\dfrac{x^2}{2\sigma^2} \right), & x > 0, \quad \sigma > 0 \\ 0, & \text{otherwise,} \end{cases}$$

then X^2/σ^2 has a chi-square distribution with 3 degrees of freedom.
27. The random variable X is said to have log-normal distribution if $\log X$ has normal distribution; that is, its probability density function (with $-\infty < u < \infty$, $\sigma > 0$) is given by

$$f_X(x) = \begin{cases} \dfrac{1}{\sqrt{2\pi}\ \sigma x} \exp\left[-\dfrac{(\log x - u)^2}{2\sigma^2} \right], & x \geq 0 \\ 0, & \text{otherwise.} \end{cases}$$

   (a) Find the mean and variance of X.
   (b) Show that if X is log-normal, so is X^r.
28. Show that if X, Y are random variables independently and uniformly distributed over the interval $(1,3)$ and if $Z = X + Y$, then

$$f_Z(z) = \begin{cases} 0, & z \leq 2 \quad \text{or} \quad z \geq 6 \\ \dfrac{z - 2}{4}, & 2 < z \leq 4 \\ \dfrac{6 - z}{4}, & 4 < z < 6. \end{cases}$$

29. Suppose that the random variables X, Y are independent and each has a uniform distribution over the interval $(0,1)$. Find the probability density function of $U = X + Y$ and $V = X/Y$.
30. Let X, Y be independent and identically distributed geometric random variables. Let $Z = \max(X,Y)$. Find the distribution of Z.
31. Find the probability density function of $Y = F_X(X)$ when X is a continuous random variable with distribution function $F_X(x)$.
32. Suppose that the random variable X has a probability density function and is independent of another random variable Y. Show that $X + Y$ has a probability density function.
33. Let X, Y be independently and identically distributed normal random variables with mean 2 and variance 4.
   (a) Show that $X + Y$, $X - Y$ are independent.

(b) Show that $(X + Y, X + 2Y)$ has a bivariate normal distribution with $\rho = 3/\sqrt{10}$.

34. Show that if the sum of n independent random variables is normally distributed then each variable is normally distributed. See Cramer (1937).

35. Let (X,Y) be a bivariate normal random variable with $m_1 = m_2 = 0$ and $\sigma_1^2 = \sigma_2^2 = 1$. Show that

(a) $P(X \geq 0, Y \geq 0) = P(X \leq 0, Y \geq 0)$

$$= \frac{1}{4} + \frac{1}{2\pi} \sin^{-1} \rho$$

(b) $P(X \leq 0, Y \geq 0) = P(X \geq 0, Y \leq 0)$

$$= \frac{1}{4} - \frac{1}{2\pi} \sin^{-1} \rho.$$

36. Let X be a uniform random variable on the interval $(1,3)$ and Y be an exponential random variable with parameter λ such that $V(X) = V(Y)$. Find λ.

37. Let X_1 and X_2 be independently and identically distributed normal random variables with mean 0 and variance σ^2.
    (a) Show that X_1/X_2 has a Cauchy distribution.
    (b) Show that $X_1^2 + X_2^2$, $X_1(X_1^2 + X_2^2)^{-1/2}$ are independent.

38. Show that if $X_1,\ldots,X_n$ are independently distributed $G(1,\lambda)$, then $Z = 2\lambda \sum_{r=1}^{n} X_i$ is distributed as χ_{2n}^2.

39. Suppose that X, Y are independent gamma random variables with parameters $\lambda, 1$ and $\lambda + \frac{1}{2}, 1$, respectively. Show that the random variable $2\sqrt{XY}$ has gamma distribution with parameters $2\lambda, 1$.

3.11 MATHEMATICAL EXPECTATION

We have seen earlier that the probability density function or mass function of a random variable X depends on certain parameters. These parameters play a very important role in statistics and probability. In this section we express these parameters in terms of certain characteristics of the random variable, called *moments*. Let $Y = g(X)$ be a Borel function of the random variable X and let $|\cdot|$ be the absolute-value symbol.

Definition 3.11.1: Mathematical expectation of a function of a discrete random variable Let X be a discrete random variable with probability mass function $p_X(x)$. The *mathematical expectation* of $g(X)$, denoted by $E(g(X))$, is defined by

$$E(g(X)) = \sum_{x:p_X(x)>0} g(x)p_X(x) \tag{3.45}$$

provided that

$$E|g(X)| = \sum_{x:p_X(x)>0} |g(x)|p_X(x) < \infty$$

If $E|g(X)|$ is not finite, we say that the expectation does not exist.

Example 3.11.1 Let X be a binomial random variable with parameters n, p, and let (a) $g(X) = X$, (b) $g(X) = X^2$. Then the following relations hold:

(a) $E(X) = \displaystyle\sum_{x=0}^{n} x\binom{n}{x}p^x(1 - p)^{n-x}$

$\qquad\quad = np \displaystyle\sum_{x=1}^{n} \frac{(n - 1)!}{(x - 1)! \, (n - x)!} \, p^{x-1}(1 - p)^{n-x}$

$\qquad\quad = np[p + (1 - p)]^{n-1} = np.$

(b) $E(X^2) = \displaystyle\sum_{x=0}^{n} x^2\binom{n}{x}p^x(1 - p)^{n-x}$

$\qquad\qquad = \displaystyle\sum_{x=0}^{n} [x(x - 1) + x]\binom{n}{x}p^x(1 - p)^{n-x}$

$\qquad\qquad = \displaystyle\sum_{x=0}^{n} x(x - 1)\binom{n}{x}p^x(1 - p)^{n-x} + \sum_{x=0}^{n} x\binom{n}{x}p^x(1 - p)^{n-x}$

$\qquad\qquad = n(n - 1)p^2 \displaystyle\sum_{x=2}^{n} \frac{(n - 2)!}{(x - 2)! \, (n - x)!} \, p^{x-2}(1 - p)^{n-x} + np$

$\qquad\qquad = n(n - 1)p^2 + np.$

Definition 3.11.2: Mathematical expectation of a function of a continuous random variable Let X be a continuous random variable with probability density function $f_X(x)$. The mathematical expectation of $g(X)$ is defined by

$$E(g(X)) = \int_{-\infty}^{\infty} g(x)f_X(x) \, dx \tag{3.46}$$

provided that

$$E|g(X)| = \int_{-\infty}^{\infty} |g(x)| f_X(x) \, dx < \infty.$$

Example 3.11.2 Let X be a normal random variable with parameters μ, σ^2. Then we have

$$E(X) = \int_{-\infty}^{\infty} x \, \frac{1}{\sigma\sqrt{2\pi}} \exp\left[-\frac{1}{2\sigma^2} (x - \mu)^2 \right] dx$$

$$= \int_{-\infty}^{\infty} (\mu + y\sigma) \, \frac{1}{\sqrt{2\pi}} e^{-(1/2)y^2} \, dy$$

$$= \mu + \sigma \int_{-\infty}^{\infty} \frac{1}{\sqrt{2\pi}} y e^{-(1/2)y^2} \, dy$$

$$= \mu,$$

where the second term vanishes because the integrand is an odd function. We also have

$$E(X^2) = \int_{-\infty}^{\infty} x^2 \, \frac{1}{\sqrt{2\pi}\,\sigma} \exp\left[-\frac{1}{2} \left(\frac{x - \mu}{\sigma} \right)^2 \right] dx$$

$$= \int_{-\infty}^{\infty} (\mu^2 + 2\mu\sigma y + \sigma^2 y^2) \, \frac{1}{\sqrt{2\pi}} e^{-(1/2)y^2} \, dy$$

$$= \mu^2 + \sigma^2 \int_{-\infty}^{\infty} y^2 \, \frac{1}{\sqrt{2\pi}} e^{-(1/2)y^2} \, dy$$

$$= \mu^2 + \sigma^2 \int_{0}^{\infty} \frac{1}{\sqrt{2\pi}} e^{-(1/2)z} z^{(3/2)-1} \, dz$$

$$= \mu^2 + \sigma^2 \, \frac{\Gamma(3/2)}{(1/2)^{3/2}\sqrt{2\pi}}$$

$$= \mu^2 + \sigma^2.$$

We have not defined the meaning of $\int_{-\infty}^{\infty} g(x) f_X(x) \, dx$. In elementary calculus, the theory of Riemann integration tells us how to evaluate integrals of the form $\int_a^b g(x) \, dx$, where $g(x)$ is a continuous function of x and a, b are specified real constants. This integral is defined as the common value of the two sums S_1, S_2, called the *upper sum* and the *lower sum*, and defined by

$$S_1 = \sup \sum_{i=1}^{n} m_i(x_i - x_{i-1})$$

$$S_2 = \inf \sum_{i=1}^{n} M_i(x_i - x_{i-1}), \tag{3.47}$$

where the interval $[a,b]$ is partitioned by points x_i, $i = 0,1,\ldots,n$, such that

$$a = x_0 < x_1 < \cdots < x_n = b,$$

and m_i, M_i denote the greatest lower bound and the least upper bound, respectively, of $g(X)$ in the subinterval $x_{i-1} < x \le x_i$ and "sup" and "inf" are, respectively, the greatest lower and least upper bounds over all possible partitions $x_0,\ldots,x_n$ of $[a,b]$ as $n \to \infty$. The case of an integral over the infinite interval is treated as a limiting case of the same integral over a finite interval:

$$\int_{-\infty}^{\infty} g(x)\, dx = \lim_{a \to -\infty, b \to \infty} \int_a^b g(x)\, dx.$$

To give a general definition of mathematical expectation, we need to evaluate integrals of the form

$$\int_a^b g(x)\, dF_X(x).$$

This involves a more general concept than that of Riemann integral, known as the *Riemann–Stieltjes integral*. If $F_X(x)$ is a monotone nondecreasing function of x and has a continuous derivative (as in the case of continuous random variable), then $dF_X(x)$ is interpreted as the differential $f_X(x)\, dx$ and the Riemann–Stieltjes integral $\int_a^b g(x)\, dF_X(x)$ becomes the Riemann integral of $g(x)f_X(x)$. However, the theory of the Riemann–Stieltjes integral is more general than that of the Riemann integral, in the sense that it deals with cases where $F_X(x)$ is a monotone nondecreasing function of x, but where $F_X(x)$ may not be differentiable or even continuous at a finite or a countably infinite number of points. For any partition $(x_0,x_1,\ldots,x_n)$ of $[a,b]$ such that $a = x_0 < x_1 < \cdots < x_n = b$ and any point t_i in the subinterval $[x_{i-1},x_i]$

$$I = \sum_{i=1}^{n} g(t_i)[F_X(x_i) - F_X(x_{i-1})] \tag{3.48}$$

is called the *Riemann–Stieltjes sum* of g with respect to F_X. We say that g is *Riemann–Stieltjes integrable* with respect to F_X over the interval $[a,b]$ if as $n \to \infty$ and $\max(x_i - x_{i-1}) \to 0$, the sum I tends to a finite limit C independent of the choice of the partition $(x_0, x_1, \ldots, x_n)$ of $[a,b]$ and the choice of t_i in the subinterval $[x_{i-1}, x_i]$:

$$C = \lim_{n \to \infty} I = \int_a^b g(x) \, dF_X(x).$$

If $a = -\infty$, $b = +\infty$, we define the improper integral

$$\int_{-\infty}^{\infty} g(x) \, dF_X(x) = \lim_{a \to -\infty, b \to +\infty} \int_a^b g(x) \, dF_X(x).$$

We have remarked earlier that if X is a continuous random variable with probability density function $f_X(x)$, then $E(g(X))$ is the improper Riemann integral of $g(X)f_X(x)$. However, if X is a discrete random variable with probability mass function

$$p_X(x_i) = \begin{cases} p_i, & i = 1, \ldots, n \\ 0, & \text{otherwise,} \end{cases}$$

then $F_X(x)$ is a step function—a function that increases only in jumps and is constant between jumps (see Figure 3.6). Then we obtain for $x_i \in [a,b]$, $i = 1, \ldots, n$,

$$\int_a^b g(x) \, dF_X(x) = \sum_{i=1}^{n} g(x_i) p_i.$$

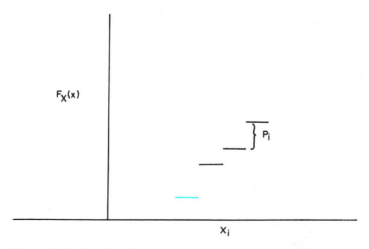

Figure 3.6 Step function.

Thus the separate definitions given above for the expectation of $g(X)$ for the continuous and discrete random variables can now be combined in terms of the Riemann–Stieltjes integral:

$$E(g(X)) = \int g(x)\, dF_X(x).$$

Since $F_X(x)$ is the distribution function of the random variable X, it is proved in real analysis that the mathematical expectation of any Borel function $g(X)$ exists if $\int_{-\infty}^{\infty} |g(x)|\, dF_X(x)$ is finite.

If $F_X(x)$ is continuous except for jumps p_i at the jump points x_i, $i = 1,\ldots,n$ and has continuous derivative between jumps, then $F_X(x)$ can be written as the sum of two functions, $F_1(x)$ and $F_2(x)$, where $F_1(x)$ is a step function with the same jumps as $F_X(x)$ and $F_2(x)$ is a continuous function that is differentiable except at the jump points of $F_X(x)$. Then we have

$$\int_a^b g(x)\, dF_X(x) = \int_a^b g(x)\, dF_2(x) + \sum_{i=1}^n g(x_i)p_i.$$

We will now state without proof some important properties of the Riemann–Stieltjes integral.

1. If $g(X)$ is Riemann–Stieltjes integrable with respect to $F_X(x)$ over the interval $[a,b]$, then for any constants C_1, C_2,

$$\int_a^b C_1 g(x)\, d[C_2 F_X(x)] = C_1 C_2 \int_a^b g(x)\, dF_X(x).$$

2. If g_1, g_2 are Riemann–Stieltjes integrable with respect to F_X over the interval $[a,b]$, then for any constants C_1, C_2,

$$\int_a^b [C_1 g_1(x) + C_2 g_2(x)]\, dF_X(x) = C_1 \int_a^b g_1(x)\, dF_X(x)$$
$$+ C_2 \int_a^b g_2(x)\, dF_X(x).$$

An immediate extension of this is that if g_i, $i = 1,\ldots,n$, are Riemann–Stieltjes integrable with respect to $F_X(x)$ over a, b, then for constants $C_1,\ldots,C_n$,

$$\int_a^b \left[\sum_{i=1}^n C_i g_i(x) \right] dF_X(x) = \sum_{i=1}^n C_i \int_a^b g_i(x)\, dF_X(x).$$

3. If g is Riemann–Stieltjes integrable with respect to F_1 and F_2 over the interval $[a,b]$, then for any constants C_1, C_2,

$$\int_a^b g(x)[C_1 \, dF_1(x) + C_2 \, dF_2(x)] = C_1 \int_a^b g(x) \, dF_1(x)$$

$$+ C_2 \int_a^b g(x) \, dF_2(x).$$

An extension of this is that if g is Riemann–Stieltjes integrable with respect to F_i, $i = 1,\ldots,n$ over the interval $[a,b]$, then for constants $C_1,\ldots,C_n$,

$$\int_a^b g(x)\left[\sum_{i=1}^n C_i \, dF_i(x) \right] = \sum_{i=1}^n C_i \int_a^b g(x) \, dF_i(x).$$

From properties 1–3 above, it then follows that:

(a) For any constant C, $E(C) = C$.
(b) $E(Cg(X)) = CE(g(X))$.
(c) For any two Borel functions g_1, g_2 of the random variable X with distribution function $F_X(x)$ and any two constants C_1, C_2,

$E(C_1 g_1(X) + C_2 g_2(X))$

$$= \int_{-\infty}^\infty [C_1 g_1(x) + C_2 g_2(x)] \, dF_X(x)$$

$$= C_1 \int_{-\infty}^\infty g_1(x) \, dF_X(x) + C_2 \int_{-\infty}^\infty g_2(x) \, dF_X(x)$$

$$= C_1 E(g_1(X)) + C_2 E(g_2(X)).$$

(d) If $g_1(x) \le g_2(x)$ for all x, then $E(g_1(X)) \le E(g_2(X))$. This follows from the fact that

$$E(g_1(X) - g_2(X)) = \int_{-\infty}^\infty [g_1(x) - g_2(x)] \, dF_X(x) \ge 0.$$

Because of these properties of the mathematical expectation, we call it a linear operator.

3.12 MOMENTS

Let X be a random variable with distribution function $F_X(x)$.

Definition 3.12.1: Raw moment of order k For any positive integer k, the mathematical expectation of $g(X) = X^k$,

$$m_k = E(X^k) = \int_{-\infty}^{\infty} x^k \, dF_X(x), \tag{3.49}$$

is called the *raw moment of order k* of the random variable X.

Definition 3.12.2: Central moment of order k For any positive integer k, the mathematical expectation of $g(X) = (X - m_1)^k$,

$$\mu_k = \int_{-\infty}^{\infty} (x - m_1)^k \, dF_X(x), \tag{3.50}$$

where $m_1 = E(X)$, is called the *central moment of order k* of the random variable X.

The relation between the two types of moments is given by

$$\begin{aligned}
\mu_k &= E(X - m_1)^k \\
&= \int_{-\infty}^{\infty} \left[\sum_{i=0}^{k} (-1)^i \binom{k}{i} m_1^i x^{k-i} \right] dF_X(x) \\
&= \sum_{i=0}^{k} (-1)^i \binom{k}{i} m_1^i m_{k-i}.
\end{aligned} \tag{3.51}$$

In particular, $\mu_2 = m_2 - m_1^2$. We usually call μ_2 the *variance*, and write it as $V(X)$. The raw moment m_1 is called the *mean* of the random variable X.

Some invariant properties of the variance of a random variable X are as follows:

1. If X is translated to $Y = X + b$ (translation), then $V(X) = V(Y)$, since from the properties of the operator E, we have

$$\begin{aligned}
V(Y) &= E[Y - E(Y)]^2 \\
&= E[X + b - E(X) - b]^2 \\
&= E[X - E(X)]^2 \\
&= V(X).
\end{aligned}$$

In other words, the variance remains invariant (unchanged) under translation.

2. If X is transformed to $Y = aX + b$ (affine transformation), then $V(Y) = a^2V(X)$, for

$$V(Y) = E[aX + b - aE(X) - b]^2$$
$$= E(a^2[X - E(X)]^2)$$
$$= a^2V(X).$$

The variance is not invariant under affine transformation.

It is interesting to note that for any Borel function $Y = g(X)$ of the random variable X,

$$\int_{-\infty}^{\infty} y \, dF_Y(y) = \int_{-\infty}^{\infty} g(x) \, dF_X(x).$$

This follows from the fact that $Y = g(X)$ is also a random variable with distribution function $F_Y(y)$, and hence

$$E(g(X)) = E(Y) = \int_{-\infty}^{\infty} y \, dF_Y(y). \tag{3.52}$$

On the other hand, from the definition of mathematical expectation,

$$E(g(X)) = \int_{-\infty}^{\infty} g(x) \, dF_X(x).$$

A useful representation of the expectation of a random variable X is given by

$$E(X) = -\int_{-\infty}^{0} F(x) \, dx + \int_{0}^{\infty} [1 - F(x)] \, dx. \tag{3.53}$$

We leave the proof of this representation to the reader.

As has been indicated, the existence of the moment of order k depends on the condition that

$$\int_{-\infty}^{\infty} |x|^k \, dF_X(x) < \infty.$$

Thus it follows that if the moment of order k exists, the moments of orders lower than k also exist. Furthermore, if the moment of order k exists, the following relation holds:

$$\lim_{a \to \infty} a^k P(|X| > a) \le \lim_{a \to \infty} \int_{|x| > a} |x|^k \, dF_X(x) = 0.$$

Hence the existence of the moments depends on the probability that the random variable takes on large absolute values.

Definition 3.12.3: Moment-generating function of a random variable   Let X be a random variable with distribution function $F_X(x)$. Then for real t,

$$m_X(t) = E(e^{tX}) \tag{3.54}$$

is called the *moment-generating function* of the random variable X.

In the case of continuous random variable X, we have

$$m_X(t) = \int_{-\infty}^{\infty} e^{tX} \, dF_X(x)$$

and in the case of discrete random variable X with probability mass function,

$$p_X(x_i) = \begin{cases} p_i, & i = 1,\ldots,n \\ 0, & \text{otherwise,} \end{cases}$$

we have

$$m_X(t) = \sum_{i=1}^{n} e^{tx_i} p_i.$$

Since the exponential function e^{tX} is always positive, $m_X(t)$ may be finite or infinite depending on the distribution function of the random variable X. For $t = 0$, $m_X(t)$ always exists and is equal to 1. If $m_X(t)$ exists, expanding e^{tX} by power series, we get

$$m_X(t) = 1 + m_1 t + m_2 \frac{t^2}{2!} + \cdots + m_k \frac{t^k}{k!} + \cdots.$$

Hence

$$\left. \frac{d^k m_X(t)}{dt^k} \right|_{t=0} = m_X^k(0) = m_k, \qquad k = 1,2,\ldots. \tag{3.55}$$

Thus, if the moment-generating function exists, we can find the moments of different orders by differentiation. Furthermore, the moment-generating function can be used to characterize the distribution function of the random variable. We now define the characteristic function, which is an analog of the moment-generating function, but which, unlike the moment-generating function, always exists.

Definition 3.12.4: Characteristic function of a random variable   Let X be a random variable with distribution function $F_X(x)$. Then for real t and $i = \sqrt{-1}$, the function

$$\phi_X(t) = E(e^{itX}) \tag{3.56}$$

is called the *characteristic function* of the random variable X.

In the case of continuous random variable X,

$$E(e^{itX}) = \int_{-\infty}^{\infty} e^{itX} \, dF_X(x).$$

Since $e^{itX} = \cos tX + i \sin tX$,

$$
\begin{aligned}
|e^{itX}|^2 &= (\cos tX + i \sin tX)(\cos tX - i \sin tX) \\
&= (\cos tX)^2 + (\sin tX)^2 \\
&= 1.
\end{aligned}
$$

Hence

$$E(e^{itX}) \le E|e^{itX}| \le E(1) = 1.$$

Thus the characteristic function always exists. In the case of discrete random variable X with probability mass function $p_X(x_j) = p_j$, $j = 1,\dots,n$, $\sum_{i=1}^{n} p_i = 1$, we have

$$\phi_X(t) = \sum_{j=1}^{n} e^{itx_j} p_j.$$

Since $|e^{itX}| = 1$, $\sum_{j=1}^{n} e^{itx_j} p_j$ is absolutely convergent.

We now state some properties of the characteristic function $\phi_X(t)$.

1. $\phi_X(0) = E(1) = 1$.
2. $|\phi_X(t)| \le E|e^{itX}| = 1$.
3. $\phi_X(-t) = \overline{\phi_X(t)}$, where $\overline{\phi_X(t)}$ denotes the complex conjugate of $\phi_X(t)$.

Since $e^{itX} = \cos tX + i \sin tX$, we have

$$\phi_X(t) = E(\cos tX) + iE(\sin tX)$$

$$
\begin{aligned}
\phi_X(-t) &= E(\cos tX - i \sin tX) \\
&= E(\cos tX) - iE(\sin tX) \\
&= \overline{\phi_X(t)}.
\end{aligned}
$$

Every characteristic function $\phi_X(t)$ must satisfy the three conditions above. However, the converse is not always true: A function $\phi_X(t)$ that satisfies the three conditions above is not necessarily a characteristic function of a random variable. An example of such a function is $\phi_X(t) = -t^4$. The necessary and sufficient condition for a function $\phi_X(t)$ to be a characteristic function was given by Bochner (1933). We give here the statement of Bochner's theorem without proof.

Theorem 3.12.1 Let $\phi_X(t)$ be defined for $-\infty < t < \infty$ and $\phi_X(0) = 1$. The necessary and sufficient conditions for $\phi_X(t)$ to be a characteristic function of a random variable X are:

   (1) $\phi_X(t)$ is a continuous function of t.

   (2)   For reals $t_1,\ldots,t_k$ and complex numbers $C_1,\ldots,C_k$,

$$\sum_{i,j}^{k} \phi_X(t_i - t_j)C_i\overline{C}_j \geq 0 \qquad \text{for } k = 1,2,\ldots .$$

Now let us consider a real random variable X whose moment of order k exists.

$$\phi_X^k(t) = \frac{d^k}{dt^k}\,\phi_X(t)$$

$$= \frac{d^k}{dt^k}\int_{-\infty}^{\infty} e^{itX}\,dF_X(x) = \int_{-\infty}^{\infty} (ix)^k e^{itX}\,dF_X{}^{(x)}$$

$$= E(i^k X^K e^{itX}).$$

Thus we have

$$m_k = \frac{\phi_X^k(0)}{i^k}, \qquad k = 1,2,\ldots . \tag{3.57}$$

The central moments of different orders of a random variable X can also be obtained from the moment-generating or characteristic function of $Y = X - E(X)$. To see this we need the following theorem.

Theorem 3.12.2 Let X be a random variable and let $Y = aX + b$, where a, b are reals and $a \neq 0$. Then the following relations hold:

$$\phi_Y(t) = e^{itb}\phi_X(ta)$$

$$m_Y(t) = e^{tb}m_X(ta).$$

Proof

$$\phi_Y(t) = E(e^{itY})$$

$$= E(e^{it(aX+b)})$$

$$= e^{itb}E(e^{itaX})$$

$$= e^{itb}\phi_X(ta),$$

$$m_X(t) = E(e^{tY})$$

$$= E(e^{t(aX+b)})$$

$$= e^{tb}E(e^{taX})$$

$$= e^{tb}m_X(ta).$$

Now, taking $a = 1$, $b = E(X)$; $Y = X - E(X)$; then

$$\phi_{X-E(X)}(t) = e^{-itE(X)}\phi_X(t)$$

$$m_{X-E(X)}(t) = e^{-tE(X)}m_X(t).$$

Thus

$$E(X - E(X))^k = \frac{1}{i^k}\frac{d^k}{dt^k}[e^{itE(X)}\phi_X(t)]$$

$$= \frac{d^k}{dt^k}[e^{tE(X)}m_X(t)].$$

Example 3.12.1 Let X be a binomial random variable with parameters n, p and let $q = 1 - p$. Then we have

$$\phi_X(t) = E(e^{itX})$$

$$= \sum_{j=0}^{n} e^{itj}\binom{n}{j}p^jq^{n-j}$$

$$= \sum_{j=0}^{n}\binom{n}{j}(pe^{it})^jq^{n-j}$$

$$= (q + pe^{it})^n.$$

Similarly, we have $m_X(t) = (q + pe^t)^n$. Hence $m_1 = np$, $\mu_2 = npq$.

Example 3.12.2 Let X be a Poisson random variable with parameter λ. Then we have

$$\phi_X(t) = \sum_{j=0}^{\infty} e^{itj}\frac{e^{-\lambda}\lambda^j}{j!}$$

$$= \sum_{j=0}^{\infty} e^{-\lambda}\frac{(e^{it})^j}{j!}$$

$$= e^{\lambda(e^{it}-1)}.$$

Similarly, we have $m_X(t) = e^{\lambda(e^t-1)}$. Hence $m_1 = \lambda$, $\mu_2 = \lambda$.

Example 3.12.3 Let X be a normal random variable with parameters μ, σ^2 and let $Y = (X - \mu)/\sigma$; then

$$\phi_X(t) = \int_{-\infty}^{\infty} e^{itX}\frac{1}{\sqrt{2\pi}\,\sigma}\exp\left[-\frac{1}{2\sigma^2}(x - \mu)^2\right]dx$$

$$= e^{it\mu}\int_{-\infty}^{\infty} e^{it\sigma y}\frac{1}{\sqrt{2\pi}}e^{-(1/2)y2}\,dy.$$

Now expanding $e^{it\sigma y}$ by Taylor series, we get

$$\Phi_X(t) = e^{it\mu} \int_{-\infty}^{\infty} \sum_{k=0}^{\infty} \frac{(it\sigma y)^k}{k!} \frac{1}{\sqrt{2\pi}} e^{-(1/2)y^2} \, dy$$

$$= e^{it\mu} \sum_{k=0}^{\infty} \frac{(it\sigma)^k}{k!} \int_{-\infty}^{\infty} y^k \frac{1}{\sqrt{2\pi}} e^{-(1/2)y^2} \, dy$$

$$= e^{it\mu} \sum_{m=0}^{\infty} \frac{(it\sigma)^{2m}}{(2m)!} \int_{-\infty}^{\infty} y^{2m} \frac{1}{\sqrt{2\pi}} e^{-(1/2)y^2} \, dy$$

$$= \exp(it\mu - \tfrac{1}{2} t^2\sigma^2).$$

Note that

$$\int_{-\infty}^{\infty} y^n \frac{1}{\sqrt{2\pi}} e^{-(1/2)y^2} \, dy$$

$$= \begin{cases} 0 & \text{if } n \text{ is odd} \\ \Gamma\left(\dfrac{2k+1}{2}\right) 2^{k+(1/2)} = \dfrac{\Gamma(2k+1)}{\Gamma(k+1)2^k} & \text{if } n = 2k, \quad k = 1,2,\dots \, . \end{cases}$$

We now leave it to the reader to verify that

$$m_X(t) = \exp(t\mu + \tfrac{1}{2} t^2\sigma^2).$$

The interchange of the summation and the integration in finding $\phi_X(t)$ is justified because of the fact that the infinite series

$$\sum_{n=0}^{\infty} \frac{(it\sigma y)^n}{n!} e^{-(1/2)y^2}$$

is dominated by the integrable function $\exp(|t\sigma y| - \tfrac{1}{2} y^2)$. It is easy to show that $m_1 = \mu$, $\mu_2 = \sigma^2$, all moments of order $2k + 1$, $k = 1,2,\dots$, are zero, and

$$\mu_{2k} = 1 \cdot 3 \cdots (2k-1)\sigma^{2k}, \qquad k = 1,2,\dots \, .$$

Example 3.12.4: Laplace distribution   Let X be a random variable with probability density function (α being a positive constant)

$$f_X(x) = \begin{cases} \dfrac{1}{2\alpha} \exp\left(-\left|\dfrac{x-a}{\alpha}\right|\right), & -\infty < x < \infty, \ -\infty < a < \infty, \ \alpha > 0 \\ 0, & \text{otherwise.} \end{cases}$$

This is known as the *Laplace probability density function* with parameters a, α. Let $Y = X - a$. Since $\exp(-|y|/\alpha)$ is an even function of y and $e^{ity} = \cos ty + i \sin ty$, we get

$$\phi_X(t) = \int_{-\infty}^{\infty} e^{itx} \frac{1}{2\alpha} \exp\left(-\left|\frac{x-a}{\alpha}\right|\right) dx$$

$$= e^{ita} \int_{-\infty}^{\infty} e^{ity} \frac{1}{2\alpha} \exp\left(-\frac{|y|}{\alpha}\right) dy$$

$$= e^{ita} \int_{-\infty}^{\infty} (\cos ty + i \sin ty) \frac{1}{2\alpha} \exp\left(-\frac{|y|}{\alpha}\right) dy$$

$$= \frac{e^{ita}}{\alpha} \int_{0}^{\infty} \cos ty \, e^{-y/\alpha} \, dy$$

$$= \frac{e^{ita}}{\alpha} \frac{\exp(-y/\alpha)[t \sin ty - (1/\alpha) \cos ty]}{(1/\alpha^2) + t^2} \bigg|_{0}^{\infty}$$

$$= e^{ita} \frac{1}{1 + t^2\alpha^2}.$$

We have pointed out earlier that although the characteristic function always exists, the moment-generating function may or may not exist, depending on the distribution function. We will now give an example of a random variable whose moments of different orders do not exist.

Example 3.12.5: Cauchy distribution   The random variable X has Cauchy distribution if its probability density function is given by, for $\lambda > 0$, $-\infty < u < \infty$,

$$f_X(x) = \begin{cases} \dfrac{1}{\pi} \dfrac{1}{\lambda^2 + (x - u)^2} & \text{if } -\infty < x < \infty \\ 0 & \text{otherwise.} \end{cases}$$

The distribution function of this variable is

$$F_X(x) = \int_{-\infty}^{x} \frac{1}{\pi} \frac{dt}{\lambda^2 + (t - u)^2}$$

$$= \frac{1}{\pi} \left[\arctan\left(\frac{t - u}{\lambda}\right) \right]_{-\infty}^{x}$$

$$= \frac{1}{2} + \frac{1}{\pi} \arctan\left(\frac{x - u}{\lambda}\right).$$

Its characteristic function is

$$\phi_X(t) = \int_{-\infty}^{\infty} \frac{1}{\pi} e^{itx} \frac{1}{\lambda^2 + (x - u)^2} \, dx$$

$$= \frac{1}{\pi} \int_{-\infty}^{\infty} \exp[it(u + y\lambda)] \frac{1}{1 + y^2} \, dy$$

$$= \frac{1}{\pi} e^{itu} \int_{-\infty}^{\infty} e^{it\lambda y} \frac{1}{1 + y^2} \, dy$$

$$= \exp(itu - |t\lambda|).$$

Thus it follows that $\phi_X(t)$ is not differentiable at $t = 0$, which implies that moments of all orders do not exist.

3.13   PROBABILITY-GENERATING FUNCTIONS

Let X be a discrete random variable that assumes only integral values $0,1,...,n$ (n may be ∞) and let $p_X(i) = p_i$, $i = 0,1,...$, such that

$$\sum_{i=1}^{\infty} p_i = 1.$$

Then the following function can be defined.

Definition 3.13.1: Probability-generating function of a discrete random variable   The function

$$P_X(t) = \sum_{k=0}^{\infty} t^k p_k, \qquad -1 < t < 1$$

is called the *probability-generating function* of the integral-valued discrete random variable X, where X takes values $0,1,...$ with probabilities $p_0, p_1,...$, respectively.

Since

$$\sum_{k=0}^{\infty} p_k = 1,$$

$P_X(t)$ converges absolutely, at least for $|t| < 1$. Let

$$\frac{d^k}{dt^k} P_X(t) \bigg|_{t=1} = P_X^k(1) \qquad \text{for } k = 1,2,... \, .$$

Then the mean and the variance of the random variable X are given by

$$E(X) = P_X^1(1)$$

$$V(X) = P_X^2(1) + P_X^1(1) - [P_X^1(1)]^2$$

since

$$P_X^1(1) = \sum_{k=0}^{\infty} kp_k = E(X)$$

$$V(X) = E(X^2) - E^2(X)$$
$$= E(X)(X - 1) + E(X) - [E(X)]^2$$

and

$$E[X(X - 1)] = \sum_{k=0}^{\infty} k(k - 1)p_k = P_X^2(1).$$

Example 3.13.1   Let X be a Poisson random variable with parameter λ. Then we have

$$P_X(t) = \sum_{k=0}^{\infty} t^k \frac{e^{-\lambda}\lambda^k}{k!}$$

$$= \sum_{k=0}^{\infty} \frac{e^{-\lambda}(\lambda t)^k}{k!}$$

$$= e^{\lambda(t-1)}.$$

Hence $E(X) = \lambda$, $V(X) = \lambda$.

3.14   MATHEMATICAL EXPECTATION OF A FUNCTION OF A BIVARIATE RANDOM VARIABLE

Let (X,Y) be a bivariate random variable with distribution function $F_{X,Y}(x,y)$. Let $g(X,Y)$ be a single-valued function (Borel function) of X, Y. Then the following definitions may be used.

Definition 3.14.1: Mathematical expectation of a function of a bivariate random variable   Let (X,Y) be a bivariate continuous random variable. The *mathematical expectation* of $g(X,Y)$ is given by

$$E(g(X,Y)) = \int_{-\infty}^{\infty} \int_{-\infty}^{\infty} g(x,y)f_{X,Y}(x,y) \, dx \, dy \tag{3.58}$$

provided that $E|g(X,Y)| < \infty$.

Definition 3.14.2: Joint raw moment of order $l + k$ Let $g(X,Y) = X^l Y^k$, where l, k are nonnegative integers. Then $m_{lk} = E(X^l Y^k)$ is called the *joint raw moment* of X,Y of order $l + k$.

Definition 3.14.3: Joint central moment of order $l + k$ Let $E(X) = \alpha$, $E(Y) = \beta$ and let $g(X,Y) = (X - \alpha)^l (Y - \beta)^k$, where l, k are nonnegative integers. Then

$$\mu_{lk} = E(X - \alpha)^l (Y - \beta)^k$$

is called the *joint central moment* of X, Y of order $l + k$.

Definition 3.14.4: Covariance   For any two random variables X, Y, the moment μ_{11} is called the *covariance* of X and Y. We will also denote μ_{11} as $\mathrm{cov}(X,Y)$.

Theorem 3.14.1   Let (X,Y) be a bivariate random variable; then (a) $E(X + Y) = E(X) + E(Y)$, and (b) $E(XY) = E(X)E(Y)$ if X, Y are independent.

Proof of (a)

Continuous case.   Let $f_{X,Y}(x,y)$, $f_X(x)$, $f_Y(y)$ be the joint probability density function of X, Y, the marginal probability density function of X, and the marginal probability density function of Y, respectively. Then we have

$$E(X + Y) = \int_{-\infty}^{\infty} \int_{-\infty}^{\infty} (x + y) f_{X,Y}(x,y) \, dx \, dy$$

$$= \int_{-\infty}^{\infty} \int_{-\infty}^{\infty} x f_{X,Y}(x,y) \, dx \, dy + \int_{-\infty}^{\infty} \int_{-\infty}^{\infty} y f_{X,Y}(x,y) \, dx \, dy$$

$$= \int_{-\infty}^{\infty} x f_X(x) \, dx + \int_{-\infty}^{\infty} y f_Y(y) \, dy$$

Discrete case.   Let $p_{X,Y}(x_i,y_j) = p_{ij}$, $p_X(x_i) = \Sigma_j \, p_{ij} = p_{i\cdot}$, $p_Y(y_j) = \Sigma_i \, p_{ij} = p_{\cdot j}$ be the joint probability mass function of X,Y, the marginal probability mass function of X, and the marginal probability mass function of Y, respectively. Then we have

$$E(X + Y) = \sum_I \sum_j (x_i + y_j) p_{ij}$$

$$= \sum_i x_i p_{i\cdot} + \sum_j y_j p_{\cdot j}$$

$$= E(X) + E(Y).$$

Proof of (b)
  Continuous case.   If X, Y are independent, then $f_{X,Y}(x,y) = f_X(x)f_Y(y)$. Thus we have

$$E(XY) = \int_{-\infty}^{\infty} \int_{-\infty}^{\infty} xy f_X(x)f_Y(y) \, dx \, dy$$

$$= \int_{-\infty}^{\infty} xf_X(x) \, dx \int_{-\infty}^{\infty} yf_Y(y) \, dy$$

$$= E(X)E(Y).$$

Discrete case

$$E(XY) = \sum_i \sum_j x_i x_j p_{ij}$$

$$= \sum_i \sum_j x_i y_j p_{i \cdot} p_{\cdot j}$$

$$= \sum_i x_i p_{i \cdot} \sum_j y_j p_{\cdot j}$$

$$= E(X)E(Y). \qquad \text{Q.E.D.}$$

Definition 3.14.5: Correlation coefficient   For any two random variables X, Y,

$$\rho(X,Y) = \frac{\operatorname{cov}(X,Y)}{\sqrt{V(X)V(Y)}}$$

is called the *coefficient of correlation* between X, Y.

The correlation coefficient gives us a numerical measure of the linear association between X and Y. Some important characteristic properties of ρ are given by the following theorem.

Theorem 3.14.2   For any two random variables X, Y:
  (a) $\rho(X,Y) = 0$ if X,Y are independent.
  (b) $|\rho(X,Y)| \le 1$.
  (c) $\rho(X,Y) = 1$ if and only if X, Y are linearly related with unit probability.
  (d) $\rho(X,Y) = -1$ if and only if X and $-Y$ are linearly related with unit probability.

Proof

(a) $\rho(X,Y) = \dfrac{E[(X - E(X))(Y - E(Y))]}{\sqrt{V(X)V(Y)}}$

$= \dfrac{E(X - E(X))E(Y - E(Y))}{\sqrt{V(X)V(Y)}}$

$= 0.$

(b) For any real t, define $h(t)$ by

$$h(t) = E[Y - E(Y) - t(X - E(X))]^2$$
$$= t^2 V(X) + V(Y) - 2t \, \text{cov}(X,Y). \qquad (3.59)$$

Then, since the square of any random variable is always positive, $h(t) \geq 0$ for all t. Taking, in particular, $t = \rho\sqrt{V(Y)/V(X)}$, we get, from Eq. (3.59),

$$V(Y) - \rho^2 V(Y) = V(Y)(1 - \rho^2) \geq 0.$$

Hence $\rho^2 \leq 1$.

(c) If $X = aY + b$, with $a > 0$, we have

$$\text{cov}(X,Y) = E[(X - E(X))(Y - E(Y))]$$
$$= E[a(Y - E(Y))(Y - E(Y))]$$
$$= aV(Y), \qquad V(X) = a^2 V(Y).$$

Hence we have

$$\rho(X,Y) = \dfrac{\text{cov}(X,Y)}{\sqrt{V(X)V(Y)}}$$

$$= \dfrac{aV(Y)}{aV(Y)}$$

$$= 1.$$

(d) The proof in this case is similar to that of (c).   Q.E.D.

Let $X_1, X_2, \ldots, X_n$ be n random variables having the joint distribution function $F_{X_1, \ldots, X_n}(x_1, \ldots, x_n)$ and let $g(X_1, \ldots, X_n)$ be a single-valued function (Borel function) of $X_1, \ldots, X_n$; then the mathematical expectation of $g(X_1, \ldots, X_n)$ is obtained in the same way as in the case of bivariate random variables, by replacing the joint bivariate probability density or mass function by the corresponding joint probability density or mass function of $X_1, \ldots, X_n$. In other words, if $X_1, \ldots, X_n$ are jointly continuous with probability density function $f_{X_1, \ldots, X_n}(x_1, \ldots, x_n)$, then we have

$$E[g(X_1, \ldots, X_n)] = \int_{-\infty}^{\infty} \cdots \int_{-\infty}^{\infty} g(x_1, \ldots, x_n) f_{X_1, \ldots, X_n}(x_1, \ldots, x_n) \, dx_1, \ldots, dx_n$$

(3.60)

provided $E|g(X_1, \ldots, X_n)| < \infty$. If $X_1, \ldots, X_n$ are jointly discrete random variables with probability mass function $p_{X_1, \ldots, X_n}(x_1, \ldots, x_n)$, then we have

$$E[g(X_1, \ldots, X_n)] = \sum_{x_1} \cdots \sum_{x_n} g(x_1, \ldots, x_n) p_{X_1, \ldots, X_n}(x_1, \ldots, x_n)$$

provided that $E|g(X_1, \ldots, X_n)| < \infty$.

Let

$$Y = \sum_{i=1}^{n} a_i X_i$$

where the a_i are real constants. Then it is evident that Y is a random variable. The moment-generating function and the characteristic function of Y are given, for real t, by

$$m_Y(t) = E\left(\exp\left[t \sum_{j=1}^{n} a_i X_i \right] \right)$$

$$= E\left(\exp\left[\sum_{j=1}^{n} (ta_j) X_j \right] \right)$$

$$\phi_Y(t) = E\left(\exp\left[it \sum_{i=1}^{n} a_i X_i \right] \right)$$

$$= E\left(\exp\left[i \sum_{j=1}^{n} (ta_j) X_j \right] \right).$$

Note: Given the joint probability density function $f_{X_1, \ldots, X_n}(x_1, \ldots, x_n)$ and the joint probability mass function $p_{X_1, \ldots, X_n}(x_1, \ldots, x_n)$, the marginal probability density function and the marginal probability mass functions of (X_i, X_j) and X_i are given by

$$f_{X_i X_j}(x_i, x_j) = \int_{-\infty}^{\infty} \cdots \int_{-\infty}^{\infty} f_{X_1, \ldots, X_n}(x_1, \ldots, x_n) \prod_{k \neq i, k \neq j} dx_k$$

$$f_{X_i}(x_i) = \int_{-\infty}^{\infty} \cdots \int_{-\infty}^{\infty} f_{X_1, \ldots, X_n}(x_1, \ldots, x_n) \prod_{k \neq i} dx_k$$

$$p_{X_i X_j}(x_i, x_j) = \sum_{x_1, \ldots, x_n} p_{X_1, \ldots, X_n}(x_1, \ldots, x_n)$$

but not x_i, x_j

$$p_{X_i}(x_i) = \sum_{x_1, \ldots, x_n} p_{X_1, \ldots, X_n}(x_1, \ldots, x_n).$$

but not x_i

Linear functions of random variables play a very important role in probability and statistics. We will establish several results about linear functions in the following theorem.

Theorem 3.14.3   Let $X_1, \ldots, X_n$ be n random variables; then

(a) $E\left(\sum_{i=1}^{n} a_i X_i\right) = \sum_{i=1}^{n} a_i E(X_i).$

(b) $V\left(\sum_{i=1}^{n} a_i X_i\right) = \sum_{i=1}^{n} a_i^2 V(X_i) + \sum_{i} \sum_{\substack{j \\ i \neq j}} a_i a_j \, \text{cov}(X_i X_j).$

(c) $m_Y(t) = \prod_{i=1}^{n} m_{X_i}(t a_i), \; \phi_Y(t) = \prod_{i=1}^{n} \phi_{X_i}(t a_i)$ if $X_1, \ldots, X_n$

are independent.

(d)   If $Z = X_1 + \cdots X_n$, then

$$P_Z(t) = E(t^Z) = \prod_{i=1}^{n} P_{X_i}(t).$$

We will give the proof of Theorem 3.14.3 for the case of continuous random variables only. The proof for the discrete case is analogous.

Proof (Continuous Case)

(a) $\displaystyle E\left(\sum_{i=1}^{n} a_i X_i\right) = \int_{-\infty}^{\infty} \cdots \int_{-\infty}^{\infty} \left(\sum_{i=1}^{n} a_i x_i\right) f_{X_1,\dots,X_n}(x_1,\dots,x_n) \prod_{i=1}^{n} dx_i$

$\displaystyle \qquad = \sum_{i=1}^{n} a_i \int_{-\infty}^{\infty} x_i f_{X_i}(x_i)\, dx_i$

$\displaystyle \qquad = \sum_{i=1}^{n} a_i E(X_i).$

(b) $\displaystyle V\left(\sum_{i=1}^{n} a_i X_i\right) = E\left(\sum_{i=1}^{n} a_i x_i - \sum_{i=1}^{n} a_i E(X_i)\right)^2$

$\displaystyle \qquad = E\left(\sum_{i=1}^{n} a_i[X_i - E(X_i)]\right)^2$

$\displaystyle \qquad = E\left(\sum_{i=1}^{n} a_i^2[X_i - E(x_i)]^2 + \sum_{\substack{i=1 \\ i \neq j}}^{n} \sum_{j=1}^{n} a_i a_j [X_i - E(X_i)]\right.$

$\displaystyle \qquad \qquad \left. \times [X_j - E(X_j)]\right)$

$\displaystyle \qquad = \int_{-\infty}^{\infty} \cdots \int_{-\infty}^{\infty} \left\{\sum_{i=1}^{n} a_i^2[X_i - E(X_i)]^2 \right.$

$\displaystyle \qquad \qquad \left. + \sum_{\substack{i=1 \\ i \neq j}}^{n} \sum_{j=1}^{n} a_i a_j [x_i - E(X_i)][x_j - E(X_j)]\right\}$

$\displaystyle \qquad \qquad \times f_{X_1,\dots,X_n}(x_1,\dots,x_n) \prod_{i=1}^{n} dx_i$

$\displaystyle \qquad = \sum_{i=1}^{n} a_i^2 \int_{-\infty}^{\infty} [x_i - E(X_i)]^2 f_{X_i}(x_i)\, dx_i + \sum_{i} \sum_{\substack{i \neq j}}^{j} a_i a_j$

$\displaystyle \qquad \qquad \times \int_{\infty}^{\infty} \int_{\infty}^{\infty} [x_i - E(X_i)][x_j - E(X_j)]$

$\displaystyle \qquad \qquad \times f_{X_i,X_j}(x_i,x_j)\, dx_i\, dx_j$

$\displaystyle \qquad = \sum a_i^2 V(X_i) + \sum_{i} \sum_{\substack{i \neq j}}^{j} a_i a_j \operatorname{cov}(X_i,X_j).$

Note that if $X_1,\ldots,X_n$ are independent, $\text{cov}(X_i, X_j) = 0$ for all $i \neq j$. Hence we have

$$V\left(\sum_{i=1}^{n} a_i X_i\right) = \sum_{i=1}^{n} a_i^2 V(X_i).$$

(c) If $X_i,\ldots,X_n$ are independent, then

$$m_Y(t) = \int_{-\infty}^{\infty} \cdots \int_{-\infty}^{\infty} \exp\left[\sum_{j=1}^{n} (a_j t) x_j\right] \prod_{j=1}^{n} f_{X_j}(x_j)\, dx_j$$

$$= \prod_{j=1}^{n} \int_{-\infty}^{\infty} \exp(a_j t x_j) f_{X_j}(x_j)\, dx_j$$

$$= \prod_{j=1}^{n} m_{X_j}(a_j t),$$

and similarly,

$$\phi_Y(t) = \prod_{j=1}^{n} \phi_{X_j}(t a_j).$$

(d) If $X_1,\ldots,X_n$ are independent discrete random variables and $Z = X_1 + \cdots + X_n$, we have

$$P_Z(t) = \sum_{x_1} \sum_{\substack{x_n \\ x_1 \cdots x_n:\, \sum_{j=1}^{n} x_j = t}} t^{\sum_{j=1}^{n} x_j} \prod_{j=1}^{n} p_{X_j}(x_j)$$

$$= \prod_{j=1}^{n} \left[\sum_{x_j} t^{x_j} p_{X_j}(x_j)\right]$$

$$= \prod_{j=1}^{n} P_{X_j}(t). \qquad \text{Q.E.D.}$$

Example 3.14.1 Let $X_1,\ldots,X_k$ be independently distributed binomial random variables with parameters n_i, $p(i = 1,\ldots,k)$, and let

$$Y = \sum_{i=1}^{k} X_i.$$

Then with $q = 1 - p$ and

$$n = \sum_{i=1}^{k} n_i,$$

we have

$$m_Y(t) = \prod_{i=1}^{k} m_{X_i}(t)$$

$$= \prod_{i=1}^{k} (q + pe^t)^{n_i}$$

$$= (q + pe^t)^n$$

and

$$\phi_Y(t) = \prod_{i=1}^{k} (q + pe^{it})^{n_i} = (q + pe^{it})^n.$$

Hence we have $E(Y) = np$, $V(Y) = npq$.

Example 3.14.2 Let (X,Y) be a bivariate normal random variable with probability density function given in Example 3.4.2. Then

$$\text{cov}(X,Y) = \int_{-\infty}^{\infty} \int_{-\infty}^{\infty} (x - m_1)(y - m_2) f_{X,Y}(x,y) \, dx \, dy$$

$$= \int_{-\infty}^{\infty} \int_{-\infty}^{\infty} \sigma_1 \sigma_2 x_1 x_2 \frac{1}{2\pi(1 - \rho^2)^{1/2}}$$

$$\times \exp\left[-\frac{1}{2(1 - \rho^2)} (x_1^2 + x_2^2 - 2\rho x_1 x_2) \right] dx_1 \, dx_2$$

$$= \sigma_1 \sigma_2 \int_{-\infty}^{\infty} x_1 \frac{1}{\sqrt{2\pi}} \exp\left(-\frac{1}{2} x_1^2 \right) \left\{ \int_{-\infty}^{\infty} x_2 \frac{1}{\sqrt{2\pi(1 - \rho^2)}} \right.$$

$$\left. \times \exp\left[\frac{-1}{2(1 - \rho^2)} (x_2 - \rho x_1)^2 \right] dx_2 \right\} dx_1$$

$$= \sigma_1 \sigma_2 \rho.$$

Hence we have $\rho(X,Y) = \rho$.

It may be noted here that if $\rho(X,Y) = 0$, then

$$f_{X,Y}(x,y) = \frac{1}{\sigma_1 \sqrt{2\pi}} \exp\left[-\frac{1}{2} \left(\frac{x - m_1}{\sigma_1} \right)^2 \right]$$

$$\times \frac{1}{\sigma_2 \sqrt{2\pi}} \exp\left[-\frac{1}{2} \left(\frac{y - m_2}{\sigma_2} \right)^2 \right].$$

In other words, if $\rho(X,Y) = 0$, then X, Y are independent, although this is not true in general.

Example 3.14.3   Let $X_1,\ldots,X_n$ be independent and identically distributed normal random variables with mean μ and variance σ^2, and let $\overline{X} = \sum_{j=1}^{n} X_j/n$; then

$$\phi_{\overline{X}}(t) = \prod_{j=1}^{n} \phi_{X_j}\left(\frac{t}{n}\right)$$

$$= \prod_{j=1}^{n} \exp\left(i\,\frac{t}{n}\,\mu - \frac{1}{2}\frac{t^2}{n^2}\,\sigma^2\right)$$

$$= \exp\left(it\mu - \frac{1}{2}t^2\,\frac{\sigma^2}{n}\right)$$

and similarly,

$$m_{\overline{X}}(t) = \exp\left(t\mu + \frac{1}{2}t^2\,\frac{\sigma^2}{n}\right).$$

From this it follows that $E(\overline{X}) = m$, $V(\overline{X}) = \sigma^2/n$, and the characteristic function and the moment-generating function of $\overline{X}$ are the same as the characteristic function and the moment-generating function of a normal random variable with mean μ and variance σ^2/n.

Given the distribution function $F_X(x)$, the characteristic function is uniquely determined by

$$\phi_X(t) = \int_{-\infty}^{\infty} e^{itx}\,dF_X(x).$$

The converse of this fact [namely, given the characteristic function $\phi_X(t)$, the distribution function $F_X(x)$ is uniquely determined by $\phi_X(t)$] is given by the following theorem.

Theorem 3.14.4   If $\phi_X(t)$ is the characteristic function of the random variable X with probability distribution function $F_X(x)$, and if a, b $(a < b)$ are the points of continuity of $F_X(x)$, the following relation holds:

$$F_X(b) - F_X(a) = \lim_{T\to\infty} \frac{1}{2\pi} \int_{-T}^{T} \frac{e^{-ita} - e^{-itb}}{it}\,\phi_X(t)\,dt. \tag{3.61}$$

Note: Since $a < b$, we can write for some real c and $h > 0$, $b = c + h$, $a = c - h$,

$$F_X(c + h) - F_X(c - h) = \lim_{T\to\infty} \frac{1}{\pi} \int_{-T}^{T} \frac{\sin th}{t}\,e^{-itc}\phi_X(t)\,dt.$$

We will prove the theorem for the continuous random variable only. The proof for the discrete random variable goes in the same way, except that the integral is replaced by a summation. To prove this theorem, we first prove two lemmas.

Lemma 3.14.1 For reals α, T, with $T > 0$, let

$$I(\alpha,T) = \frac{2}{\pi} \int_0^T \frac{\sin \alpha t}{t} \, dt.$$

Then we have

$$\lim_{T \to \infty} I(\alpha,T) = \lim_{T \to \infty} \frac{2}{\pi} \int_0^T \frac{\sin \alpha t}{t} \, dt$$

$$= \begin{cases} 1 & \text{if } \alpha > 0 \\ 0 & \text{if } \alpha = 0 \\ -1 & \text{if } \alpha < 0 \end{cases}$$

and the convergence is uniform for $|\alpha| \geq \delta$, where δ is any small positive real number.

Proof

$$\frac{2}{\pi} \int_0^T \frac{\sin \alpha t}{t} \, dt = \frac{2}{\pi} \int_0^{\alpha T} \frac{\sin t}{t} \, dt.$$

Let us define

$$F(y) = \int_0^\infty e^{-xy} \frac{\sin x}{x} \, dx.$$

The integrand is continuous on $[0,\infty]$ and the integral converges uniformly on $[0,\infty]$, and hence $F(y)$ is continuous for y belonging to any finite interval $[0,d]$. So we have

$$\lim_{y \to 0+} F(y) = F(0) = \int_0^\infty \frac{\sin x}{x} \, dx.$$

Now

$$\frac{dF(y)}{dy} = F'(y) = -\int_0^\infty e^{-xy} \sin x \, dx$$

and since

$$|e^{-xy} \sin x| \leq e^{-cx}$$

for $y \geq c$, $F'(y)$ converges uniformly on $[c,\infty]$ for $c > 0$. Now integrating by parts, we get

$$F'(y) = -\frac{1}{1 + y^2}.$$

Hence we have

$$F(y) = K - \tan^{-1}y$$

if $y > 0$, where K is the constant of integration. To evaluate K, let us observe that $F(y) \to 0$ as $y \to \infty$. For $\epsilon > 0$ and x_0 belonging to $[\epsilon,b]$, we get from the mean-value theorem of Riemann integration

$$\left| \int_\epsilon^b e^{-xy} \frac{\sin x}{x} \, dx \right| \leq \frac{e^{-\epsilon y}}{\epsilon} \left| \int_{x_0}^b \sin x \, dx \right|$$

$$\leq 2 \frac{e^{-\epsilon y}}{\epsilon}.$$

Hence

$$\int_\epsilon^b e^{-xy} \frac{\sin x}{x} \, dx \longrightarrow 0$$

as $y \to \infty$. Furthermore, since

$$\int_0^\epsilon e^{-xy} \frac{\sin x}{x} \, dx \leq \epsilon$$

it follows that $F(y) \to 0$ as $y \to \infty$. Thus

$$F(y) = \frac{\pi}{2} - \tan^{-1}y$$

and hence $F(0) = \pi/2$, since for $\alpha < 0$,

$$\int_0^T \frac{\sin \alpha x}{x} \, dx = -\int_0^T \frac{\sin \beta x}{x} \, dx, \qquad \beta > 0 \ (\alpha = -\beta).$$

The proof is complete.   Q.E.D.

Lemma 3.14.2 Let

$$I(T,x,a,b) = \frac{1}{2\pi} \int_{-T}^T \frac{\sin t(x - a) - \sin t(x - b)}{t} \, dt.$$

Then the following relation holds:

$$\lim_{T\to\infty} I(T,x,a,b) = \begin{cases} 1 & \text{if } a < x < b \\ \frac{1}{2} & \text{if } x = a \text{ or } x = b \\ 0 & \text{if } x < a \text{ or } b < x \end{cases}$$

and the convergence is uniform for $|x - a| \geq \delta$, $|x - b| \geq \delta$ for $\delta > 0$.

Proof. It is easy to see that

$$I(T,x,a,b) = \tfrac{1}{2}[I(x - a, T) - I(x - b, T)].$$

Hence by Lemma 3.14.1 we get the result.   Q.E.D.

Proof of Theorem 3.14.4.   Let

$$\psi(t) = \frac{e^{-ita} - e^{-itb}}{it}\, \phi_X(t).$$

Then we have

$$F(b) - F(a) = \lim_{T\to\infty} \frac{1}{2\pi} \int_{-T}^{T} \psi(t)\, dt.$$

Since $\phi(-t) = \overline{\phi(t)}$, we get $\overline{\psi(t)} = \psi(-t)$. Hence, denoting by $\text{Im}(\psi(t))$, $\text{Re}(\psi(t))$ the imaginary and the real parts of $\psi(t)$, respectively, we get

$$\int_{-T}^{T} \text{Im}[\psi(t)]\, dt = 0$$

for all T. Hence

$$F(b) - F(a) = \lim_{T\to\infty} \frac{1}{2\pi} \int_{-T}^{T} \text{Re}[\psi(t)]\, dt$$

$$= \lim_{T\to\infty} \frac{1}{2\pi}$$

$$\times \int_{-T}^{T}\left[\int_{-\infty}^{\infty}\left(\frac{\sin t(x - a) - \sin t(x - b)}{t}\right) dF_X(x)\right] dt.$$

$$(3.62)$$

The order of integration in the last integral can be changed, since the integral with respect to x converges absolutely and the integral with respect to t has finite limits of integration. Hence the last integral in Eq. (3.62) becomes

$$\lim_{T\to\infty} \int_{-\infty}^{\infty} I(T,x,b,a)\, dF_X(x). \tag{3.63}$$

By Lemma 3.14.2, we can write Eq. (3.63) as

$$\frac{1}{2}[F(a + 0) - F(a - 0)] + F(b - 0) - F(a - 0)$$
$$+ \frac{1}{2}[F(b + 0) - F(b - 0)]$$
$$= \frac{1}{2}[F(b - 0) + F(b + 0)] - \frac{1}{2}[F(a - 0) + F(a + 0)]$$
$$= F(b) - F(a).$$

Hence Theorem 3.14.4 is proved.

If the characteristic function $\phi_X(t)$ of the random variable X is absolutely integrable over the interval $(-\infty,\infty)$, the probability density function of X is given by

$$f_X(x) = \lim_{h \to 0} \frac{F(x + h) - F(x - h)}{2h}$$

$$= \lim_{h \to 0} \mathop{Lt}_{T \to \infty} \frac{1}{2\pi} \int_{-T}^{T} \frac{\sin th}{ht} e^{-itx}\phi_X(t) \, dt$$

$$= \frac{1}{2\pi} \int_{-\infty}^{\infty} e^{-itx}\phi_X(t) \, dt.$$

We discuss below some examples that will illustrate application of Theorem 3.14.4.

Example 3.14.4 Let $X_1,\ldots,X_k$ be independently distributed binomial random variables with parameters n_i, p, with $i = 1,\ldots,k$, and let

$$Y = \sum_{i=1}^{k} X_i$$

$$\phi_{X_i}(t) = (q + pe^{it})^{n_i}, \qquad i = 1,\ldots,k.$$

Hence we have

$$\phi_Y(t) = (q + pe^{it})^{\sum_{i=1}^{k} n_i},$$

which is the characteristic function of a binomial random variable with parameters $\sum_{i=1}^{k} n_i, p$. Hence, by Theorem 3.14.4 (and by the uniqueness of characteristic function) Y is a binomial random variable with parameters $\sum_{i=1}^{k} n_i, p$.

Example 3.14.5 Let $X_1,\ldots,X_n$ be independently and identically distributed normal random variables with mean μ and variance σ^2, and let

$$\overline{X} = \sum_{i=1}^{n} \frac{X_i}{n}.$$

We have seen in Example 3.12.3 that

$$\phi_{\overline{X}}(t) = \exp\left(it\mu - \frac{1}{2} t^2 \frac{\sigma^2}{n}\right).$$

Hence $\overline{X}$ is normal with mean μ and variance σ^2/n.

3.15 CONDITIONAL MATHEMATICAL EXPECTATION

Let (X,Y) be a bivariate random variable having a joint probability density or mass function. In Section 3.6 we defined the conditional probability density and the conditional mass function of X given $Y = y$ and of Y given $X = x$. We will now define the conditional expectation of any Borel function $g(X)$ of X given $Y = y$ and that of any Borel function $g(Y)$ given $X = x$.

Definition 3.15.1: Conditional mathematical expectation (discrete random variables) Let (X,Y) be a bivariate discrete random variable with joint probability mass function $p_{X,Y}(x_i,y_j)$, and let the conditional probability mass function of X given $Y = y_j$ and that of Y given $X = x_i$ be $p_{X|Y}(x_i|y_j)$ and $p_{Y|X}(y_j|x_i)$, respectively. Then the *conditional mathematical expectations* of Borel functions $g(X)$, $h(Y)$ are given by

$$E(g(X)|Y = y_j) = \sum g(x_i)p_{X|Y}(x_i|y_j)$$
$$x_i : p_X(x_i) > 0$$

$$E(h(Y)|X = x_i) = \sum h(y_j)p_{Y|X}(y_j|x_i)$$
$$y_j : p_Y(y_j) > 0$$

provided that $E(|g(X)||Y = y_j) < \infty$, $E(|h(Y)||X = x_i) < \infty$.

Note: The terms $p_X(x_i)$, $p_Y(y_j)$ are the marginal probability mass functions of X and Y, respectively.

Definition 3.15.2: Conditional mathematical expectation (continuous random variables) Let (X,Y) be a continuous bivariate random variable with the joint probability density function $f_{X,Y}(x,y)$ and let the conditional probability density function of X given $Y = y$ and that of Y given $X = x$ be

$f_{X|Y}(x|y)$, $f_{Y|X}(y|x)$, respectively. Then the *conditional mathematical expectations* of $g(X)$ and $h(Y)$ are given by

$$E(g(X)|Y = y) = \int_{-\infty}^{\infty} g(x)f_{X|Y}(x|y)\, dx$$

$$E(h(Y)|X = x) = \int_{-\infty}^{\infty} h(y)f_{Y|X}(y|x)\, dy$$

provided that $E(|g(X)||Y = y) < \infty$, $E(|h(Y)||X = x) < \infty$. $E(X|Y = y)$ and $E(Y|X = x)$ are of considerable importance in statistical theory.

Definition 3.15.3: Regression curve   The set of points $\{E(X|Y = y), y\}$ is called the regression curve of X on Y, and the set of points $\{x, E(Y|X = x)\}$ is called the *regression curve of Y* on X. If the regression curve is a straight line, we call it a linear regression.

Example 3.15.1   Let (X, Y) be a bivariate discrete random variable with the joint probability mass function given in Example 3.4.1. From Example 3.6.1 we have

$$p_{X|Y}(x|y) = \frac{(n - y)!}{x!\,(n - x - y)!} \left(\frac{p_1}{1 - p_2}\right)^x \left(1 - \frac{p_1}{1 - p_2}\right)^{n-x-y}$$

if $x = 0, 1, \ldots, n$. Hence $E(X|Y = y) = (n - y)p_1/(1 - p_2)$ and the regression of X on Y is linear.

Example 3.15.2   Let (X, Y) be a bivariate normal random variable with the probability density function given in Example 3.4.2. In Example 3.6.2 we have shown that the conditional distribution of X given $Y = y$ is normal with mean $m_1 + \rho(\sigma_1/\sigma_2)(y - m_2)$ and variance $\sigma_1^2(1 - \rho^2)$. Thus

$$E(X|Y = y) = m_1 + \rho\frac{\sigma_1}{\sigma_2}(y - m_2)$$

$$V(X|Y = y) = \sigma_1^2(1 - \rho^2)$$

and the regression of X on Y is linear. Similarly, one can show that the regression of Y on X is also linear.

3.16 TWO IMPORTANT INEQUALITIES

3.16.1 Chebychev's Inequality

Let X be a random variable with finite mean m and finite variance σ^2. Then for every $\epsilon > 0$, the following relation holds:

$$P\left\{\left|\frac{X - m}{\sigma}\right| \geq \epsilon\right\} \leq \frac{1}{\epsilon^2}$$

or equivalently,

$$P\{|X - m| \geq \epsilon\} \leq \frac{\sigma^2}{\epsilon^2}.$$

This was named after the great Russian probabilist P. L. Chebychev, and it gives a relation between the variance and the dispersion of probability for any random variable X. We will give the proof for the continuous case only. The proof for the discrete case is analogous.

Proof. Let $f_X(x)$ be the probability density function of the random variable X. Then we have

$$E(X - m)^2 = \int_{-\infty}^{\infty} (x - m)^2 f_X(x) \, dx$$

$$= \int_{|(x-m)/\sigma| \geq k} (x - m)^2 f_X(x) \, dx$$

$$+ \int_{|(x-m)/\sigma| < k} (x - m)^2 f_X(x) \, dx$$

$$\geq \int_{|(x-m)/\sigma| \geq k} (x - m)^2 f_X(x) \, dx$$

$$\geq \sigma^2 k^2 \int_{|(x-m)/\sigma| \geq k} f_X(x) \, dx$$

$$= \sigma^2 k^2 P\left\{\left|\frac{X - m}{\sigma}\right| \geq k\right\}.$$

Hence

$$P\left\{\left|\frac{X - m}{\sigma}\right| \geq \epsilon\right\} \leq \frac{1}{\epsilon^2}. \qquad \text{Q.E.D.}$$

3.16.2 Schwartz' Inequality

For any two random variables X, Y with finite second moments (order 2),

$$E^2(XY) \le E^2(X)E^2(Y).$$

The equality holds if and only if $X = aY$ for some constant a with probability 1, that is, $P\{\omega \in \Omega : X(\omega) \ne aY(\omega)\} = 0$, where Ω is the sample space.

Proof. For any real t, let

$$h(t) = E(X - tY)^2$$
$$= E(X^2) + t^2 E(Y^2) - 2tE(XY).$$

Then $h(t) \ge 0$ for all t. Regarding $h(t) = 0$ as a quadratic equation in t, we establish that $h(t) \ge 0$ if the discriminant of the equation $h(t) = 0$ is less than or equal to 0. In other words,

$$4(E^2(XY) - E(X^2)E(Y^2)) \le 0$$

or

$$E^2(XY) \le E(X^2)E(Y^2).$$

If $E^2(XY) = E(X^2)E(Y^2)$, then $E(X - tY)^2 = 0$, which implies that $X = tY$ with probability 1. Conversely, if $X = tY$, then $h(t) = 0$, which implies that $E^2(XY) = E(X^2)E(Y^2)$.   Q.E.D.

Schwartz's inequality is a restatement of the fact that $\rho^2(X,Y) \le 1$, where $\rho(X,Y)$ is the correlation coefficient of X and Y, as proved in Theorem 3.14.2. This follows by replacing X and Y in Schwartz's inequality by $X - EX$ and $Y - EY$, respectively. Moreover, it follows from Schwartz's inequality that $\rho^2(X,Y) = 1$ if and only if X and Y are linearly related with probability 1 (i.e., there exist two constants a and b such that $X = aY + b$ with probability 1); $\rho(X,Y) = 1$ or -1 according as $a > 0$ or $a < 0$. This generalizes the results of parts (c) and (d) of Theorem 3.14.2 by establishing the sufficiency of the conditions therein.

EXERCISES

40. Find the moment-generating function, the characteristic function, and the first and second central moments of the following distributions:
    (a) Uniform distribution over the interval (a,b)
    (b) Gamma distribution with parameters λ, r
    (c) Geometric distribution

(d) Negative binomial distribution with probability mass function

$$p_X(x) = \begin{cases} \binom{r + x - 1}{x} p^r (1 - p)^x, & x = 0,1,\ldots \\ 0 & \text{otherwise.} \end{cases}$$

41. Let the characteristic function of the random variable X be given by

$$\phi_X(t) = \tfrac{1}{16}(1 + 2e^{it} + e^{2it})^2.$$

Find (a) $P(1 < X < 4)$, and (b) $P(X > 2)$.

42. Determine whether or not the following are characteristic function and why.

(a) $\dfrac{1 - t^2}{1 + t^2}$

(b) $\cos t$

(c) $\log(e + |t|)$

(d) e^{-t^4}

43. Suppose that X, Y are independent random variables with distribution functions $F_X(x)$, $F_Y(y)$ and characteristic functions $\phi_X(t)$, $\phi_Y(t)$, respectively. Show that the characteristic function of $Z = XY$ is given by

$$\int_{-\infty}^{\infty} \phi_Y(tx)\, dF_X(x) = \int_{-\infty}^{\infty} \phi_X(ty)\, dF_Y(y).$$

44. Let X be a random variable with characteristic function $\phi_X(t)$. Let $R(t)$ be the real part of $\phi_X(t)$. Show that:

(a) $R(t) = E(\cos(tX))$.

(b) $R^2(t) \le \tfrac{1}{2}[1 + R(2t)]$.

45. If $\phi_X(t)$ is the characteristic function of the random variable X and

$$\int_{-\infty}^{\infty} |\phi_X(t)|\, dt \le \infty,$$

what can be inferred?

46. (Bienayme–Chebychev inequality) Let $g(X)$ be a monotone function of a random variable X. Show that for every $k > 0$, $P\{g(X) \ge k\} \le E(g(X))/k$.

47. In a sequence of n independent Bernoulli trials (with probability p of success), let E_n denote the probability of an even number of successes. Prove the recursion formula

$$E_n = (1 - p)E_{n-1} + (1 - E_{n-1})p.$$

From this derive the probability-generating function of the random variable representing even number of successes in n trials.

48. A population of N distinct elements in equal proportions is sampled with replacement. If S_r denotes the sample size necessary for acquisition of r distinct elements and if

$$P_n(r) = P(S_r = n),$$

show that

$$P_{n+1}(r) = \frac{r-1}{N} P_n(r) + \frac{N-r+1}{N} P_n(r-1).$$

Hence or otherwise, derive the probability-generating function of $P_n(r)$.

49. Obtain the probability-generating function of q_n: the probability that in n independent tosses of an unbiased coin no run of three heads occurs.

50. Let $U, V_1, \ldots, V_n$ and $W_1, \ldots, W_n$ be mutually independent random variables each with mean zero and variance $\sigma^2 < \infty$. Let X be independent of each of these random variables and have probability mass function

$$p_X(j) = \begin{cases} p_j, & j = 1, \ldots, N \\ 0, & \text{otherwise} \end{cases}$$

with $\sum_1^N p_j = 1$. Y, Z are random variables defined by the equations

$$Y = U + \sum_{j=1}^{X} V_j, \qquad Z = U + \sum_{j=1}^{X} W_j.$$

(a) Show that X, YZ are uncorrelated but not independent.
(b) Find the correlation coefficient between Y and Z.

51. Suppose that the random variables $X_1, X_2, \ldots$ (ad infinitum) are non-negative and independent. Define

$$S_n = X_1 + \cdots + X_n$$

with $S_0 = 0$.

(a) Show that if $x > 0$,

$$P\{S_n > x\} \geq \sum_{m=1}^{n-1} P\{S_{m-1} \leq x \text{ and } X_m \geq x\}.$$

(b) Hence deduce that

$$P\{S_n > x\} \geq \frac{\alpha_n}{1 + \alpha_n},$$

where

$$\alpha_n = \sum_{m=1}^{n-1} P\{X_m > x\}.$$

52. Suppose that the random variables $X_1,\ldots,X_n$ are mutually independent and are identically normally distributed with zero mean and unit variance. Let

$$Y_1 = \rho Y_n + X_1$$

$$Y_2 = \rho Y_1 + X_2$$

$$Y_3 = \rho Y_2 + X_3$$

$$\vdots$$

$$Y_n = \rho Y_{n-1} + X_n,$$

where $|\rho| < 1$ (constant). Prove that as $n \to \infty$,

$$E\left(\frac{X_1 X_2 + X_2 X_3 + \cdots + X_n X_1}{n}\right) \longrightarrow \rho.$$

53. Let X_1, X_2, X_3, X_4 be independently, identically distributed normal random variables with zero mean and unit variance. Let $a_1 > 0$, $a_2 > 0$ and

$$R_1 = a_1 \sqrt{X_1^2 + X_2^2}, \qquad R_2 = a_2 \sqrt{X_3^2 + X_4^2}.$$

Find $P(R_1 > R_2)$.

54. Let X be a random variable with probability density function $f(x)$ which is symmetric about 0 [i.e., $f(x) = f(-x)$ for all x]. Show that

$$E(X^k) = \begin{cases} 0 & \text{if } k \text{ is odd} \\ \text{finite} & \text{if } k \text{ even.} \end{cases}$$

BIBLIOGRAPHY

Apostol, T. (1957). *Mathematical Analysis*, Addison-Wesley, Reading, Mass.

Bochner, S. (1933). Montone Funktionen, stieltjessche Integrale und harmonische Analyse, *Math. Ann.*, 108.

Cramér, H. (1937). *Random Variables and Probability Distributions*, Cambridge Tracts in Mathematics and Mathematical Physics, 36, Cambridge University Press, London.

Fisz, M. (1963). *Probability Theory and Mathematical Statistics*, Wiley, New York.

Kolmogorov, A. N. (1933). *Grundbegriffe der Wahrscheinlichkeitsrechnung*, Springer-Verlag, Berlin.

Parzen, E. (1960). *Modern Probability Theory*, Wiley, New York.

Pearson, K. (1922). *Tables of Incomplete Γ-Function*, Cambridge University Press, London.

Pearson, K. (1934). *Tables of Incomplete Beta Function*, Cambridge University Press, London.

4

Stochastic Convergence and Limit Theorems

As indicated in Chapter 1, a statistical inquiry is directed at studying some specific characteristics of a target population that describe a particular variant of a natural phenomenon. Ideally speaking, if we could observe all the units of the target population, it would be possible to appraise the true value of the characteristics under study without any error. In real situations, however, the number of units making up the target population are usually very large, sometimes infinitely large, and as such it would be very costly and time consuming and often impossible to observe all its units. Sometimes all the units in a population may not even be discernible: for example, the units in a continuous production process or all the drops in a mass of liquid. Moreover, we shall very often be interested in obtaining a quick answer to our specific inquiry and the purpose of the inquiry may be foiled if too long a time is taken to observe all the units in the target population, even granting that there are enough funds for such an observation. There is thus a practical need to observe only a few units of a target population and to utilize the information contained in these few units to derive a quick answer to the inquiry at hand. Herein lies the utility of the science of statistics. The few units of a target population so observed are said to constitute a sample of the population, and the method adopted for drawing a sample is called the *sampling method*, which physically describes a random experiment yielding the sample.

It should be evident that any particular sampling method is capable of yielding more than one sample from a population when it is repeated, and as such, following a particular sampling method, there would be more than one possible answer (one for each of the samples) for any statistical inquiry.

Statisticians are aware that none of these answers may be able to give the true value of the specific characteristic of the target population under study, and their main concern when faced with such a situation is to secure some sort of guarantee that the answer based on any of the samples is, in a certain sense, not far from the true value. They choose to ensure this by drawing a sample from the target population by using some random mechanism (random sampling), so that the probability of the departure of the answer based on the sample from the true value of the characteristic by a fixed amount may be calculated by applying the laws of probability and be small enough. The dynamics of this approach is brought out in greater detail in chapters which follow. What we would like to emphasize at this stage is the fact that random sampling plays a dominant role in statistical theory, which can be viewed as an applied branch of probability in the sense that it uses the laws of probability to derive answers to any statistical inquiry.

A sequence of random variables $X_1, X_2, \ldots, X_n$ is said to form a random sample of size n of a random variable X if $X_1, X_2, \ldots, X_n$ are independent and have the same distribution as the random variable X. Taking the distribution of X as describing the target population, the main problem the statistician faces is to process the information contained in a particular realization of the random sample $X_1, X_2, \ldots, X_n$ in order to obtain suitable answers for statistical inquiry. This is done by constructing some specific functions of the random sample $X_1, X_2, \ldots, X_n$ intended to give the statistician the relevant part of the information contained in the sample with reference to the inquiry at hand; the answer to the inquiry is then built up in terms of these functions. Any such function of the random sample is technically called a *statistic*.

It is quite apparent from what has been said above that the probability distributions of statistics play a very important role in statistical theory, since it is only in terms of these distributions that it is possible to calculate the probabilities of departures of the answer to the statistical inquiry (which is a function of these statistics) from the true value of the desired characteristic of the target population in the inquiry.

In this chapter we consider only linear functions of random variables $X_1, X_2, \ldots, X_n$, and prove some useful probabilistic results concerning these linear functions. Two important series of results of probability theory that play a dominant role in statistics are the laws of large numbers and limit theorems. Throughout this chapter we use the following notation. For a random sample $X_1, X_2, \ldots, X_n$ we shall always write

$$S_n = X_1 + X_2 + \cdots + X_n \quad \text{and} \quad \overline{X}_n = \frac{S_n}{n}.$$

Hence we have

$$E(\overline{X}_n) = E(X) = m$$

$$V(\overline{X}_n) = \frac{1}{n^2} \sum_{i=1}^{n} V(X_i) = \frac{1}{n} V(X),$$

where the use of the symbol m is arbitrary. Thus if $V(X)$ is finite, $V(\overline{X}_n) \to 0$ as $n \to \infty$. Therefore, from the Chebychev's inequality, for large n, $P\{|\overline{X}_n - m| > \epsilon\}$ will be close to zero, for any $\epsilon > 0$. Thus for large n, the probability that $\overline{X}_n$ differs significantly from m is 0. We will see later in this chapter that we can make a stronger statement about $\overline{X}_n$, namely, that as $n \to \infty$, $\overline{X}_n$ converges to m with probability 1. In general, for any sequence of random variables $\{X_n\}$, the laws of large numbers specify conditions under which one can assert that $\overline{X}_n$ converges weakly or strongly to an unknown constant, which is, in most cases, the mean.

We have seen in Chapter 3 that if X_i, $i = 1,\ldots,n$, are independently distributed normal random variables with means m_i and variances σ_i^2, then for all n, S_n is also normally distributed with mean

$$m = \sum_{i=1}^{n} m_i$$

and variance

$$\sigma^2 = \sum_{i=1}^{n} \sigma_i^2.$$

An important question arises: What can we say about S_n if the X_i are not normal? The answer comes from the central limit theorems, which assert that under very general conditions with respect to the distribution of X_i, $i = 1,\ldots,n$, the random variable $Y_n = (S_n - m)/\sigma$ is approximately normally distributed with mean 0 and variance 1. In other words,

$$\lim_{n\to\infty} P(Y_n \le y) = \int_{-\infty}^{y} \frac{1}{\sqrt{2\pi}} e^{-1/2x^2} \, dx.$$

Before we start our detailed discussion of the laws of large numbers and the limit theorems, we first define a few types of convergence that are useful in the probability theory.

4.1 TWO TYPES OF CONVERGENCE

Definition 4.1.1: Convergence in probability (weak convergence) A sequence of random variables $\{X_n\}$ *converges in probability* to a random variable or constant X if for every $\epsilon > 0$,

$$\lim_{n\to\infty} P\{|X_n - X| \ge \epsilon\} = 0$$

or, equivalently,

$$\lim_{n \to \infty} P\{|X_n - X| < \epsilon\} = 1.$$

It is to be understood that Definition 4.1.1 tells us nothing about the convergence of the sequence $\{X_n\}$ to X in the sense of mathematical analysis. In other words, if $\{X_n\}$ converges to X in probability, it is not necessary that for $\epsilon > 0$ there exist an integer n_0 such that for all $n \geq n_0$, $|X_n - X| \leq \epsilon$. This definition asserts only that given $\epsilon > 0$ and $\delta > 0$, there exists an integer n_0 such that $P\{|X_{n_0+i} - X| \leq \epsilon\} > 1 - \delta$ for $i = 1, 2, \dots$. This type of convergence is also known as *stochastic convergence*.

Definition 4.1.2: Convergence with probability 1 (strong convergence) A sequence of random variables $\{X_n\}$ converges to X, which may be a random variables or a constant, with probability 1, if

$$P\left(\lim_{n \to \infty} X_n = X\right) = 1$$

or, equivalently, if for every $\epsilon > 0$,

$$\lim_{n \to \infty} P\left\{\bigcup_{k \geq n} |X_k - X| > \epsilon\right\} = 0.$$

In other words, if $\{X_n\}$ converges with probability 1, then for every $\epsilon > 0$ and $\delta > 0$ there exists an integer n_0 such that the probability of the simultaneous occurrence of the events

$$|X_{n_0+i} - X| < \epsilon, \qquad i = 1, 2, \dots$$

is greater than $1 - \delta$. Since for any n, $|X_n - X| > \epsilon$ implies that

$$\bigcup_{k \geq n} |X_k - X| > \epsilon,$$

we have

$$P\{|X_n - X| > \epsilon\} \leq P\left\{\bigcup_{k \geq n} |X_n - X| > \epsilon\right\},$$

and hence convergence with probability 1 implies convergence in probability. Convergence in probability is sometimes called *weak convergence*, and convergence with probability 1, *strong convergence*.

We now define some laws of large numbers for a sequence of random variables $\{X_n\}$. Let $Y_n = \overline{X}_n - E(\overline{X}_n)$, $n = 1, 2, \dots$.

Definition 4.1.3: Weak (strong) law of large numbers A sequence of random variables $\{X_n\}$ is said to obey the *weak* or the *strong law of large numbers* according as the sequence $\{Y_n\}$ converges to zero weakly or strongly.

We will limit our discussion to sequences of independent random variables. The reader is referred to Loève (1963) or Chung (1968) for results concerning dependent random variables.

Let $\{X_n\}$ be a sequence of independent and identically distributed Bernoulli random variables with $P(X_i = 1) = p$, $P(X_i = 0) = 1 - p = q$, and let

$$S_n = \sum_{i=1}^{n} X_i.$$

Then S_n is a binomial random variable with

$$P(S_n = j) = \begin{cases} \binom{n}{j} p^j q^{n-j}, & j = 0,1,\ldots,n \\ 0, & \text{otherwise.} \end{cases}$$

Denoting $\overline{X}_n = S_n/n$, we have

$$p_{\overline{x}_n}\left(\frac{j}{n}\right) = \begin{cases} \binom{n}{j} p^j q^{n-j}, & j = 0,1,\ldots,n \\ 0, & \text{otherwise.} \end{cases}$$

Theorem 4.1.1: Bernoulli's law of large numbers Let $\{X_n\}$ be a sequence of independent and identically distributed Bernoulli random variables with $P(X_i = 1) = p$, $p(X_i = 0) = q = 1 - p$, $0 < p < 1$. For every $\epsilon > 0$,

$$\lim_{n\to\infty} P\{|\overline{X}_n - p| < \epsilon\} = 1.$$

Proof.

$$E(\overline{X}_n) = p$$

$$V(\overline{X}_n) = \frac{1}{n^2} \sum_{i=1}^{n} V(X_i)$$

$$= \frac{npq}{n^2}$$

$$= \frac{pq}{n}.$$

By Chebychev's inequality, for every $\epsilon > 0$,

$$P\left\{ \left| \frac{\overline{X}_n - p}{\sqrt{pq/n}} \right| > \epsilon \right\} \leq \frac{1}{\epsilon^2}$$

and hence

$$P\{|\overline{X}_n - p| \geq \epsilon\} \leq \frac{pq}{n\epsilon^2}.$$

Thus

$$\lim_{n \to \infty} P\{|\overline{X}_n - p| \geq \epsilon\} = 0. \qquad \text{Q.E.D.}$$

Since in practice the unknown p has to be estimated empirically, Theorem 4.1.1 asserts that in order to have an agreement between p and its estimate $\overline{X}_n$ we need to have a large number of observations. However, this theorem does not tell us anything about the convergence of $\overline{X}_n$ to p in a mathematical sense.

Theorem 4.1.2: Chebychev's law of large numbers   Let $\{X_n\}$ be a sequence of independent random variables with $E(X_i) = m_i$, $V(X_i) = \sigma_i^2$, $i = 1,\ldots,n$, and let $\sigma_i^2 \leq c < \infty$; then for $\epsilon > 0$

$$\lim_{n \to \infty} P\left\{ \left| \overline{X}_n - \frac{1}{n} \sum_{i=1}^n m_i \right| < \epsilon \right\} = 1.$$

Proof.   Since the X_i are independent,

$$V(\overline{X}_n) = \sum_{i=1}^n \frac{\sigma_i^2}{n^2} \leq \frac{c}{n}.$$

Applying Chebychev's inequality to the random variable $\overline{X}_n$, we get, for $\epsilon > 0$,

$$P\left\{ \left| \overline{X}_n - \frac{1}{n} \sum_{i=1}^n m_i \right| \geq \epsilon \right\} \leq \sum_{i=1}^n \frac{\sigma_i^2}{n^2 \epsilon^2}$$

$$\leq \frac{c}{n\epsilon^2}.$$

Hence

$$\lim_{n \to \infty} P\left\{ \left| \overline{X}_n - \frac{1}{n} \sum_{i=1}^n m_i \right| \geq \epsilon \right\} = 0.$$

Equivalently,

$$\lim_{n\to\infty} P\left\{ \left| \overline{X}_n - \frac{1}{n} \sum_{i=1}^{n} m_i \right| < \epsilon \right\} = 1.$$

Example 4.1.1: Poisson's law of large numbers   Let $\{X_n\}$ be a sequence of independent Bernoulli random variables with $P(X_i = 1) = p_i$, $P(X_i = 0) = 1 - p_i$, $i = 1,2,...$: then for every $\epsilon > 0$,

$$\lim_{n\to\infty} P\left\{ \left| \overline{X}_n - \frac{1}{n} \sum_{i=1}^{n} p_i \right| < \epsilon \right\} = 1.$$

Proof.

$$V(X_i) = p_i(1 - p_i) < \tfrac{1}{4} \qquad \text{for all } i.$$

$$E(\overline{X}_n) = \frac{1}{n} \sum_{i=1}^{n} p_i$$

$$V(\overline{X}_n) = \sum_{i=1}^{n} \frac{p_i(1 - p_i)}{n^2} \le \frac{1}{4n}.$$

Hence, by Chebychev's law of large numbers, we have

$$\lim_{n\to\infty} P\left\{ \left| \overline{X}_n - \frac{1}{n} \sum_{i=1}^{n} p_i \right| < \epsilon \right\} = 1. \qquad \text{Q.E.D.}$$

In Chebychev's law of large numbers we assumed that the $V(X_i)$ for all i are uniformly bounded by a constant c. However, if $E(|X_i|) < \infty$ for all i but the $V(X_i)$ are not uniformly bounded, a similar result is available for independent identically distributed random variables. This is given in the following theorem.

Theorem 4.1.3: Khinchin's theorem   Let $\{X_n\}$ be a sequence of independent identically distributed random variables with finite mean m; then for every $\epsilon > 0$,

$$\lim_{n\to\infty} P\{|\overline{X}_n - m| \le \epsilon\} = 1.$$

Proof.   For any fixed $\delta > 0$ let us define two sequences of random variables $\{Y_n\}$, $\{Z_n\}$ such that

$$Y_i = X_i, \qquad Z_i = 0, \qquad \text{if } |X_i| < n\delta$$
$$Y_i = 0, \qquad Z_i = X_i, \qquad \text{if } |X_i| \ge n\delta,$$

then, evidently, $X_i = Y_i + Z_i$. Since $E(|X_i|) < \infty$, defining

$$M = \int_{-\infty}^{\infty} |X_i| \, dF_{X_i}(x_i)$$

$$m(n) = E(Y_i) = \int_{-n\delta}^{n\delta} x_i \, dF_{X_i}(x_i),$$

we conclude that both M and $m(n)$ are finite for all n. Furthermore, since $\lim_{n\to\infty} m(n) = m$, for large n we then have $|m(n) - m| < \epsilon$. Now, for all $i = 1,2,\ldots$, we have

$$V(Y_i) = \int_{-n\delta}^{n\delta} x_i^2 \, dF_{X_i}(x_i) - m^2(n)$$

$$\leq n\delta \int_{-n\delta}^{n\delta} |x_i| \, dF_{X_i}(x_i)$$

$$\leq n\delta M$$

and

$$P(Z_i \neq 0) = \int_{|x_i| \geq n\delta} dF_{X_i}(x_i) \leq \frac{1}{n\delta} \int_{|x_i| \geq n\delta} |x_i| \, dF_{X_i}(x_i).$$

Since $E(X_i)$ is finite, we can choose n large enough that

$$P(Z_i \neq 0) \leq \frac{\delta}{n}$$

and hence

$$P\left(\sum_{i=1}^{n} Z_i \neq 0\right) \leq \sum_{i=1}^{n} P(Z_i \neq 0) \leq \delta. \tag{4.1}$$

Writing

$$\overline{X}_n - m = \frac{1}{n}\sum_{i=1}^{n} Y_i - m + \frac{1}{n}\sum_{i=1}^{n} Z_i,$$

we get

$$P\{|\overline{X}_n - m| \geq 2\epsilon\} \leq P\left\{\left|\frac{1}{n}\sum_{i=1}^{n} Y_i - m\right| > 2\epsilon\right\} + P\left(\sum_{i=1}^{n} Z_i \neq 0\right).$$

$$\tag{4.2}$$

By Chebychev's inequality,

$$P\left\{\left|\frac{1}{n}\sum_{i=1}^{n} Y_i - m\right| \geq 2\epsilon\right\}$$

$$= P\left\{\left|\frac{1}{n}\sum_{i=1}^{n} Y_i - m(n) + m(n) - m\right| \geq 2\epsilon\right\}$$

$$\leq P\left\{\left|\frac{1}{n}\sum_{i=1}^{n} Y_i - m(n)\right| \geq \epsilon\right\}$$

$$\leq \frac{M\delta}{\epsilon^2}. \tag{4.3}$$

Thus from Eqs. (4.1) to (4.3) we get

$$P\{|\overline{X}_n - m| \geq 2\epsilon\} \leq \frac{M\delta}{\epsilon^2} + \delta. \tag{4.4}$$

Since δ, ϵ are arbitrary, by proper choice of δ, ϵ we can make the right-hand side of Eq. (4.4) very small, and hence we prove the theorem.

Q.E.D.

If $\{X_n\}$ is a sequence of independent random variables, then for every $\epsilon > 0$,

$$\lim_{n\to\infty} P\left\{\left|\overline{X}_n - \frac{1}{n}\sum_{i=1}^{n} E(X_i)\right| \geq \epsilon\right\} = 0,$$

provided that

$$\frac{1}{n^2} V\left(\sum_{i=1}^{n} X_i\right) \longrightarrow 0$$

as $n \to \infty$. The proof of this fact follows trivially from Chebychev's inequality, and this result is due to Markov.

We now discuss the strong law of large numbers. As in the case of weak laws of large numbers, we first consider the sequence of independent Bernoulli random variables and then consider the general case. The starting point of all investigations of the strong laws of large numbers is Borel's law of large numbers. This is due to Emile Borel (1909) and deals with the strong convergence of a sequence of independent and identically distributed Bernoulli random variables.

Theorem 4.1.4: Borel's strong law of large numbers Let $\{X_n\}$ be a sequence of independent, identically distributed Bernoulli random variables with $P\{X_i = 1\} = p$ for all i; then

$$P\left\{\lim_{n\to\infty} \overline{X}_n = p\right\} = 1. \tag{4.5}$$

To prove Theorem 4.1.4, we need the following lemmas.

Lemma 4.1.1 If for any integer m and a sequence of random variables $\{X_n\}$,

$$\sum_{n=1}^{\infty} P\left\{|X_n| \geq \frac{1}{m}\right\} \leq \infty, \tag{4.6}$$

then

$P\{X_n$ does not converge to zero$) = 0$.

Proof. Let

$$E_{n,m} = \bigcup_{k=1}^{\infty} \left[|X_{n+k}| \geq \frac{1}{m}\right]$$

$$E_m = \bigcap_{n=1}^{\infty} E_{n,m}$$

$$E = \bigcup_{m=1}^{\infty} E_m.$$

Then E represents the event that X_n does not converge to 0. We have

$$P\{E_{n,m}\} = P\left\{\bigcup_{k=1}^{\infty} |X_{n+k}| \geq \frac{1}{m}\right\} \leq \sum_{k=1}^{\infty} P\left\{|X_{n+k}| \geq \frac{1}{m}\right\}.$$

Thus, in view of Eq. (4.6), we have for any fixed m,

$$\lim_{n\to\infty} P\{E_{n,m}\} = 0.$$

Since E_m implies every one of the events $E_{n,m}$ for $n = 1,2,...$, it follows that for some integer $n = N$, we get, for every m,

$$P\{E_m\} = P\left\{\bigcap_{n=1}^{\infty} E_{n,m}\right\} < P(E_{N,m}).$$

Thus, letting $N \to \infty$, we get

$$P(E_m) = 0. \tag{4.7}$$

Now E is an event which asserts that a value of m can be found such that

$$|X_{n+k}| \geq \frac{1}{m}$$

for all $n = 1, 2, \ldots$ and for at least one value of k. Since

$$P(E) < \sum_{m=1}^{\infty} P(E_m)$$

we obtain from Eq. (4.7) that

$$P(E) = 0,$$

which proves the lemma.   Q.E.D.

Lemma 4.1.2   Let X be a random variable for which $E(X - E(X))^4 < \infty$; then for any $\epsilon > 0$

$$P\{|X - E(X)| > \epsilon\} \leq \frac{1}{\epsilon^4} E(X - E(X))^4.$$

We give the proof for the continuous random variable only. The proof for the discrete case is analogous.

Proof.

$$
\begin{aligned}
E[X - E(X)]^4 &= \int_{-\infty}^{\infty} [x - E(X)]^4 \, dF_X(x) \\
&\geq \int_{|x - E(X)| \geq \epsilon} [x - E(X)]^4 \, dF_X(x) \\
&\geq \epsilon^4 \int_{|x - E(X)| \geq \epsilon} dF_X(x) \\
&= \epsilon^4 P\{|X - E(X)| \geq \epsilon\}.
\end{aligned}
$$

Hence, for $\epsilon > 0$, we have

$$P\{|X - E(X)| \leq \epsilon\} \leq \frac{1}{\epsilon^4} E[X - E(X)]^4.   \text{Q.E.D.}$$

Proof of Borel's strong law.   In view of Lemma 4.1.1, it is sufficient to show that for any integer m,

$$\sum_{n=1}^{\infty} P\left\{|\overline{X}_n - p| \geq \frac{1}{m}\right\} < \infty.$$

By Lemma 4.1.2 we have

$$P\left\{|\bar{X}_n - p| \geq \frac{1}{m}\right\} \leq m^4 \, E(\bar{X}_n - p)^4 = \frac{m^4}{n^4} \, E(S_n - np)^4. \qquad (4.8)$$

We have seen in Chapter 3 that S_n is a binomial random variable with parameters n, p, and hence

$$E(S_n - np)^4 = npq(p^3 + q^3) + 3pq\,(n^2 - n),$$

so that

$$E(S_n - np)^4 \leq \frac{3n^2}{4}.$$

Thus, from Eq. (4.8), we have

$$P\left\{|\bar{X}_n - p| \geq \frac{1}{m}\right\} \leq \frac{3m^4}{4n^2}$$

and hence

$$\sum_{n=1}^{\infty} P\left\{|\bar{X}_n - p| \geq \frac{1}{m}\right\} \leq \frac{3m^4}{4} \sum_{n=1}^{\infty} \frac{1}{n^2} < \infty. \qquad \text{Q.E.D.}$$

We now discuss Kolmogorov's law of large numbers for the sequence of independent, but not necessarily identically distributed, random variables $\{X_n\}$.

Theorem 4.1.5: Kolmogorov's law of large numbers Let $\{X_n\}$ be a sequence of independent random variables. Then $\{X_n\}$ obeys the strong law of large numbers, that is,

$$P\left\{\lim_{n \to \infty} \bar{X}_n = \lim_{n \to \infty} \frac{1}{n} \sum_{i=1}^{n} E(X_i)\right\} = 1$$

if

$$\sum_{n=1}^{\infty} \frac{V(X_n)}{n^2} < \infty.$$

Corollary If $V(X_n) < C$ for all n, then $\{X_n\}$ obeys the strong law of large numbers.

To prove Theorem 4.1.5, we need the following result.

Lemma 4.1.3: Hajek–Renyi inequality If the sequence of independent random variables $\{Y_n\}$ is such that $E(Y_n) = 0$, $V(Y_n) = \sigma^2 < \infty$ for all n, and if $\{C_n\}$ is a nonincreasing sequence of positive constants, then for any positive integers, m, n with $m < n$ and $\epsilon > 0$,

$$P\left\{\max_{m \leq k \leq n} C_k|Y_1 + \cdots + Y_k| \geq \epsilon\right\} < \frac{1}{\epsilon^2}\left(C_m^2 \sum_{i=1}^{m} \sigma_i^2 + \sum_{i=m+1}^{n} C_i^2\sigma_i^2\right).$$

Proof. Let

$$S_k = \sum_{i=1}^{k} Y_i, \qquad k = 1,2,\ldots$$

and

$$Y = \sum_{k=m}^{n-1} S_k^2(C_k^2 - C_{k+1}^2) + C_n^2 S_n^2.$$

Then we have

$$E(S_k^2) = \sum_{j=1}^{k} \sigma_j^2$$

and thus

$$E(Y) = C_m^2 \sum_{k=1}^{m} \sigma_k^2 + \sum_{k=m+1}^{n} C_k^2\sigma_k^2. \tag{4.9}$$

Let E_i, $i = m, m+1,\ldots,n$ be the events that

$$C_j|S_j| \begin{cases} <\epsilon & \text{if } m \leq j < i \\ \geq\epsilon & \text{if } j = i. \end{cases}$$

Then the events E_i are mutually exclusive. Since the event $\max_{m \leq i \leq n} C_i|S_i| > \epsilon$ implies that there exists at least one k such that $C_k|S_k| \geq \epsilon$, we get

$$P\left\{\max_{m<i<n} C_i|S_i| > \epsilon\right\} = \sum_{i=1}^{n} P(E_i). \tag{4.10}$$

Defining E_0 to be the event that $C_j|S_j| < \epsilon$ for $m \leq j \leq n$, we get, from the definition of conditional expectation and from the fact that Y is a non-negative random variable,

$$E(Y) = \sum_{i=0}^{n} E(Y|E_i)P(E_i)$$

$$\geq \sum_{i=1}^{n} E(Y|E_i)P(E_i).$$

Now, for $j \geq i$, we have

$$S_j = S_i + Y_{i+1} + \cdots + Y_j$$

and hence

$$E(S_j^2|E_i) = E(S_i^2 + (Y_{i+1} + \cdots + Y_j)^2 + 2S_i(Y_{i+1} + \cdots + Y_j)|E_i)$$

$$\geq E(S_i^2|E_i) + 2E(S_i(Y_{i+1} + \cdots + Y_j)|E_i).$$

Since the occurrence of the event E_i imposes restriction only on random variables $Y_1, \ldots, Y_i$, it follows that under the condition that E_i has occurred, the random variables $Y_{i+1}, Y_{i+2}, \ldots$ in the sequence $\{Y_n\}$ remain independent of each other and of S_i. Hence, for $j > i$, $E(S_i Y_j) = 0$, and thus

$$E(S_j^2|E_i) \geq E(S_i^2|E_i)$$

$$= E\left(S_i^2 \Big| S_i^2 \geq \frac{\epsilon^2}{C_i^2}\right)$$

$$\geq \frac{\epsilon^2}{C_i^2}.$$

Now we have

$$E(Y|E_i) = \sum_{j=m}^{n-1} E(S_j^2|E_i)(C_j^2 - C_{j+1}^2) + C_n^2 E(S_n^2|E_i)$$

$$\geq \sum_{j=i}^{n-1} E(S_j^2|E_i)(C_j^2 - C_{j+1}^2) + C_n^2 E(S_n^2|E_i)$$

$$\geq \frac{\epsilon^2}{C_i^2}\left[\sum_{j=i}^{n-1} (C_j^2 - C_{j+1}^2) + C_n^2\right]$$

$$= \epsilon^2. \tag{4.11}$$

From Eqs. (4.10) and (4.11) we get

$$E(Y) \geq \epsilon^2 \sum_{i=m}^{n} P(E_i).$$

Hence from Eqs. (4.9) and (4.10) we get the result.   Q.E.D.

Proof of Kolmogorov's law of large numbers.   Taking $C_j = 1/j$ and $Y_i = X_i - E(X_i)$, $\overline{Y}_i = S_i/i$ in the Hajek–Renyi inequality, we get

$$P\left\{\max_{m \leq j \leq n} |\overline{Y}_j| > \epsilon\right\} < \frac{1}{\epsilon^2}\left(\frac{1}{m^2} \sum_{i=1}^{m} \sigma_i^2 + \sum_{i=m+1}^{n} \frac{\sigma_i^2}{i^2}\right).$$

Now it is clear that

$$P\left\{\max_{m\le j}|\overline{Y}_j|\ge\epsilon\right\} = \lim_{n\to\infty}P\left\{\max_{m\le j\le n}|\overline{Y}_j|\ge\epsilon\right\}$$

$$\le \frac{1}{\epsilon^2}\left(\frac{1}{m^2}\sum_1^m\sigma_i^2 + \sum_{i=m+1}^{\infty}\frac{\sigma_i^2}{i^2}\right).$$

Thus we get

$$\lim_{m\to\infty}P\left\{\max_{m\le j}|\overline{Y}_j|\ge\epsilon\right\} = 0,$$

or, equivalently, $P\{\lim_{m\to\infty}\overline{Y}_m = 0\} = 1.$ Q.E.D.

On the basis of Kolmogorov's law of large numbers we obtain the following theorem for the sequence of independent and identically distributed random variables $\{X_n\}$, which is also due to Kolmogorov.

Theorem 4.1.6: Kolmogorov's theorem   Let $\{X_n\}$ be a sequence of independent and identically distributed random variables [distributed as the random variable X with the distribution function $F_X(x)$]. Then

$$P\left\{\lim_{n\to\infty}\overline{X}_n = \mu\right\} = 1$$

if and only if $E(X_i) = \mu < \infty.$

To prove this theorem we need the following lemma.

Lemma 4.1.4: Borel–Cantelli lemma   Let $\{E_n\}$ be a sequence of events and let

$$E = \bigcap_{n=1}^{\infty}\bigcup_{j=n}^{\infty}E_j.$$

Then:

1. If $\Sigma_{i=1}^{\infty}P(E_i) < \infty$, then $P(E) = 0$.
2. If the sequence of events $\{E_n\}$ are independent and if

$$\sum_{i=1}^{\infty}P(E_i) = \infty,$$

then $P(E) = 1$.

Proof. Since E is the set of all elements that belong to infinitely many of the events E_n, we have

$$E \subset \bigcup_{j=n}^{\infty} E_j, \qquad n = 1,2,\dots \ .$$

Hence

$$P(E) \le P\left(\bigcup_{j=n}^{\infty} E_j\right) \le \sum_{j=n}^{\infty} P(E_j). \tag{4.12}$$

Since

$$\sum_{j=1}^{\infty} P(E_j) < \infty$$

it follows that for sufficiently large n, the right-hand side of Eq. (4.12) is smaller than any given $\delta > 0$, and hence that

$$\lim_{n\to\infty} \sum_{j=n}^{\infty} P(E_j) = 0,$$

which implies that $P(E) = 0$. To prove part 2, we now suppose that the $\{E_n\}$ are independent. The complementary event $\overline{E}$ of E is given by

$$\overline{E} = \bigcup_{n=1}^{\infty} \bigcap_{j=n}^{\infty} \overline{E}_j.$$

The independence of the E_n implies the independence of the events in the sequence $\{\overline{E}_n\}$. Hence we get

$$1 - P(E) = P(\overline{E})$$

$$= P\left(\bigcup_{n=1}^{\infty} \bigcap_{j=n}^{\infty} \overline{E}_j\right)$$

$$\le \sum_{n=1}^{\infty} P\left(\bigcap_{j=n}^{\infty} \overline{E}_j\right)$$

$$= \sum_{n=1}^{\infty} \prod_{j=n}^{\infty} [1 - P(E_j)]. \tag{4.13}$$

Since

$$\sum_{j=1}^{\infty} P(E_j) = \infty$$

the infinite product on the right side of Eq. (4.13) is divergent to zero for each n. Hence $P(E) = 1$.

Proof of Kolmogorov's theorem.   Suppose that $E(X_i) < \infty$ for all i, and let $E(X_i) = E(X)$. This implies that

$$E|X| = \int_{-\infty}^{\infty} |X| \, dF_X(x) < \infty.$$

Therefore, we have

$$\sum_{n=1}^{\infty} P\{|X| \geq n\} = \sum_{n=1}^{\infty} \sum_{k \geq n} P\{k \leq |X| < k + 1\}$$

$$= \sum_{k=1}^{\infty} kP\{k \leq |X| < k + 1\}$$

$$\leq \sum_{k=0}^{\infty} \int_{k \leq |x| < k+1} |x| \, dF_X(x)$$

$$= \int_{-\infty}^{\infty} |x| \, dF_X(x)$$

$$< \infty.$$

Let us now define a sequence of truncated random variables $\{Y_n\}$ such that

$$Y_n = \begin{cases} X_n & \text{if } |X_n| < n \\ 0 & \text{if } |X_n| \geq n. \end{cases}$$

Then

$$V(Y_n) \leq E(Y_n^2)$$
$$= \int_{-n}^{n} x^2 \, dF_X(x)$$
$$\leq \sum_{k=0}^{n-1} (k + 1)^2 P\{k < |x| < k + 1\}.$$

Since

$$\sum_{n=k}^{\infty} \frac{1}{n^2} < \frac{1}{k^2} + \frac{1}{k} < \frac{2}{k}$$

we have

$$\sum_{n=1}^{\infty} \frac{V(Y_n)}{n^2} \le \sum_{n=1}^{\infty} \sum_{k=0}^{n-1} \frac{(k+1)^2}{n^2} P\{k \le |X| < k + 1\}$$

$$\le \sum_{k=1}^{\infty} P\{k - 1 \le |X| < k\} k^2 \sum_{n=k}^{\infty} \frac{1}{n^2}$$

$$< \sum_{k=1}^{\infty} P\{k - 1 \le |X| < k\} k^2 \frac{2}{k}$$

$$= 2 \sum_{k=1}^{\infty} k P\{k - 1 \le |X| < k\}$$

$$< \infty.$$

Thus from Kolmogorov's law of large numbers we get

$$P\left\{ \lim_{n \to \infty} \left[\frac{1}{n} \sum_{i=1}^{n} Y_i - \frac{1}{n} \sum_{i=1}^{n} E(Y_i) \right] = 0 \right\} = 1. \tag{4.14}$$

In other words, $\{Y_n\}$ satisfies the strong law of large numbers. Since

$$E(Y_n) = \int_{-n}^{n} x \, dF_X(x)$$

it is evident that $E(Y_n) \to E(X)$ as $n \to \infty$. In other words, given $\epsilon > 0$ there exists an integer $N(\epsilon)$ such that for $n \ge N(\epsilon)$,

$$|E(Y_n) - E(X)| < \epsilon.$$

Now

$$\left| \frac{1}{n} E\left(\sum_{i=1}^{n} Y_i \right) - E(X) \right| = \left| \frac{1}{n} \sum_{i=1}^{n} E[Y_i - E(X)] \right|$$

$$= \left| \frac{1}{n} \sum_{i<N(\epsilon)} E[Y_i - E(X)] \right.$$

$$\left. + \frac{1}{n} \sum_{i \ge N(\epsilon)}^{n} E[Y_i - E(X)] \right|$$

$$\le \left| \frac{1}{n} \sum_{i<N(\epsilon)} E[Y_i - E(X)] \right| + \epsilon.$$

Letting $n \to \infty$ first and then $\epsilon \to 0$, we have, from Eq. (4.14),

$$\lim_{n \to \infty} \frac{1}{n} \sum_{i=1}^{n} Y_i = E(X).$$

Now we prove that $\{X_n\}$ and $\{Y_n\}$ are equivalent in the sense that

$$P\{Y_n \neq X_n \text{ for some } n \geq N(\epsilon)\} \longrightarrow 0 \quad \text{as} \quad N(\epsilon) \longrightarrow \infty,$$

which will imply that $\{X_n\}$ obeys strong law of large numbers if $\{Y_n\}$ does so and the limits are the same.

Now

$$
\begin{aligned}
P\{X_n \neq Y_n \text{ for any } n \geq N(\epsilon)\} &\leq \sum_{n \geq N(\epsilon)} P\{X_n \neq Y_n\} \\
&= \sum_{n \geq N(\epsilon)} P\{|X_n| \geq n\} \\
&\leq \sum_{n \geq N(\epsilon)} (n - N(\epsilon) + 1) \\
&\qquad \times P\{n \leq |X_n| < n + 1\} \\
&\leq \sum_{n \geq N(\epsilon)} nP\{n \leq |X_n| < n + 1\} \\
&\leq \int_{|x| \geq N(\epsilon)} |x|\, dF_X(x) \longrightarrow 0 \\
&\qquad\qquad\qquad \text{as } N(\epsilon) \longrightarrow \infty.
\end{aligned}
$$

This is due to the fact that since $\int_{-\infty}^{\infty} |x|\, dF_X(x) < \infty$, $\int_{|x| \geq N(\epsilon)} |x|\, dF_X(x)$ can be made arbitrarily small by choosing $N(\epsilon)$ sufficiently large.

To prove the necessity let E_i, $i = 1,2,\ldots$, be the event that $|X_i| \geq i$. Since

$$\frac{X_n}{n} = \overline{X}_n - \frac{n-1}{n}\overline{X}_{n-1}$$

and $\lim_{n \to \infty} \overline{X}_n = E(X)$, we get

$$\lim_{n \to \infty} \frac{X_n}{n} = 0. \tag{4.15}$$

Thus the probability that infinitely many of the events E_i happen is zero. Furthermore, as the X_n are independent, the events E_n are also independent. Now by the Borel–Cantelli lemma, we have

$$\sum_{n=1}^{\infty} P\{|X_n| \geq n\} = \sum_{n=1}^{\infty} P(E_n) < \infty.$$

Denoting

$$P_j = P(|X| \geq j), \qquad j = 0,1,2,\ldots$$

we get

$$E|X| \leq 1(1 - p_1) + 2(p_1 - p_2) + 3(p_2 - p_3) + \cdots$$
$$= 1 + p_1 + p_2 + p_3 + \cdots$$
$$= 1 + \sum_{i=1}^{\infty} P(E_i)$$
$$< \infty.$$

Now from the argument in the sufficiency case it follows that $E(X) = \mu$. Q.E.D.

4.2 LIMIT THEOREMS

Let $\{X_n\}$ be a sequence of independent random variables with the distribution functions $\{F_{X_n}(x)\}$ and let

$$S_n = \sum_{i=1}^{n} X_i$$

$$Y_n = \frac{S_n - E(S_n)}{\sqrt{V(S_n)}}.$$

Our aim is to investigate conditions under which, for every x,

$$\lim_{n \to \infty} P(Y_n < x) = \int_{-\infty}^{x} \frac{1}{\sqrt{2\pi}} e^{-(1/2)y^2} \, dy. \tag{4.16}$$

All theorems considered in this connection will go by the generic name of central limit theorems. In general, when a sum of random variables satisfy Eq. (4.16) we say that $\{X_n\}$ satisfies a central limit theorem. These theorems are called central limit theorems because they are central to the distribution theory necessary for statistical inference.

The developments in this section depend heavily on the following notion of convergence in distribution of a sequence of random variables $\{X_n\}$.

Definition 4.2.1 A sequence of random variables $\{X_n\}$ with distribution functions $\{F_{X_n}(x)\}$ is said to *converge in distribution* to a random variable X with distribution function $F_X(x)$ if

$$\lim_{n \to \infty} F_{X_n}(x) = F_X(x)$$

at all points of continuity x of $F_X(x)$.

The distribution function $F_X(x)$ is called the *limiting distribution function* of $\{F_{X_n}(x)\}$. From Eq. (4.16) it is clear that all limit theorems that we discuss in this section deal with convergence in distribution. To prove these theorems we first need to prove a theorem that gives us the relation between convergence in distribution and convergence in terms of characteristic functions. The use of characteristic functions has proven to be a very powerful tool in proving these limit theorems. Furthermore, there is a one-to-one correspondence between convergence in distribution and convergence in terms of characteristic function.

Theorem 4.2.1: Lévy–Cramér theorem Let $\{X_n\}$ be a sequence of random variables and let, for reals t and x, $\phi_{X_n}(t)$, $F_{X_n}(x)$ be, respectively, the characteristic function and the distribution function of the random variable X_n. Then $\{F_{X_n}(x)\}$ converges to $F_X(x)$ if and only if for every t, $-\infty < t < \infty$, the sequence of characteristic functions $\{\phi_{X_n}(t)\}$ converges to the function $\phi_X(t)$, which is the characteristic function of the limit distribution $F_X(x)$, and convergence is uniform with respect to t in every finite interval $T_1 \leq t \leq T_2$.

To prove this theorem we need Helly's lemma and the Lebesgue dominated-convergence theorem.

Lemma 4.2.1: Helly's lemma Every sequence of distribution functions is weakly compact. In other words, every sequence of distribution functions contains a subsequence that converges weakly to a nondecreasing function (not necessarily a distribution function) at all points of continuity of the latter function.

Proof. Let $\{F_{X_n}(x) = F_n(x)\}$ be a sequence of distribution functions and let D be a set of all rational numbers $\{r_k\}$. Since $0 \leq F_n(r_1) \leq 1$ for all n, by the Bolzano–Weistrass theorem, $\{F_n(r_1)\}$ contains a subsequence $\{F_{n_1}(r_1)\}$ that converges weakly. Similarly, since $0 \leq \{F_{n_1}(r_2)\} \leq 1$ for all n_1, $\{F_{n_1}(r_2)\}$ contains a subsequence $\{F_{n_2}(r_2)\}$ which converges weakly, and so on. Evidently, the subsequence $\{F_{n_2}(r_1)\}$ of $\{F_{n_1}(r_1)\}$ converges weakly. Thus the diagonal sequence $\{F_{nn}(\cdot)\}$ of the subsequences $\{F_{nk}(\cdot)\}$, $k = 1,2,...,$ converges weakly for all r_k in D.

Let us call the limiting distribution function $F_D(\cdot)$. Clearly, it is defined for all r_k in D and it is bounded and nondecreasing. We now define, for any real x,

$$F(x) = \text{Upper bound } F_D(r_i), \ r_i \leq x.$$

By definition $F(x)$ is continuous to the right, bounded, and nondecreasing. Let x (real) be a point of continuity of $F(\cdot)$. Given x there exists a sequence of rationals $\{(r_i', r_i'')\}$ such that $r_i' < x < r_i''$ for all i with $r_i' \uparrow x$, $r_i'' \downarrow x$ and $\lim_{i \to \infty} (F(r_i'') - F(r_i')) = 0$. For any element F_{nn} in the diagonal sequence $\{F_{nn}(\cdot)\}$ that converges to $F_D(\cdot)$, we have

$$F_{nn}(r_i') \le F_{nn}(x) \le F_{nn}(r_i''),$$

$$F_D(r_i'') \le \lim \inf F_{nn}(x)$$

$$\le \lim \sup F_{nn}(x)$$

$$\le F_D(r_i'').$$

Since $F_D(r_i'') - F_D(r_i')$ can be made arbitrarily small by choosing i sufficiently large, we see that $\lim_{n \to \infty} F_{nn}(x)$ exists and is equal to $F(x)$. Thus the subsequence $\{F_{nn}(\cdot)\}$ converges to $F(\cdot)$ at all continuity points of $F(\cdot)$.   Q.E.D.

Theorem 4.2.2: Lebesque's dominated-convergence theorem   Let the sequence of functions f_n be such that each improper integral $\int_a^\infty f_n(x)\, dx$ exists and let $\{f_n\}$ converge pointwise on $[a, +\infty)$ to a function f for which $\int_a^\infty f(x)\, dx$ exists. Suppose that there exists a nonnegative function g such that $\int_a^\infty g(x)\, dx$ exists and for all n, $|f_n(x)| \le g(x)$ for each $x \ge a$. Then

$$\lim \int_a^\infty f_n(x)\, dx = \int_a^\infty f_n(x)\, dx = \int_a^\infty f(x)\, dx.$$

The reader is referred to Apostol (1957) for the proof of this theorem.

Proof of the Lévy–Cramér theorem.   By definition

$$\phi_{X_n}(t) = \int_{-\infty}^\infty e^{itx}\, dF_{X_n}(x)$$

$$\phi_X(t) = \int_{-\infty}^\infty e^{itx}\, dF_X(x).$$

Choose two continuity points $a < 0$, $b > 0$ of the distribution function $F_X(x)$. Then the following equations hold:

$$\phi_{X_n}(t) = \int_{-\infty}^a e^{itx}\, dF_{X_n}(x) + \int_a^b e^{itx}\, dF_{X_n}(x) + \int_b^\infty e^{itx}\, dF_{X_n}(x)$$

$$= I_{n_1} + I_{n_2} + I_{n_3}$$

$$\phi_X(t) = \int_{-\infty}^{a} e^{itx} \, dF_X(x) + \int_{a}^{b} e^{itx} \, dF_X(x) + \int_{b}^{\infty} e^{itx} \, dF_X(x)$$

$$= I_1 + I_2 + I_3,$$

where $I_1, I_2, I_3, I_{n_1}, I_{n_2}, I_{n_3}$ are arbitrary designations. Now integrating by parts yields

$$I_{n_2} - I_2 = \int_{a}^{b} e^{itx} [dF_{X_n}(x) - dF_X(x)]$$

$$= e^{itx} [F_{X_n}(x) - F_X(x)]_a^b - it \int_{a}^{b} [F_{X_n}(x) - F_X(x)] e^{itx} \, dx.$$

Hence

$$|I_{n_2} - I_2| \leq |F_{X_n}(b) - F_X(b)| + |F_{X_n}(a) - F_X(a)|$$

$$+ |t| \int_{a}^{b} |F_{X_n}(x) - F_X(x)| \, dx.$$

Since $F_{X_n}(x)$ converges to $F_X(x)$ at all continuity points of $F_X(x)$, given $\epsilon > 0$ we can have, for sufficiently large n,

$$|F_{X_n}(b) - F_X(b)| < \frac{\epsilon}{9}, \qquad |F_{X_n}(a) - F_X(a)| < \frac{\epsilon}{9}.$$

For any t in the interval $T_1 < t < T_2$, where T_1, T_2 are arbitrary constants and $K = \max(|T_1|, |T_2|)$, we have

$$|t| \int_{a}^{b} |F_{X_n}(x) - F_X(x)| \, dx \leq K \int_{a}^{b} |F_{X_n}(x) - F_X(x)| \, dx.$$

By the Lebesque theorem and by the fact that $|F_{X_n}(x) - F_X(x)|$ is uniformly bounded in every interval, we get

$$\lim_{n \to \infty} \int_{a}^{b} |F_{X_n}(x) - F_X(x)| \, dx = \int_{a}^{b} \lim_{n \to \infty} |F_{X_n}(x) - F_X(x)| \, dx.$$

Thus for sufficiently large n we can have

$$|t| \int_{a}^{b} |F_{X_n}(x) - F_X(x)| \, dx < \frac{\epsilon}{9}$$

and hence $|I_{n_2} - I_2| < \epsilon/3$. Furthermore,

$$|I_{n_1} - I_1| \leq |I_{n_1}| + |I_1|$$

$$\leq \left| \int_{-\infty}^{a} e^{itx} \, dF_{X_n}(x) \right| + \left| \int_{-\infty}^{a} e^{itx} \, dF_X(x) \right|$$

$$\leq \int_{-\infty}^{a} dF_{X_n}(x) + \int_{-\infty}^{a} dF_X(x)$$

$$= F_{X_n}(a) + F_X(a).$$

Choosing a such that $|a|$ is large, we can make both $F(a)$ and $F_n(a)$ smaller than $\epsilon/6$. Therefore, for n large and all t under consideration, we have $|I_{n_1} - I_1| < \epsilon/3$. By a similar argument we get, for large n and all t, $|I_{n_3} - I_3| < \epsilon/3$. Thus for large n and all t,

$$|\phi_{X_n}(t) - \phi_X(t)| < \epsilon$$

or, equivalently, for all t,

$$\lim_{n \to \infty} \phi_{X_n}(t) = \phi_X(t),$$

which proves the necessity part of the theorem.

We now prove the sufficiency part. By Helly's lemma the sequence of distribution functions $\{F_{X_n}(x)\}$ has a subsequence $\{F_{X_{n_k}}(x)\}$, which converges to a nondecreasing right-continuous, bounded function $F_X(x)$ at all points of continuity of $F_X(x)$. To claim that $F_X(x)$ is a distribution function we must show that $F_X(-\infty) = 0$, $F_X(\infty) = 1$. To do this, let us assume that

$$\alpha = F(\infty) - F(-\infty) < 1. \tag{4.17}$$

By hypothesis $\phi_{X_n}(t) \to \phi_X(t)$ and $\phi_{X_n}(0) = 1$. Hence $\phi_X(0) = 1$. Since $\phi_X(t)$ is continuous in t, we conclude that in the small neighborhood $(-\tau, \tau)$ of 0,

$$\frac{1}{2\tau} \left| \int_{-\tau}^{\tau} \phi_X(t)\, dt \right| > 1 - \frac{\epsilon}{2} > \alpha + \frac{\epsilon}{2}, \tag{4.18}$$

where $\alpha + \epsilon > 1$. From Eq. (4.17) it follows that we can choose $a > 4/\epsilon\tau$ such that $-a$, a are continuity points of $F_X(x)$ and for all $k > K$,

$$\alpha_k = F_{X_{n_k}}(a) - F_{X_{n_k}}(-a) < \alpha + \frac{\epsilon}{4}.$$

From Eq. (4.18) and from the fact that $\phi_{X_n}(t) \to \phi_X(t)$, we get for sufficiently large k,

$$\frac{1}{2\tau} \left| \int_{-\tau}^{\tau} \phi_{X_{n_k}}(t)\, dt \right| > \alpha + \frac{\epsilon}{2}.$$

Now

$$\int_{-\tau}^{\tau} \phi_{X_{n_k}}(t)\, dt = \int_{-\tau}^{\tau} \left[\int_{-\infty}^{\infty} e^{itx}\, dF_{X_{n_k}}(x) \right] dt$$

$$= \int_{-\infty}^{\infty} \left(\int_{-\tau}^{\tau} e^{itx}\, dt \right) dF_{X_{n_k}}(x).$$

Since

$$\left| \int_{-\tau}^{\tau} e^{itx} \, dt \right| \le 2\tau$$

we have

$$\begin{aligned}
\left| \int_{-\tau}^{\tau} e^{itx} \, dt \right| &= \left| \left[\frac{e^{itx}}{ix} \right]_{-\tau}^{\tau} \right| \\
&= \frac{2}{|x|} |\sin \tau x| \\
&\le \frac{2}{|x|} \\
&\le \frac{2}{a}
\end{aligned}$$

if $|x| > a$. Now

$$\begin{aligned}
\frac{1}{2\tau} \left| \int_{-\tau}^{\tau} \phi_{X_{n_k}}(t) \, dt \right| &\le \frac{1}{2\tau} \left| \int_{|x|<a} \left(\int_{-\tau}^{\tau} e^{itx} \, dt \right) dF_{X_{n_k}}(x) \right| \\
&\quad + \frac{1}{2\tau} \left| \int_{|x|>a} \left(\int_{-\tau}^{\tau} e^{itx} \, dt \right) dF_{X_{n_k}}(x) \right| \\
&\le \left| \int_{|x|\le a} dF_{X_{n_k}}(x) \right| + \frac{1}{a\tau} \left| \int_{|x|>a} dF_{X_{n_k}}(x) \right| \\
&\le \alpha_k + \frac{1}{a\tau} \\
&\le \alpha_k + \frac{\epsilon}{4} \\
&< \alpha + \frac{\epsilon}{4} + \frac{\epsilon}{4} \\
&= \alpha + \frac{\epsilon}{2}.
\end{aligned} \qquad (4.19)$$

Thus Eq. (4.19) contradicts Eq. (4.18). Hence $F_X(x)$ is a distribution function and its characteristic function is $\phi_X(t)$.

The last thing we want to show is that the sequence $\{F_{X_n}(x)\}$ converges to $F_X(x)$. If this is not so, there exists another subsequence $\{F_{X_{n_k}}(x)\}$ of $\{F_{X_n}(x)\}$ which converges to $\overline{F}_X(x)$ where $\overline{F}_X(x)$ and $F_X(x)$ are different. By the argument above, $\overline{F}_X(x)$ is a distribution function and by the necessary part of the theorem the characteristic function of $F_X(x)$ is $\phi_X(t)$. Thus from the fact that the characteristic function uniquely determines its distribution function, we conclude that $\overline{F}_X(x)$ and $F_X(x)$ are identical. Thus every subsequence of $\{F_{X_n}(x)\}$ converges to the same distribution function $F_X(x)$, and hence $\{F_{X_n}(x)\}$ converges to $F_X(x)$.   Q.E.D.

Theorem 4.2.3: De Moivre–Laplace theorem   Let $\{X_n\}$ be a sequence of independent and identically distributed Bernoulli random variables with $P(X_n = 1) = p, 0 < p < 1, q = 1 - p$, and let

$$S_n = \sum_{i=1}^{n} X_i.$$

Then for every x,

$$\lim_{n \to \infty} P\left\{ Y_n = \frac{S_n - np}{\sqrt{npq}} \le x \right\} = \frac{1}{\sqrt{2\pi}} \int_{-\infty}^{x} e^{(1/2)y^2} \, dy.$$

Proof.   We have seen in Chapter 3 that S_n is a binomial random variable with parameters n, p, that is,

$$P(S_n = j) = \begin{cases} \binom{n}{j} p^j q^{n-j}, & j = 0,1,\ldots,n \\ 0, & \text{otherwise} \end{cases}$$

with $E(S_n) = np$, $V(S_n) = npq$, and $\phi_{S_n}(t) = (q + pe^{it})^n$. Thus

$$\phi_{Y_n}(t) = \exp\left(-\frac{inpt}{\sqrt{npq}} \right)\left[q + p \exp\left(\frac{it}{\sqrt{npq}} \right) \right]^n$$

$$= \left[q \exp\left(-\frac{ipt}{\sqrt{npq}} \right) + p \exp\left(\frac{itq}{\sqrt{npq}} \right) \right]^n.$$

We now expand the exponential function e^{ix} in the neighborhood of $x = 0$ by Taylor series for k terms with the remainder in the Peano form, to get

$$e^{ix} = \sum_{j=0}^{k} \frac{(ix)^j}{j!} + O(x^k).$$

Thus we have

$$q \exp\left(-\frac{ipt}{\sqrt{npq}}\right) + p \exp\left(\frac{itq}{\sqrt{npq}}\right) = (p + q) - \left(\frac{p + q}{2}\right)\frac{t^2}{n}$$
$$+ O\left(\frac{t^2}{n}\right)$$
$$= 1 - \frac{t^2}{2n} + O\left(\frac{t^2}{n}\right),$$

where for every t,

$$\lim_{n\to\infty} nO\left(\frac{t^2}{n}\right) = 0.$$

Thus

$$\log \phi_{Y_n}(t) = n \log\left[1 - \frac{t^2}{2n} + O\left(\frac{t^2}{n}\right)\right].$$

Since for every fixed t and for sufficiently large n,

$$\left|-\frac{t^2}{2n} + O\left(\frac{t^2}{n}\right)\right| < 1,$$

we get

$$\lim_{n\to\infty} \log \phi_{Y_n}(t) = \lim_{n\to\infty}\left[-\frac{t^2}{2} + nO\left(\frac{t^2}{n}\right)\right]$$
$$= -\frac{t^2}{2},$$

or, equivalently,

$$\lim_{n\to\infty} \phi_{Y_n}(t) = e^{-t^2/2},$$

which is the characteristic function of a normal random variable with mean 0 and variance 1. Thus by the Lévy–Cramér theorem we get the proof of the De Moivre–Laplace theorem.   Q.E.D.

From Theorem 4.2.3 it follows that for $y_1 < y_2$,

$$\lim_{n\to\infty} P(y_1 \le Y_n \le y_2) = \lim_{n\to\infty} P(y_1 \sqrt{npq} + np \le Y_n \le y_2 \sqrt{npq} + np)$$
$$= \int_{y_1}^{y_2} \frac{1}{\sqrt{2\pi}} e^{(1/2)y^2} \, dy.$$

For moderately large values of n a better approximation of $P(y_1 \le y_n \le y_2)$ is obtained from

$$\int_{y_1}^{y_2} \frac{1}{\sqrt{2\pi}} e^{-(1/2)y^2} \, dy$$

by replacing y_1 and y_2 by $y_1 + (1/2\sqrt{npq})$ and $y_2 + (1/2\sqrt{npq})$, respectively.

Example 4.2.1   Suppose that an unbiased die is tossed 100 times. Let us associate a Bernoulli random variable X_i with the ith toss by $X_i = 1$ if the outcome of the ith toss is an even number. Then, clearly, the X_i are independent and identically distributed Bernoulli random variables with $P(X_i = 1) = \frac{1}{2}$, $P(X_i = 0) = \frac{1}{2}$. Let $S_n = \Sigma_{i=1}^{n} X_i$. Then

$$P(40 \le S_n \le 60) = \int_{-2.1}^{2.1} \frac{1}{\sqrt{2\pi}} e^{(1/2)y^2} \, dy = 0.9642.$$

The general limit theorem for the sequence $\{X_n\}$ of independent and identically distributed random variables (not necessarily Bernoulli random variables) with finite mean and variance is due to Lindeberg and Lévy and is given below.

Theorem 4.2.4: Lindeberg–Lévy theorem   Let $\{X_n\}$ be a sequence of independent random variables having the same distribution function as the random variable X with $E(X) = m$, $V(X) = \sigma^2 < \infty$. Then for every x,

$$\lim_{n \to \infty} F_{Y_n}(x) = \int_{-\infty}^{x} \frac{1}{\sqrt{2\pi}} e^{(1/2)y^2} \, dy,$$

where

$$Y_n = \frac{1}{\sigma\sqrt{n}} \sum_{i=1}^{n} (X_i - m).$$

Proof. $E(S_n) = nm$, $V(S_n) = n\sigma^2$. The characteristic function of Y_n is

$$\phi_{Y_n}(t) = \left[\phi_{X-m}\!\left(\frac{t}{\sigma\sqrt{n}}\right) \right]^n.$$

Expanding $\phi_{X-m}(t/\sigma\sqrt{n})$ in the neighborhood of $t = 0$ by MacLaurin series, we get

$$\phi_{X-m}\!\left(\frac{t}{\sigma\sqrt{n}}\right) = 1 + \frac{it}{\sigma\sqrt{n}} E(X - m) + \frac{i^2 t^2}{2\sigma^2 n} E(X - m)^2 + O\!\left(\frac{t^2}{n}\right)$$

$$= 1 - \frac{t^2}{2n} + O\!\left(\frac{t^2}{n}\right).$$

Thus

$$\lim_{n\to\infty} \log \phi_{Y_n}(t) = -\frac{t^2}{2},$$

so for sufficiently large n,

$$\phi_{Y_n}(t) = e^{-t^2/2}.$$

Since $e^{-t^2/2}$ is the characteristic function of a normal random variable with mean 0 and variance 1, the proof of the Lindeberg–Lévy theorem follows from the Lévy–Cramér theorem.

If the random variables of the sequence $\{X_n\}$ are independent but not identically distributed, the Lindeberg–Lévy theorem does not hold even if all the random variables in the sequence have finite variances. Although it is beyond the scope of this book to include limit theorems for independent random variables, for completeness we include here without proof the Liapunov theorem, which gives a sufficient condition for the sequence of independent random variables $\{X_n\}$ to satisfy the central limit theorem, and the Lindeberg–Feller theorem, which gives the necessary and sufficient condition for the sequence $\{X_n\}$ of independent random variables to satisfy the central limit theorem. For proofs of these theorems the reader is referred to Gnedenko (1962), Loève (1963), Fisz (1963), and Parzen (1967) or to the original papers of Liapunov (1901), Lindeberg (1922), and Feller (1935).

Theorem 4.2.5: Liapunov's theorem   Let $\{X_n\}$ be a sequence of independent random variables with $\mu_n = E(X_n) < \infty$, $\sigma_n^2 = V(X_n) < \infty$, $\beta_n = E|X_n - \mu_n|^3 < \infty$, and let

$$B_n = \left(\sum_{i=1}^{n} \beta_i\right)^{1/3}, \quad C_n = \left(\sum_{i=1}^{n} \sigma_i^2\right)^{1/2}, \quad Y_n = \frac{\sum_{i=1}^{n} (X_i - \mu_i)}{C_n}.$$

Then if $\lim_{n\to\infty} B_n/C_n = 0$, then for every x,

$$\lim_{n\to\infty} P(Y_n \leq x) = \int_{-\infty}^{x} \frac{1}{\sqrt{2\pi}} e^{(1/2)y^2} \, dy.$$

Theorem 4.2.6: Lindeberg–Feller theorem   Let $\{X_n\}$ be a sequence of independent random variables with distribution functions $\{F_{X_n}(x)\}$ and

$$\mu_n = E(X_n) < \infty, \quad \sigma_n^2 = V(X_n) < \infty, \quad C_n = \left(\sum_{i=1}^{n} \sigma_i^2\right)^{1/2}.$$

Then for every x the relations

$$\lim_{n\to\infty} \max_{1\le i\le n} \frac{\sigma_i}{C_n} = 0$$

$$\lim_{n\to\infty} P(Y_n \le x) = \int_{-\infty}^{x} \frac{1}{\sqrt{2\pi}} e^{(1/2)y^2} \, dy$$

hold good if and only if for every $\epsilon > 0$,

$$\lim_{n\to\infty} \frac{1}{C_n^2} \sum_{i=1}^{n} \int_{|x-\mu_i|>\epsilon C_n} (x - \mu_i)^2 \, dF_{X_i}(x) = 0.$$

EXERCISES

1. Demonstrate by an example that the limit of a convergent sequence of distribution functions may not always be a distribution function.
2. Show by an example that convergence in distribution does not necessarily imply convergence in probability.
3. Let $\{X_n\}$ ($n = 1,2,...$) be a sequence of independent random variables converging to a constant c in distribution. Show that $\{X_n\}$ converges to c in probability.
4. (Slutsky's theorem) Let $\{X_n\}$ ($n = 1,2,...$) and $\{Y_n\}$ ($n = 1,2,...$) be two sequences of random variables such that X_n converges to a random variable X in distribution and $\{Y_n\}$ converges to C (constant) in probability. Show that $\{X_n + Y_n\}$ converges to $X + C$ in distribution and $X_n Y_n$ converges to CX in distribution.
5. Let $\{X_n\}$ converge to X in distribution and $\{Y_n\}$ converge to Y in distribution. Show that $F_X(x) = F_Y(x)$ for all x.
6. Prove that if $\{X_n\}$ converges to X in probability and $\{X_n\}$ converges to Y in probability, then $P(X = Y) = 1$.
7. Let $\{X_n\}$ ($n = 1,2,...$) be a sequence of identically distributed random variables that have finite second moments equal to σ^2. Let $\{Y_n\}$ ($n = 1,2,...$) be a sequence of identically distributed random variables that are independent of the X_n and take the values ± 1 each with probability $\frac{1}{2}$. Let $S_n = Y_1 X_1 + \cdots + Y_n X_n$. Show that $n^{-1/2} S_n$ is asymptotically normal and find its asymptotic mean and variance.
8. (Law of iterated logarithm) Let $\{X_n\}$ ($n = 1,2,...$) be a sequence of independent and identically distributed binomial random variables with parameters n, p, and let $q = 1 - p$, $Y_n = (X_n - np)/\sqrt{npq}$.

Show that

$$P\left\{\limsup_{n\to\infty} \frac{Y_n}{\sqrt{2\log\log n}} = 1\right\} = 1.$$

9. Let $\{X_n\}$ $(n = 1,2,\dots)$ be a sequence of independent random variables. Show that the probability of convergence of $S_n = \Sigma_{i=1}^n X_i$ is either zero or one.

10. Let $\{X_n\}$ be a sequence of independent and identically distributed Poisson random variables with parameter $\lambda = 3$ and let $Y_{100} = X_1 + \cdots + X_{100}$. Find $P(100 \le Y_{100} \le 200)$.

11. Let $\{X_n\}$ be a sequence of independent and identically distributed Cauchy random variables with probability density function

$$f_X(x) = \begin{cases} \dfrac{1}{\pi(1 + x^2)}, & -\infty < x < \infty \\ 0, & \text{otherwise.} \end{cases}$$

Let $\overline{X}_n = (1/n)(X_1 + \cdots + X_n)$. Show that there does not exist a finite constant c to which $\overline{X}_n$ converges in probability.

12. Let $h(X)$ be a nonnegative and increasing function of a nonnegative random variable X and $E(h(X)) < \infty$. Show that for every $\epsilon > 0$,

$$P\{X - E(X) > \epsilon\} \le \frac{E(h(X))}{h(\epsilon)}.$$

13. Let $X_1,\dots,X_n$ be a random sample of size n from a random phenomenon (the X_i take nonnegative values only) with finite mean μ. Show that the probability that $\Sigma_{i=1}^n X_i$ exceeds some constant α is not greater than $n\mu/\alpha$.

14. Prove the inequality (4.12).

15. Let $\{F_n(x)\}$ be a sequence of distribution functions $F_n(x)$ defined by

$$F_n(x) = \begin{cases} 0 & \text{if } x \le -n \\ \dfrac{x + n}{2n} & \text{if } -n < x < n \\ 1 & \text{if } x \ge n. \end{cases}$$

Is $\lim_{n\to\infty} F_n(x)$ a distribution function?

16. Examine whether the weak law of large numbers holds for the sequence $\{X_n\}$ of independent random variables X_n, $n = 1,2,\dots$, defined by

$$P(X_n = x) = \begin{cases} \frac{1}{2} & \text{if } x = -n^{1/4} \\ \frac{1}{2} & \text{if } x = n^{1/4} \\ 0 & \text{otherwise.} \end{cases}$$

17. Let $\{X_n\}$ be a sequence of independent random variables X_n, $n = 1,2,\ldots$, defined by

$$P\{X_n = x\} = \begin{cases} \frac{1}{2}(1 - 2^{-n}) & \text{if } x = 1 \\ \frac{1}{2}(1 - 2^{-n}) & \text{if } x = -1 \\ \dfrac{1}{2^{n+1}} & \text{if } x = 2^n \\ \dfrac{1}{2^{n+1}} & \text{if } x = -2^n. \end{cases}$$

   (a) Verify that the weak law of large numbers can be applied to the sequence $\{X_n\}$.
   (b) Examine whether the Lindeberg–Feller condition holds for the sequence $\{X_n\}$.
18. Let $\{X_n\}$ be a sequence of independent random variables X_n with $E(X_n) = n$. Show that for any real numbers α, β ($\beta > \alpha$),

$$\lim_{n\to\infty} P\left\{\alpha \le \frac{X_n - n}{\sqrt{n}} \le \beta\right\} = \frac{1}{\sqrt{2\pi}} \int_\alpha^\beta e^{-(1/2)x^2}\, dx.$$

BIBLIOGRAPHY

Apostol, T. (1957). *Mathematical Analysis*, Addison-Wesley, Reading, Mass.

Borel, E. (1909). Les Probabilités dénombrables et leurs applications arithmetiques, *Rend. Cire. Mat. Palermo*, Vol. 27.

Chung, K. L. (1968). *A Course in Probability Theory*, Harcourt, Brace & World, New York.

Cramér, H. (1946). *Mathematical Methods of Statistics*, Princeton University Press, Princeton, N.J.

Cramér, H. (1970). *Random Variables and Probability Distributions*, 3rd ed., Cambridge University Press, London.

Feller, W. (1935). Über den zentralen Granzwertsatz der Wahrscheinlichkeitsrechnung, *Math. Z.*, Vol. 40.

Feller, W. (1966). *An Introduction to Probability Theory and Its Applications*, Vol. 2, Wiley, New York.

Fisz, M. (1963). *Probability Theory and Mathematical Statistics*, Wiley, New York.

Giri, N. (1974). *Introduction to Probability and Statistics*, Part II, *Statistics*, Marcel Dekker, New York.

Gnedenko, B. V. (1962). *Theory of Probability*, Chelsea, New York.

Liapunov, A. (1901). Nouvelle forme du théorème sur la limite de la probabilité, *Mem. Acad. St. Petersbourg*, Vol. 12.

Lindeberg, J. W. (1922). Über das Exponentialgesetz in der Washrscheinfichkeits-rechnung, *Math. Z.*, Vol. 15.

Loève, M. (1963). *Probability Theory*, 3rd ed., Van Nostrand Reinhold, New York.

Parzen, E. (1967). *Modern Probability Theory and Its Applications*, Wiley, New York.

5
Concepts of Statistics

5.1 INDUCTIVE INFERENCE

Experimentation and inference making are the twin essential features of general scientific methodology. A research scientist faced with the problem of studying some characteristics of a natural phenomenon performs an experiment and collects a mass of data relevant to the problem at hand. If the experiment could be repeated indefinitely under similar circumstances, the totality of data so collected would describe the totality of all possible outcomes of repeated occurrences of the phenomenon, and would in this sense yield a physical description of the phenomenon. Calling this totality of data population, and calling the mass of data collected in a single experiment sample, the problem of inference the scientist faces is that of drawing conclusions about the characteristics of the population under study on the basis of the information contained in a sample of observations from a single experiment. An inference of this type is called an *inductive inference*. Inductive inference thus proceeds from the particular (the sample) to the general (the population).

As an example of inductive inference, let us consider a process of manufacturing items that can be classified as defective or nondefective according to some accepted standard. The manufacturer will be happy if this production process does not yield more than 5% defective items. To determine the percentage of defective items, the manufacturer performs the following experiment. He or she draws out, at random, 100 produced items in a row as they come out of the production process and checks each to see whether or not it is defective. The manufacturer must now draw conclusions about the performance of the entire production process on the basis of the number of defective items observed in the sample of 100 items drawn. It is obvious that if an infinite number of such samples of 100 items each could be drawn (in other words, if the experiment could be repeated indefinitely), the totality of these items would demonstrate the relevant characteristic of the production process, namely, the percentage of defective items produced.

5.2   THE NATURE OF A STATISTICAL PROBLEM

It was not long ago that by statistics we meant the collection of data of some general description and their presentation in organized form by means of graphs and charts. For example, statistics of land holdings have long been gathered by governments for the collection of revenue. Today's statistician would, however, be inclined to claim that the present-day science of statistics is more broad-based and is generally concerned with the problems of inductive inference in relation to stochastic models describing random phenomena.

To give a general description of such problems, we note that a random phenomenon will, in general, have more than one variant, each variant being characterized by a distribution function (sometimes called its law of randomness). Let any particular variant of a phenomenon be denoted by θ and the distribution describing it by the distribution function $F_X(x|\theta)$. The density or mass function corresponding to F describes the relative frequencies of all possible outcomes of repetitions of the variant θ of the phenomenon. Let Ω be the set of all values of θ, corresponding to all possible variants of the phenomenon. Then the stochastic model for the phenomenon is a specification of the set

$$\left(\Omega, \frac{\partial}{\partial x} F_X(x|\theta)\right) \tag{5.1}$$

where for each $\theta \in \Omega$, $\partial F_X(x|\theta)/\partial x$ gives the law of randomness for the variant θ of the phenomenon. The set Ω is the index set for the distribution function or the corresponding probability mass function in the case of discrete phenomenon or the corresponding probability density function in the absolutely continuous case.

For example, the stochastic model for the production process described in Section 5.1 could appropriately be $(\Omega, p_X(x|\theta))$, where Ω is the unit interval $(0,1)$ of the real line, for every $\theta \in \Omega$ the equations

$$
p_X(x|\theta) = \begin{cases} \binom{100}{x} \theta^x (1 - \theta)^{100-x}, & x = 0,1,\ldots,100 \\ 0, & \text{otherwise} \end{cases} \tag{5.2}
$$

represent the probability of observing x defectives in a sample of 100 items drawn at random from the production process, and θ represents the probability that a single item drawn at random from the production process is defective. An example of a continuous stochastic model is $(\Omega, f_X(x|\theta))$ where $\Omega = \{\theta : \theta = (\mu, \sigma^2), -\infty < \mu < \infty, \sigma^2 > 0\}$ and

$$
f_X(x|\theta) = \begin{cases} \dfrac{1}{\sigma\sqrt{2\pi}} \exp\left[-\dfrac{1}{2}\left(\dfrac{x - \mu}{\sigma}\right)^2 \right], & -\infty < x < \infty \\ 0, & \text{otherwise.} \end{cases} \tag{5.3}
$$

Sometimes the research scientist engaged in studying a random phenomenon will be concerned with only a single variant of the phenomenon describing the population. If $\theta = \theta_0$ is this variant, we shall call θ_0 the true variant of the phenomenon under study, and the distribution $F_X(x|\theta_0)$ in the stochastic description (5.1) of the phenomenon the true distribution under study. For example, the production process described in Section 5.1 will be characterized by a specific value of θ in the stochastic description (5.2). In sum, the study of a random phenomenon is equivalent to the study of the population defined by a true distribution embedded in the set of distributions occurring in the stochastic model of the phenomenon.

It should be noted that when faced with the problem of studying a random phenomenon the scientist may not have any knowledge of the true variant of the phenomenon he encounters. A statistical problem arises when the scientist is interested in some specific behavior of the unknown true variant of the phenomenon described by the true distribution. In other words, the statement of a statistical problem is a statement regarding the true distribution of the phenomenon under study, and can formally be represented as follows:

1. Let $(\Omega, \partial F_X(x|\theta)/\partial x)$ be the stochastic model for the phenomenon.
2. Let $\theta_0 \in \Omega$ represent the true variant of the phenomenon.
3. The statement of a statistical problem is a statement $g(\theta_0)$.

For example, in the production process of Section 5.1, if θ_0 is the true variant of the process described by the stochastic model (5.2), the producer is interested in knowing if $\theta_0 \leq 0.05$; the corresponding statistical problem is the statement $g(\theta_0)$ such that

$$g(\theta_0) = \begin{cases} 1 & \text{if } \theta_0 \leq 0.05 \\ 2 & \text{if } \theta_0 > 0.05. \end{cases}$$

There are two situations for the problem given by the two values of $g(\theta_0)$; the producer is interested in finding out which of these situations hold for a specific production process.

The nature of a statistical problem depends on the nature of the statement or the nature of the function g of the true variant of the phenomenon under study. Three of the standard statistical problems are outlined below.

Testing Problems

If in the statement of a statistical problem Ω is initially partitioned into two known subsets Ω_1 and Ω_2 such that $\Omega = \Omega_1 \cup \Omega_2$ and $\Omega_1 \cap \Omega_2 = \emptyset$ (the null set), and if we define g as

$$g(\theta_0) = \begin{cases} 1 & \text{if } \theta_0 \in \Omega_1 \\ 2 & \text{if } \theta_0 \in \Omega_2 \end{cases}$$

the statistical problem is called a *problem in the testing of a hypothesis*. The physical description of a testing-type problem is that we are interested in knowing which of the two situations, $\theta_0 \in \Omega_1$ and $\theta_0 \in \Omega_2$, described by the two values of the function g is true for the phenomenon under study. In statistical terminology, each of the situations above ($\theta_0 \in \Omega_1$ and $\theta_0 \in \Omega_2$) is called a *statistical hypothesis*. The problem considered in the example in Section 5.1 is an instance of a testing-type problem.

Estimation Problems

If in the statement of a statistical problem g is defined as

$$g(\theta_0) = \theta_0$$

the problem is called a *problem in estimation*. The physical description of an estimation problem is that we are interested in knowing the exact value of the true variant of the phenomenon under study, or, in other words, the true distribution involved. As an instance of an estimation problem, in the production-process example in Section 5.1 the producer may be interested in knowing the exact value of the fraction defective θ_0 of the process described by the model (5.2), instead of knowing in which of the two intervals (0,0.05) and (0.05,1) lies θ_0, and that would be an estimation-type problem.

Multiple Classification Problems

If in the statement of a statistical problem Ω is initially partitioned into k ($k > 2$, is finite), known subsets $\Omega_1, \Omega_2, \ldots, \Omega_k$, such that $\Omega = \Omega_1 \cup \Omega_2 \cup \cdots \cup \Omega_k$; if $\Omega_i \cap \Omega_j = \emptyset$ (the null set) for $i \neq j = 1, 2, \ldots, k$; and if the function g is defined as

$$g(\theta_0) = \begin{cases} 1 & \text{if } \theta_0 \in \Omega_1 \\ 2 & \text{if } \theta_0 \in \Omega_2 \\ \vdots \\ k & \text{if } \theta_0 \in \Omega_k \end{cases}$$

the statistical problem is called a *problem in multiple classification*. The physical description of a multiple classification problem is that we are interested in knowing which of the k situations above is true for the phenomenon under study.

A statistical problem can also be classified under one of two very broad headings, outlined below.

Parametric

A statistical problem for which Ω is a subset of a finite-dimensional Euclidean space (in other words, for which each element $\theta \in \Omega$ is a finite vector whose coordinates are real numbers) is called a *parametric problem*. A parametric problem arises when we have prior knowledge of the functional form of the distribution functions F in the stochastic representation (5.1) of the phenomenon under study, and the unknown element in a distribution is the value of θ entering in the common form F of the distributions. In statistical terminology θ is called a *parameter* of the distribution F. For example, the production-process example of Section 5.1 assumed that the distribution of the number of defective items of the process followed a binomial distribution with parameter θ, and the problem was then one of testing the hypothesis $\theta \leq 0.05$.

Nonparametric Problems

Any statistical problem that is not parametric in nature is called a *nonparametric problem*. Such problems are characterized by the fact that we have no prior knowledge of the functional form F of the distribution functions in the underlying model (5.1) or that it would be unrealistic to make an assumption regarding the functional form of distributions F. For example, we may be interested in knowing if the students of the University of Montreal are as intelligent as the students of the McGill University. If we take the IQ of a student as the measure of his intelligence, the problem can be posed as testing the hypothesis $F = G$, where F and G are the distributions of the IQ's of all the students attending the University of Montreal and McGill University, respectively. We make no assumptions regarding the functional forms of the distributions F and G.

5.3 THE NATURE OF STATISTICAL INFERENCE

After a statistical problem of the form given in Section 5.2, parametric or nonparametric in nature, has been set up, the next step is to perform an experiment with a view to collecting some information in light of which the statistician intends to make his inference about the problem at hand. The result of the experiment is called a *sample*. If there are n observations arising from the experiment, we shall call the sample a sample of size n and write $(x_1, x_2, \ldots, x_n)$. Every time the experiment is repeated, we get a sample, which varies from experiment to experiment; in other words, the outcome of an experiment is uncertain. There are two types of uncertainty attached to the outcome of an experiment: (1) One type is due to the basic randomness of the outcome (whenever one throws a die the outcome is random, and thus uncertain in the sense that any of the numbers 1, 2, 3, 4, 5, 6 may appear). (2) The other type of uncertainty arises when it is not known which law of randomness governs the outcomes of the experiment. For example, if the die is fair, the numbers 1, 2, 3, 4, 5, 6 will be equally likely; on the other hand, if the die is not fair, the probabilities of the appearance of these numbers will no longer be equal. There is a basic law of randomness applicable to the outcomes of the experiment, and this law is identical with the true distribution for the problem under study.

It should be noted that if the experiment could be repeated indefinitely, the distribution of the observations in the samples so generated would be given by the true distribution underlying the problem. The statistician is allowed to perform only one experiment and obtain a sample x, which is thus a sample from the population defined by the true distribution of the problem. The sample generated by the experiment gives the statistician some partial information about this true distribution. The problem of statistical inference that the statistician faces is to use this information in the best possible manner to draw inductive conclusions about the problem at hand. There are two important aspects of the problem: design of the experiment and choice of a statistical method.

5.3.1 Design of the Experiment

The first of these aspects of inference making concerns what manner of experimentation the statistician should adopt for generating the sample.

It should be evident that the quantity and quality of information contained in a sample that is relevant for making inferences about the problem under study will substantially depend on the manner in which the experiment is performed (i.e., on the particular method of sampling). For example, let us suppose that we are interested in estimating the average income of all families in the city of Montreal. Let the total number of families in the city of Montreal be N and let us decide to draw a sample of $n < N$ families therefrom. Consider the two following methods of sampling: (1) The n families are selected at random from the population of N families; and (2) before drawing the sample we divide the population of N families into k well-defined income groups of sizes $N_1,...,N_k$ such that $\Sigma_1^k N_i = N$, and thereafter select n_i families from the ith group, $i = 1,...,k$ such that $n_i/n = N_i/N$, where $\Sigma_1^k n_i = n$. It is obvious that the second method of sampling is more informative than the first for estimating the average income. Generally speaking, the desirability of a method of sampling will depend on how far the sample it generates is representative of the population under study in terms of the problem at hand. The two methods of sampling just discussed are called simple random sampling and stratified random sampling, respectively.

Simple random sampling is a method of selecting a sample of n items out of N similar items such that each of the $\binom{N}{n}$ samples has an equal chance of being chosen. A simple random sample is drawn item by item. The items are numbered from 1 to N. A series of n random numbers is then drawn from the table of random numbers. The items that bear these numbers are selected.

Stratified random sampling is a method in which the entire population of N items is divided into k nonoverlapping subpopulations or strata of $N_1,...,N_k$ items such that $\Sigma_1^k N_i = N$. The items within each stratum are chosen so as to be as nearly alike as possible and the strata are defined so that items from different strata are as unlike as possible. From the ith stratum a sample of n_i items is drawn by using simple random sampling such that $\Sigma_1^k n_i = n$. An optimum choice of n_i is

$$\frac{n_i}{n} = \frac{N_i}{N}, \qquad i = 1,...,k.$$

The full advantage of the stratified random sampling is obtained if the values of N_i are known.

These two sampling methods are quite simple in principle and are widely used in statistical work. It is needless to mention that there are other sampling methods that are also widely used in statistics. We will not make any attempt to describe them here. The reader is referred to Cochran (1963) for descriptions and critiques of these methods.

Design for different methods of sampling adopted in practice can be classified under two broad headings:

Single-Experiment Design (Fixed-Sample-Size Design)

According to this design, the experimenter decides in advance of experimentation (or sampling) the number of units of the population to be included in the sample (in other wirds, the size of the sample). If n is the prescribed sample size, the sample consists of n observations on the n units included in the sample, each of which may be a vector with a fixed number of components. Considering each of these observations as the outcome of a subexperiment, we may say that the sample consisting of n observations is the outcome of a single experiment consisting of n subexperiments. What makes an experiment a single experiment or gives it a fixed-sample-size design is the fact that a prescribed number of subexperiments are performed generating a sample of fixed size. Examples of fixed-sample-size designs are (i) recording the daily road accidents in a city for 100 randomly selected days (sample size is 100); (ii) measuring the heights and weights of 150 randomly selected students in a school (sample size is 150 and each observation is a vector consisting of two components, height and weight); (iii) assigning five varieties of feeds, each given only once at random on any one of five specified days, to each of 10 animals of approximately the same age and measuring the gains in their body weights over a specified period of time (sample size is 10 and each observation is a vector consisting of five components corresponding to gains in weight for the five varieties).

Sequential-Experiment Design

Instead of prescribing the number of subexperiments (the size of the sample) in advance of experimentation, as in a fixed-sample-size design, the experimenter may take observations or perform subexperiments in a particular sequence, and at each stage of experimentation (or sampling) decide, on the basis of the information obtained from the observations already collected, whether to stop experimentation and take appropriate action in relation to the problem at hand, or to make further observations (continue experimentation). The number of units to be sampled at each stage of sampling (experimentation) may be one or more. Strictly, sequential sampling refers to the case where this number is one. The main features of sequential sampling are (1) the possibility of continuation of subexperiments in a sequence characterized by a stopping rule that indicates when sampling or experimentation should stop, and (2) the nature of the action that should be taken regarding the problem at hand after sampling is stopped according to the rule above.

As an example of sequential sampling, consider the following problem. A producer must install electric switches in his production plant and has to decide whether to accept or reject a particular lot of switches offered to him for purchase. He has his own specifications for acceptance of a switch for installation in his production unit. Calling a switch that does not meet his specifications defective, he decides to examine the switches picked up one by one at random from the lot. In other words, he performs sub-experiments in a sequence, each subexperiment being the random selection from the lot of one switch, which he examines to find out if it is defective. Let m_n be the number of defective switches counted by him after he performs n subexperiments (at the nth stage of sampling). He must now define his stopping rule and his actions (rejection or acceptance of the lot) after stopping experimentation. He argues that he would be prepared to accept the lot if the proportion of defective switches in the lot is reasonably small— say, smaller than a fraction p_1. On the other hand, he would certainly like to reject the lot if the proportion of defective switches in the lot is large— say, larger than a fraction p_2. Naturally, $p_1 < p_2$. He then can decide on the following scheme of stopping rule and rule of action. He chooses two constants, $A \geq 1$ and $B \leq 1$, and decides to continue sampling as long as

$$\lambda_1 = \frac{\log B - n \log[(1 - p_2)/(1 - p_1)]}{\log(p_2/p_1) - \log[(1 - p_2)/(1 - p_1)]}$$

$$< m_n < \frac{\log A - n \log[(1 - p_2)/(1 - p_1)]}{\log(p_2/p_1) - \log[(1 - p_2)/(1 - p_1)]} = \lambda_2$$

and to terminate sampling and decide to reject or accept the lot according as $m_n > \lambda_2$ or $m_n < \lambda_1$.

If performing a subexperiment (taking observations) is costly, it is expected that a sequential-sampling design properly formulated will be superior to a fixed-sample-size design, since the former takes into account the information collected at each stage of sampling. We shall bring out the implications of this observation more clearly when we consider sequential sampling in a later chapter.

It should be evident from the definition of a sequential-sampling design that sometimes it may not terminate in the taking of a definite final action. To prevent such an unhappy situation, or for other reasons, sequential-sampling designs are sometimes truncated. In truncated sequential-sampling designs the total number of subexperiments, which in a nontruncated sequential-sampling design may be infinitely large, is not allowed to exceed a certain preassigned integer N, and some well-defined rule of action is prescribed when experimentation terminates at the Nth subexperiment.

It is worth pointing out that statistically the significant difference between a fixed-sample-size design and a sequential-sampling design is that
in the former case the sample size is a fixed integer, whereas in the latter
it is a random variable taking integral values.

5.3.2 Choice of a Statistical Method

The second aspect of inference making arises after an experiment is performed
giving rise to a sample that characterizes the true variant of the phenomenon
under study. This is the choice of the manner in which the information
contained in the sample is to be used to make an inference about the problem
at hand (i.e., the statistical method to be used in making an inference). Having
a clear understanding of the nature of the inference involved, the statistician
must, on the basis of the information contained in the sample, choose an
action from the set of all possible actions available. Mathematically, this
implies that if $x = (x_1,...,x_n)$ is the sample observed by the statistician by
means of the experiment, the nature of the inference made by the statistician
is described by the choice of a function d defined over the sample space $\mathcal{X}$
consisting of all possible values of the sample x when the experiment is
repeated indefinitely, such that $d(x)$ is an element of the set A of all possible
actions. In classical statistics the choice of any particular d is referred to as a
statistical method used for making an inference. If

$$D = \{d : d(x) \in A \text{ for every } x \in \mathcal{X}\} \tag{5.4}$$

D represents the set of all statistical methods that can be used in making an
inference about a problem. A complete statement of a statistical problem,
by including the problem of inference involved, can be summarized as

$$(\Omega, F, g, A, D) \tag{5.5}$$

which is obtained by extending the statistical problem by the inclusion of the
sets A and D. We shall sometimes refer to the p problem represented by
(5.5) as the statistical inference problem associated with the statistical problem. In (5.5), $\theta \in \Omega$ represents the true variant of the phenomenon under
study, $g(\theta)$ gives the statement of the problem, and the problem of inference
refers to the choice of a $d \in D$.

5.4 TRANSITIONS IN THE HISTORY OF STATISTICAL METHODOLOGY

If one surveys the growth of the subject of statistical inference since the
inception of the science of statistics (which is relatively young, barely 60
years old), one notices at least three distinct transitional stages.

R. A. Fisher and his associates laid the foundations of the science of statistics. Fisher was responsible for developing a large number of powerful statistical methods, which even now form the basic framework of the subject of statistical inference. He introduced many important concepts essential to making inferences about statistical problems. The techniques of analysis of variance (with its associated tests of significance), regression analysis, and discriminant analysis (multiple classification), as well as the concepts of likelihood function and sufficiency, are only a few instances of Fisher's contributions to the general body of statistical methodology. The statistical methods developed by Fisher, however, suffered from one general shortcoming: They were based primarily on intuitive appeal and lacked a coherent and coordinated logical justification for universal acceptance. The justification for these methods must lie along the following lines: It should be possible to define for every suggested method a performance characteristic that describes the consequences of the acceptance of the method in the different situations that can arise for the problem at hand, so that given any two competing methods for attacking the problem it would be possible to compare their relative effectiveness in the light of their performance characteristics. Fisher's work lacks a definition of any such performance characteristics for the statistical methods he developed, although this does not in any way detract from importance of his pioneering contributions to the development of statistics.

At a later stage, it was the school of workers inspired by J. Neyman and E. S. Pearson who gave a new orientation to the study evolving statistical methods by giving concrete shape to the basic principles. They argued that a statistical method should be evaluated by the consequences it produces in various hypothetical situations that may arise. This principle was clearly emphasized for the first time by Neyman and Pearson in their theory of the testing of hypotheses, where they considered the suitabilities of a test criterion in the light of its performance as described by its power function. The choice between different tests of a hypothesis was thus based on the relative values that their power functions could take in different situations (i.e., under different alternative hypotheses). Any criterion of choice between different tests for a statistical hypothesis based on the performance of their power functions was indeed a new and completely satisfying approach, in line with the general principle of choice used in making decisions in scientific experiments, where a utility function is associated with each decision and any criterion of choice between different decisions is based on the relative performances of their utility functions.

Neyman and Pearson, however, could not extend this principle of evaluating a statistical method by its consequences in various possible circumstances to the derivation of appropriate methods for statistical problems other than the testing of hypotheses (e.g., problems of estimation and multiple classification). It was left to A. Wald to extend the foregoing principle for general decision making when faced with any type of statistical problem. This was made possible by identifying a statistical problem as a two-person zero-sum game with Nature and Statistician as the two players (a statistical game) and defining optimum decisions (statistical methods) in terms of the optimum strategies of the game. This novel approach for evaluating optimum methods for various types of statistical problems was proposed and formulated by Wald in a series of brilliant papers, culminating in his book *Statistical Decision Functions* (1950). The highlight of Wald's contribution was the establishment of an equivalence between a statistical problem and a statistical game; thereafter the derivation of optimum statistical methods for the problem in terms of optimal strategies for the statistician in the statistical game corresponding to the problem. Statistics has thereby developed a very broad perspective as an inductive science applied to the task of decision making for all types of statistical problems. By classical statistics we mean the development of statistics that took place prior to Wald's formulation of a game-theoretical approach to the solution of statistical problems.

5.5   DATA REDUCTION AND SUFFICIENCY

Let X be a random variable (or vector) whose distribution depends on $\theta \in \Omega$. As indicated in Section 1.2, the basic problem of statistical inference is to draw valid conclusions about the characteristic θ of the population characterized by the distribution function $F_X(x|\theta)$, on the basis of the information contained in the sample from a single experiment. Let $(x_1,\ldots,x_n)$ denote the observations in the sample. Every time the experiment is repeated we get a sample, which varies from one repetition to the other. In other words, the sample $(x_1,\ldots,x_n)$ is a value of the random vector $(X_1,\ldots,X_n)$, where the X_i' are independent and identically distributed random variables. We call $(x_1,\ldots,x_n)$ a *random sample of size n*. In actual practice, for statistical inference it may suffice to use a suitable function of $(X_1,\ldots,X_n)$ instead of the samples themselves. In other words, all the information contained in the sample may not be necessary for making the inference. For example, in the experiment to determine the proportion of defectives among the manufactured items (Section 5.2) the sample of 100 items contains information about the total number of defective items and also the order in which the defective items occurred. But the order in which the defective items occurred is not relevant to the problem at hand.

By selecting a suitable function of the sample observations we eliminate some irrelevant information contained in the sample. To find out what kind of information (i.e., what kind of function) is relevant to the inference problem at hand is an important problem in statistics, the solution of which involves the principle of sufficiency.

Definition 5.5.1: Statistic   A random variable (real- or vector-valued) that is a function of the random variables (vectors) $X_1,...,X_n$ is called a *statistic*.

Example 5.5.1   In the example of Section 5.2, let X_i be 1 if the ith item in the sample of 100 items is defective, and 0 if the ith item is not defective. The random variable $\sum_1^{100} X_i$ is a statistic.

Definition 5.5.2: Sufficient Statistic   Let X be a random variable (or vector) whose distribution depends on $\theta \in \Omega$. A real- or vector-valued function $T(X_1,...,X_n)$ of the random sample $X_1,...,X_n$ of size n from the distribution of X is called a *sufficient statistic* for θ if the conditional distribution of $X_1,...,X_n$ given $T = t$ is independent of θ [except possibly for a set of values of t in A such that $P_\theta(A) = 0$].

Thus $T(X_1,...,X_n)$ discards information contained in the sample that is statistically of no importance for making the inference about θ. The concept of sufficient statistics was introduced by R. A. Fisher (1922). Later Neyman (1935) provided a simple criterion, popularly called the *Fisher–Neyman factorization theorem*, to characterize the sufficient statistic.

As a word of caution we may add that when we say that a function is independent of θ, we mean not only that θ does not appear in the expression of the function, but also that the domain of the function does not involve θ.

Example 5.5.2   Let $X_1,...,X_n$ be a random sample of size n with the joint probability mass function

$$\prod_1^n P_{X_i}(x_i|\theta) = \prod_1^n \binom{N}{x_i}\theta^{\Sigma x_i}(1 - \theta)^{Nn - \Sigma x_i}, \qquad 0 < \theta < 1.$$

It is well known that $\Sigma_1^n X_i$ is a binomial random variable with parameters Nn, θ (see Chapter 3). Let

$$T(X_1,...,X_n) = \sum_1^n X_i, \qquad t = \sum_1^n x_i.$$

Clearly,

$$P(X_1 = x_1,\ldots,X_n = x_n | T = t) = \prod_1^n \binom{N}{x_i} \Big/ \binom{Nn}{\Sigma x_i} \tag{5.6}$$

is independent of θ. Hence $\Sigma_1^n X_i$ is sufficient for θ.

Example 5.5.3 Let $(X_1,\ldots,X_n)$ be a random sample of size n from a normal distribution with mean μ and variance σ^2. Let

$$\overline{X} = \frac{\Sigma_1^n X_i}{n}$$

and

$$S^2 = \sum_1^n (X_i - \overline{X})^2.$$

From Chapter 6 $\overline{X}$ and S^2 are independent in distribution, $n\overline{X}$ is normally distributed with mean $n\mu$ and variance $n\sigma^2$, and S^2/σ^2 has chi-square distribution with $n - 1$ degrees of freedom. The joint probability density function of $\overline{X}$, S^2 is given by

$$\frac{\sqrt{n}}{2^{n/2}\sqrt{\pi}\,\sigma^n\,\Gamma[(n-1)/2]}\,\exp\{-1/2\sigma^2[s^2 + n(\overline{x} - \mu)^2]\}(s^2)^{[(n-1)/2]-1}. \tag{5.7}$$

Now the conditional density of $(X_1,\ldots,X_n)$ given $S^2 = s^2$, $\overline{X} = \overline{x}$, is given by [note that $\Sigma(x_i - \mu)^2 = s^2 + n(\overline{x} - \mu)^2$]

$$\frac{\Gamma[(n-1)/2]}{(\pi)^{(n-1)/2}}\,(s^2)^{-(n-3)/2} \tag{5.8}$$

and is independent of $\theta = (\mu,\sigma^2)$. Hence $(\overline{X},S^2)$ is sufficient for θ.

If $T(X_1,\ldots,X_n) = (X_1,\ldots,X_n)$, then obviously the conditional distribution of $(X_1,\ldots,X_n)$ given T is always independent of the parameter under consideration. Thus the sample itself is always sufficient. In Examples 5.5.2 and 5.5.3 it may be noted that the dimension of the sufficient statistic is less than the dimension of the sample. Reduction of the dimension of the sample without losing information about the parameter is desirable for statistical inference. Thus a question arises as to how great a reduction in the dimension of the sample is possible without losing sufficiency.

Definition 5.5.3: Minimal sufficient statistics A sufficient statistic $T(X_1,\ldots,X_n)$ is said to be *minimal* if the sample cannot be reduced beyond T without losing sufficiency.

Explicit procedures for obtaining a minimal sufficient statistic are given by Lehmann and Scheffé (1950) and Bahadur (1954). These are beyond the scope of this book. However, it has been established that the sufficient statistic obtained through the following Fisher–Neyman factorization theorem is minimal. It is easy to conclude that the sufficient statistics obtained in Examples 5.5.2 and 5.5.3 are minimal. In what follows and throughout this book we will use the term sufficient statistic to mean minimal sufficient statistic. As a word of caution we add that our claim that the sample can be replaced by the sufficient statistic is subject to certain mild restriction on the conditional probability density or mass function.

Theorem 5.5.1: Fisher–Neyman factorization theorem (continuous chase) Let $X_1,\ldots,X_n$ be a random sample of size n from the probability density function $f_X(x|\theta)$, $\theta \in \Omega$. The random variable (or vector) $T(X_1,\ldots,X_n)$ is sufficient for θ if and only if there exists a factorization

$$\prod_1^n f_{X_i}(x_i|\theta) = f_T(t|\theta)H(x_1,\ldots,x_n), \qquad (5.9)$$

where $f_T(t|\theta)$ depends on θ and is the probability density function of T and $H(x_1,\ldots,x_n)$ does not depend on θ.

Proof. Suppose that Eq. (5.9) holds. Let $T = (T_1,\ldots,T_q)$ be a q-dimensional vector and let $Y = (Y_1,\ldots,Y_{n-q})$. Assume that there exists a one-to-one transformation

$$T_i = U_i(X_1,\ldots,X_n), \qquad i = 1,\ldots,q$$
$$Y_i = U_{q+i}(X_1,\ldots,X_n), \qquad i = 1,\ldots,n - q \qquad (5.10)$$

having the inverse transformation

$$X_i = W_i(T,y), \qquad i = 1,\ldots,n.$$

Let J denote the Jacobian of this transformation and J^* denote the Jacobian of the inverse transformation. Writing

$$t = (t_1,\ldots,t_q), \qquad y = (y_1,\ldots,y_{n-q}),$$

where

$$t_i = U_i(x_1,\ldots,x_n)$$
$$y_i = U_{q+i}(x_1,\ldots,x_n)$$
$$x_i = W_i(t,y),$$

we get from Eq. (5.9)

$$\prod_{i=1}^{n} f_{X_i}(W_i|\theta)|J| = f_T(t|\theta)H(W_1,\ldots,W_n)|J|. \tag{5.11}$$

Since the left-hand side of Eq. (5.11) is the joint probability density function of T, Y, and since $f_T(t|\theta)$ is the joint probability density function of T, $H(W_1,\ldots,W_n)$ is the conditional probability density function of T, Y given $T = t$. By assumption, $H(W_1,\ldots,W_n)$ is independent of θ. Hence T is sufficient for θ.

To prove the converse let the joint probability density function of T, Y be given by

$$f_{T,Y}(t,y|\theta) = f_T(t|\theta)f_{Y|T}(y|t), \tag{5.12}$$

where $f_{Y|T}(y|t)$ is the conditional probability density function of Y given T and is independent of θ. From (5.12) we have

$$f_{X_1\ldots X_n}(x_1,\ldots,x_n|\theta)|J^*|$$
$$= f_T(U_1,\ldots,U_q|\theta)f_{Y|T}(U_{q+1},\ldots,U_n|U_1,\ldots,U_q)|J^*|. \tag{5.13}$$

But

$$f_{X_1\ldots X_n}(x_1,\ldots,x_n|\theta) = \prod_{1}^{n} f_{X_i}(x_i|\theta).$$

Since $f_{Y|T}(y|t)$ by assumption does not depend on θ, $f_{Y|T}(U_{q+1},\ldots, U_n|U_1,\ldots,U_q)|J^*|$ is a function $H(x_1,\ldots,x_n)$ that does not depend on θ. Hence the theorem.   Q.E.D.

If $X_1,\ldots,X_n$ are discrete random variables, the proof is identical with the one given above, except that no Jacobian is involved and the integral is replaced by the sum. Regularity conditions for the existence of one-to-one transformations from $(X_1,\ldots,X_n)$ to (T,Y) are given by Tukey (1958).

Example 5.5.4   Let $X_1,\ldots,X_n$ be a random sample of size n from the Poisson distribution with parameter θ. Then

$$\prod_{i=1}^{n} f_{X_i}(x_i|\lambda) = \frac{\lambda^{\Sigma x_i}e^{-n\lambda}}{\Pi_1^n(x_i)!}$$

$$= \frac{\lambda^{\Sigma x_i}e^{-n\lambda}}{(\Sigma x_i)!}\frac{(\Sigma x_i)!}{\Pi_1^n(x_i)!}. \tag{5.14}$$

Since the first factor on the right-hand side of Eq. (5.14) is the probability density function (pdf) of

$$T(X_1,\ldots,X_n) = \sum_1^n X_i$$

and the second factor is independent of λ, from Theorem 5.5.1 we conclude that

$$T(x) = \sum_1^n X_i$$

is sufficient for λ.

Example 5.5.5 Consider once again Example 5.5.3. The joint pdf of $X_1,\ldots,X_n$ is given by

$$\frac{1}{(2\pi)^{n/2}\sigma^n} \exp\left\{ -\frac{1}{2\sigma^2} \left[\sum_1^n (x_i - \bar{x})^2 + n(\bar{x} - \mu)^2 \right] \right\}$$

$$= \frac{\sqrt{n}}{2^{n/2}\sqrt{\pi}\,\sigma^n\,\Gamma[(n-1)/2]} \exp\left\{ -\frac{1}{2\sigma^2} [s^2 + n(\bar{x} - \mu)^2] \right\}$$

$$\times (s^2)^{(n-3)/2} H(x_1,\ldots,x_n), \qquad (5.15)$$

where

$$H(x_1,\ldots,x_n) = \frac{\Gamma[(n-1)/2]}{\pi^{(n-1)/2}}(s^2)^{-(n-3)/2}.$$

Hence by Theorem 5.5.1 we conclude that $(\bar{X},S^2)$ is sufficient for (μ,σ^2).

To apply the Fisher–Neyman factorization theorem we must first of all know the pdf of the sufficient statistics. Sometimes this poses extra difficulty in the case where the pdf of the sufficient statistic is not known. The following theorem will establish that the factorization theorem holds good even if it is not a pdf of the sufficient statistic.

Theorem 5.5.2 Let $X_1,\ldots,X_n$ be a random sample of size n from a distribution with pdf $f_X(x|\theta)$, $\theta \in \Omega$. The random variable (or vector) $T(X_1,\ldots,X_n)$ is sufficient for θ if and only if we can find two nonnegative functions $g_T(t|\theta)$ (not necessarily a pdf) and $K(X_1,\ldots,X_n)$ such that

$$\prod_1^n f_{X_i}(x_i|\theta) = g_T(t|\theta)K(x_1,\ldots,x_n), \qquad (5.16)$$

where $g_T(t|\theta)$ depends on $(x_1,\ldots,x_n)$ only through t and depends on θ, and where $K(X_1,\ldots,X_n)$ is independent of θ.

Proof.   As for Theorem 5.5.1, we shall provide the proof for continuous random variables only. Suppose that Eq. (5.16) holds. Using the same notation as in Theorem 5.5.1, we get

$$\prod_1^n f_{X_i}(W_i|\theta)|J| = g_T(t|\theta)K(W_1,\ldots,W_n)|J|. \tag{5.17}$$

The left-hand side of Eq. (5.17) is the joint pdf of (T,Y). Hence the marginal pdf of T is given by

$$f_T(t|\theta) = g_T(t|\theta) \int_{-\infty}^{\infty} \cdots \int_{-\infty}^{\infty} K(W_1,\ldots,W_n)|J|dy_1,\ldots,dy_{n-q}$$

$$= g_T(t|\theta)h_T(t), \tag{5.18}$$

where

$$h_T(t) = \int_{-\infty}^{\infty} \cdots \int_{-\infty}^{\infty} K(W_1,\ldots,W_n)|J|dy_1,\ldots,dy_{n-q} \tag{5.19}$$

Obviously, $h_T(t)$ is independent of θ and $h_T(t) > 0$. Writing

$$g_T(t|\theta) = \frac{f_T(t|\theta)}{h_T(t)},$$

we get for Eq. (5.17),

$$\prod_1^n f_{X_i}(x_i|\theta) = f_T(t|\theta) \frac{K(x_1,\ldots,x_n)}{h_T(U_1,\ldots,U_q)}$$

$$= f_T(t|\theta)H(x_1,\ldots,x_n),$$

where

$$H(X_1,\ldots,X_n) = \frac{K(X_1,\ldots,X_n)}{h_T(U_1,\ldots,U_q)}$$

and is independent of θ. Hence, by Theorem 5.5.1, T is sufficient for θ.

Conversely, if T is sufficient for θ, Eq. (5.16) can be obtained by taking $g_T(t|\theta)$ to be the pdf of T.   Q.E.D.

Example 5.5.6   Let $(X_1,\ldots,X_n)$ be a random sample of size n from a population with pdf

$$f_X(x|\lambda) = \begin{cases} \lambda x^{\lambda-1}, & 0 < x < 1, \quad \lambda > 0 \\ 0, & \text{otherwise.} \end{cases} \tag{5.20}$$

Then the joint pdf of $X_1,\ldots,X_n$ is

$$\prod_1^n f_{X_i}(x_i|\lambda) = \lambda^n(x_1 \cdots x_n)^{\lambda-1}$$

if $0 < x_i < 1$ for all i. Taking $K(X_1,\ldots,X_n) = 1$, from Theorem 5.5.2 we conclude that $T(X_1,\ldots,X_n) = X_1 \cdot X_2 \cdots X_n$ is sufficient for λ.

Example 5.5.7 Let $(X_1,\ldots,X_n)$ be a random sample of size n from the waiting-time distribution with pdf

$$f_X(x|\lambda) = \begin{cases} \dfrac{1}{\lambda} e^{-x/\lambda}, & x > 0, \quad \lambda > 0 \\ 0, & \text{otherwise.} \end{cases} \tag{5.21}$$

The joint pdf of $X_1,\ldots,X_n$ is

$$\prod_1^n f_{X_i}(x_i|\lambda) = \frac{1}{\lambda^n} \exp\left(-\frac{1}{\lambda}\sum_1^n x_i\right).$$

Taking $K(X_1,\ldots,X_n) = 1$, from Theorem 5.5.2 we conclude that

$$T(X_1,\ldots,X_n) = \sum_1^n X_i$$

is sufficient for λ.

Let X be a vector-valued random variable with pdf $f_X(x|\theta)$, $\theta \in \Omega$ and let $(X_1,\ldots,X_n)$ be a sample of size n from $f_X(x|\theta)$. Then $T(X_1,\ldots,X_n)$ is sufficient for θ if and only if

$$\prod_1^n f_{X_i}(x_i|\theta) = g_T(t|\theta)K(x_1,\ldots,x_n), \tag{5.22}$$

where $g_T(t|\theta)$ is positive, depends on θ, and depends on $X_1,\ldots,X_n$ only through T, and where $K(X_1,\ldots,X_n)$ is positive and is independent of θ.

5.6 THE EXPONENTIAL FAMILY OF DISTRIBUTIONS

An important class of distributions that permit considerable reduction of the dimension of the sample through sufficiency is the exponential family. This class is defined by the probability density function (for continuous random variable)

$$f_X(x|\theta) = c(x)K(\theta) \exp\left[\sum_{j=1}^m Q_j(\theta)T_j(x)\right] \tag{5.23}$$

where c, K, Q_j, T_j are known functions. For a discrete random variable $f_X(x|\theta)$ is to be interpreted as a probability mass function.

Example 5.6.1: Binomial distribution Let X be a binomial random variable with probability mass function

$$P_X(x) = \binom{n}{x} \theta^x (1 - \theta)^{n-x} \qquad (0 < \theta < 1)$$

$$= \binom{n}{x} (1 - \theta)^n \exp\left(x \log \frac{\theta}{1 - \theta} \right). \tag{5.24}$$

The binomial distribution obviously belongs to the exponential family.

Example 5.6.2: Normal distribution Let X be normally distributed with pdf

$$f_X(x|\mu,\sigma^2) = \frac{1}{\sigma\sqrt{2\pi}} \exp\left[-\frac{1}{2\sigma^2} (x - \mu)^2 \right]$$

$$= \frac{1}{\sqrt{2\pi}} \left[\frac{1}{\sigma} \exp\left(-\frac{\mu^2}{2\sigma^2} \right) \right] \exp\left(-\frac{1}{2\sigma^2} x^2 + \frac{\mu}{\sigma^2} x \right). \tag{5.25}$$

Clearly, the normal distribution belongs to the exponential family.

Similarly, various other distributions (gamma, Poisson, multinomial, etc.) belong to the exponential family.

EXERCISES

1. Let the events $E_1, E_2,...$ be pairwise disjoint and exhaustive and let $P(E_i) > 0$ for all i. Show that for any event $A \in \cup_i E_i$,

$$P(A) = \sum_i P(A|E_i)P(E_i).$$

2. (Stratified sampling theorem) Let a population of N individuals be divided into k strata, N_i being the size of the ith stratum. A stratified random sampling is used to draw a sample of size n, n_i individuals being chosen from the ith stratum. Let

$$\sum_1^k n_i = n, \qquad \sum_1^k N_i = N.$$

Let

$$Y_{ij} = \text{income of } j\text{th individual in } i\text{th stratum}$$

$$W_i = N_i/N \text{ be } i\text{th stratum weight}$$

$$f_i = n_i/N \text{ be sampling fraction for } i\text{th stratum}$$

$$\overline{Y}_i = \text{true average income for } i\text{th stratum}$$

$$\overline{y}_i = \text{sample average income for } i\text{th stratum}$$

$$S_i^2 = \sum_1^{N_i} (Y_{ij} - \overline{Y}_i)^2/N_i - 1 = i\text{th stratum variance}$$

and let

$$\overline{y} = \sum_1^k W_i \overline{y}_i, \qquad \overline{Y} = \sum_1^k W_i \overline{Y}_i.$$

Show that:
(a) $E(\overline{y}) = \overline{Y}$.
(b) $\text{var}(\overline{y}) = \sum_1^k W_i^2 (S_i^2/n_i)(1 - f_i)$.
(c) Find $\text{var}(\overline{y})$ when $n_i/N_i \to 0$ for all i.
3. In Exercise 2 let C_i denote the cost per unit for taking the sample from the ith stratum. The total cost for taking the sample is

$$C = c_o + \sum_1^k C_i n_i$$

where c_o is the overhead cost. Show that given C, $\text{var}(\overline{y})$ is minimum if n_i is proportional to $N_i S_i/\sqrt{c_i}$.
4. (Neyman allocation) Show that for stratified random sampling, $\text{var}(\overline{y})$ is minimized for fixed n if

$$\frac{n_i}{n} = \frac{N_i S_i}{\sum_1^k N_i S_i}.$$

5. Let $(X_1,\ldots,X_n)$ be a random sample of size n from the uniform distribution with pdf

$$f_X(x|\theta) = \begin{cases} \dfrac{1}{\theta}, & 0 < x < \theta, \quad 0 < \theta < \infty \\ 0, & \text{otherwise.} \end{cases}$$

Show that $\max(X_1,\ldots,X_n)$ is sufficient for θ.

6. Let $X_1, \ldots, X_n$ be a random sample of size n from the uniform distribution over the interval (α, β), $\alpha < \beta$.
   (a) Show that $\max(X_1, \ldots, X_n)$ is a sufficient statistic for β if α is known.
   (b) Show that $\min(X_1, \ldots, X_n)$ is sufficient for α if β is known.
   (c) Show that $(\min(X_1, \ldots, X_n), \max(X_1, \ldots, X_n))$ is sufficient for (α, β) if α, β unknown.

7. Prove the Fisher–Neyman factorization theorem for discrete random variables.

8. Prove Theorem 5.5.2 for the sequence $(X_1, \ldots, X_n)$ of discrete random variables where the X_i are independent and identically distributed.

9. Let X be a Weibull random variable with pdf

$$f_X(x|\alpha, \beta) = \begin{cases} \beta\alpha x^{\beta-1} \exp(-\alpha x^\beta), & x > 0, \quad \alpha > 0, \quad \beta > 0 \\ 0, & \text{otherwise.} \end{cases}$$

If β is known, find the sufficient statistic for α on the basis of a random sample of size n.

10. Let $(X_1, \ldots, X_n)$ be a sequence of independently distributed normal random variables with unit variance and $E(X_j) = j\mu$, μ being unknown. Find the sufficient statistics for μ.

11. Let $(X_1, \ldots, X_n)$ be a random sample of size n from the exponential family of distributions. Show that $(\Sigma_1^n T_1(X_i), \ldots, \Sigma_1^n T_m(X_i))$ is sufficient for θ.

12. Let $(X_1, \ldots, X_n)$ be a random sample of size n from the normal distribution with mean 0, variance σ^2. Find the sufficient statistics for σ^2.

13. Let $(X_1, \ldots, X_n)$ be a random sample of size n from the geometric distribution with pdf

$$P_X(x|\theta) = \begin{cases} (1 - \theta)^x\theta, & x = 0,1,2,\ldots \\ 0, & \text{otherwise.} \end{cases}$$

Show that $\Sigma_1^n X_i$ is sufficient for θ. Does $P_X(x|\theta)$ belong to the exponential family?

14. Let $(X_1, \ldots, X_n)$ be a random sample from a population with beta distribution with parameter $p = \theta > 0$ and $q = 2$. Find the sufficient statistics for θ.

15. Let $(X_1, \ldots, X_n)$ be a sample of size n from a Cauchy distribution

$$f_X(x|\theta) = \begin{cases} \dfrac{1}{\pi[1 + (x - \theta)^2]}, & -\infty < x < \infty, \quad -\infty < \theta < \infty \\ 0, & \text{otherwise.} \end{cases}$$

Does θ have a sufficient statistic?

16. Let X_1, X_2, X_3 be independent identically distributed Bernoulli random variables with $P(X_i = 1) = p$. Show that $3X_1 + X_2 + 2X_3$ is not sufficient for p.

BIBLIOGRAPHY

Bahadur, R. R. (1954). Sufficiency and statistical decision functions, *Ann. Math. Stat.*, Vol. 25, pp. 423–462.

Cochran, W. G. (1963). *Sampling Techniques*, Wiley, New York.

Fisher, R. A. (1922). On the mathematical foundations of theoretical statistics, *Philos. Trans. Roy. Soc.* (London), Ser. A, Vol. 222, pp. 309–368.

Lehmann, E., and Scheffé, H. (1950). Completeness, similar regions and unbiased estimation. *Sankhya*, Vol. 10, pp. 305–340.

Neyman, J. 1932). Sur un teorema concernente: la considette statistiche sufficiente, *Giorn. Ist. Ital. AH*, Vol. 6, pp. 320–334.

Neyman, J., and Pearson, E. S. (1928). On the use and interpretation of certain test criteria for purposes of statistical inference, *Biometrika*, Vol. 20A, pp. 175–240.

Tukey, J. R. (1958). A smooth invertibility theorem, *Ann. Math. Stat.*, Vol. 29, pp. 581–584.

Wald, A. (1950). *Statistical Decision Functions*, Wiley, New York.

6
Univariate Distributions

The appropriateness of a stochastic (probabilistic) model to the explanation of a natural phenomenon admitting of more than one variant was brought out in Chapter 1. Generally speaking, each variant of a natural random phenomenon generates a random variable (which may be vector-valued) whose different values correspond to different outcomes when the particular variant of the phenomenon is repeated. The probability distributions of the random variables generated by different variants of the phenomenon define a class of probability distributions. A specification of this class of distributions offers a stochastic model for the phenomenon.

It may be argued that nature seldom acts erratically and that the class of probability distributions defining the different variants of a natural random phenomenon usually follow a regular pattern describing the stochastic model of the phenomenon. This is actually the situation in the description of most random phenomena. Some random variables occur very frequently in such descriptions. The descriptions of these random variables are thus of special interest and give rise to what are known as *standard distributions*. These distributions are sometimes given special names to highlight their importance as stochastic descriptions of random phenomena. The distributions covered here are classified into two major classes:

1. *Univariate distributions*: the binomial, Poisson, geometric, hypergeo-metric, negative binomial, uniform, normal, log-normal, chi-square

(central and noncentral), beta, Cauchy, Student's t (central and non-central), F (central and noncentral) Laplace, Rayleigh (chi), Weibull, and others.

2. *Multivariate distributions*: the multinomial and multivariate normal distributions, elliptically symmetric distributions, and the distribution of quadratic forms.

Most of the distributions listed under class 1 were discussed in Chapter 3. We introduce in this chapter only those distributions under class 1 that have not been covered in Chapter 3 (the hypergeometric, the negative binomial, the noncentral χ^2, the noncentral t, and the noncentral F distributions). In the class of univariate discrete distributions, the Poisson distribution is perhaps the most useful because of its wide applicability in the description of natural random phenomena. In Chapter 3 we have shown that the Poisson distribution can be obtained as a limiting case of binomial distributions for rare events, and we have given examples of a few natural phenomena that closely follow the Poisson distribution. Because of its wide applicability we first study this distribution in greater detail and formulate several mathematical conditions, the realization of which will give rise to it.

In addition to the standard distributions defined above, we cover in this chapter a few other distributions, which we call sampling distributions of statistics. A set of random variables $X_1,\ldots,X_N$ (vector or scalar) is said to form a random sample of size N from a distribution $F_X(x)$ of a random variable X if $X_1,\ldots,X_N$ are independent and have the same distribution as the random variable X. The probability distribution of any statistic $g(X_1,\ldots,X_N)$ of $X_1,\ldots,X_N$ is known as the sampling distribution of g. Under this class we will cover the distribution of the sample mean, the sample variances and covariances, and their functions.

6.1 STANDARD UNIVARIATE DISTRIBUTIONS

6.1.1 Further Characterization of the Poisson Distribution

As emphasized earlier, the Poisson distribution plays an important role in the description of random phenomena generating repetitions of a discrete event in a continuous interval of any kind. For example, we may observe the arrival of automobiles at a busy parking lot between certain hours. The arrival of an automobile at the parking lot is a discrete event, which takes place at a single instant of time within the continuous time period specified. Similarly, the number of incoming telephone calls at a telephone exchange within a specified period is an instance of such a phenomenon. The unit of measurement of the interval for a phenomenon need not be always time. For example, in assessing the quality of manufactured goods, say cloth, we may observe the number of defects in one yard of cloth; in this case the occurrence of a defect is a discrete event taking place at any specific location within the continuous space defined by one yard of the cloth.

Any random phenomenon generating repetitions of a discrete event E in a continuous interval can be described by the Poisson distribution if the phenomenon can be identified with a process having the following structure:

1. The continuous interval in which the phenomenon is defined can be subdivided into a large number of equal subintervals of length $h(h > 0$ is very small).
2. The probability of exactly one occurrence of the discrete event E in any of these subintervals of length h is equal to $\lambda h + r_1(h)$, where λ is a positive constant and $r_1(h)/h \to 0$ as $h \to 0$.
3. The probability of two or more occurrences of the event E in an interval of length h is equal to $r_2(h)$, where $r_2(h)/h \to 0$ as $h \to 0$.
4. The occurrence of the event E in any one interval is independent of its occurrence in any other interval.

If these assumptions are satisfied, the number of occurrences of the event E in any continuous interval of length t follows a Poisson distribution with parameter λt. In other words, if $X(t)$ denotes the number of times the discrete event E occurs in any continuous interval of length t, then

$$P(X(t) = k) = \begin{cases} \dfrac{e^{-\lambda t}(\lambda t)^k}{k!} & \text{if } k = 0,1,\ldots \\ 0 & \text{otherwise.} \end{cases} \tag{6.1}$$

To prove this, let us divide the continuous interval of length t into n subintervals of length $h = t/n$. The event that E occurs k times in the continuous interval of length t is equivalent to the event that E has occurred once in each of exactly k of these n subintervals. Hence, if h is small, the probability that E has occurred k times is approximately equal to (cf. the binomial distribution)

$$\binom{n}{k}(\lambda h)^k(1 - \lambda h)^{n-k} = \binom{n}{k}\left(\frac{\lambda t}{n}\right)^k\left(1 - \frac{\lambda t}{n}\right)^{n-k} \tag{6.2}$$

By Example 3.2.4 we get

$$P(X(t) = k) = \lim_{n \to \infty} \binom{n}{k}\left(\frac{\lambda t}{n}\right)^k\left(1 - \frac{\lambda t}{n}\right)^{n-k}$$

$$= \begin{cases} \dfrac{e^{-\lambda t}(\lambda t)^k}{k!}, & k = 0,1,\ldots \\ 0, & \text{otherwise.} \end{cases}$$

Hence $E(X(t)) = \lambda t$. Consequently, λ is the mean rate of occurrence of the discrete event E in any continuous interval of unit length.

6.1.2 Polya and Hypergeometric Distributions

Consider an urn containing N balls of which $a = Np$ are white and $b = Nq$, where $p + q = 1$, are black. A sample of n balls is drawn from this urn according to the following scheme (known as the Polya scheme): The first ball is drawn at random and its color is noted, and before the second drawing from the urn we replace the one we have drawn and add s balls of the same color as the first one. This is continued n times. Let X denote the number of white balls in the sample of size n. Then X is a discrete random variable with probability mass function

$$p_X(k) =$$
$$\begin{cases} \binom{n}{k} \dfrac{a(a + s) \cdots [a + (k - 1)s]b(b + s) \cdots [b + (n - k - 1)s]}{N(N + s) \cdots (N + (n - 1)s)} \\ \quad \text{if } k = 0,\ldots,n \\ 0 \quad \text{otherwise.} \end{cases}$$

$$(6.3)$$

The distribution of the random variable X with the probability mass function above is known as the Polya distribution. It is clear that

$$\sum_{k=0}^{n} p_X(k) = 1.$$

An example of a random phenomenon that generates the Polya distribution is any epidemic disease, where the occurrence of the event increases the probability of its reoccurrence. Here we have

$$E(X) = \sum_{k=0}^{n} k p_X(k)$$
$$= \frac{na}{N} \sum_{l=0}^{n-1} \binom{n-1}{l}$$
$$\times \frac{(a + s) \cdots [a + ls]b(b + s) \cdots [b + (n - l - 2)s]}{(N + s) \cdots [N + (n - 1)s]}.$$

Now suppose that we have an urn of $N + s$ balls of which $a + s$ are white and b are black and we have drawn a sample of $(n - 1)$ balls according to the Polya scheme. The probability that the sample contains l white balls is given by

$$\binom{n-1}{l} \frac{(a+s)\cdots(a+ls)b(b+s)\cdots[b+(n-l-2)s]}{(N+s)\cdots[N+(n-1)s]}.$$

Thus

$$\sum_{l=0}^{n-1} \binom{n-1}{l} \frac{(a+s)\cdots(a+ls)b(b+s)\cdots[b+(n-l-2)s]}{(N+s)\cdots[N+(n-1)s]} = 1.$$

Hence

$$E(X) = np.$$

By a similar argument we get

$$E(X(X-1)) = \frac{n(n-1)a(a+s)}{N(N+s)} = \frac{n(n-1)p(p+s/N)}{1+s/N}$$

so that

$$\begin{aligned}
V(X) &= E(X(X-1) + E(X) - E^2(X) \\
&= \frac{n(n-1)p(p+s/N)}{1+s/N} + np(1-np) \\
&= npq\,\frac{1+ns/N}{1+s/N}.
\end{aligned}$$

If $s = -1$, we have the case of sampling without replacement and the Polya distribution reduces to the hypergeometric distribution. Defining

$$(a)_{(k)} = a(a-1)\cdots(a-k+1)$$
$$(a)^{[k]} = a(a+1)\cdots(a+k-1),$$

the probability-generating function of the hypergeometric distribution is given by

$$\begin{aligned}
P_X(t) &= \sum_{k=0}^{n} t^k \binom{n}{k} \frac{(a)_{(k)}(b)_{(n-k)}}{(N)_{(n)}} \\
&= \frac{(b)_{(n)}}{(N)_{(n)}} \sum_{k=0}^{n} \frac{(-n)^{[k]}(-Np)^{[k]}}{(Nq-n+1)^{[k]}} \frac{t^k}{k!}.
\end{aligned} \qquad (6.4)$$

If n is very large, then

$$P_X(t) = \frac{(b)_{(n)}}{(N)_{(n)}} F(-n,-np;(Nq-n+1),t),$$

where $F(\alpha,\beta;\delta,t)$ is the hypergeometric series and is given by

$$F(\alpha,\beta: \delta,t) = 1 + \frac{\alpha\beta}{\delta} t + \frac{\alpha(\alpha + 1)\beta(\beta + 1)}{\delta(\delta + 1)} \frac{t^2}{2!} + \cdots . \tag{6.5}$$

The name *hypergeometric distribution* has been derived from the fact that its probability-generating function can be expressed in terms of the hypergeometric series.

6.1.3 Negative Binomial Distribution

Consider a natural random phenomenon that leads to two alternatives every time the phenomenon occurs. The sex (male or female) of a newborn baby is one such phenomenon. Without any loss of generality, we can identify the phenomenon with Bernoulli trials with two outcomes, to be called success and failure to represent the two alternatives of a single occurrence of the phenomenon. Let p be the probability of success in a single trial, different values of p defining different variants of the phenomenon. Obviously, the probability of failure in a single trial is $1 - p = q$. If we define X as the number of failures encountered in a sequence of independent repeated Bernoulli trials to achieve the rth success for the first time, the random variable X is called a negative binomial random variable with parameters r, p. It is easy to verify that its probability mass function is given by

$$p_X(x) = \begin{cases} \binom{r + x - 1}{x} p^r q^x = \binom{-r}{x} p^r(-q)^x & \text{if } x = 0,1,\ldots \\ 0, & \text{otherwise.} \end{cases} \tag{6.6}$$

Its characteristic function is given by

$$\phi_X(t) = \sum_{x=0}^{\infty} e^{itx} \binom{r + x - 1}{x} p^r q^x = p^r(1 - qe^{it})^{-r}.$$

Hence

$$E(X) = \frac{rq}{p}, \qquad V(X) = \frac{rq}{p^2}.$$

6.1.4 Rayleigh Distributions

Let X be a positive random variable with pdf

$$f_X(x|\theta) = \theta^{-2}x \exp\left(\frac{x^2}{2\theta}\right), \qquad x > 0, \quad \theta > 0.$$

It is known as the pdf of the Rayleigh distribution. It gives the pdf of a random variable representing the waiting time for the failure of certain types of equipments.

Let $\{X(t), t > 0\}$ be a family of random variables representing a narrowband noise voltage. Then $X(t)$ can be written (Rice, 1945) as

$$X(t) = X_c(t) \cos \omega t + X_s(t) \sin \omega t,$$

where ω is a known frequency and $X_c(t)$, $X_s(t)$ are independently distributed normal random variables with the same mean zero and the same unit variance. The envelope of $X(t)$ is defined by

$$R(t) = [X_c^2(t) + X_s^2(t)]^{1/2}$$

and $R(t)$ has the Rayleigh distribution with parameter $\theta = 1$. A Rayleigh distribution with $\theta = 1$ is also known as the *chi distribution* with 2 degrees of freedom. A chi distribution with n degrees of freedom is defined by the distribution of the positive square root of the chi-square distribution with n degrees of freedom.

6.1.5   Noncentral Chi-Square Distribution

Let X be a normal random variable with mean μ and variance σ^2, and let $Y = X^2/\sigma^2$. The distribution function of Y is given by ($y > 0$)

$$F_Y(y) = P(-\sigma\sqrt{y} \le X \le \sigma\sqrt{y})$$
$$= F_X(\sigma\sqrt{y}) - F_X(-\sigma\sqrt{y}).$$

Hence the probability density function of Y is

$$f_Y(y) = \frac{\{f_X(\sigma\sqrt{y}) + f_X(-\sigma\sqrt{y})\}\sigma}{2\sqrt{y}}$$

$$= \frac{1}{2\sqrt{2\pi}} y^{-1/2} \exp\left[-\frac{1}{2}\left(\sqrt{y} - \frac{\mu}{\sigma}\right)^2 \right]$$

$$+ \frac{1}{2\sqrt{2\pi}} y^{-1/2} \exp\left[-\frac{1}{2}\left(\sqrt{y} + \frac{\mu}{\sigma}\right)^2 \right]$$

$$= \frac{1}{2\sqrt{2\pi}} \exp\left[-\frac{1}{2}\left(y + \frac{\mu^2}{\sigma^2}\right) \right] y^{-1/2}$$

$$\times \left[\exp\left(\frac{\mu}{\sigma}\sqrt{y}\right) + \exp\left(-\frac{\mu}{\sigma}\sqrt{y}\right) \right]$$

$$= \frac{1}{2\sqrt{2\pi}} \exp\left[-\frac{1}{2}\left(y + \frac{\mu^2}{\sigma^2}\right) \right] y^{-1/2} \cosh\left(\frac{\mu}{\sigma}\sqrt{y}\right)$$

$$= \frac{1}{\sqrt{2\pi}} \exp\left[-\frac{1}{2}\left(y + \frac{\mu^2}{\sigma^2}\right)\right] y^{(1/2)-1}$$

$$\times \sum_{j=0}^{\infty} \frac{[(\mu^2/\sigma^2)y]^j}{(2j)!}.$$

The distribution of Y is known as the noncentral chi-squre distribution $\chi_1^2(\lambda)$ with noncentrality parameter $\lambda = \mu^2/\sigma^2$ and with one degree of freedom. Obviously, if $\lambda = 0$ ($\mu = 0$), $\chi_1^2(\lambda)$ reduces to the central chi-square distribution χ_1^2 with one degree of freedom. The probability density function of a noncentral chi-square random variable $Y = \chi_n^2(\lambda)$ with parameter λ and with n degrees of freedom is given by

$$f_Y(y) = \frac{1}{\sqrt{\pi}} 2^{-n/2} \exp\left[\frac{1}{2}(y + \lambda)\right] y^{(n/2-1)} \sum_{j=0}^{\infty} \frac{(\lambda y)^j \Gamma(j + \frac{1}{2})}{(2j)! \Gamma(j + n/2)}. \quad (6.7)$$

If $\lambda = 0$, $f_Y(y)$ gives the distribution of the central chi-square χ_n^2 (or simply the chi-square) with n degrees of freedom.

Let $X_1,\dots,X_n$ be independently distributed normal random variables with $E(X_i) = \mu_i$, $V(X_i) = \sigma_i^2$, $i = 1,\dots,n$. Then

$$Z = \sum_{i=1}^{n} \frac{X_i^2}{\sigma_i^2}$$

is distributed as the noncentral chi-square with n degrees of freedom and noncentrality parameter

$$\mu^2 = \sum_{i=1}^{n} \frac{\mu_i^2}{\sigma_i^2}.$$

To show this, let us denote $Y_i = X_i/\sigma_i$, $i = 1,\dots,n$. Clearly, $Y_1,\dots,Y_n$ are independently distributed normal random variables with variance 1 and $E(Y_i) = \mu_i/\sigma_i = \mu_i$, where μ_i is an arbitrary designation.

Let O be an orthogonal $n \times n$ matrix, given by

$$O = \begin{bmatrix} \frac{\mu_1}{\mu} & \cdots & \frac{\mu_n}{\mu} \\ O_{21} & \cdots & O_{2n} \\ O_{n1} & \cdots & O_{nn} \end{bmatrix}$$

with $OO' = O'O = I$, and let

$$Z_i = \sum_{j=1}^{n} O_{ij} Y_j, \qquad i = 1,\dots,n$$

with $O_{1j} = \mu_j/\mu$, $j = 1,\ldots,n$. Now

$$E(Z_1) = \frac{\mu_1^2 + \cdots + \mu_n^2}{\mu} = \mu$$

$$E(Z_i) = O_{i1}\mu_1 + \cdots + O_{in}\mu_n$$

$$= \mu\left(O_{i1}\frac{\mu_1}{\mu} + \cdots + O_{in}\frac{\mu_n}{\mu}\right)$$

$$= 0, \qquad i = 2,\ldots,n$$

$$v(Z_i) = \sum_{j=1}^{n} O_{ij}^2 V(Y_i) + \sum_{j}\sum_{\substack{j' \\ j \neq j'}} O_{ij}O_{ij'} \, \mathrm{cov}(Y_j Y_{j'})$$

$$= \sum_{j=1}^{n} O_{ij}^2$$

$$= 1, \qquad i = 1,\ldots,n$$

$$\mathrm{cov}(Z_i Z_{i'}) = \sum_{j=1}^{n} O_{ij}O_{i'j}V(Y_j) + \sum_{j=1}^{n}\sum_{j'=1}^{n} O_{ij}O_{i'j'} \, \mathrm{cov}(Y_J Y_{j'})$$

$$= \sum_{j=1}^{n} O_{ij}O_{i'j}$$

$$= 0 \qquad \text{if } i \neq i'.$$

Furthermore,

$$\sum_{i=1}^{n} Z_i^2 = \sum_{i=1}^{n}\left(\sum_{j=1}^{n} O_{ij}Y_j\right)\left(\sum_{j'=1}^{n} O_{ij'}Y_{j'}\right)$$

$$= \sum_{j=1}^{n}\sum_{j'=1}^{n}\left(\sum_{i=1}^{n} O_{ij}O_{ij'}\right)Y_j Y_{j'}$$

$$= \sum_{j=1}^{n} Y_j^2.$$

Since $\mathrm{cov}(Z_i Z_{i'}) = 0$ for all $i \neq i'$, and since the Z_i are normally distributed, it follows that $Z_1,\ldots,Z_n$ are independently normally distributed random variables with variance 1 and $E(Z_1) = \mu$, $E(Z_j) = 0$, $j = 2,\ldots,n$. Hence, letting $X = Z_1^2$, $Y = \Sigma_{i=2}^{n} Z_i^2$, we get that X has the noncentral chi-square distribution with one degree of freedom and noncentrality parameter μ^2, Y has chi-square distribution with $n - 1$ degrees of freedom, and X

and Y are independent. Hence the joint probability density of X and Y is given by

$$\frac{\exp[\frac{1}{2}(\mu^2 + x + y)]}{\sqrt{2\pi}\, 2^{(n-1)/2}\Gamma[(n-1)/2]} \sum_{j=0}^{\infty} \frac{(\mu^2)^j}{(2j)!} x^{j+(1/2)-1} y^{[(n-1)/2]-1}.$$

Thus from results in Section 3.8 it follows that the probability density function of $Z = X + Y$ is given by

$$f_Z(z) = \exp\frac{[-\frac{1}{2}(\mu^2 + z)]}{\sqrt{\pi}\, 2^{n/2}} z^{(n/2)-1} \sum_{j=0}^{\infty} \frac{(\mu^2)^j z^j \Gamma(j + \frac{1}{2})}{(2j)!\Gamma[(n/2) + j]}$$

$$\times \int_0^1 \frac{\Gamma[(n/2) + j]\mu^{j+(1/2)-1}}{\Gamma[j + \frac{1}{2}]\Gamma[(n-1)/2]} (1 - u)^{(n-1)/2-1}\, du$$

$$= \exp\frac{[-\frac{1}{2}(\mu^2 + z)]z^{(n/2)-1}}{\sqrt{\pi}\, 2^{n/2}} \sum_{j=0}^{\infty} \frac{(\mu^2)^j z^j \Gamma(j + \frac{1}{2})}{(2j)!\Gamma[(n/2) + j]},$$

where $u = x/z$.

Using the fact that the characteristic function of χ_n^2 is $(1 - 2it)^{-n/2}$ and $\Gamma(2k + 1)\sqrt{\pi} = 2^{2k}\Gamma(k + \frac{1}{2})\Gamma(k + 1)$, we obtain the characteristic function of the noncentral chi-square $Y = \chi_n^2(\lambda)$ as follows.

$$\phi_Y(t) = \frac{1}{\sqrt{\pi}\, 2^{n/2}} e^{-\lambda/2} \sum_{j=0}^{\infty} \frac{\lambda^j \Gamma(j + \frac{1}{2})}{(2j)!\Gamma(j + n/2)} \int_0^{\infty} e^{-(1/2)y} y^{n/2+j-1} e^{ity}\, dy$$

$$= \frac{1}{\sqrt{\pi}} e^{-\lambda/2} \sum_{j=0}^{\infty} \frac{\lambda^j}{(2j)!} 2^j (1 - 2it)^{-(n/2)-j} \Gamma(j + \frac{1}{2})$$

$$= e^{-\lambda/2} \sum_{j=0}^{\infty} \frac{(\lambda/2)^j}{j!} (1 - 2it)^{-n/2-j}$$

$$= (1 - 2it)^{-n/2} e^{it\lambda/(1-2it)}. \tag{6.8}$$

From this it follows that if $Y_1,\ldots,Y_k$ are independently and identically distributed noncentral $\chi_{n_i}^2(\lambda_i)$, then

$$\Phi_{\Sigma_1^k Y_i}(t) = (1 - 2it)^{-\Sigma_{i=1}^k (n_i/2)} \exp\left(\frac{it}{1 - 2it} \sum_{i=1}^k \lambda_i\right),$$

and hence from the uniqueness of the characteristic function we get that $\Sigma_{i=1}^k Y_i$ is a noncentral chi-square random variable with noncentrality $\Sigma_{i=1}^k \lambda_i$ and $\Sigma_{i=1}^k n_i$ degrees of freedom. From Eq. (6.8) it follows that $E(\chi_n^2(\lambda)) = n + \lambda$, $v(\chi_n^2(\lambda)) = 2n + 4\lambda^2$.

The distribution of X^2 with $\mu = 0$ is known as the Rayleigh distribution in theoretical physics and its probability density function is given by

$$
f_{X^2}(y) = \begin{cases} \dfrac{1}{\sigma\sqrt{2\pi}} \exp\left(\dfrac{1}{2}\dfrac{y}{\sigma^2}\right) y^{(1/2)} - 1 & \text{if } y > 0 \\ 0 & \text{otherwise} \end{cases}
$$

6.1.6 Noncentral *t* Distribution

Let the random variable X distributed normally with mean μ and variance σ^2 and the random variable Y/σ^2 having a chi-square distribution with n degrees of freedom be independent and let $t = \sqrt{n}X/\sqrt{Y}$. The probability density function of t is given by

$$
f_t(y) = \frac{n^{n/2}}{\sqrt{\pi}\,\Gamma(n/2)} \frac{e^{-\lambda^2/2}}{(n + y^2)^{(n+1)/2}} \sum_{j=0}^{\infty} \Gamma\left(\frac{n + j + 1}{2}\right) \frac{\lambda^j}{j!} \left(\frac{\sqrt{2}y^2}{\sqrt{n + y^2}}\right)^j,
$$

$$(6.9)$$

where $\lambda = \mu/\sigma$. The distribution function corresponding to this probability density function is known as the noncentral t distribution with n degrees of freedom and with noncentrality parameter λ.

To prove it, let us write $t = X_1/\sqrt{X_2}$, where $X_1 = \sqrt{n}X/\sigma$, $X_2 = Y/\sigma^2$. Since X_1, X_2 are independent, the joint probability density function of X_1, X_2,

$$
f_{X_1, X_2}(x_1, x_2) = \frac{1}{\sqrt{2n\pi}} \frac{(\frac{1}{2})^{n/2}}{\Gamma(n/2)} \exp\left[-\frac{1}{2n}(x_1 - \lambda\sqrt{n})^2\right]
$$
$$
\times \exp\left(-\frac{1}{2}x_2\right) x_2^{n/2 - 1}
$$

By results in Section 3.9, the probability density function of t is given by

$$
f_t(y) = \int_0^\infty \frac{1}{\sqrt{2n\pi}} \frac{(\frac{1}{2})^{n/2}}{\Gamma(n/2)} \exp\left[-\frac{1}{2n}(y\sqrt{x_2} - \lambda\sqrt{n})^2\right]
$$
$$
\times \exp\left(-\frac{1}{2}x_2\right) x_2^{(n-1)/2} \, dx_2
$$
$$
= \frac{(\frac{1}{2})^{n/2}}{\sqrt{2n\pi}\,\Gamma(n/2)} e^{-(1/2)\lambda^2} \sum_{j=0}^{\infty} \frac{(y\lambda/\sqrt{n})^j}{j!}
$$
$$
\times \int_0^\infty \exp\left[-\frac{1}{2}x_2\left(1 + \frac{y^2}{n}\right)\right] x_2^{[(n+1+j)/2]-1} \, dx_2
$$

$$= \frac{(\frac{1}{2})^{n/2}}{\sqrt{2n\pi}\,\Gamma(n/2)}\,e^{-(1/2)\lambda^2}\sum_{j=0}^{\infty}\frac{(y\lambda/\sqrt{n})^j\,\Gamma[(n+1+j)/2]}{j!(\frac{1}{2}(1+y^2/n))^{(n+1+j)/2}}$$

$$= \frac{n^{n/2}e^{-\lambda^2/2}}{\sqrt{\pi}(n+y^2)^{(n+1)/2}}\sum_{j=0}^{\infty}\frac{\lambda^j\Gamma[(n+1+j)/2]}{j!\,\Gamma(n/2)}\left(\frac{\sqrt{2}y^2}{\sqrt{n+y^2}}\right)^j.$$

If $\lambda = 0$, the distribution of t is known as *Student's t distribution* with n degrees of freedom and is given by

$$f_t(y) = \begin{cases} \dfrac{1}{\sqrt{n\pi}}\dfrac{\Gamma[(n+1)/2]}{\Gamma(n/2)}\left(1+\dfrac{y^2}{n}\right)^{-(n+1)/2} & \text{if } -\infty < y < \infty \\ 0 & \text{otherwise.} \end{cases}$$

$$(6.10)$$

Student's t distribution was derived independently in Chapter 3.

6.1.7   Noncentral *F* Distribution

Let the random variable X, distributed as $\chi_m^2(\lambda)$, and the random variable Y, distributed as χ_n^2, be independent and let

$$F = \frac{n}{m}\frac{\chi_m^2(\lambda)}{\chi_n^2}.$$

Then the distribution of F is known as the noncentral F distribution with noncentrality parameter λ and degrees of freedom m, n. Its probability density function is given by

$$f_F(z) =$$
$$\begin{cases} \dfrac{m}{n}e^{-\lambda/2}\displaystyle\sum_{j=0}^{\infty}\frac{(\lambda/2)^j}{j!}\frac{\Gamma[j+(m+n)/2][(m/n)z]^{(m/2)+j-1}}{\Gamma(j+m/2)\Gamma(n/2)[1+(m/n)z]^{[(m+n)/2]+j}} \\ \qquad \text{if } z > c \\ 0, \qquad \text{otherwise.} \end{cases}$$

$$(6.11)$$

To prove it, we have the joint probability density function of X and Y given by (with X and Y independent by assumption)

$$f_{X,Y}(x,y) = \frac{1}{\sqrt{\pi}\,\Gamma(n/2)2^{(m+n)/2}}\exp\left[-\frac{1}{2}(x+y+\lambda)\right]x^{(m/2)-1}y^{(n/2)-1}$$
$$\times\sum_{j=0}^{\infty}\frac{(\lambda x)^j\Gamma(j+\frac{1}{2})}{(2j)!\,\Gamma(j+m/2)}.$$

By results in Section 3.9 and from the fact that $\sqrt{\pi}\Gamma(2j + 1) = \Gamma(j + \frac{1}{2})\Gamma(j + 1)2^{2j}$, we have

$f_F(z)$

$$= \frac{m}{n\sqrt{\pi}2^{(m+n)/2}\Gamma(n/2)} e^{-\lambda/2} \sum_{j=0}^{\infty} \frac{\lambda^j \Gamma(j + \frac{1}{2})}{(2j)!\, \Gamma(j + m/2)}$$

$$\times \int_0^{\infty} \left(\frac{m}{n} z\right)^{(m/2)+j-1} \exp\left[-\frac{1}{2} y\left(1 + \frac{m}{n} z\right)\right] dy$$

$$= \begin{cases} \dfrac{m}{n} e^{-\lambda/2} \displaystyle\sum_{j=0}^{\infty} \frac{(\lambda/2)^j}{j!} \frac{\Gamma[(m + n)/2 + j][(m/n)z]^{(m/2)+j-1}}{\Gamma[(m/2) + j]\Gamma(n/2)[1 + (m/n)z]^{(m+n)/2+j}} \\ \qquad \text{if } z > 0 \\ 0, \qquad \text{otherwise.} \end{cases}$$

When $\lambda = 0$ the distribution of Z is called the central F distribution (or simply the F distribution) with degrees of freedom m, n which was derived independently in Chapter 3.

6.2   SAMPLING DISTRIBUTION
6.2.1   Univariate Normal Case: Distribution of Sample Mean and Sample Variance

Let $X_1,\ldots,X_N$ be a sample of size N from the univariate normal population with mean μ and variance σ^2, and let

$$\overline{X} = \sum_{i=1}^{N} \frac{X_i}{N}$$

$$S = \sum_{i=1}^{N} (X_i - \overline{X})(X_i - \overline{X}) = \sum_{i=1}^{N} (X_i - \overline{X})^2. \tag{6.12}$$

Then $\overline{X}$ and S^2/N are called the sample mean and sample variance, respectively. We are interested in finding the joint distribution of $\overline{X}$, S.

Theorem 6.2.1.1   Let $X_1,\ldots,X_N$ be independent and identically distributed univariate normal random variables with mean μ and variance σ^2, and let $\overline{X}$ and S be as defined in Eq. (6.12). Then $\sqrt{N}\overline{X}$, S are independent and $\sqrt{N}\overline{X}$ is normally distributed with mean $\sqrt{N}\mu$ and variance σ^2, and S/σ^2 is chi-square distributed with $N - 1$ degrees of freedom.

Proof. Let O be an orthogonal matrix of dimension of $N \times N$ and be of the form

$$O = \begin{bmatrix} \dfrac{1}{\sqrt{N}} & \cdots & \dfrac{1}{\sqrt{N}} \\ O_{21} & \cdots & O_{2N} \\ \vdots & \cdots & \vdots \\ O_{N1} & \cdots & O_{NN} \end{bmatrix}. \tag{6.13}$$

The first row of O is the equiangular vector of unit length. Since $OO' = I$ we get (with $O_{1j} = 1/\sqrt{N}$, $j = 1,...,N$)

$$\sum_{j=1}^{N} O_{ij}O_{i'j} = \begin{cases} 1 & \text{if } i = i' \\ 0 & \text{if } i \neq i' \end{cases} \tag{6.14}$$

and from $O'O = I$ we get

$$\sum_{i=1}^{N} O_{ij}O_{ij'} = \begin{cases} 1 & \text{if } j = j' \\ 0 & \text{if } j \neq j'. \end{cases} \tag{6.15}$$

Let

$$Y_i = \sum_{j=1}^{N} O_{ij}X_j, \qquad i = 1,...,N.$$

Then

$$E(Y_1) = \frac{\mu}{\sqrt{N}} + \cdots + \frac{\mu}{\sqrt{N}} = \sqrt{N}\mu$$

$$E(Y_i) = \sum_{j=1}^{N} O_{ij}\mu = \sqrt{N}\mu \sum_{j=1}^{N} O_{ij} \frac{1}{\sqrt{N}} = 0, \qquad i = 2,...,N.$$

Since $X_1,...,X_N$ are independent, we get

$$V(Y_i) = \sum_{j=1}^{N} O_{ij}^2 v(X_j) + \sum_{\substack{j=1 \\ j \neq j'}}^{N} \sum_{j'=1}^{N} O_{ij}O_{ij'} \, \mathrm{cov}(X_j X_{j'})$$

$$= \sum_{j=1}^{N} O_{ij}^2 \sigma^2 = \sigma^2$$

$$\text{cov}_{i \neq i'}(Y_i Y_{i'}) = \sum_{j=1}^{N} O_{ij} O_{i'j} V(X_j) + \sum_{j'=1}^{N} \sum_{\substack{j=1 \\ j \neq j'}}^{N} O_{ij} O_{i'j'} \text{cov}(X_j X_{j'})$$

$$= \sigma^2 \sum_{j=1}^{N} O_{ij} O_{i'j}$$

$$= 0$$

in view of relations (6.15). Thus $Y_1,\ldots,Y_n$ are independent and are normally distributed with $E(Y_1) = \sqrt{N}\mu$, $E(Y_i) = 0$, $i = 2,\ldots,N$, $v(Y_i) = \sigma^2$ for all i. Furthermore,

$$\sum_{j=1}^{N} Y_j^2 = \sum_{i=1}^{N} \sum_{j=1}^{N} O_{ij} X_j \sum_{j'=1}^{N} O_{ij'} X_{j'}$$

$$= \sum_{j=1}^{N} \sum_{j'=1}^{N} \left(\sum_{i=1}^{N} O_{ij} O_{ij'} \right) X_j X_{j'}$$

$$= \sum_{j=1}^{N} X_j^2.$$

in view of relations (6.14) and (6.15). Equivalently, we have

$$\sum_{i=2}^{N} Y_j^2 = \sum_{j=1}^{N} X_j^2 - N\overline{X}^2 = \sum_{j=1}^{N} (X_j - \overline{X})^2.$$

Since $Y_1,\ldots,Y_N$ are independent, it follows that $Y_1 = \sqrt{N}\overline{X}$ and

$$\sum_{i=2}^{N} Y_i^2 = \sum_{i=1}^{N} (X_j - \overline{X})^2$$

are independent and that $\sqrt{N}\overline{X}$ is normal with mean $\sqrt{N}\mu$ and variance σ^2 and that

$$\frac{S^2}{\sigma^2} = \sum_{j=2}^{N} \left(\frac{Y_j}{\sigma} \right)^2$$

being the sum of the squares of $N - 1$ independent unit normal random variables, is chi-square distributed with $N - 1$ degrees of freedom.   Q.E.D.

Corollary 6.2.1.1.1 $\sqrt{N}(\overline{X} - \mu)/\sqrt{(S/N - 1)}$ has Student's t distribution with $N - 1$ degrees of freedom.

Proof. The proof follows trivially from results in Section 6.1.6.
                                                                    Q.E.D.

EXERCISES

1. Let X, Y be two independent beta random variables with pdf $B(p_1, q_1)$, $B(p_2, q_2)$, respectively. Find the joint distribution of $Y_1 = X$, $Y_2 = Y(1 - X)$.

2. Let X, Y be two independent gamma random variables with pdf $G(1, n)$, $G(1, n + \frac{1}{2})$, respectively. Show that $Z = 2\sqrt{XY}$ is distributed $G(1, 2n)$.

3. Let X, Y be independent where X is $N(\mu, \sigma^2)$ and Y is $N(\theta, \tau^2)$. Let $Z = X + Y$. Find the conditional pdf of X given $Z = z$.

4. Let $X_1, \ldots, X_n$ be independently and identically distributed $N(\mu, \sigma^2)$. Define

$$\overline{X} = \frac{1}{n} \sum_{i=1}^{n} X_i; \ Z_i = X_i - \overline{X}, \qquad i = 1, \ldots, n.$$

   (a) Show that $\overline{X}$ is independent of $(Z_2, \ldots, Z_n)$.
   (b) Using (a) show that $\overline{X}$, $S^2 = \sum_{i=1}^{n} (X_i - \overline{X})^2$ are independent.

5. Let $X_1, \ldots, X_n$ be independently distributed random variables and let $S_m = \sum_{i=1}^{m} X_i$, $m < n$. Show that the joint distribution of X_i and S_m does not depend on $i \leq m$.

6. Let $X_1, \ldots, X_n$ be independently and identically distributed normal random variables with $E(X_i) = \mu_i$, $\mathrm{var}(X_i) = 1$ for $i = 1, \ldots, n$. Let $\delta^2 = \sum_{i=1}^{m} \mu_i^2$. Show that the pdf of $U = \sum_{i=1}^{n} X_i^2$ is given by $f_U(i) = \sum_{i=1}^{\infty} p_V(i) f(x_{n+2i}^2)$, where $p_v(i)$ is the pdf of a Poisson random variable with parameter $\frac{1}{2} \delta^2$, $f(\chi_{n+2i}^2)$ is the pdf of a central chi-square with $n + 2i$ degrees of freedom and V is independent of χ_{n+2i}^2. (*Note*: This is also known as the pdf of a noncentral chi-square random variable with n degrees of freedom and noncentrality parameter δ^2.)

7. Let $X_1, \ldots, X_n$ be independently, identically distributed $N(\mu, \sigma^2)$. Show that $U = \sum_j \sum_j (X_i - X_j)^2$ and $W = \sum_{i=1}^{n} X_i$ are independent. Find the distribution of U.

8. Show that if X, Y are independent random variables (univariate) such that $X + Y$ is independent of $X - Y$, then both X and Y are normal random variables.

9. Show that if X, Y are independent and identically distributed normal random variables with mean 0 and variance 1, then:
   (a) $X^2 + Y^2$ and X/Y are independent random variables.
   (b) X/Y has Cauchy distribution.

10. Show that if X, Y are two independent chi-square random variables with m, n degrees of freedom, respectively, the random variables $X + Y$ and X/Y are independent.

11. (Logistic distribution) A random variable X is said to have logistic distribution function $F_X(x)$ if

$$F_X(x) = (1 + e^{-(ax+b)})^{-1}, \qquad -\infty < x < \infty$$

where $a > 0$, b are real constants. Show that the corresponding probability density function is

$$f_X(x) = aF_X(x)[1 - F_X(x)].$$

12. (Pearson probability distribution) Distributions where the probability density functions $f_X(x)$ satisfy the differential equation

$$\frac{df_X(x)}{dx} = \frac{(x - a)f_X(x)}{b_0 + b_1x + b_2x^2},$$

where a, b_0, b_1, b_2 are real constants, are called Pearson probability distributions.

(a) Show that if $\mu_k = E(X - E(X))^k$, then

$$a = \frac{-\mu_3(\mu_4 + 3\mu_2^2)}{A}$$

$$b_0 = \frac{-\mu_2(4\mu_2\mu_4 - 3\mu_3^2)}{A}$$

$$b_1 = \frac{-\mu_3(\mu_4 + 3\mu_2^2)}{A}$$

$$b_2 = \frac{-(2\mu_2\mu_4 - 3\mu_3^2 - 6\mu_2^3)}{A},$$

where $A = 10\mu_4\mu_2 - 18\mu_2^3 - 12\mu_3^2$.

(b) Show that normal, gamma, and beta distributions belong to the family of Pearson probability distributions.

13. (One-parameter exponential family) Let θ be a real parameter. The distribution of the random variable X having probability density function

$$f_X(x|\theta) = C(\theta)e^{Q(\theta)T(x)}h(x)$$

is said to belong to the exponential family of distributions. Show that the gamma distribution, the normal distribution with known variance, the normal distribution with known mean, and the binomial and Poisson distributions belong to the exponential family.

14. (Infinitely divisible distribution) A random variable X is said to be infinitely divisible if for every n, X can be written as

$$X = X_1 + \cdots + X_n$$

where the random variables $X_1,\ldots,X_n$ are independent and have the same distribution function $F_n(X)$ (depending on n). The distribution function of an infinitely divisible random variable is called the infinitely divisible distribution function. Show that normal, gamma, Cauchy, and Poisson distributions are infinitely divisible distributions.

15. Let $X_1,\ldots,X_n$ be a sample of size n from a normal population with mean v and variance σ^2, and let $Y_1,\ldots,Y_m$ be a sample of size m from a normal population with mean v and variance σ_2^2. Assume that $(X_1,\ldots,X_n)$ are independent of $(Y_1,\ldots,Y_m)$. Write

$$\overline{X} = \sum_1^n \frac{X_i}{n}, \qquad S_1 = \sum_1^n (X_i - \overline{X})^2$$

$$\overline{Y} = \sum_1^m \frac{Y_i}{m}, \qquad S_2 = \sum_1^m (Y_i - \overline{Y})^2.$$

(a) Show that $\overline{X} - \overline{Y}$ is independent of $S_2 + S_2$.
(b) Find the distribution of $(S_1/\sigma_1^2) + (S_2/\sigma_2^2)$.
(c) Find the distribution of $(m/n)(S_1/S_2)$.

16. If Q is a quadratic form of rank q $(\leq p)$ in $X_1,\ldots,X_p$, show that Q can be expressed as

$$Q = \sum_{i=1}^q \pm Y_i^2,$$

where Y_i is a linear function of $X_1,\ldots,X_p$ for $i = 1,\ldots,q$.

17. Let $X_1,\ldots,X_n$ be a random sample of size n from a random phenomenon specified by the Cauchy probability density function

$$f_X(x) = \begin{cases} \dfrac{1}{\pi(1 + x^2)}, & -\infty < x < \infty \\ 0 & \text{otherwise.} \end{cases}$$

Show that the probability density function of the random variable

$$Z = \frac{1}{n}\sum_{i=1}^n X_i$$

is given by

$$f_Z(z) = \begin{cases} \dfrac{1}{\pi(1 + z^2)}, & -\infty < z < \infty \\ 0 & \text{otherwise.} \end{cases}$$

18. Let $X_1,\ldots,X_n$ be a random sample of size n from a random phenomenon specified by the uniform probability density function

$$f_X(x) = \begin{cases} \dfrac{1}{a}, & \text{if } 0 < x < a \\ 0 & \text{otherwise} \end{cases}$$

and let

$$Z = \exp\left(\frac{U}{n}\right), \qquad U = \sum_{i=1}^{n} \log X_i.$$

Show that the pdf of Z and U are given by

$$f_U(u) = \begin{cases} \dfrac{(n\log a - u)^{n-1}\exp[-(n\log a - u)]}{\Gamma(n)}, \\ \qquad \text{if } n\log a - u \geq 0 \\ 0, \qquad \text{otherwise}; \end{cases}$$

$$f_Z(z) = \begin{cases} \dfrac{n^n z^{n-1}}{a^n \Gamma(n)}\log\left(\dfrac{a}{z}\right)^{n-1}, & \text{if } \log a \geq \log z, \\ 0, & \text{otherwise}. \end{cases}$$

19. Let X, Y be two independent normal random variables with means m_1, m_2 and variances σ_1^2, σ_2^2, respectively, and let

$$U = \frac{m_1 - m_2 Z}{(\sigma_1^2 + \sigma_2^2 Z^2)^{1/2}}, \qquad Z = \frac{X}{Y}.$$

Show that U is normally distributed with mean zero and unit variance.

20. Shots are fired at a vertical circular target of unit radius. The distribution of the horizontal and the vertical deviations of the shots from the center are normal, with zero means, unit variances, and correlation coefficient ρ. Show that the probability of heating the target is

$$\sqrt{\frac{1 - \rho^2}{\pi}} \int_{1-\rho}^{1+\rho} \left(1 - \exp\left[\frac{x}{2(1 - \rho^2)}\right]\right) \frac{dx}{x[\rho^2 - (1 - x^2)]^{1/2}}.$$

BIBLIOGRAPHY

Bickel, P. J., and Doksum, K. A. (1977). *Mathematical Statistics*, Holden-Day, San Francisco.

Rao, C. R. 1973). *Linear Statistical Inference and Its Application*, Wiley, New York.

Rice, S. O. (1945). Mathematical analysis of random noise, *Bell System Tech. Jour.* Vol. 24, p. 81.

7
Multivariate Distributions

Multivariate distributions and related techniques are becoming increasingly popular with the progress of computing facilities. At present most statistical data are multidimensional. In this chapter we discuss multivariate, continuous and discrete, distributions. The continuous multivariate distributions include, among others, the multivariate normal, the beta distribution, and the multivariate t distribution. The multinomial distribution is a discrete multivariate distribution. A justification of using different types of multivariate distributions is that, in many situations given a set of data, the assumption that they arise from a multivariate normal population does not always hold. For example, the stock market data seem to follow closely the multivariate t distribution. No doubt, the multivariate normal distribution plays a dominant role in the multivariate statistical analysis. It is mainly due to the fact that among all known multivariate distributions, it is the simplest one to handle and many well-known optimum statistical techniques of univariate normal have been successfully extended to the multivariate normal case.

7.1 PROPERTIES OF MULTIVARIATE DISTRIBUTIONS

In this section we define multivariate distributions and describe some of their properties. By multivariate distribution we mean the distribution of a p-dimensional ($p \geq 2$) vector random variable $X = (X_1,\ldots,X_p)'$ whose elements are unidimensional random variables X_i with distribution function $F_{X_i}(x_i)$. Writing $X = (X_1,\ldots,X_p)'$ the distribution function of the vector random variable X is defined by

$$F_X(x) = P(X_1 \leq x_1,\ldots,X_p \leq x_p). \tag{7.1}$$

In the notation of Chapter 3 we can also write

$$F_X(x) = F_{X_1,\ldots,X_p}(x_1,\ldots,x_p). \tag{7.2}$$

If each X_i is a discrete random variable, X is called a p-dimensional discrete random variable and its probability mass function is given by

$$\begin{aligned} P_X(x) &= P_{X_1,\ldots,X_p}(x_1,\ldots,x_p) \\ &= P(X_1 = x_1,\ldots,X_p = x_p). \end{aligned} \tag{7.3}$$

If $F_X(x)$ is continuous in $x_1,\ldots,x_p$ for $-\infty < x_i < \infty$ for all i, and if there exists a nonnegative continuous function

$$f_X(x) = f_{X_1,\ldots,X_p}(x_1,\ldots,x_p) \text{ in } x_1,\ldots,x_p$$

such that

$$F_X(x) = \int_{-\infty}^{x_1} \cdots \int_{-\infty}^{x_p} f_X(x)\, dx_1 \cdots dx_{p'}, \tag{7.4}$$

then X is called a continuous random variable with probability density function $f_X(x)$. If the components $X_1,\ldots,X_p$ of X are independent, then

$$F_X(x) = \prod_{i=1}^{P} F_{X_i}(x_i)$$

or, equivalently,

$$p_X(x) = \prod_{i=1}^{P} p_{X_i}(x_i), \qquad f_X(x) = \prod_{i=1}^{P} f_{X_i}(x_i).$$

From Eq. (7.4) it follows that if $f_X(x)$ is given, the marginal density function of any subset of the components of X is obtained by integrating (or, in the case of a mass function, taking the sum of) the original probability density function over the domain of the variables, not in the subset. In other words, for $q < p$,

$$f_{X_1,\ldots,X_q}(x_1 \cdots x_q) = \int_{-\infty}^{\infty} \cdots \int_{-\infty}^{\infty} f_X(x)\, dx_{q+1} \cdots dx_p. \tag{7.5}$$

In multivariate analysis it is often necessary to find the conditional probability density function or the distribution function of a subset of the components of X given that the variates of another subset of the components of X have assumed constant specified values or have been constrained to lie in some subregion of the space described by their variable values. Such a probability density function (distribution function) is called a conditional probability density function (conditional distribution function). It is customary to denote the conditional distribution of $X_1,\ldots,X_q$ when $X_{q+1} = x_{q+1},\ldots,X_p = x_p$ as

$$f_{X_1,\ldots,X_q|X_{q+1},\ldots,X_p}(x_1,\ldots,x_q|x_{q+1},\ldots,x_p) = \frac{f_X(x)}{f_{X_{q+1},\ldots,X_p}(x_{q+1},\ldots,x_p)}, \tag{7.6}$$

where $f_{X_{q+1},\ldots,X_p}(x_{q+1},\ldots,x_p)$ is the marginal distribution of $f_{X_1,\ldots,X_q}$. It is to be noted that the conditional distribution (7.6) is defined only if the denominator $f_{X_{q+1},\ldots,X_p}(x_{q+1},\ldots,x_p) \neq 0$.

The fact that Eq. (7.6) can be derived from the notion of conditional probability of two events follows from the explanation below in the case of bivariate random variable (X,Y) with probability density function $f_{X,Y}(x,y)$. For $x_1 < x_2$, $y_1 < y_2$,

$$P(x_1 \leq X \leq x_2|y_1 \leq Y \leq y_2) = \frac{p(x_1 \leq X \leq x_2, y_1 \leq Y \leq y_2)}{P(y_1 \leq Y \leq y_2)}$$

$$= \frac{\int_{x_1}^{x_2} \int_{y_1}^{y_2} f_{X,Y}(x,y)\, dx\, dy}{\int_{y_1}^{y_2} f_Y(y)\, dy},$$

where $f_Y(y)$ denotes the marginal probability density function of Y and $P(y_1 \leq Y \leq y_2) > 0$. Let $y_1 = y$, $y_2 = Y + \Delta y$, $\Delta y > 0$. By the mean-value theorem, there exist y^* and $y(x)$ with $y \leq y^* \leq y + \Delta y$, $y \leq y(x) \leq y + \Delta y$ such that

$$\int_y^{y+\Delta y} f_Y(y)\, dy = f_Y(y^*)\, \Delta y$$

$$\int_y^{y+\Delta y} f_{X,Y}(x,y)\, dy = f_{X,Y}(x,y(x))\, \Delta y.$$

For y such that $f_Y(y) > 0$, define $P(x_1 \leq X \leq x_2|Y = y)$, the probability that X lies in the closed interval $[x_1,x_2]$ given that $Y = y$. Hence, from above, as Δy tends to zero, we get

$$P(x_1 \leq X \leq x_2|Y = y) = \int_{x_1}^{x_2} f_{X|Y}(x|y)\, dx,$$

where

$$f_{X|Y}(x|y) = \frac{f_{X,Y}(x,y)}{f_Y(y)}.$$

For a given y, $f_{X|Y}(x|y)$ is called the conditional probability density function of X given $Y = y$. Thus

$$f_{X_1,\ldots,X_q|X_{q+1},\ldots,X_p}(x_1,\ldots,x_q|x_{q+1},\ldots,x_p)$$

$$= \int_{-\infty}^{x_1} \cdots \int_{-\infty}^{x_q} F_{X_1,\ldots,X_q|X_{q+1},\ldots,X_p}(x_1,\ldots,x_q|x_{q+1},\ldots,x_p) \, dx_1 \cdots dx_q.$$

Since

$$\int_{-\infty}^{\infty} \cdots \int_{-\infty}^{\infty} F_{X_1,\ldots,X_q|X_{q+1},\ldots,X_p}(x_1,\ldots,x_q|x_{q+1},\ldots,x_p) \, dx_1 \cdots dx_q = 1$$

we get

$$F_{X_1,\ldots,X_q|X_{q+1},\ldots,X_p}(\infty,\ldots,\infty|x_{q+1},\ldots,x_p) = 1.$$

We now define the notion of mathematical expectation of a random vector and of a random matrix. By a random matrix we mean a matrix whose elements are random variables. For example, if X is a random p-vector (column), then XX' is a random matrix of dimension $p \times p$ with $X_i X_j$ as the (i,j)th element.

Definition 7.1.1: Mathematical expectation of a random matrix   The *mathematical expectation of a random matrix* of dimension $p \times q$,

$$Z = \begin{bmatrix} Z_{1p} & \cdots & Z_{1q} \\ \vdots & \cdots & \vdots \\ Z_{p1} & \cdots & Z_{pq} \end{bmatrix},$$

is defined by

$$E(Z) = \begin{bmatrix} E(Z_{11}) & \cdots & E(Z_{1q}) \\ \vdots & \cdots & \vdots \\ E(Z_{p1}) & \cdots & E(Z_{pq}) \end{bmatrix}.$$

Since a column vector is a matrix of dimension $p \times 1$, we have

$$E(X) = (E(X_1),\ldots,E(X_p))'.$$

From this definition it may be verified that if Z is a random matrix of dimension $m \times n$ and if A, B, C are matrices of real constants of dimension $l \times m$, $n \times q$, and $l \times q$, respectively, then

$$E(AZB + C) = AE(Z)B + C.$$

Definition 7.1.2: Mean and covariance of a multivariate random variable For any random p-vector X (column), $\mu = E(X)$, $\Sigma = E(X - \mu)(X - \mu)'$ are, respectively, called the *mean* and the *covariance matrix* of X.

Definition 7.1.3: Characteristic function of a multivariate random variable For every real vector $t = (t_1,...,t_p)'$, the *characteristic function* of any random p-vector X is defined by

$$\phi_X(t) = E(e^{it'X})$$

$$= \int_{-\infty}^{\infty} \cdots \int_{-\infty}^{\infty} e^{it'x} f_X(x) \, dx_1 \cdots dx_p$$

if X is a continuous random variable

$$\sum_{x_1} \cdots \sum_{x_p} e^{it'x} p_X(x)$$

if X is a discrete random variable.

Since $E(|e^{it'X}|) = 1$, $\Phi_X(t)$ always exists.

For the necessary background on vectors and matrices, the reader is referred to Appendix A.

7.2 BIVARIATE NORMAL DISTRIBUTION

Before we start discussing the p-variate normal distribution, it is advantageous for comprehensions to discuss the bivariate normal distribution in some details. A two-dimensional vector $X = (X_1, X_2)'$ with values in R^2 is said to have a bivariate normal distribution if its probability density function (pdf) can be written as

$$f_X(x) = \frac{1}{2\pi\sigma_1\sigma_2(1 - \rho^2)^{1/2}} \exp\left\{ -\frac{1}{2(1 - \rho^2)} \left[\left(\frac{x_1 - \mu_1}{\sigma_1}\right)^2 \right.\right.$$

$$\left.\left. + \left(\frac{x_2 - \mu_2}{\sigma_2}\right)^2 - 2\rho\left(\frac{x_1 - \mu_1}{\sigma_1}\right)\left(\frac{x_2 - \mu_2}{\sigma_2}\right) \right] \right\}, \tag{7.7}$$

where $-\infty < \mu_i < \infty$, $\sigma_i^2 > 0$, $i = 1,2, -1 < \rho < 1$. In Example 3.4.2 we have verified that it is an honest density function, that is,

$$\int_{-\infty}^{\infty} \int_{-\infty}^{\infty} f_X(x) \, dx_1 \, dx_2 = 1.$$

Theorem 7.2.1 Let $X = (X_1, X_2)$ be a random vector with pdf given in (7.7). Then $E(X_i) = \mu_i$, $E(X_i - \mu_i)^2 = \sigma_i^2$, $i = 1,2$; $\rho = \text{cov}(X_1, X_2)/\sigma_1\sigma_2$.

Proof. Define

$$y_i = \frac{x_i - \mu_i}{\sigma_i}, \qquad i = 1,2.$$

Since $dx_i/dy_i = \sigma_i$, we get

$$E(X_1 - \mu_1) = \int_{-\infty}^{\infty} \int_{-\infty}^{\infty} (x_1 - \mu_1) f_X(x) \, dx_1 \, dx_2$$

$$= \sigma_1 \int_{-\infty}^{\infty} \int_{-\infty}^{\infty} \frac{y_1}{2\pi(1 - \rho^2)^{1/2}}$$

$$\times \exp\left[-\frac{1}{2(1 - \rho^2)} (y_1^2 + y_2^2 - 2\rho y_1 y_2) \right] dy_1 \, dy_2$$

$$= \sigma_1 \int_{-\infty}^{\infty} y_1 \frac{1}{\sqrt{2\pi}} \exp\left(-\frac{1}{2} y_1^2 \right)$$

$$\times \left\{ \int_{-\infty}^{\infty} \frac{1}{\sqrt{2\pi(1 - \rho^2)}} \right.$$

$$\left. \times \exp\left[-\frac{1}{2(1 - \rho^2)} (y_2 - \rho y_1)^2 \right] dy_2 \right\} dy_1$$

$$= \sigma_1 \int_{-\infty}^{\infty} y_1 \frac{1}{\sqrt{2\pi}} \exp\left(-\frac{1}{2} y_1^2 \right) dy_1$$

$$= 0$$

$$E(X_1 - \mu_2)^2 = \int_{-\infty}^{\infty} \int_{-\infty}^{\infty} (x_1 - \mu_1)^2 f_X(x) \, dx_1 \, dx_2$$

$$= \sigma_1^2 \int_{-\infty}^{\infty} \int_{-\infty}^{\infty} \frac{y_1^2}{2\pi(1 - \rho^2)^{1/2}}$$

$$\times \exp\left[-\frac{1}{2(1 - \rho^2)} (y_1^2 + y_2^2 - 2\rho y_1 y_2) \right] dy_1 \, dy_2$$

$$= \sigma_1^2 \int_{-\infty}^{\infty} y_1^2 \frac{1}{\sqrt{2\pi}} \exp\left(-\frac{1}{2} y_1^2 \right)$$

$$\times \left\{ \int_{-\infty}^{\infty} \frac{1}{\sqrt{2\pi(1 - \rho^2)}} \right.$$

$$\left. \times \exp\left[-\frac{1}{2(1 - \rho^2)} (y_2 - \rho y_1)^2 \right] dy_2 \right\} dy_1$$

$$= \sigma_1^2 \int_{-\infty}^{\infty} y_1^2 \frac{1}{\sqrt{2\pi}} \exp\left(-\frac{1}{2} y_1^2 \right) dy_1$$

$$= \sigma_1^2$$

Hence $E(X_1) = \mu_1$, $\text{var}(X_1) = \sigma_1^2$. Interchanging the roles of X_1, X_2, we get $E(X_2) = \mu_2$, $\text{var}(X_2) = \sigma_2^2$. Now

$$
\begin{aligned}
E(X_1 - \mu_1)(X_2 - \mu_2) &= \sigma_1\sigma_2 \int_{-\infty}^{\infty} \int_{-\infty}^{\infty} \frac{y_1 y_2}{2\pi(1 - \rho^2)^{1/2}} \\
&\quad \times \exp\left[-\frac{1}{2(1 - \rho^2)} \right. \\
&\quad \left. \times (y_1^2 + y_2^2 - 2\rho y_1 y_2) \right] dy_1 \, dy_2 \\
&= \sigma_1\sigma_2 \int_{-\infty}^{\infty} y_1 \frac{1}{\sqrt{2\pi}} \left(-\frac{1}{2} y_1^2 \right) \\
&\quad \times \left\{ \int_{-\infty}^{\infty} y_2 \frac{1}{\sqrt{2\pi(1 - \rho^2)}} \right. \\
&\quad \left. \times \exp\left[-\frac{1}{2(1 - \rho^2)} (y_2 - \rho y_1)^2 \right] dy_2 \right\} dy_1 \\
&= \sigma_1\sigma_2 \int_{-\infty}^{\infty} \rho y_1^2 \frac{1}{\sqrt{2\pi}} \\
&\quad \times \exp\left(-\frac{1}{2} y_1^2 \right) dy_1 = \rho\sigma_1\sigma_2. \qquad \text{Q.E.D.}
\end{aligned}
$$

Note: ρ is the coefficient of correlation between components of X. If $\rho = 0$, the pdf of X can be written as

$$
\begin{aligned}
f_X(x) &= \frac{1}{\sqrt{2\pi} \, \sigma_1} \exp\left[-\frac{1}{2} \left(\frac{x_1 - \mu_1}{\sigma_1} \right)^2 \right] \\
&\quad \times \frac{1}{\sqrt{2\pi} \, \sigma_2} \exp\left[-\frac{1}{2} \left(\frac{x_2 - \mu_2}{\sigma_2} \right)^2 \right].
\end{aligned}
$$

which implies that X_1, X_2 are independent $N(\mu_1,\sigma_1^2)$, $N(\mu_2,\sigma_2^2)$, respectively. The following theorem will assert that the marginal distributions of the components of a bivariate normal are normals even if $\rho \neq 0$.

Theorem 7.2.2 Let $f_X(x)$ be given in (7.7).

(a) The marginal pdf of X_1 is $N(\mu_1,\sigma_1^2)$ and the marginal of X_2 is $N(\mu_2,\sigma_2^2)$.

(b) The conditional pdf of X_1 given $X_2 = x_2$ is

$$
N\left(\mu_1 + \rho \frac{\sigma_1}{\sigma_2} (x_2 - \mu_2), \, \sigma_1^2(1 - \rho^2) \right).
$$

Proof. The marginal pdf of X_1 is

$$f_{X_1}(x_1) = \int_{-\infty}^{\infty} f_X(x) \, dx_2$$

$$= \frac{1}{\sqrt{2\pi\sigma_1^2}} \exp\left[-\frac{1}{2}\left(\frac{x_1 - \mu_1}{\sigma_1}\right)^2\right] \int_{-\infty}^{\infty} \frac{1}{\sqrt{2\pi(1 - \rho^2)}}$$

$$\times \exp\left[-\frac{1}{2(1 - \rho^2)}(y_2 - \rho y_1)^2\right] dy_2$$

$$= \frac{1}{\sqrt{2\pi\sigma_1^2}} \exp\left[-\frac{1}{2}\left(\frac{x_1 - \mu_1}{\sigma_1}\right)^2\right],$$

which is a normal with mean μ_1 and variance σ_1^2. Similarly, we can show that X_2 is $N(\mu_2, \sigma_2^2)$.

(b) $\quad f_{X_1|X_2}(x_1|x_2) = \dfrac{f_X(x)}{f_{X_2}(x_2)}$

$$= \frac{1}{\sqrt{2\pi\sigma_1^2(1 - \rho^2)}} \exp\left\{\frac{-1}{2(1 - \rho^2)\sigma_1^2}\right.$$

$$\times \left.\left[x_1 - \mu_1 - \rho\frac{\sigma_1}{\sigma_2}(x_2 - \mu_2)\right]^2\right\}. \quad \text{Q.E.D.}$$

We have observed that if X is a bivariate normal, the marginal distribution of each component of X is a univariate normal. The following example will show that the converse is not always true.

Example 7.2.1   Let

$$\phi_1(x_1,x_2) = \frac{1}{2\pi(1 - \rho_1^2)^{1/2}} \exp\left[-\frac{1}{2(1 - \rho_1^2)}(x_1^2 + x_2^2 - 2\rho_1 x_1 x_2)\right]$$

$$\phi_2(x_1,x_2) = \frac{1}{2\pi(1 - \rho_2^2)^{1/2}} \exp\left[-\frac{1}{2(1 - \rho_2^2)}(x_1^2 + x_2^2 - 2\rho_2 x_1 x_2)\right]$$

where $-1 < \rho_i < 1$, $i = 1,2$; and let

$$\phi(x_1,x_2) = \alpha\phi_1(x_1,x_2) + (1 - \alpha)\phi_2(x_1,x_2)$$

with $0 < \alpha < 1$. Since

$$\int_{-\infty}^{\infty} \int_{-\infty}^{\infty} \phi(x_1,x_2) \, dx_1 \, dx_2 = \alpha \int_{-\infty}^{\infty} \int_{-\infty}^{\infty} \phi_1(x_1,x_2) \, dx_1 \, dx_2 + (1 - \alpha)$$

$$\times \int_{-\infty}^{\infty} \phi_2(x_1,x_2) \, dx_1 \, dx_2$$

$$= \alpha + (1 - \alpha) = 1,$$

ϕ is a pdf of $X = (X_1, X_2)'$. The marginal pdf of X_i, $i = 1, 2$, is

$$\alpha \frac{1}{\sqrt{2\pi}} \exp\left(-\frac{1}{2} x_i^2\right) + (1 - \alpha) \frac{1}{\sqrt{2\pi}} \exp\left(-\frac{1}{2} x_i^2\right) = \frac{1}{\sqrt{2\pi}} e^{-(1/2)x_i^2},$$

which is a $N(0,1)$. But $\phi(x_1, x_2)$ is not a bivariate normal. *Note*: A bivariate normal distribution with $-1 < \rho < 1$, that is, $|\Sigma| > 0$ with

$$\Sigma = E(X - \mu)(X - \mu)' = \begin{pmatrix} \sigma_1^2 & \rho\sigma_1\sigma_2 \\ \rho\sigma_1\sigma_2 & \sigma_2^2 \end{pmatrix},$$

is called *nondegenerate*; others are called *degenerate*.

Let $\mu = (\mu_1, \mu_2)' = E(X)$, $\Sigma = \text{cov}(X)$. Since

$$|\Sigma| = \sigma_1^2 \sigma_2^2 (1 - \rho^2),$$

$$\Sigma^{-1} = (1 - \rho^2)^{-1} \begin{vmatrix} \dfrac{1}{\sigma_1^2} & \dfrac{-\rho}{\sigma_1\sigma_2} \\ \dfrac{-\rho}{\sigma_1\sigma_2} & \dfrac{1}{\sigma_2^2} \end{vmatrix}$$

$$(x - \mu)'\Sigma^{-1}(x - \mu) = (1 - \rho^2)^{-1}\left[\left(\frac{x_1 - \mu_2}{\sigma_1}\right)^2 + \left(\frac{x_2 - \mu_2}{\sigma_2}\right)^2 \right.$$

$$\left. - 2\rho\left(\frac{x_1 - \mu_1}{\sigma_2}\right)\left(\frac{x_2 - \mu_2}{\sigma_2}\right)\right].$$

Hence we can write the pdf of X in the vector and matrix notation

$$f_X(x) = \frac{1}{2\pi|\Sigma|^{1/2}} \exp\left[-\frac{1}{2}(x - \mu)'\Sigma^{-1}(x - \mu)\right].$$

7.3 MULTIVARIATE NORMAL DISTRIBUTION

A p-dimensional random variable $X = (X_1, \ldots, X_p)'$ is said to have a multivariate normal distribution if its probability density function can be written as [with $x = (x_1, \ldots, x_p)'$]

$$f_X(x) = \begin{cases} \dfrac{1}{(2\pi)^{p/2}|\Sigma|^{1/2}} \exp\left[-\dfrac{1}{2}(x - \mu)'\Sigma^{-1}(x - \mu)\right] \\ \qquad \text{if } -\infty < x_i < \infty \text{ for all } i \\ 0 \qquad \text{otherwise,} \end{cases} \tag{7.8}$$

where $\mu = (\mu_1,\ldots,\mu_p)'$ is a vector of constant components μ_i and

$$\Sigma = \begin{pmatrix} \sigma_1^2 & \sigma_{12} & \cdots & \sigma_{1p} \\ \sigma_{21} & \sigma_2^2 & \cdots & \sigma_{2p} \\ \vdots & & \cdots & \vdots \\ \sigma_{p1} & \sigma_{p2} & \cdots & \sigma_p^2 \end{pmatrix}$$

is a positive definite symmetric matrix of constants $\sigma_{ij} = \sigma_{ji}$.

Since Σ is positive definite, $|\Sigma| > 0$ and $(x - \mu)'\Sigma^{-1}(x - \mu) \geq 0$ for all x. Hence $f_X(x) \geq 0$ and is bounded for all values of X. To show that it is a true probability density function, we proceed as follows. By Corollary A.2.6.6.2 (Appendix A) we can find a nonsingular matrix A such that $A\Sigma A' = I$. Let $Y = AX$ and $v = A\mu = (v_1,\ldots,v_p)'$. The Jacobian of the transformation $X \to Y = AX$ is mod $|A^{-1}| = |AA'|^{-1/2}$ (see Exercise 20 in Appendix A). The absolute-value symbol is written here as "mod" to avoid confusion with the determinant symbol. Now

$$\int_{-\infty}^{\infty} \cdots \int_{-\infty}^{\infty} \frac{1}{(2\pi)^{p/2}|\Sigma|^{1/2}} \exp\left[-\frac{1}{2}(x - \mu)'\Sigma^{-1}(x - \mu) \right] dx_1 \cdots dx_p$$

$$= \int_{-\infty}^{\infty} \cdots \int_{-\infty}^{\infty} \frac{1}{(2\pi)^{p/2}|A\Sigma A'|^{1/2}}$$

$$\times \exp\left[-\frac{1}{2}(y - A\mu)'(A\Sigma A')^{-1}(y - A\mu) \right] dy_1 \cdots dy_p$$

$$= \int_{-\infty}^{\infty} \cdots \int_{-\infty}^{\infty} \frac{1}{(2\pi)^{p/2}} \exp\left[-\frac{1}{2}(y - v)'(y - v) \right] dy_1 \cdots dy_p$$

$$= \prod_{i=1}^{P} \int_{-\infty}^{\infty} \frac{1}{\sqrt{2\pi}} \exp\left[-\frac{1}{2}(y_i - v_i)^2 \right] dy_i$$

$$= 1$$

since

$$\int_{-\infty}^{\infty} \frac{1}{\sqrt{2\pi}} \exp\left[-\frac{1}{2}(y_i - v_i)^2 \right] dy_i = 1 \qquad \text{for all } i.$$

By results in Section 3.8 the probability density function of Y [$y = (y_1,\ldots,y_p)'$] is

$$f_Y(y) = \begin{array}{ll} \dfrac{1}{(2\pi)^{p/2}} \exp\left[-\dfrac{1}{2}\sum_{i=1}^{P}(y_i - v_i)^2 \right] & \text{if } -\infty < y_i < \infty \\ 0 & \text{otherwise.} \end{array} \qquad (7.9)$$

From the above it follows that $Y_1,...,Y_p$ are independently normally distributed with variance 1 and means $v_1,...,v_p$. In matrix and vector notations,

$$E(Y) = v = (v_1,...,v_p)'$$
$$E(Y - v)(Y - v)' = I.$$

In other words,

$$E(AX) = AE(X) = v = A\mu$$
$$E(AX - A\mu)(AX - A\mu)' = AE(X - \mu)(X - \mu)'A' = I$$

or, equivalently,

$$E(X) = \mu$$
$$E(X - \mu)(X - \mu)' = A^{-1}(A')^{-1} = (A'A)^{-1} = \Sigma$$

since $A\Sigma A' = I$ implies that $\Sigma = (A'A)^{-1}$.

Thus if X has probability density function given by Eq. (7.8), then

$$E(X) = \mu \quad \text{and} \quad E(X - \mu)(X - \mu)' = \Sigma.$$

We can conclude that the distribution of a normal random vector X is completely characterized by the parameters μ and Σ, which represent the mean and covariance matrices of X, respectively.

Definition 7.3.1: Covariance matrix $E(X - \mu)(X - \mu)'$ is called the *covariance matrix* of the random vector X and is sometimes denoted by $\text{cov}(X)$. If $\Sigma = I$, then $f_X(x)$ can be written as

$$f_X(x) = \frac{1}{(2\pi)^{p/2}} \exp\left[-\frac{1}{2}(x - \mu)'(x - \mu) \right]$$
$$= \prod_{i=1}^{P} \frac{1}{\sqrt{2\pi}} \exp\left[-\frac{1}{2}(x_i - \mu_i)^2 \right].$$

Thus if $\text{cov}(X) = I$, then the components of X are independently distributed normal random variables with mean μ_i and variance 1. This further implies that if $X_1,..,X_p$ are p univariate normal random variables with $\text{cov}(X_i,X_j) = 0$ for all $i \neq j = 1,...,p$, then $X_1,...,X_p$ are independent. A standard notation for the p-variate normal distribution with mean μ and covariance matrix Σ is $N_p(\mu,\Sigma)$.

An important characteristic property of the multivariate normal distribution is that any linear combination of its components is normally distributed. We establish this in the following two theorems.

Theorem 7.3.1   Let $(X = X_1,\ldots,X_p)'$ be normally distributed with mean μ and positive definite covariance matrix Σ. Then the random vector $Y = CX$, where C is a nonsingular matrix of dimension $p \times p$, is normally distributed with mean $C\mu$ and positive definite covariance matrix $C\Sigma C'$.

Proof.   The Jacobian of the transformation $X \to Y$ is mod $|C^{-1}| = |CC'|^{-1/2}$. By results in Section 3.8 the probability density function of Y is

$$f_Y(y) = \frac{1}{(2\pi)^{p/2}|\Sigma|^{1/2}} \exp\left[-\frac{1}{2}(C^{-1}y - \mu)'\Sigma^{-1}(C^{-1}y - \mu)\right]|CC'|^{-1/2}$$

$$= \frac{1}{(2\pi)^{p/2}|C\Sigma C'|^{1/2}} \exp\left[-\frac{1}{2}(y - C\mu)'(C\Sigma C')^{-1}(y - C\mu)\right],$$

where $y = (y_1,\ldots,y_p)'$. Hence Y is $N_p(C\mu, C\Sigma C')$.   Q.E.D.

Theorem 7.3.2   Let X be distributed as $N_p(\mu,\Sigma)$ and let $Y = AX$, where A is a matrix of dimension $q \times p(q < p)$ and is of rank q. Then Y is distributed as $N_q(A\mu, A\Sigma A')$.

Proof.   Let $Y = AX$. Then

$$E(Y) = AE(X) = A\mu$$

$$E(Y - A\mu)(Y - A\mu)' = E(AX - A\mu)(AX - A\mu)' = A\Sigma A'$$

and by Theorem A.2.6.7, $A\Sigma A'$, which is of dimension $q \times q$, is positive definite. Let C be a nonsingular matrix of dimension $p \times p$ such that $C = \binom{A}{B}$, where B is a matrix of dimension $(p - q) \times p$ and is of rank $p - q$, and let $W = BX$. Thus, by Theorem 7.3.1, $\binom{Y}{W}$ is distributed as

$$N_p\left[\begin{bmatrix} A\mu \\ B\mu \end{bmatrix}, \begin{bmatrix} A\Sigma A' & A\Sigma B' \\ B\Sigma A' & B\Sigma B' \end{bmatrix}\right].$$

Hence, applying Theorem 7.3.4 (to be proved later), Y is distributed as $N_q(A\mu, A\Sigma A')$.   Q.E.D.

One effective way of choosing the matrix C in the Theorem 7.3.2 would be as follows: Assume without any loss of generality that the first q columns of A are linearly independent. Write $A = (A_{11}, A_{12})$, where the submatrix A_{12} contains the remaining $p - q$ columns of A. Then choose B as $(0,I)$, where I is of dimension $(p - q) \times (p - q)$. Then

$$C = \begin{bmatrix} A_{11} & A_{12} \\ 0 & I \end{bmatrix}$$

and

$$|C| = \begin{vmatrix} A_{11} & A_{12} \\ 0 & I \end{vmatrix} = |A_{11}| \neq 0.$$

Another characteristic property of the normal distribution is that if X is distributed as $N_p(\mu,\Sigma)$, the distribution of any subvector (the marginal distribution of the subvector) is also normal. In particular, each component of X is normally distributed. This result is demonstrated by the two following theorems.

Theorem 7.3.3 Let X be distributed as $N_p(\mu,\Sigma)$ and let the vectors X, μ and the matrix Σ be partitioned as

$$X = (X'_{(1)},X'_{(2)})', \quad X_{(1)} = (X_1,\ldots,X_q)', \quad X_{(2)} = (X_{q+1},\ldots,X_p)'$$

$$\mu = (\mu'_{(1)},\mu'_{(2)})', \quad \mu_{(1)} = (\mu_1,\ldots,\mu_q)', \quad \mu_{(2)} = (\mu_{q+1},\ldots,\mu_p)'$$

$$\Sigma = \begin{bmatrix} \Sigma_{11} & \Sigma_{12} \\ \Sigma_{21} & \Sigma_{22} \end{bmatrix}, \quad \Sigma_{11} = \begin{pmatrix} \sigma_1^2 & \cdots & \sigma_{1q} \\ \vdots & \cdots & \vdots \\ \sigma_{q1} & \cdots & \sigma_q^2 \end{pmatrix}, \quad \Sigma_{12} = \Sigma'_{21}. \qquad (7.10)$$

If $\Sigma_{12} = \Sigma'_{21} = 0$, then $X_{(1)}$, $X_{(2)}$ are independently normally distributed as $N_q(\mu_{(1)}, \Sigma_{11})$, $N_{p-q}(\mu_{(2)},\Sigma_{22})$.

Proof. Since $\Sigma_{12} = \Sigma'_{21} = 0$, and Σ_{11}, Σ_{22} are (symmetric) positive definite, we have

$$\Sigma^{-1} = \begin{bmatrix} \Sigma_{11}^{-1} & 0 \\ 0 & \Sigma_{22}^{-1} \end{bmatrix}.$$

Now, for

$$x = (x'_{(1)},x'_{(2)})', \quad x_{(1)} = (x_1,\ldots,x_q)', \quad x_{(2)} = (x_{q+1},\ldots,x_p)',$$

we have

$$(x - \mu)'\Sigma^{-1}(x - \mu) = (x_{(1)} - \mu_{(1)})'\Sigma_{11}^{-1}(x_{(1)} - \mu_{(1)})$$
$$+ (x_2 - \mu_{(2)})'\Sigma_{22}^{-1}(x_{(2)} - \mu_{(2)}).$$

Hence

$$f_X(x) = \frac{1}{(2\pi)^{p/2}|\Sigma|^{1/2}} \exp\left[-\frac{1}{2} (x - \mu)'\Sigma^{-1}(x - \mu) \right]$$

$$= \frac{1}{(2\pi)^{q/2}|\Sigma_{11}|^{1/2}} \exp\left[-\frac{1}{2} (x_{(1)} - \mu_{(1)})'\Sigma_{11}^{-1}(x_{(1)} - \mu_{(1)}) \right]$$

$$\times \frac{1}{(2\pi)^{(p-q)/2}|\Sigma_{22}|^{1/2}} \exp\left[-\frac{1}{2} (x_{(2)} - \mu_2)'\Sigma_{22}^{-1}(x_{(2)} - \mu_{(2)}) \right]$$

$$= N_q(\mu_{(1)},\Sigma_{11}), \quad N_{p-q}(\mu_{(2)},\Sigma_{22}). \qquad \text{Q.E.D.}$$

Theorem 7.3.4 Let X be distributed as $N_p(\mu, \Sigma)$ and let X, μ, Σ be partitioned as in Eq. (7.10).

(a) $X_{(1)}$, $X_{(2)} - \Sigma_{21}\Sigma_{11}^{-1}X_{(1)}$ are independently distributed:

$$N_q(\mu_{(1)}, \Sigma_{11}), \quad N_{p-q}(\mu_{(2)} - \Sigma_{21}\Sigma_{11}^{-1}\mu_{(1)}, \Sigma_{22} - \Sigma_{21}\Sigma_{11}^{-1}\Sigma_{12}).$$

(b) The conditional probability density function of $X_{(2)}$ given $X_{(1)} = x_{(1)}$ is normal with mean $\mu_{(2)} + \Sigma_{21}\Sigma_{11}^{-1}(x_{(1)} - \mu_{(1)})$ and covariance $\Sigma_{22} - \Sigma_{21}\Sigma_{11}^{-1}\Sigma_{12}$.

Proof. Let

$$Y_{(1)} = X_{(1)}, \qquad Y_{(2)} = X_{(2)} - \Sigma_{21}\Sigma_{11}^{-1}X_{(1)}.$$

Then

$$Y = (Y'_{(1)}, Y'_{(2)})' = CX$$

where

$$C = \begin{pmatrix} I_1 & 0 \\ -\Sigma_{21} & I_2 \end{pmatrix}$$

with I_1 a identity matrix of dimension $q \times q$ and I_2 a identity matrix of dimension $(p - q) \times (p - q)$. By Theorem 7.3.1, Y is distributed as $N_p(C\mu, C\Sigma C')$. Since

$$C\mu = \begin{pmatrix} \mu_{(1)} \\ \mu_{(2)} - \Sigma_{21}\Sigma_{11}^{-1}\mu_{(1)} \end{pmatrix}, \qquad C\Sigma C' = \begin{pmatrix} \Sigma_{11} & 0 \\ 0 & \Sigma_{22} - \Sigma_{21}\Sigma_{11}^{-1}\Sigma_{12} \end{pmatrix},$$

we conclude (by Theorem 7.3.3.) that $X_{(1)}$, $X_{(2)} - \Sigma_{21}\Sigma_{11}^{-1}X_{(1)}$ are independent and

$$X_{(1)} \text{ is } N_q(\mu_{(1)}, \Sigma_{11}),$$
$$X_{(2)} - \Sigma_{21}\Sigma_{11}^{-1}X_{(1)} \text{ is } N_{p-q}(\mu_{(2)} - \Sigma_{21}\Sigma_{11}^{-1}\mu_{(1)}, \Sigma_{22} - \Sigma_{21}\Sigma_{11}^{-1}\Sigma_{12}).$$

(b) From part (a) the pdf $f_X(x)$ can be written as

$$f_{X_{(1)}, X_{(2)}}(x_{(1)}, x_{(2)}) = \frac{1}{(2\pi)^{q/2}|\Sigma_{11}|^{1/2}} \exp\left[-\frac{1}{2}(x_{(1)} - \mu_{(1)})'\Sigma_{11}^{-1}(x_{(1)} - \mu_{(1)}) \right]$$

$$\times \frac{1}{(2\pi)^{(p-q)/2}|\Sigma_{22} - \Sigma_{21}\Sigma_{11}^{-1}\Sigma_{12}|^{1/2}}$$

$$\times \exp\{-\tfrac{1}{2}(x_{(2)} - \mu_{(2)} - \Sigma_{21}\Sigma_{11}^{-1}[x_{(1)} - \mu_{(1)}])'$$

$$\times (\Sigma_{22} - \Sigma_{21}\Sigma_{11}^{-1}\Sigma_{12})^{-1}$$

$$\times (x_{(2)} - \mu_{(2)} - \Sigma_{21}\Sigma_{11}^{-1}[x_{(1)} - \mu_{(1)}])\}.$$

From the definition of the conditional probability density function, we get

$$f_{X_{(2)}|X_{(1)}}(x_{(2)}|x_{(1)}) = \frac{f_{X_{(1)},X_{(2)}}(x_{(1)},x_{(2)})}{f_{X_{(1)}}(x_{(1)})}$$

$$= \frac{1}{(2\pi)^{(p-q)/2}|\Sigma_{22} - \Sigma_{21}\Sigma_{11}^{-1}\Sigma_{12}|^{1/2}}$$

$$\times \exp\{-\tfrac{1}{2}(x_{(2)} - \mu_{(2)} - \Sigma_{21}\Sigma_{11}^{-1}[x_{(1)} - \mu_{(1)}])'$$

$$\times (\Sigma_{22} - \Sigma_{21}\Sigma_{11}^{-1}\Sigma_{12})^{-1}$$

$$\times (x_{(2)} - \mu_{(2)} - \Sigma_{21}\Sigma_{11}^{-1}[x_{(1)} - \mu_{(1)}])\}. \qquad \text{Q.E.D.}$$

$$(7.11)$$

Definition 7.3.2: Regression surface The matrix $\Sigma_{21}\Sigma_{11}^{-1}$ is called the *matrix of regression coefficients* of $X_{(2)}$ on $X_{(1)} = x_{(1)}$. The quantity

$$E(X_{(2)}|X_{(1)} = x_{(1)}) = \mu_{(2)} + \Sigma_{21}\Sigma_{11}^{-1}(x_{(1)} - \mu_{(1)})$$

is called the *regression surface* of $X_{(2)}$ on $x_{(1)}$.

The regression terminology is due to Galton (1889). He introduced it in his studies of the association between diameter of seeds of parents and daughters of sweet peas and between heights of fathers and sons. He observed that the daughters of dwarf peas are less dwarfish and the daughters of giant peas are less giant than their respective parents, and that the heights of sons of either unusually short or tall fathers tend more closely to the average height than their deviant father's values did to the mean for their generation. Galton called this a *regression to mediocrity* and the parameters of the linear relationship as regression parameters.

Definition 7.3.3: Partial correlation coefficient Let $\sigma_{ij.1...q}$ be the (i,j)th element of the matrix $\Sigma_{22} - \Sigma_{21}\Sigma_{11}^{-1}\Sigma_{12}$ of dimension $(p - q) \times (p - q)$. Then

$$\rho_{ij.1...q} = \frac{\sigma_{ij.1...q}}{\sqrt{\sigma_{ii.1...q}\sigma_{jj.1...q}}} \qquad (7.12)$$

is called the *partial correlation coefficient* of the components $X_{q+i}X_{q+j}$ of X when $X_1,...,X_q$ are held constant.

The partial correlation is the correlation between two variables when the combined effect of some other variables are eliminated. Let $\beta_{(i)}$ denote the ith row of the matrix Σ_{21} of dimension $(p - q) \times q$. From the relation

$$E(X_{(2)}|X_{(1)} = x_{(1)}) = \mu_{(2)} + \Sigma_{21}\Sigma_{11}^{-1}(x_{(1)} - \mu_{(1)}) \qquad (7.13)$$

it follows that

$$E(X_{q+i} - \mu_{q+i}|X_{(1)} = x_{(1)}) = \beta_{(i)}\Sigma_{11}^{-1}(x_{(1)} - \mu_{(1)}).$$ (7.14)

The covariance between X_{q+i} and $\beta_{(i)}\Sigma_{11}^{-1}X_{(1)}$ is given by

$$E(X_{q+i} - \mu_{q+i})(\beta_{(i)}\Sigma_{11}^{-1}[X_{(1)} - \mu_{(1)}])'$$
$$= E(X_{q+i} - \mu_{q+i})(X_{(1)} - \mu_{(1)})'\Sigma_{11}^{-1}\beta_{(i)}'$$
$$= \beta_{(i)}\Sigma_{11}^{-1}\beta_{(i)}'$$

and

$$V(X_{q+i}) = \sigma_{q+i}^2$$
$$V(\beta_{(i)}\Sigma_{11}^{-1}X_{(1)}) = E(\beta_{(i)}\Sigma_{11}^{-1}(X_{(1)} - \mu_{(1)})(X_{(1)} - \mu_{(1)})'\Sigma_{11}^{-1}\beta_{(i)}')$$
$$= \beta_{(i)}\Sigma_{11}^{-1}\Sigma_{11}\Sigma_{11}^{-1}\beta_{(i)}')$$
$$= \beta_{(i)}\Sigma_{11}^{-1}\beta_{(i)}'.$$

The correlation between X_{q+i} and $\beta_{(i)}\Sigma_{11}^{-1}X_{(1)}$ is given by

$$\rho = \frac{\sqrt{\beta_{(i)}\Sigma_{11}^{-1}\beta_{(i)}'}}{\sigma_{q+i}}.$$ (7.15)

Definition 7.3.4: Multiple correlation coefficient The term ρ, as defined in Eq. (7.15), is called the *multiple correlation coefficient* between the component X_{q+i} of X_2 and the linear function $\beta_{(i)}\Sigma_{11}^{-1}X_{(1)}$ (or simply the multiple correlation of X_{q+i} on $X_{(1)}$). (*Note*: The multiple correlation coefficient is defined to be positive.)

Theorem 7.3.5 Of all linear combinations $\alpha X_{(1)}$, the one that minimizes the variance of $X_{q+i} - \alpha X_{(1)}$ and maximized the correlation between X_{q+1} and $\alpha X_{(1)}$ is the linear function $\beta_{(i)}\Sigma_{11}^{-1}X_{(1)}$.

Proof. Let $\beta = \beta_{(i)}\Sigma_{11}^{-1}$. Then

$$V(X_{q+i} - \alpha X_{(1)}) = E(X_{q+i} - \mu_{q+i} - \alpha(X_{(1)} - \mu_{(1)}))^2$$
$$= E(X_{q+i} - \mu_{q+i} - \beta(X_{(1)} - \mu_{(1)})$$
$$+ (\beta - \alpha)(X_{(1)} - \mu_{(1)}))^2$$
$$= E(X_{q+i} - \mu_{q+i} - \beta(X_{(1)} - \mu_{(1)}))^2$$
$$+ E((\beta - \alpha)(X_{(1)} - \mu_{(1)}))^2$$
$$+ 2E(X_{q+i} - \mu_{q+i} - \beta(X_{(1)} - \mu_{(1)}))$$
$$\times ((\beta - \alpha)(X_{(1)} - \mu_{(1)}))'.$$

But

$$E\beta(X_{(1)} - \mu_{(1)})(X_{q+i} - \mu_{q+i}) = \beta\beta'_{(i)} = \beta_{(i)}\Sigma_{11}^{-1}\beta'_{(i)}$$

$$\begin{aligned}
E(X_{q+i} - \mu_{q+i} - \beta(X_{(1)} - \mu_{(1)}))(X_{(1)} - \mu_{(1)})' &= \beta_{(i)} - \beta\Sigma_{11} \\
&= \beta_{(i)} - \beta_{(i)} \\
&= 0
\end{aligned}$$

$$\begin{aligned}
E((\beta - \alpha)(X_{(1)} - \mu_{(1)}))^2 \\
= E((\beta - \alpha)(X_{(1)} - \mu_{(1)})(X_{(1)} - \mu_{(1)})'(\beta - \alpha)') \\
= (\beta - \alpha)E((X_{(1)} - \mu_{(1)})(X_{(1)} - \mu_{(1)})'(\beta - \alpha)') \\
= (\beta - \alpha)\Sigma_{11}(\beta - \alpha)'
\end{aligned}$$

$$\begin{aligned}
E(X_{q+i} - \mu_{q+i} - \beta(X_{(1)} - \mu_{(1)}))^2 \\
= \sigma_{q+i}^2 - 2E((X_{q+i} - \mu_{q+i})(X_{(1)} - \mu_{(1)})'\beta') \\
\quad + \beta E((X_{(1)} - \mu_{(1)})(X_{(1)} - \mu_{(1)})'\beta') \\
= \sigma_{q+i}^2 - 2\beta_{(i)}\Sigma_{11}^{-1}\beta'_{(i)} + \beta_{(i)}\Sigma_{11}^{-1}\beta'_{(i)} \\
= \sigma_{q+i}^2 - \beta_{(i)}\Sigma_{11}^{-1}\beta'_{(i)}. \tag{7.16}
\end{aligned}$$

Hence

$$V(X_{q+i} - \alpha X_{(1)}) = \sigma_{q+i}^2 - \beta_{(i)}\Sigma_{11}^{-1}\beta'_{(i)} + (\beta - \alpha)\Sigma_{11}(\beta - \alpha)'.$$

Since Σ_{11} is positive definite, $(\beta - \alpha)\Sigma_{11}(\beta - \alpha)' \geq 0$ and is 0 if $\beta - \alpha = 0$. Thus $\beta X_{(1)}$ is the linear function such that $X_{q+i} - \beta X_{(1)}$ has the minimum variance. We now consider the correlation between X_{q+i} and $\alpha X_{(1)}$ and show that this correlation is maximum if $\alpha = \beta$.

For any scalar c, $c\alpha X_{(1)}$ is a linear function of $X_{(1)}$. Hence

$$E(X_{q+i} - \beta(X_{(1)} - \mu_{(1)}))^2 \leq E(X_{q+i} - \mu_{q+i} - c\alpha(X_{(1)} - \mu_{(1)}))^2. \tag{7.17}$$

Dividing both sides of Eq. (7.17) by $\sigma_{q+i}\sqrt{E(\beta(X_{(1)} - \mu_{(1)}))^2}$ and choosing

$$c = \left[\frac{E(\beta(X_{(1)} - \mu_{(1)}))^2}{E(\alpha(X_{(1)} - \mu_{(1)}))^2} \right]^{1/2}$$

we get from Eq. (7.17), after some obvious simplifications,

$$\frac{E(X_{q+1} - \mu_{q+i})(\beta(X_{(1)} - \mu_{(1)}))}{\sigma_{q+i}\sqrt{E(\beta(X_{(1)} - \mu_{(1)}))^2}}$$

$$\geq \frac{E(X_{q+1} - \mu_{q+i})(\alpha(X_{(1)} - \mu_{(1)}))}{\sigma_{q+i}\sqrt{E(\alpha(X_{(1)} - \mu_{(1)}))^2}}. \qquad \text{Q.E.D.}$$

Example 7.3.1 Let $X = (X_1, X_2, X_3)'$ be distributed as $N_3(\mu, \Sigma)$, where

$$\mu = \begin{pmatrix} 1 \\ 2 \\ 3 \end{pmatrix}, \quad \Sigma = \begin{pmatrix} 4 & 1 & 0 \\ 1 & 8 & 0 \\ 0 & 0 & 9 \end{pmatrix}.$$

Then X_1 is $N(1,4)$, X_2 is $N(2,8)$, X_3 is $N(3,9)$;

$$\begin{pmatrix} X_1 \\ X_2 \end{pmatrix} \quad \text{is} \quad N_2\left(\begin{pmatrix} 1 \\ 2 \end{pmatrix}, \begin{pmatrix} 4 & 1 \\ 1 & 8 \end{pmatrix} \right),$$

The condition pdf of X_1 given $X_2 = x_2$ is $N(\frac{3}{4} + \frac{1}{8}x_2, \frac{31}{8})$.

Example 7.3.2 Let $X = (X_1, X_2, X_3)'$ be distributed as $N_3(\mu, \Sigma)$, where

$$\mu = (\mu_1, \mu_2, \mu_3)', \quad \Sigma = \begin{pmatrix} \sigma_1^2 & \sigma_{12} & \sigma_{13} \\ \sigma_{12} & \sigma_1^2 & \sigma_{23} \\ \sigma_{13} & \sigma_{23} & \sigma_3^2 \end{pmatrix}.$$

The distribution of

$$\begin{pmatrix} X_1 - X_3 \\ X_1 - 2X_2 + X_3 \end{pmatrix} = AX,$$

where

$$A = \begin{pmatrix} 1 & 0 & -1 \\ 1 & -2 & 1 \end{pmatrix},$$

is $N_2(A\mu, A\Sigma A')$ with

$A\mu = (\mu_1 - \mu_3, \mu_1 - 2\mu_2 + \mu_3)'$

$A\Sigma A' =$

$$\begin{pmatrix} \sigma_1^2 - 2\sigma_{13} + \sigma_3^2 & \sigma_1^2 - 2\sigma_{12} + 2\sigma_{23} - \sigma_3^2 \\ \sigma_1^2 - 2\sigma_{13} + 2\sigma_{23} - \sigma_3^2 & \sigma_1^2 - 4\sigma_{12} + 2\sigma_{13} + 4\sigma_2^2 - 4\sigma_{23} + \sigma_3^2 \end{pmatrix}.$$

Example 7.3.3 Let $X = (X_1, X_2, X_3)'$ be distributed as $N_3(\mu, \Sigma)$, where $\mu = (\mu_1, \mu_2, \mu_3)'$ and

$$\Sigma = \begin{pmatrix} 2 & 3 & -1 \\ 3 & 7 & 1 \\ -1 & 1 & 10 \end{pmatrix}.$$

The multiple correlation coefficient of X_3 on $(X_1, X_2)'$ is

$$\rho = \sqrt{\dfrac{(-1,1)\begin{pmatrix} 2 & 3 \\ 3 & 7 \end{pmatrix}^{-1}\begin{pmatrix} -1 \\ 1 \end{pmatrix}}{10}} = \sqrt{\dfrac{3}{10}} = 0.548.$$

The regression of X_3 on $X_1 = x_1$ and $X_2 = x_2$ is

$$E(X_3|X_1 = x_1, X_2 = x_2) = \mu_3 + (-1,1)\begin{pmatrix} 2 & 3 \\ 3 & 7 \end{pmatrix}^{-1}\begin{pmatrix} x_1 - \mu_1 \\ x_2 - \mu_2 \end{pmatrix}$$

$$= \mu_3 + (-2,1)\begin{pmatrix} x_1 - \mu_1 \\ x_2 - \mu_2 \end{pmatrix}$$

$$= \mu_3 - 2(x_1 - \mu_1) + (x_2 - \mu_2).$$

7.4 ELLIPTICALLY SYMMETRIC DISTRIBUTIONS

In analyzing multivariate data, the role of multivariate normal distribution is utmost important. In actual practice, however, the assumption of multinormality does not always hold and the verification of this assumption is cumbersome, if not, impossible. However, somtimes, optimum statistical procedures derived under multinormality assumption remain also optimum (to be defined later) when the underlying distribution is a member of the family of elliptically symmetric distributions. This family of distributions is frequently used in filtering and stochastic control (Chu, 1973), random signal input (McGraw and Wagner, 1968), and stock market data analysis (Zellner, 1986). This family included the multivariate normal, the multivariate Cauchy, the Pearsonian type II and IV distributions, and the Laplace distributions, among others. The family of spherically symmetric distributions is a particular case of this family. The contaminated normal, the compound normal, and the multivariate t distribution are some examples of spherically symmetric distributions.

Definition 7.4.1: Elliptically symmetric distribution   A random vector $X = (X_1,...,X_p)'$ with values $x = (x_1,...,x_p)'$ in R^p is said to have a distribution belonging to the family of elliptically symmetric distributions with the location parameter $\mu = (\mu,...,\mu_p)'$ and the scale parameter Σ, a $p \times p$ positive definite matrix, if its pdf is a function of the quadratic form $(x - \mu)'\Sigma^{-1}(x - \mu)$ and is given by

$$f_X(x) = |\Sigma|^{-1/2}q((x - \mu)'\Sigma^{-1}(x - \mu)), \tag{7.18}$$

where q is a function on $[0,\infty)$ satisfying

$$\int_{-\infty}^{\infty} \cdots \int_{-\infty}^{\infty} q(y'y) \, dy_1 \cdots dy_p = 1$$

with $y = (y_1,\ldots,y_p)'$.

We denote the family of elliptically symmetric distributions with parameters (μ,Σ) by $E_p(\mu,\Sigma)$.

Definition 7.4.2: Spherically symmetric distributions A random vector $X = (X_1,\ldots,X_p)'$ is said to have a distribution belonging to the family of spherically symmetric distributions if X and OX have the same distribution for all $p \times p$ orthogonal matrix O. In other words, the pdf of X depends on X only through $X'X$.

Theorem 7.4.1 If $X = (X_1,\ldots,X_p)'$ is distributed as $E_p(\mu,\Sigma)$, then $\Sigma^{-1/2}(X - \mu)$ has a spherically symmetric distribution.

Proof. Let $Y = \Sigma^{-1/2}(X - \mu)$, where $\Sigma^{1/2}$ is a $p \times p$ symmetric matrix such that $\Sigma^{1/2}\Sigma^{1/2} = \Sigma$. The Jacobian of the transformation $X \to Y$ is $|\Sigma|^{1/2}$. Hence the pdf of Y is

$$f_Y(y) = |\Sigma|^{-1/2}q(y'y)|\Sigma|^{1/2}$$
$$= q(y'y)$$

and $q(y'y) = q((Oy)'(Oy))$.   Q.E.D.

7.4.1 Some Examples of Symmetric Distributions

Example 7.4.1: Contaminated normal A p-variate random vector $X = (X_1,\ldots,X_p)'$ has a contaminated normal distribution if its pdf is given by

$$f_X(x) = \frac{\alpha}{(2\pi\sigma_1^2)^{p/2}} \exp\left(-\frac{1}{2\sigma_1^2} x'x\right)$$
$$+ \frac{1 - \alpha}{(2\pi\sigma_2^2)^{p/2}} \exp\left(-\frac{1}{2\sigma_2^2} x'x\right), \tag{7.19}$$

where $0 < \alpha < 1$ and $\sigma_i^2 > 0$, $i = 1,2$. Since the Jacobian of the transformation $X \to Y = OX$, with O a $p \times p$ orthogonal matrix, is unity, the pdf of Y is

$$f_Y(y) = \frac{\alpha}{(2\pi\sigma_1^2)^{p/2}} \exp\left(\frac{-1}{2\sigma_1^2} y'y\right) + \frac{1 - \alpha}{(2\pi\sigma_2^2)^{p/2}} \exp\left(-\frac{1}{2\sigma_2^2} y'y\right),$$

Hence it is spherically symmetric.

Example 7.4.2: Compound normal A p-variate random vector $X = (X_1,\ldots,X_p)'$ has a compound normal distribution if its pdf is given by

$$f_X(x) = \int_0^\infty \frac{1}{(2\pi z)^{p/2}} \exp\left(-\frac{1}{2z} x'x\right) f_Z(z)\, dz, \tag{7.20}$$

where Z is a positive random variable with pdf $f_Z(z)$. It is spherically symmetric.

Example 7.4.3: Multivariate t distribution A p-variate random vector $X = (X_1,\ldots,X_p)'$ has a p-variate t distribution with n degrees of freedom if its pdf is given by

$$f_X(x) = \frac{\Gamma(\tfrac{1}{2}(n + p))}{\Gamma(n/2)(n\pi)^{p/2}} (1 + x'x)^{-(n+p)/2}. \tag{7.21}$$

It is spherically symmetric. Multivariate t distribution with 1 degree of freedom is called the multivariate Cauchy distribution for which no moment exists.

Example 7.4.4: Multivariate t distribution with parameters μ,Σ A p-variate random vector $X = (X_1,\ldots,X_p)'$ is distributed as a multivariate t distribution with parameters (μ,Σ) and n degrees of freedom if its pdf is given by

$$f_X(x) = \frac{\Gamma(\tfrac{1}{2}(n + p))}{\Gamma(n/2)(n\pi)^{p/2}} (\det \Sigma)^{-1/2}$$
$$\times \left[1 + (x - \mu)' \frac{\Sigma^{-1}}{n} (x - \mu)\right]^{-1/2(n+p)}. \tag{7.22}$$

It is elliptically symmetric.

The following theorems will establish that the elliptically symmetric distributions possess most of the characteristic properties of the multivariate normal distribution, which is also a member of this class.

Theorem 7.4.2 Let X be distributed as $E_p(\mu,\Sigma)$. The characteristic function of X is given by

$$E(\exp(it'x)) = \exp(it'\mu)\psi(t'\Sigma t), \tag{7.23}$$

where $t = (t_1,\ldots,t_p)' \in R^p$ and ψ is some function on $[0,\infty)$.

Proof. Let $Y = X - \mu$, $Z = \Sigma^{-1/2}Y$, $\alpha = \Sigma^{1/2}t$, where $\Sigma^{1/2}$ is a symmetric matrix satisfying $\Sigma^{1/2}\Sigma^{1/2} = \Sigma$.

$$
\begin{aligned}
\phi_X(t) &= E(\exp(it'X)) \\
&= \exp(it'\mu)E(\exp(it'Y)) \\
&= \exp(it'\mu)|\Sigma|^{-1/2} \int_{-\infty}^{\infty} \cdots \int_{-\infty}^{\infty} \exp(it'y)q(y'\Sigma^{-1}y)\, dy_1 \cdots dy_p \\
&= \exp(it'\mu) \int_{-\infty}^{\infty} \cdots \int_{-\infty}^{\infty} \exp(i\alpha'z)q(z'z)\, dz_1 \cdots dz_p.
\end{aligned}
$$

Let O be a $p \times p$ orthogonal matrix, given by

$$
O = \begin{pmatrix} \dfrac{\alpha_1}{\sqrt{\alpha'\alpha}} & \cdots & \dfrac{\alpha_p}{\sqrt{\alpha'\alpha}} \\ O_{21} & \cdots & O_{2p} \\ O_{p1} & \cdots & O_{pp} \end{pmatrix},
$$

where O_{ij}'s are arbitrary. Since

$$
\sum_{j=1}^{p} O_{ij}\alpha_j = \sqrt{\alpha'\alpha} \left(\sum_{j=1}^{p} O_{ij} \frac{\alpha_j}{\sqrt{\alpha'\alpha}} \right)
$$

we get

$$
O\alpha = ((\alpha'\alpha)^{1/2}, 0 \ldots, 0)'.
$$

Let $OZ = U = (U_1, \ldots, U_p)'$. Since $q(z'z) = q(u'u)$ is symmetric, we get

$$
\begin{aligned}
\phi_X(t) &= \exp(it'\mu) \int_{-\infty}^{\infty} \cdots \int_{-\infty}^{\infty} \exp[i(\alpha'\alpha)^{1/2}u_1]q(u'u)\, du_1 \cdots du_p \\
&= \exp(it'\mu)\psi(\alpha'\alpha) \\
&= \exp(it'\mu)\psi(t'\Sigma t). \qquad \text{Q.E.D.}
\end{aligned}
$$

It follows from this theorem that

$$
E(X) = \mu, \qquad \text{cov}(X) = -2\psi'(0)\Sigma. \tag{7.24}
$$

Theorem 7.4.3 Let X be $E_p(\mu, \Sigma)$. Then $Y = CX + b$, where $b = (b_1, \ldots, b_p)' \in R^p$ and C is a $p \times p$ nonsingular matrix, is distributed as $E_P(C\mu + b, C\Sigma C')$.

Proof. The Jacobian of the transformation $X \to Y = CX + b$ is $|C|^{-1}$. Hence, from (7.18),

$$f_Y(y) = f_X(C^{-1}(y - b))|C|^{-1}$$
$$= |C\Sigma C'|^{-1/2}q((y - C\mu - b)'(C\Sigma C')^{-1}(y - C\mu - b)),$$

which is $E_p(C\mu + b, C\Sigma C')$.

Theorem 7.4.4 Let $X = (X'_{(1)}, X'_{(2)})'$, $X_{(1)} = (X_1, \ldots, X_q)'$, $X_{(2)} = (X_{q+1}, \ldots, X_p)'$, $p > q$. Let μ be similarly partitioned as $\mu = (\mu'_{(1)}, \mu'_{(2)})'$ and Σ be partitioned as

$$\Sigma = \begin{pmatrix} \Sigma_{11} & \Sigma_{12} \\ \Sigma_{21} & \Sigma_{22} \end{pmatrix},$$

where Σ_{11} is the upper left-hand-corner submatrix of Σ of dimension $q \times p$.

(a) The marginal pdf of $X_{(1)}$ is $E_q(\mu_{(1)}, \Sigma_{11})$.   (7.25)
(b) The conditional pdf of $X_{(2)}$ given $X_{(1)} = x_{(1)}$ is

$$E_{p-q}(\mu_{(2)} + \Sigma_{21}\Sigma_{11}^{-1}(x_{(1)} - \mu_{(1)}), \Sigma_{22.1}), \qquad \text{where}$$
$$\Sigma_{22.1} = \Sigma_{22} - \Sigma_{21}\Sigma_{11}^{-1}\Sigma_{12}. \tag{7.26}$$

Proof. Since Σ is positive definite, by Theorem A.2.7.4 there exists a lower-triangular nonsingular $p \times p$ matrix T in the block form

$$T = \begin{pmatrix} T_{11} & 0 \\ T_{21} & T_{22} \end{pmatrix},$$

where T_{11} is the upper left-hand $q \times q$ submatrix of T, satisfying $T\Sigma T' = I$. This implies that $T_{11}\Sigma_{11}T'_{11} = I$. Let $Y = T(X - \mu)$. Since the Jacobian of the transformation $X \to Y$ is $|T|^{-1}$, we get, using (7.18),

$$f_Y(y) = f_X(T^{-1}y + \mu)|T|^{-1}$$
$$= q(y'y).$$

Partition $Y = (Y'_{(1)}, Y'_{(2)})'$ as X. The pdf of $Y_{(1)}$ is

$$f_{Y_{(1)}}(y_{(1)}) = \int_{-\infty}^{\infty} \cdots \int_{-\infty}^{\infty} q(y'_{(1)}y_{(1)} + y'_{(2)}y_{(2)}) \, dy_{q+1} \cdots dy_p$$
$$= \bar{q}(y'_{(1)}y_{(1)}) \qquad \text{(say).} \tag{7.27}$$

Since

$$\int_{-\infty}^{\infty} \cdots \int_{-\infty}^{\infty} \overline{q}(y_{(1)}'y_{(1)})\, dy_1 \cdots dy_q = \int_{-\infty}^{\infty} \cdots \int_{-\infty}^{\infty} \overline{q}(y'y)\, dy_1 \cdots dy_p = 1,$$

$\overline{q}(y_{(1)}'y_{(1)})$ is a pdf and $\overline{q}$ is on $[0,\infty)$. From (7.27),

$$\begin{aligned}
f_{X_{(1)}}(x_{(1)}) &= f_{Y_{(1)}}(T_{11}(x_{(1)} - \mu_{(1)}))|T_{11}|^{-1} \\
&= |\Sigma_{11}|^{-1/2}\overline{q}((x_{(1)} - \mu_{(1)})'\Sigma_{11}^{-1}(x_{(1)} - \mu_{(1)})),
\end{aligned}$$

which is $E_q(\mu_{(1)},\Sigma_{11})$.

(b) Let $U = X_{(2)} - \mu_{(2)} - \Sigma_{21}\Sigma_{11}^{-1}(X_{(1)} - \mu_{(1)})$. Then

$$(x - \mu)'\Sigma^{-1}(x - \mu) = (x_{(1)} - \mu_{(1)})'\Sigma_{11}^{-1}(x_{(1)} - \mu_{(1)}) + u'\Sigma_{22.1}^{-1}u.$$

From (7.18)

$$f_{X_{(1)}U}(x_{(1)},u) = |\Sigma_{11}|^{-1/2}|\Sigma_{22.1}|^{-1/2}q(x_{(1)} - \mu_{(1)})\Sigma_{11}^{-1}(x_{(1)} - \mu_{(1)}) + u'\Sigma_{22.1}^{-1}u).$$
$$(7.28)$$

Let

$$\begin{aligned}
W_{(1)} &= \Sigma_{11}^{-1/2}(X_{(1)} - \mu_{(1)}) \\
W_{(2)} &= \Sigma_{22.1}^{-1/2}U = (W_{q+1},\ldots,W_p)'.
\end{aligned}$$

From (7.28) the joint pdf of $W_{(1)}$ and $W_{(2)}$ is given by

$$f_{W_{(1)},W_{(2)}}(w_{(1)},w_{(2)}) = q(w_{(1)}'w_{(1)} + w_{(2)}'w_{(2)}).$$

The marginal pdf of $W_{(1)}$ is

$$\begin{aligned}
f_{W_{(1)}}(w_{(1)}) &= \int_{-\infty}^{\infty} \cdots \int_{-\infty}^{\infty} q(w_{(1)}'w_{(1)} + w_{(2)}'w_{(2)})\, dw_{q+1} \cdots dw_p. \\
&= \overline{q}(w_{(1)}'w_{(1)}) \qquad \text{(say)}.
\end{aligned}$$

Hence the conditional pdf of $W_{(2)}$ given $W_{(1)} = w_{(1)}$ is

$$\overline{\overline{q}}(w_{(2)}'w_{(2)}) = \frac{q(w_{(1)}'w_{(1)} + w_{(2)}'w_{(2)})}{\overline{q}(w_{(1)}'w_{(1)})}.$$

Thus the conditional pdf of $X_{(2)}$ given $X_{(1)} = x_{(1)}$ is

$$\begin{aligned}
|\Sigma_{22.1}|^{-1/2}&\overline{\overline{q}}(x_{(2)} - \mu_{(2)} - \Sigma_{21}\Sigma_{11}^{-1}(x_{(1)} - \mu_{(1)}))' \\
&\times \Sigma_{22.1}^{-1}(x_{(2)} - \mu_{(2)} - \Sigma_{21}\Sigma_{11}^{-1}(x_{(1)} - \mu_{(1)})). \qquad \text{Q.E.D.}
\end{aligned}$$

7.5 MULTINOMIAL DISTRIBUTION

We have seen that a binomial distribution arises when the outcome of a random experiment admits of two classifications, which are called success and failure. A multinomial distribution is a generalization of the binomial distribution when the outcome of a random experiment admits of $p > 2$ classifications. Let p_i be the probability of the outcome of a single experiment belonging to the ith classification, $i = 1,\ldots,p$. Obviously,

$$\sum_{i=1}^{p} p_i = 1.$$

Let there be n independent repetitions of a trial and let x_i, $i = 1,\ldots,p$ be the number of outcomes belonging to the ith classification. Obviously,

$$\sum_{i=1}^{p} x_i = n.$$

The probability mass function of $X = (X_1,\ldots,X_p)'$ can easily be worked out and is given by

$$p_X(x) = P(X_1 = x_1,\ldots,X_p = x_p)$$

$$= \begin{cases} \dfrac{n!}{x_1! \cdots x_p!} p_1^{x_1} \cdots p_p^{x_p} & \text{if } 0 \leq x_i \leq n \text{ for all } i \text{ and } \sum_{i=1}^{p} x_i = n \\ 0, & \text{otherwise.} \end{cases}$$

$$(7.29)$$

Since $\sum_{i=1}^{p} x_i = n$ we can write Eq. (7.29) as

$$P(X_1 = x_1,\ldots,X_p = x_p) = \frac{n!}{x_1! \cdots x_{p-1}!(n - x_1 - \cdots - x_{p-1})!}$$

$$\times p_1^{x_1} \cdots p_{p-1}^{x_{p-1}}(1 - p_1 - \cdots - p_{p-1})^{n - x_1 - \cdots x_{p-1}}.$$

Furthermore,

$$\sum_{x_1} \cdots \sum_{x_p} p_X(x) = (p_1 + \cdots + p_p)^n = 1$$

$$\sum_{i=1}^{p} x_i = n.$$

The p-dimensional discrete random variable X with the probability mass function above is said to have *multinomial distribution*. Means and covariances for this distribution are obtained from the following identities:

$$np_i = p_i \frac{\partial}{\partial p_i} (p_1 + \cdots + p_p)^n$$

$$= p_i \frac{\partial}{\partial p_i} \sum_{x_i} \cdots \sum_{x_p} \frac{n!}{x_1! \cdots x_p!} p_1^{x_1} \cdots p_p^{x_p}$$

$$= \sum_{x_1} \cdots \sum_{x_p} x_i p_X(x) = E(X_i)$$

$$n(n-1)p_i p_j = p_j \frac{\partial}{\partial p_j} p_i n (p_1 + \cdots p_p)^{n-1}$$

$$= p_j \frac{\partial}{\partial p_j} \left[p_i \frac{\partial}{\partial p_i} (p_1 + \cdots + p_p)^n \right]$$

$$= p_j \frac{\partial}{\partial p_j} \sum_{x_1} \cdots \sum_{x_p} x_i \frac{n!}{x_1! \cdots x_p!} p_1^{x_1} \cdots p_p^{x_p}$$

$$= \sum_{x_1} \cdots \sum_{x_p} x_j x_i \frac{n!}{x_1! \cdots x_p!} p_1^{x_1} \cdots p_p^{x_p}$$

$$= E(X_i X_j),$$

and similarly,

$$n(n-1)p_i^2 = E(X_i(X_i - 1)).$$

Hence

$$\text{cov}(X_i X_j) = -np_i p_j (i \neq j), \qquad V(X_i) = np_i(1 - p_i).$$

Example 7.5.1   Let $X = (X_1, \ldots, X_4)'$ be distributed as multinomial with pdf

$$p_{X_1,\ldots,X_4}(x_1,\ldots,x_4) = \frac{n!}{x_1! \, x_2! \, x_3! (n - x_1 - x_2 - x_3)!}$$

$$\times (1 - p_1 - p_2 - p_3)^{n - x_1 - x_2 - x_3} \prod_{i=1}^{3} p_i^{x_i}$$

with $x_1 + x_2 + x_3 + x_4 = n$, $0 \le p_i \le 1$, $i = 1,\ldots,4$, $p_1 + p_2 + p_3 + p_4 = 1$. Then

$$
\underset{X_1,\ldots,X_3}{p}(x_1,\ldots,x_3) = \frac{n!}{x_1!\, x_2!(n - x_1 - x_2)!}\, p_1^{x_1} p_2^{x_2}(1 - p_1 - p_2)^{n - x_1 - x_2}
$$

$$
\times \sum_{x_3=0}^{n-x_1-x_2} \frac{(n - x_1 - x_2)!}{x_3!(n - x_1 - x_2 - x_3)!}
$$

$$
\times \left(\frac{p_3}{1 - p_1 - p_2}\right)^{x_3} \left(1 - \frac{p_3}{1 - p_1 - p_2}\right)^{n - x_1 - x_2 - x_3}
$$

$$
= \frac{n!}{x_1!\, x_2!(n - x_1 - x_2)!}\, p_1^{x_1} p_2^{x_2}(1 - p_1 - p_2)^{n - x_1 - x_2}
$$

with $p_1 + p_2 + p_3 = 1$, $x_1 + x_2 + x_3 = n$. The expression under the summation sign is equal to 1 since

$$
\sum_{j=0}^{m} \binom{m}{j} \theta^j (1 - \theta)^{m-j} = 1 \qquad \text{for } 0 \le \theta \le 1.
$$

Proceeding in the same way, we get

$$
\underset{X_1,X_2}{p}(x_1,x_2) = \frac{n!}{x_1!(n - x_1)!}\, p_1^{x_1}(1 - p_1)^{n - x_1}.
$$

7.6 DISTRIBUTION OF QUADRATIC FORMS

For the definition and properties of quadratic forms, the reader is referred to Appendix A. In this subsection we prove some important theorems regarding the distribution of quadratic forms.

Theorem 7.6.1 Let $X = (X_1,\ldots,X_p)'$ be distributed as $N_p(\mu,\Sigma)$. Then $X'\Sigma^{-1}X$ is distributed as $\chi_p^2(\mu'\Sigma^{-1}\mu)$.

Proof. Since Σ is positive definite, by Corollary A.2.6.6.2 in Appendix A there exists a nonsingular matrix C such that $C\Sigma C' = I$. Let $Y = CX$. Then

$$
E(Y) = C\mu = v = (v_1,\ldots,v_p)',
$$

where $v = (v_1,\ldots,v_p)$ is an arbitrary designation, and

$$
E(Y - v)(Y - v)' = E(CX - C\mu)(CX - C\mu)' = C\Sigma C' = I.
$$

By Theorem 7.3.1, $Y_1,...,Y_P$ are independently distributed normal random variables with means $v_1,...,v_p$, respectively, and with variance 1. Thus, by results in Section 6.1.5, $Y_1^2,...,Y_p^2$ are independent with $\chi_1^2(\lambda_i)$ distribution, $\lambda_i = v_i^2$, $i = 1,...,p$. Hence

$$X'\Sigma^{-1}X = (CX)'(C\Sigma C')^{-1}CX = Y'Y = \sum_i^p Y_i^2$$

is distributed as $\chi_p^2(\lambda)$, where

$$\lambda = \sum_{i=1}^p v_i^2 = v'v = (C\mu)'(C\mu) = \mu'C'C\mu = \mu'\Sigma^{-1}\mu. \qquad \text{Q.E.D.}$$

Corollary 7.6.1.1 If X is distributed as $N_p(0,\Sigma)$, then $X'\Sigma^{-1}X$ is distributed as χ_p^2.

Proof. Here $\lambda = 0$, and the proof follows trivially from the main theorem.   Q.E.D.

Theorem 7.6.2 Let X be distributed as $N_p(0,\Sigma)$ and let $Y = AX$, where A is a matrix of dimension $q \times p$ and is of rank $q(p > q)$, then:

(a) $Y'(A\Sigma A')^{-1}Y$ is distributed as χ_q^2.
(b)   If $X_1,...,X_p$ are subject to the restriction $AX = 0$, then $X'\Sigma^{-1}X$ is distributed as χ_{p-q}^2.

Proof.

(a)   By Theorem 7.3.2, Y is distributed as $N_q(0,A\Sigma A')$, and hence, by Corollary 7.6.1.1, $Y'(A\Sigma A')^{-1}Y$ is distributed as χ_q^2.

(b)   Let $C = \binom{A}{B}$ be a $p \times p$ nonsingular matrix such that the matrix B is of dimension $(p - q) \times p$ and is chosen in such a way that the random variable $Y = AX$ is independent of $Z = BX$, that is,

$$E(AX - A\mu)(BX - B\mu)' = AE(X - \mu)(X - \mu)'B'$$
$$= A\Sigma B' = 0.$$

Now

$$C\Sigma C' = \begin{pmatrix} A\Sigma A' & 0 \\ 0 & B\Sigma B' \end{pmatrix}$$

and $C\Sigma C'$ is nonsingular. Thus $A\Sigma A'$, $B\Sigma B'$ are nonsingular. By Theorem 7.3.1, $CX = \binom{Y}{Z}$ is distributed as $N_p(0,C\Sigma C')$. But

$$X'\Sigma^{-1}X = (CX)'(C\Sigma C')^{-1}CX = Y'(A\Sigma A')^{-1}Y + Z'(B\Sigma B')^{-1}Z.$$

Thus the value of $X'\Sigma^{-1}X$, subject to the restriction $Y = 0$, is $X'\Sigma^{-1}X = Z'(B\Sigma B')^{-1}Z$. By Theorem 7.3.2, Z is distributed as $N_{p-q}(0,B\Sigma B')$. Hence under the restriction $Y = 0$, $X'\Sigma^{-1}X$ is distributed as χ^2_{p-q}.   Q.E.D.

Theorem 7.6.3: Cochran's theorem Let $X = (X_1,\ldots,X_p)'$ be distributed as $N_p(O,I)$ and let $X'X = Q_1 + \cdots + Q_k$, where Q_i is a quadratic form of rank p_i. Then the necessary and sufficient condition for $Q_1,\ldots,Q_k$ to be independently distributed chi-square random variables with degrees of freedom $p_1,\ldots,p_k$, respectively, is that $\Sigma_1^k p_i = p_.$.

Proof. Suppose that Q_i, $i = 1,\ldots,k$ are independently distributed chi-square random variables with $p_1,\ldots,p_k$ degrees of freedom. Then $\Sigma_1^k Q_i$ is chi-square distributed with $\Sigma_1^k p_i$ degrees of freedom. Since

$$X'X = \sum_{i=1}^{p} X_i^2$$

is chi-square distributed with p degrees of freedom and

$$\sum_{i=1}^{p} X_i^2 = \sum_{1}^{k} Q_i,$$

we get $\Sigma_1^p p_i = p$. This proves the necessity of the condition.

Since Q_i is a quadratic form of rank p_i, by Exercise 16 of Chapter 6, Q_i can be expressed as

$$Q_i = \sum_{j=1}^{p_i} \pm Y_{ij}^2,$$

where the Y_{ij} are linear functions of $X_1,\ldots,X_p$. Hence

$$X'X = Y'\Delta Y, \tag{7.30}$$

where Y is a vector of dimension $\Sigma_1^k p_i$,

$$Y = (Y_{11},\ldots,Y_{1p_1},\ldots,Y_{k1},\ldots,Y_{kp_k})'$$

and Δ is a diagonal matrix of dimension $p \times p$ with diagonal elements ± 1. Let $Y = AX$ be the transformation that transforms the positive definite quadratic form $X'X$ to $Y'\Delta Y$ and let $\Sigma_1^k p_i = p$. Since

$$X'X = Y'\Delta Y = X'A'\Delta AX$$

we get $A'\Delta A = I$. This implies that A is nonsingular. Thus the transformed quadratic form $Y'\Delta Y$ is positive definite. Hence $\Delta = I$ and $A'A = I$. By Theorem 7.3.1, Y is distributed as $N_p(0,I)$. So Q_i, being the sum of squares of p_i independent $N_1(0,1)$, is chi-square distributed with p_i degrees of freedom.   Q.E.D.

Corollary 7.6.3.1 Let X be distributed as $N_p(0,\Sigma)$. If $X'\Sigma^{-1}X = Q_1 + \cdots + Q_k$, where $Q_i, i = 1,\ldots,k$ are quadratic forms of rank p_i, the necessary and sufficient condition for $Q_1,\ldots,Q_k$ to be independently distributed chi-square random variables with degrees of freedom $p_1,\ldots,p_k$, respectively, is that $\Sigma_1^k p_i = p$.

Proof. Let $Q_i = X'AX$ be a quadratic form of rank p_i, and let Q_i be transformed to the quadratic form

$$Q_i = Y'(C')^{-1}AC^{-1}Y$$

by means of the nonsingular transformation $Y = CX$ satisfying $C\Sigma C' = I$. Then Q_i is a quadratic form in Y and is of rank p_i. The proof of the corollary follows from Theorem 7.6.3.   Q.E.D.

7.7 MULTIVARIATE NORMAL CASE: DISTRIBUTION OF SAMPLE MEAN AND SAMPLE COVARIANCE MATRICES

Let $X_\alpha = (X_{\alpha 1},\ldots,X_{\alpha p})'$, $\alpha = 1,\ldots,N$, be a sample of size N from $N_p(\mu,\Sigma)$, and let

$$\overline{X} = \sum_{\alpha=1}^{N} \frac{X_\alpha}{N}, \qquad S = \sum_{\alpha=1}^{N} (X_\alpha - \overline{X})(X_\alpha - \overline{X})'. \tag{7.31}$$

Then $\overline{X}$ and S/N are, respectively, the sample mean and the sample covariance matrix.

Theorem 7.7.1 Let $X_1,\ldots,X_N$ be independent and identically distributed p-variate normal random variables with mean μ and covariance matrix Σ, and let $\overline{X}$ and S be as defined in Eq. (7.31). Then $\sqrt{N}\,\overline{X}$, S are independent and $\sqrt{N}\,\overline{X}$ is distributed as $N_p(\sqrt{N}\mu,\Sigma)$ and S is distributed as $\Sigma_{\alpha=2}^{N} Y_\alpha Y_\alpha'$, where Y_α', $\alpha = 2,\ldots,N$, are independently distributed as $N_p(0,\Sigma)$.

Proof. Since $X_\alpha, \alpha = 1,\ldots,N$, are independent,

$$E(X_\alpha - \mu)(X_\beta - \mu)' = \begin{cases} 0 & \text{if } \alpha \neq \beta \\ \Sigma & \text{if } \alpha = \beta. \end{cases}$$

Let

$$Y_\beta = \sum_{\alpha=1}^{N} O_{\beta\alpha}X_\alpha, \qquad \beta = 1,\ldots,N,$$

where $O_{\beta\alpha}$ are defined in Eq. (6.31). The components of Y_β are linear combinations of the components of $X_{\alpha'}$ $\alpha = 1,\ldots,N$. Since each X_α is distributed as $N_p(\mu,\Sigma)$, Y_β also has normal distribution. Now

$$E(Y_1) = \frac{\mu}{\sqrt{N}} + \cdots + \frac{\mu}{\sqrt{N}} = \sqrt{N}\,\mu$$

$$E(Y_\beta) = \sum_{\alpha=1}^{N} O_{\beta\alpha}\mu$$

$$= \sqrt{N}\,\mu \sum_{\alpha=1}^{N} O_{\beta\alpha} \frac{1}{\sqrt{N}} = 0, \qquad \beta = 2,\ldots,N$$

$$\operatorname*{cov}_{\beta\neq\gamma}(Y_\beta Y_\gamma) = \sum_{\alpha=1}^{N} O_{\beta\alpha}O_{\gamma\alpha}\, E(X_\alpha - \mu)(X_\alpha - \mu)'$$

$$+ \sum_{\substack{\alpha=1 \\ \alpha\neq\alpha'}}^{N} \sum_{\alpha'=1}^{N} O_{\beta\alpha}O_{\gamma a'}E(X_\alpha - \mu)(X_{\alpha'} - \mu)'$$

$$= \begin{cases} \Sigma \displaystyle\sum_{\alpha=1}^{N} O_{\beta\alpha}O_{\gamma\alpha} = 0 & \text{if } \beta \neq \gamma \\ \Sigma & \text{if } \beta = \gamma. \end{cases}$$

Furthermore,

$$\sum_{\alpha=1}^{N} Y_\alpha Y_\alpha' = \sum_{\alpha=1}^{N} \sum_{\beta=1}^{N} O_{\alpha\beta}X_\beta \sum_{\gamma=1}^{N} O_{\alpha\gamma}X_\gamma'$$

$$= \sum_{\beta=1}^{N} \sum_{\gamma=1}^{N} \left(\sum_{\alpha=1}^{N} O_{\alpha\beta}O_{\alpha\gamma} \right) X_\beta X_\gamma'$$

$$= \sum_{\beta=1}^{N} X_\beta X_\beta' \tag{7.32}$$

in view of relations (6.32) and (6.33). Now by Theorem 7.3.3, $Y_1,\ldots,Y_N$ are independent and $Y_2,\ldots,Y_N$ are distributed as $N_p(0,\Sigma)$. Hence, from Eq. (7.32), $\sqrt{N}\,\bar{X}$, S are independent in distribution, and $\sqrt{N}\,\bar{X}$ is distributed as $N_p(\sqrt{N}\,\mu,\Sigma)$ and S is distributed as $\Sigma_{\alpha=2}^{N} Y_\alpha Y_\alpha'$, where $Y_2,\ldots,Y_N$ are independently distributed as $N_p(0,\Sigma)$.   Q.E.D.

This distribution of S is known as the Wishart distribution with $N - 1$ degrees of freedom and parameter Σ and is given by (if $N > p$)

$$\frac{|s|^{(N-p-2)/2} \exp(-\tfrac{1}{2} \operatorname{tr} \Sigma^{-1}s)}{2^{(N-1)p/2}\pi^{p(p-1)/4}|\Sigma|^{(N-1)/2} \prod^{p} \Gamma[(N - i)/2]}. \tag{7.33}$$

For a proof of this the reader is referred to Giri (1977). An analog of the square of Student's t distribution in the multivariate case is called *Hotelling's T^2 statistic* and is given by

$$T^2 = \frac{N}{N-1} \, \overline{X}' S^{-1} \overline{X}.$$

The probability density function of T^2 is given by

$$f_{T^2}(t^2) = \frac{\exp(-\frac{1}{2}\delta^2)}{(N-1)\Gamma[(N-p)/2]} \sum_{j=0}^{\infty}$$

$$\times \frac{(\delta^2/2)^j [t^2/N - 1]^{(p/2)+j-1} \Gamma[(N/2)+j]}{j!\Gamma[(p/2)+j][1 + t^2/(N-1)]^{(N/2)+j}}, \tag{7.34}$$

where $\delta^2 = N\mu'\Sigma^{-1}\mu$. For a proof the reader is referred to Giri (1977).

EXERCISES

1. Let $X = (X_1, X_2)'$ be a bivariate random vector with pdf

$$f_X(x) = \frac{1}{4\pi\sqrt{3}} \exp\left\{ -\frac{1}{6} [(x_1 - 2)^2 \right.$$

$$\left. + (x_2 - 3)^2 - (x_1 - 2)(x_2 - 3)] \right\}.$$

   (a) Find the marginal pdf of X_1 and X_2.
   (b) Calculate $E(X_1)$, $E(X_2)$ cov(X).
2. Let $X = (X_1, X_2)'$ with pdf

$$= \begin{cases} \dfrac{1}{\pi} \exp[-\frac{1}{2}(x_1^2 + x_2^2)] & \text{if } x_1 x_2 > 0 \\ 0, & \text{otherwise.} \end{cases}$$

   Show that the marginal distributions of X_1 and X_2 are normal but X is not a bivariate normal.
3. Let $X = (X_1, X_2, X_3)'$ be distributed as $N_3(0, \Sigma)$, where

$$\Sigma = \begin{pmatrix} 1 & \rho & 0 \\ \rho & 1 & \rho \\ 0 & \rho & 1 \end{pmatrix}.$$

   Find the value of ρ so that $X_1 + X_2 + X_3$ and $X_1 - X_2 - X_3$ are independent.

4. Let $X = (X_1,...,X_p)'$ be $N_p(\mu,\Sigma)$, and let $a = (a_1,...,a_p)'$ be a fixed vector. Show that $F = a'(X - \mu)/\sqrt{a'\Sigma a}$ is normally distributed with mean 0 and variance 1 provided that $a'\Sigma a > 0$.

5. In Exercise 4, assume that a is random vector independent of X satisfying $P(a'\Sigma a = 0) = 0$. Show that F is $N(0,1)$ and F is independent of a.

6. Let X be normally distributed with mean 0 and variance 1. Define

$$Y = \begin{cases} -X & \text{if } -1 \le X \le 1 \\ X, & \text{otherwise.} \end{cases}$$

   Show that Y is $N(0,1)$ and the joint distribution of X, Y is not a bivariate normal.

7. (Symmetric normal distribution) The random vector $X = (X_1,...,X_p)'$ is said to have a symmetric normal distribution if it is distributed as $N_p(\mu,\Sigma)$, where $\mu = (\mu_1,...,\mu_p)'$, and

$$\Sigma = \begin{pmatrix} 1 & \rho & \cdots & \rho \\ \rho & 1 & \cdots & \rho \\ \cdot & \cdot & \cdots & \cdot \\ \rho & \rho & \cdots & 1 \end{pmatrix} \sigma^2,$$

   that is, $\text{var}(X_i) = \sigma^2$, $\text{cov}(X_i,X_j) = \rho\sigma^2$ for all $i \ne j = 1,...,p$.
   Let 0 be an orthogonal $p \times p$ matrix and be defined by

$$O = \begin{pmatrix} \dfrac{1}{\sqrt{p}} & \cdots & \dfrac{1}{\sqrt{p}} \\ O_{21} & \cdots & O_{2p} \\ \cdot & \cdots & \cdot \\ O_{p1} & \cdots & O_{pp} \end{pmatrix}$$

   and let $Y = (Y_1,...,Y_p)' = OX$. Find $E(Y)$, $\text{cov}(Y)$ and show that $Y_1,...,Y_p$ are independent.

8. Let $X = (X_1,...,X_p)'$ be distributed as $N_p(0,\sigma^2 I)$. Show that for any $p \times p$ symmetric matrix A, $E(X'AX) = \sigma^2 \text{ tr } A$.

9. Let $X = (X_1,...,X_p)'$ be $N_p(\mu,\Sigma)$. Show that $E(X'AX) = \mu'A\mu + \text{tr } \Sigma A$.

10. Let $X = (X_1,X_2)'$ be $N_2(\mu,\Sigma)$, where $\mu = (\mu_1,\mu_2)'$,

$$\Sigma = \begin{pmatrix} 1 & \rho \\ \rho & 1 \end{pmatrix} \sigma^2, \qquad -1 < \rho < 1.$$

   (a) Show that $Z = (X_1 - \mu_1)/(X_2 - \mu_2)$ has pdf
   $$f_Z(z) = (1 - \rho^2)^{1/2}(\pi(1 - 2\rho z + z^2))^{-1}.$$

   (b) Find the distribution of $(X_1 + X_2, X_1 - X_2)'$.

11. Let $X = (X_1,\ldots,X_p)'$ be $N_p(\mu,\Sigma)$, where

$$\mu = (\mu_1,\ldots,\mu_p)', \qquad \Sigma = \begin{pmatrix} 1 & \frac{1}{2} & 0 & \cdots & 0 & 0 \\ -\frac{1}{2} & 1 & \frac{1}{2} & \cdots & 0 & 0 \\ 0 & -\frac{1}{2} & 1 & \frac{1}{2} & 0 & 0 \\ \cdot & \cdot & \cdot & \cdot & & \cdot \\ 0 & 0 & 0 & 0 & \frac{1}{2} & 1 \end{pmatrix} \sigma^2.$$

(a) Show that

$$P(X_1 - \mu_1 \geq 0,\ldots,X_p - \mu_p \geq 0) = \frac{1}{\Gamma(p + 2)}.$$

(b) For $q < p$,

$$P(X_1 - \mu_1 \geq 0,\ldots,X_q - \mu_q \geq 0, X_{q+1} - \mu_{q+1} \leq 0,\ldots,X_p\mu_p \leq 0)$$

$$= \frac{1}{(p + 1)\Gamma(q + 1)\Gamma(p - q + 1)}.$$

12. Prove that if X is distributed as $N_p(0,\Sigma)$, a necessary and sufficient condition for the quadratic form $X'AX$, with A a $p \times p$ symmetric matrix, to have a chi-square distribution with k degrees of freedom is that ΣA be an idempotent matrix of rank k.

13. Let X be distributed as $N_p(\mu,I)$. Show that two semipositive definite quadratic forms $X'AX$, $X'BX$ (with A, B symmetric matrices) are independently distributed if and only if $AB = 0$.

14. Let Z be a matrix of dimension $p \times k$ $(p > k)$ and is of rank k, and let X be distributed as $N_p(0,I)$.
   (a) Show that $Z(Z'Z)^{-1}Z'$ is an idempotent matrix.
   (b) Find $E(X'(I - Z(Z'Z)^{-1}Z')X)$.

15. Let $Y = (Y_1,\ldots,Y_n)'$ be normally distributed n vector with zero means and covariance matrix I ($n \times n$ identity matrix) and let X be a $n \times p$ matrix of rank $p(<n)$. Find the distribution (a) $Y'(I - X(X'X)^{-1}X')Y$ and (b) $Y'X(X'X)^{-1}X'Y$.

16. Do Exercise 15 when $E(Y) = X\beta$, $\beta = (\beta_1,\ldots,\beta_p)'$ and $E(Y - X\beta)(Y - X\beta)' = \sigma^2I$. Show that

$$\text{cov}((I - X(X'X)^{-1}X')Y) = \sigma^2(I - X(X'X)^{-1}X').$$

17. Let X be distributed as $E_p(\mu,\Sigma)$, and let

$$e_i = \frac{X_i^2}{L}, \qquad i = 1,\ldots,p, \quad L = X'X.$$

(a) Show that the joint distribution of $(e_1,\ldots,e_p)$ is Dirichlet $D(\tfrac{1}{2},\ldots,\tfrac{1}{2})$ with pdf

$$f(e_1,\ldots,e_p) = \frac{\Gamma(p/2)}{[\Gamma(\tfrac{1}{2})]^p} \left(\prod_{i=1}^{p-1} e_i^{(1/2)-1} \right) \left(1 - \sum_{i=1}^{p-1} e_i \right)^{(1/2)-1},$$

where $0 \le e_i \le 1$ and $\Sigma_1^p \, e_i = 1$.

(b) The pdf of L is

$$f_L(1) = \frac{\pi^{p/2}}{\Gamma(p/2)} \, 1^{p/2-1} q(1)$$

and L is independent of $(e_1,\ldots,e_p)$.

18. Let $X_1,\ldots,X_n$ be independently and identically distributed normal random variables with mean μ and variance σ^2.

(a) Show that

$$S^2 = X'AX \qquad \text{where} \quad X = (X_1,\ldots,X_n)',$$

$$A = \begin{pmatrix} 1 - \dfrac{1}{n} & -\dfrac{1}{n} & \cdots & -\dfrac{1}{n} \\[2mm] -\dfrac{1}{n} & 1 - \dfrac{1}{n} & \cdots & -\dfrac{1}{n} \\[2mm] \cdot & \cdot & & \cdot \\[2mm] -\dfrac{1}{n} & -\dfrac{1}{n} & & 1 - \dfrac{1}{n} \end{pmatrix}$$

is an $n \times n$ idempotent matrix of rank $n - 1$.

(b) Using Exercise 12, show that S^2/σ^2 is distributed as chi-square with $n - 1$ degrees of freedom.

BIBLIOGRAPHY

Chu, K'ai-Ching (1973). Estimation and decision for linear systems with elliptical random processes, *IEEE Transactions on Automatic Control*, pp. 499–505.

Galton, F. (1889), *Natural Inheritance*, MacMillan, London.

Giri, N. (1977). *Multivariate Statistical Inference*, Academic Press, New York.

Johnson, R. A., and Wichern, D. W. (1982). *Applied Multivariate Statistical Analysis*, Prentice-Hall, Englewood Cliffs, N.J.

McGraw, D. K., and Wagner, J. F. (1968). Elliptically symmetric distributions, *IEEE Trans. Infor.* pp. 110–120.

Rao, C. R. (1973). *Linear Statistical Inference and Its Application*. Wiley, New York.

Zellner, A. (1976). Bayesian and non-Bayesian analysis of the regression model with multivariate *T*-error term. *J.A.S.A.*, pp. 404–405.

8
Order Statistics and Related Distributions

8.1 ORDERED PARAMETERS

Let X be a random variable with distribution function $F_X(x)$. For any p, $0 < p < 1$, the value θ_p satisfying the inequalities

$$F_X(\theta_p) \geq p \tag{8.1a}$$
$$P(X \geq \theta_p) \geq 1 - p \tag{8.1b}$$

or, equivalently,

$$p - P(X = \theta_p) \leq F_X(\theta_p) \leq p \tag{8.2}$$

is called the *quantile of order p*. If, in particular, $P(X = \theta_p) = 0$, the quantile of order p is given by $F_X(\theta_p) = p$. If $p = \frac{1}{2}$, θ_p is called the *median* of the distribution F. Quantiles and their functions are called *ordered parameters*. An important function of the quantiles that measures dispersion is given by the interquantile range, which is the difference between two special quantiles. For example, $(\frac{1}{2})(\theta_{3/4} - \theta_{1/4})$ is called the *semi-interquantile range*. In this chapter we consider the sample quantiles based on a random sample of size n from the specified distribution. In statistical terminology, the sample observations, when arranged in order of their magnitudes, are called *order statistics*.

Example 8.1.1   Let X be a uniform random variable with

$$F_X(x) = \begin{cases} 0 & \text{if } x < 0 \\ x & \text{if } 0 \le x \le 1 \\ 1 & \text{if } x > 1. \end{cases}$$

Clearly, then, $\theta_p = p$.

Example 8.1.2   Let X be a normally distributed random variable with mean 0 and variance 1. From the Table 3B in Appendix B, we have

$$\theta_{1/4} = -0.676, \qquad \theta_{3/4} = 0.676, \qquad \theta_{1/2} = 0.$$

8.2   ORDER STATISTICS

Definition 8.2.1: Order statistic   Let $X_1,\dots,X_n$ be a random sample of size n from a random phenomenon specified by the distribution function F. We arrange each collection of values $x_1,\dots,x_n$ of $X_1,\dots,X_n$ in the increasing order of magnitude such that we obtain a sequence of values $y_1,\dots,y_n$ (called the *ordered sample*), where $y_1 = \min(x_1,\dots,x_n)$, y_2 is the next x_i in order of magnitude, ..., and $y_n = \max(x_1,\dots,x_n)$. The function $Y_k(k = 1,\dots,n)$ of $X_1,\dots,X_n$, which takes on value y_k in each possible sequence $x_1,\dots,x_n$, is called the kth *order statistic*. The number k is called the *rank* of Y_k. If two numbers x_i, x_j are equal, their order is irrelevant.

Definition 8.2.2: Range   The difference $Y_n - Y_1$ is called the *range* of the sample. If n is odd, the order statistic $Y_{(n+1)/2}$ is called the *sample median*. The midrange of the random sample $X_1,\dots,X_n$ is the average $(\frac{1}{2})(Y_n + Y_1)$.

Example 8.2.1   Consider a random sample of size 5 from a certain population. Suppose that the observed sample values are $x_1 = 3$, $x_2 = 2$, $x_3 = 1.5$, $x_4 = 1$, and $x_5 = 5$. Then $y_1 = 1$, $y_2 = 1.5$, $y_3 = 2$, $y_4 = 3$, and $y_5 = 5$. The sample range is $y_5 - y_1 = 5 - 1 = 4$. The median and the midrange of the sample are 2 and 3, respectively.

Example 8.2.2   Consider a random sample of size 2 from a binomial population with $n = 2$. The random variable (X_1,X_2) can take values $(0,0)$, $(0,1)$, $(0,2)$, $(1,0)$, $(1,1)$, $(1,2)$, $(2,0)$, $(2,1)$, (2.2). The order statistic Y_1 will take on the smallest value in each of the values above.

The results presented in this chapter are valid for both continuous and discrete random variables. As previously, we deal with continuous random phenomena only. The modifications needed for the discrete cases are straightforward.

Theorem 8.2.1 The joint probability density function of the order statistics $Y_1,\ldots,Y_n$ is given by

$$f_{Y_1,\ldots,Y_n}(y_1,\ldots,y_n) = \begin{cases} n! \prod_1^n f_X(y_i) & \text{if } -\infty < y_1 < \cdots < y_n < \infty \\ 0 & \text{otherwise.} \end{cases}$$

(8.3)

Proof. First we note that each of the $n!$ permutations of $X_1,\ldots,X_n$ leaves the corresponding order statistics invariant. Thus a particular realization of $Y_1,\ldots,Y_n$ is associated with one of the $n!$ realizations of $X_1,\ldots,X_n$, each corresponding to a particular permutation of the X_i. Thus by the theorem of total probability, for $-\infty < y_1 < \cdots < y_n < \infty$, we have

$$F_{Y_1,\ldots yn}(y_1,\ldots,y_n) = P\{Y_1 \le y_1,\ldots,Y_n \le y_n\}$$

$$= \sum_{(i_1,\ldots,i_n)\,\in\,P_n} P\{(Y_1 \le y_1,\ldots,Y_n \le y_n)$$

$$\cap (Y_1 = X_{i_1},\ldots,Y_n = X_{i_n})\},$$

(8.4)

where the summation extends over the set P_n of all $n!$ permutations of $(1,\ldots,n)$, that is,

$$P_n = \{(i_1,\ldots,i_n) : 1 \le i_j \le n, 1 \le j \le n, i_1 \ne i_2 \ne \cdots \ne i_n\}.$$

But obviously, the right-hand side of Eq. (8.4) is equal to

$$\sum_{(i_1,\ldots,i_n)\in P_n} P\{(X_{i_1} \le y_1,\ldots,X_{i_n} \le y_n) \cap (X_{i_1} < \cdots < X_{i_n})\}$$

$$= n! \, P\{(X_1 \le y_1,\ldots,X_n \le y_n) \cap (X_1 < \cdots < X_n)\}$$

(8.5)

since the X_i have the same distribution. Now

$$P\{(X_1 < y_1,\ldots,X_n < y_n) \cap (X_1 < \cdots < X_n)\}$$

$$= \int_{-\infty}^{y_1} \int_{x_1}^{y_2} \cdots \int_{x_{n-1}}^{y_n} f_X(x_1) \cdots f_X(x_n) \, dx_n \cdots dx_1.$$

(8.6)

Differentiating the right-hand side of Eq. (8.6) with respect to $y_1,\ldots,y_n$, it follows from Eq. (8.5) that

$$f_{Y_1,\ldots,Y_n}(y_1,\ldots,y_n) = n! \prod_1^n f_X(y_i),$$

and this completes the proof of the theorem. Q.E.D.

From Eq. (8.3), the marginal joint probability density function of $Y_1,\ldots,Y_k$, $k < n$ is

$$f_{Y_1,\ldots,Y_k}(y_1,\ldots,y_k)$$

$$= n! \prod_1^k f_X(y_i) \int_{y_k}^{\infty} f_X(y_{k+1}) \, dy_{k+1} \cdots \int_{y_{n-1}}^{\infty} f_X(y_n) \, dy_n$$

$$= n! \prod_1^k f_X(y_i) \int_{y_k}^{\infty} f_X(y_{k+1}) \, dy_{k+1} \cdots$$

$$\times \int_{y_{n-2}}^{\infty} [1 - F_X(y_{n-1})] f_X(y_{n-1}) \, dy_{n-1}$$

$$= \begin{cases} n! \prod_1^k f_X(y_i) \dfrac{[1 - F_X(y_k)]^{n-k}}{(n-k)!}, & -\infty < y_1 < \cdots < y_k < \infty \\ 0 & \text{otherwise.} \end{cases}$$

$$(8.7)$$

The marginal joint probability density function of $Y_{k+1},\ldots,Y_n$ is given by

$$f_{Y_{k+1},\ldots,Y_n}(y_{k+1},\ldots,y_n)$$

$$= n! \prod_{k+1}^n f_X(y_i) \int_{-\infty}^{y_{k+1}} f_X(y_k) \, dy_k \cdots \int_{-\infty}^{y_2} f_X(y_1) \, dy_1$$

$$= \begin{cases} n! \prod_{k+1}^n f_X(y_i) \dfrac{[F_X(y_{k+1})]^k}{k!} & \text{if } -\infty < y_{k+1} < \cdots < y_n \\ 0 & \text{otherwise.} \end{cases}$$

$$(8.8)$$

Similarly, one can verify that the joint probability density function of $Y_{j_1}, Y_{j_1+j_2}, \ldots, Y_{j_1+j_2+\cdots+j_k}$ is given by

$$\frac{n!}{\Gamma(j_1) \cdots \Gamma(j_k)\Gamma(n + 1 - j_1 - \cdots - j_k)} f_X(y_{j_1}) \cdots f_X(y_{j_1+j_2+\cdots+j_k})$$

$$\times [F_X(y_{j_1})]^{j_1-1}[F_X(Y_{j_1+j_2}) - F_X(y_{j_1})]^{j_2-1}$$

$$\times \cdots \times [1 - F_X(y_{j_1+j_2+\cdots+j_k})]^{n-j_1-\cdots-j_k}$$

$$(8.9)$$

provided that

$$-\infty < y_{j_1} < \cdots < y_{j_1+\cdots+j_k} < \infty$$

and is equal to zero otherwise. From Eqs. (8.4) to (8.6), the probability density functions of $Y_n = \max(X_1,\ldots,X_n)$, $Y_1 = \min(X_1,\ldots,X_n)$ and Y_i, $1 < i < n$, are given by

$$f_{Y_n}(y_n) = \begin{cases} n[F_X(y_n)]^{n-1}f_X(y_n) & \text{if } -\infty < y_n < \infty \\ 0 & \text{otherwise} \end{cases}$$

$$(8.10a)$$

$$f_{Y_1}(y_1) = \begin{cases} n[1 - F_X(y_1)]^{n-1}f_X(y_1) & \text{if } -\infty < y_1 < \infty \\ 0 & \text{otherwise} \end{cases} \qquad (8.10b)$$

$$f_{Y_i}(y_i) = \begin{cases} \dfrac{n!}{(i-1)!\,(n-i)!}\,[F_X(y_i)]^{i-1}[1 - F_X(y_i)]^{n-i}f_X(y_i) \\ \qquad \text{if } -\infty < y_i < \infty \\ 0 \qquad \text{otherwise} \end{cases} \qquad (8.10c)$$

Example 8.2.3 We derive here the pdf of Y_1, the pdf of Y_n, and the joint pdf of Y_1 and Y_n using the distribution functions of these variables.

$$F_{Y_n}(y_n) = P(Y_n \leq y_n)$$

$$= P(\max(X_1,\ldots,X_n) \leq y_n)$$

$$= P(X_1 \leq y_n, X_2 \leq y_n, \ldots, X_n \leq y_n)$$

$$= \prod_{i=1}^{n} P(X_i \leq y_n)$$

$$= [F_X(y_n)]^n;$$

$$F_{Y_1}(y_1) = P(Y_1 \leq y_1)$$

$$= P(\min(X_1,\ldots,X_n) \leq y_1)$$

$$= 1 - P(\min(X_1,\ldots,X_n) > y_1)$$

$$= 1 - P(X_1 > y_1, X_2 > y_1, \ldots, X_n > y_1)$$

$$= 1 - \prod_{i=1}^{n} (1 - P(X_i \leq y_1))$$

$$= 1 - (1 - F_X(y_1))^n$$

$$F_{Y_1,Y_n}(y_1,y_n) = P(Y_1 \leq y_1, Y_n \leq y_n)$$

$$= P(Y_n \leq y_n) - P(Y_1 > y_1, Y_n \leq y_n)$$

$$= \prod_{i=1}^{n} P(X_i \leq y_n) - \prod_{i=1}^{n} P(y_1 < X_i \leq y_n)$$

$$= [F_X(y_n)]^n - [F_X(y_n) - F_X(y_1)]^n.$$

Hence

$$f_{Y_n}(y_n) = \frac{\partial F_{Y_n}(y_n)}{\partial y_n} = n[F_X(y_n)]^{n-1}f_X(y_n)$$

$$f_{Y_1}(y_1) = \frac{\partial F_{Y_1}(y_1)}{\partial y_1} = n(1 - F_X(y_1))^{n-1}f_X(y_1)$$

$$f_{Y_n,Y_1}(y_n,y_1) = \frac{\partial^2 F_{Y_n,Y_1(y_n,y_1)}}{\partial y_n \, \partial y_1}$$

$$= n(n - 1)[F_X(y_n) - F_X(y_1)]^{n-2}f_X(y_n)f_X(y_1).$$

8.3 SOME RELATED DISTRIBUTIONS

From Eq. (8.6), the joint probability density function of Y_n, Y_1 is given by

$$f_{Y_n,Y_1}(y_n,y_1) = \begin{cases} n(n - 1)[F_X(y_n) - F_X(y_1)]^{n-2}f_X(y_n)f_X(y_1) \\ \quad \text{if } -\infty < y_1 < y_n < \infty \\ 0 \quad \text{otherwise.} \end{cases} \qquad (8.11)$$

The probability density function of the sample range R is

$$f_R(r) = \int_{-\infty}^{\infty} f_{Y_n,Y_1}(y_n, y_n - r) \, dy_n$$

$$= \begin{cases} n(n - 1) \int_{-\infty}^{\infty} [F_X(y_n) - F_X(y_n - r)]^{n-2}f_X(y_n - r)f_X(y_n) \, dy_n \\ \quad \text{if } r \geq 0 \\ 0 \quad \text{otherwise.} \end{cases}$$

$$(8.12)$$

Example 8.3.1 Let $Y_1,\ldots,Y_n$ denote the order statistics of a random sample $X_1,\ldots,X_n$ from a random phenomenon specified by the probability density function (exponential with $\lambda = 1$)

$$f_X(x) = \begin{cases} e^{-x} & \text{if } 0 < x < \infty \\ 0 & \text{otherwise.} \end{cases}$$

Here

$$F_X(x) = \begin{cases} 0, & x < 0 \\ 1 - e^{-x}, & 0 < x < \infty. \end{cases}$$

Hence

$$f_{Y_1}(y_1) = \begin{cases} ne^{-ny_1}, & -\infty < y_1 < \infty \\ 0 & \text{otherwise} \end{cases}$$

$$f_{Y_n}(y_n) = \begin{cases} n(1 - e^{-y_n})^{n-1}e^{-y_n}, & -\infty < y_n < \infty \\ 0, & \text{otherwise} \end{cases}$$

$$f_R(r) = \begin{cases} (n - 1)(e^r - 1)^{n-2}e^r & \text{if } r > 0 \\ 0 & \text{otherwise.} \end{cases}$$

Theorem 8.3.1 Let $Y_1,\ldots,Y_n$ denote the order statistics of a random sample $X_1,\ldots,X_n$ from a random phenomenon specified by the continuous distribution function F. The random variables $U_1 = F_X(Y_1),\ldots,U_n = F_X(Y_n)$ have the ordered n-variate Dirichlet distribution, which is given by

$$f_{U_1,\ldots,U_n}(u_1,\ldots,u_n) = \begin{cases} n! & \text{if } 0 < u_1 < \cdots < u_n < \infty \\ 0 & \text{otherwise.} \end{cases} \tag{8.13}$$

Proof. The proof follows trivially from Eq. (8.3).   Q.E.D.

From Eq. (8.7), the probability density function of $U_i = F_X(Y_i)$ is

$$f_{U_i}(u_i) = \begin{cases} \dfrac{n!}{(i - 1)! (n - i)!} u^{i-1}(1 - u)^{n-i} & \text{if } 0 < u < 1 \\ 0 & \text{otherwise,} \end{cases} \tag{8.14}$$

which is the well-known beta probability density function with parameters $i, n - i + 1$. Hence

$$E(U_i) = \frac{i}{n + 1} \tag{8.15}$$

or, equivalently,

$$E(U_{i+1}) - E(U_i) = \frac{1}{n + 1}, \qquad i = 1,\ldots,n - 1.$$

Thus, on the average the n order statistics $Y_1,\ldots,Y_n$ divide the area under $f_X(x)$ into $n + 1$ equal parts of area $1/(n + 1)$.

Theorem 8.3.2 Let $Y_1,\ldots,Y_n$ denote the order statistics of the random sample $X_1,\ldots,X_n$ from the distribution function $F_X(x)$. Let $p, 0 < p < 1$, be such that $k = [np] + 1$, where $[np]$ denotes the greatest integer in np (the integral part of np). If there is only one θ_p (pth quantile) satisfying Eq. (8.1), then Y_k converges in probability to θ_p as $n \to \infty$.

Proof. From the definition of convergence in probability, the theorem will be proved if we can show that for an arbitrary $\epsilon > 0$,

$$\lim_{n\to\infty} P\{|Y_k - \theta_p| \ge \epsilon\} = 0 \tag{8.16}$$

or, equivalently,

$$\lim_{n\to\infty} P\{Y_k \ge \theta_p + \epsilon\} = 0$$

$$\lim_{n\to\infty} P\{Y_k \le \theta_p - \epsilon\} = 0. \tag{8.17}$$

For any ordered sample $y_1,\ldots,y_n$ of $x_1,\ldots,x_n$ let us define $S_n(x)$ (empirical distribution function) for $-\infty < x < \infty$, by

$$S_n(x) = \begin{cases} 0 & \text{if } x \le y_1 \\ \dfrac{r}{n} & \text{if } x > y_1 \\ & \text{and } r \text{ is the largest index for which } x_r < x. \end{cases}$$

Obviously, $nS_n(x)$ denotes the number of elements in the sample $x_1,\ldots,x_n$ that are smaller than x. Hence

$$P(nS_n(x) = i) = \begin{cases} \binom{n}{i}[F_X(x)]^i[1 - F_X(x)]^{n-i}, & i = 0,1,\ldots,n \\ 0, & \text{otherwise.} \end{cases} \tag{8.18}$$

It can be verified that $S_n(x)$ is continuous from the left and satisfies all the properties of the distribution function of a random variable X.

Clearly, from Eq. (8.17) we see that we shall have proved the theorem if the following two relations can be proved:

$$\lim_{n\to\infty} P\left\{ S_n(\theta_p + \epsilon) < \frac{k}{n} \right\} = 0 \tag{8.19}$$

$$\lim_{n\to\infty} P\left\{ S_n(\theta_p - \epsilon) \ge \frac{k-1}{n} \right\} = 0. \tag{8.20}$$

From the assumption of uniqueness of θ_p, we have

$$\lim_{n\to\infty}\left\{ F_X(\theta_p + \epsilon) - \frac{k}{n} \right\} = F_X(\theta_p + \epsilon) - p = \delta_1 > 0 \tag{8.21}$$

$$\lim_{n\to\infty}\left\{ F_X(\theta_p - \epsilon) - \frac{k-1}{n} \right\} = F_X(\theta_p - \epsilon) - p = \delta_2 < 0, \tag{8.22}$$

where δ_1 and δ_2 are arbitrary designations. Now, from Eq. (8.21) it follows that for sufficiently large n, we can have

$$F_X(\theta_p + \epsilon) - \frac{k}{n} > \frac{1}{2}\delta_1. \tag{8.23}$$

From the inequalities (8.19) and (8.23) we get, for large n,

$$P\left\{S_n(\theta_p + \epsilon) < \frac{k}{n}\right\} \le P\{S_n(\theta_p + \epsilon) - F_X(\theta_p + \epsilon) < -\tfrac{1}{2}\delta_1\}$$

$$\le P\{|S_n(\theta_p + \epsilon) - F_X(\theta_p + \epsilon)| > \tfrac{1}{2}\delta_1\}. \tag{8.24}$$

Using Chebychev's inequality, from (8.19) and (8.24) we have

$$\lim_{n\to\infty} P\left\{S_n(\theta_p + \epsilon) < \frac{k}{n}\right\} \le \lim_{n\to\infty} \frac{F_X(\theta_p + \epsilon)\{1 - F_X(\theta_p + \epsilon)\}}{n\delta_1^2}$$

$$= \lim_{n\to\infty} \frac{1}{n\delta_1^2}$$

$$= 0, \tag{8.25}$$

which proves (8.19). Similarly, we can establish (8.20).   Q.E.D.

Theorem 8.3.3 Let $Y_1,\ldots,Y_n$ denote the order statistics of the random sample $X_1,\ldots,X_n$ from a random phenomenon specified by the distribution function $F_X(x)$. (For the sake of brevity we have omitted the parameter in the writing of the distribution and the density functions.) Let p, $0 < p < 1$, be such that $k = [np] + 1$, and let θ_p be the pth quantile of $F_X(x)$. If the probability density function $f_X(x)$ is continuous and $f_X(\theta_k) > 0$, then

$$T_k = \sqrt{n}(Y_k - \theta_p) \tag{8.26}$$

converges in distribution to a normal distribution with mean 0 and variance $p(1 - p)/[f_X(\theta_p)]^2$ as $n \to \infty$.

Proof. From Eq. (8.26), the probability density function of T_k is given by

$$f_{T_k}(t) = \frac{1}{\sqrt{n}} f_{Y_k}\left(\frac{t}{\sqrt{n}} + \theta_p\right). \tag{8.27}$$

From Eqs. (8.7) and (8.27) we have

$$f_{T_k}(t) = \frac{1}{\sqrt{n}} \frac{n!}{(k-1)!\,(n-k)!} \left[F_X\left(\frac{t}{\sqrt{n}} + \theta_p\right)\right]^{k-1}$$

$$\times \left[1 - F_X\left(\frac{t}{\sqrt{n}} + \theta_p\right)\right]^{n-k} f_X\left(\theta_p + \frac{t}{\sqrt{n}}\right). \tag{8.28}$$

Substituting the approximate values $k = np + 1$, $n - k + 1 = n - np$, and using Stirling's approximation $n! \simeq n^n e^{-n} \sqrt{2\pi n}$ (approximately equal for large n), we obtain from (8.28)

$$f_{T_k}(t) = \frac{f_X(\theta_p)}{\sqrt{2\pi[p/(1-p)]}\{1 - F_X[(t/\sqrt{n}) + \theta_p]\}} A_n, \qquad (8.29)$$

where

$$A_n = \left(\left[\frac{1}{p} F_X\left(\frac{t}{\sqrt{n}} + \theta_p\right)\right]^p \left\{\frac{1}{1-p}\left[1 - F_X\left(\frac{t}{\sqrt{n}} + \theta_p\right)\right]\right\}^{1-p} \right)^n.$$

But for large n,

$$F_X\left(\theta_p + \frac{t}{\sqrt{n}}\right) \simeq F_X(\theta_p) + \frac{t}{\sqrt{n}} f_X(\theta_p)$$

$$= p + \frac{t}{\sqrt{n}} f_X(\theta_p),$$

so that

$$\frac{1}{p} F_X\left(\theta_p + \frac{t}{\sqrt{n}}\right) \simeq 1 + \frac{t}{p\sqrt{n}} f_X(\theta_p).$$

Similarly,

$$\frac{1}{1-p}\left[1 - F_X\left(\theta_p + \frac{t}{\sqrt{n}}\right)\right] \simeq 1 - \frac{t}{(1-p)\sqrt{n}} f_X(\theta_p),$$

so that

$$f_{T_k}(t) \simeq \frac{f_X(\theta_p)}{\sqrt{2\pi p(1-p)}\{1 - [t/(1-p)\sqrt{n}]f_X(\theta_p)\}}$$

$$\times \left\{\left[1 + \frac{t}{p\sqrt{n}} f_X(\theta_p)\right]^p \left[1 - \frac{t}{(1-p)\sqrt{n}} f_X(\theta_p)\right]^{1-p}\right\}^n.$$

$$(8.30)$$

From Eq. (8.30) we have

$$f_{T_k}(t) \simeq \frac{1}{\sqrt{2\pi}\,\sigma} \exp\left(-\frac{t^2}{2\sigma^2}\right),$$

where

$$\sigma = \frac{\sqrt{p(1-p)}}{f_X(\theta_p)}.$$

Hence the theorem.   Q.E.D.

Example 8.3.2 Let

$$f_X(x) = \begin{cases} \dfrac{1}{\sqrt{2\pi}} e^{-(1/2)x^2}, & -\infty < x < \infty \\ 0, & \text{otherwise.} \end{cases}$$

The median of a random sample of size n from this distribution is asymptotically (as $n \to \infty$) normally distributed with mean 0 and variance $\pi/2n$.

EXERCISES

1. Let $Y_1,\ldots,Y_5$ denote the order statistics of a random sample of size 5 from a random phenomenon specified by the probability density function

   $$f_X(x) = \begin{cases} 3x^2, & 0 < x < 1 \\ 0, & \text{otherwise.} \end{cases}$$

   Show that Y_2/Y_4 and Y_4 are independent random variables.

2. Find the probability that the range of a random sample of size 5, from a uniform probability density function on $(0,1)$, is greater than $\frac{1}{2}$.

3. Let Y_{k+1} denote the median of a random sample of size $2k + 1$ from the normal distribution with mean μ and variance σ^2. Show that the probability density function of Y_{k+1} is symmetric about $Y_{k+1} = \mu$ and deduce that $E(Y_{k+1}) = \mu$.

4. Let $Y_1,\ldots,Y_n$ denote the order statistics of a random sample of size n from a population specified by the continuous distribution function F. Let

   $$V_1 = F_X(Y_1), \quad V_i = F_X(Y_i) - F_X(Y_{i-1}), \quad i = 2,\ldots,n.$$

   Find the joint probability density functions of (i) $V_1,\ldots,V_n$, and (ii) $V_1,\ldots,V_k$, $k < n$.

5. (Continuation of Exercise 4) Show that for $R = Y_n - Y_1$:

   (a) $F_R(r) = n \displaystyle\int_{-\infty}^{\infty} [F_X(x + r) - F_X(x)]^{n-1} \, dF_X(x).$

   (b) $E(R) = \displaystyle\int_{-\infty}^{\infty} [1 - [F_X(x)]^n - [1 - F_X(x)]^n] \, dx.$

6. Let $Y_1,\ldots,Y_n$ denote the order statistics of a random sample of size n from the continuous probability density function

   $$f_X(x) = \begin{cases} \dfrac{1}{\sigma} e^{-(x-\mu)/\sigma} & \text{if } x > \mu \\ 0 & \text{if } x \le \mu. \end{cases}$$

Let

$$Z = \sum_{i=2}^{n} \frac{Y_i - Y_1}{n - 1}, \qquad T = \frac{Y_1 - \mu}{Z}.$$

Show that the probability density function of T is given by

$$f_T(t) = \begin{cases} n\left(1 + \dfrac{nt}{n-1}\right)^{-n} & \text{if } t > 0 \\ 0 & \text{if } t \leq 0. \end{cases}$$

7. Let $Y_1, \ldots, Y_n$ denote the order statistics of a random sample of size n from a random phenomenon specified by the continuous distribution function F. Show that the variance and covariance matrix of the random vector (U_i, U_j), $1 \leq i < j \leq n$, where $U_i = F_X(Y_i)$, $U_j = F_X(Y_j)$ is given by

$$\begin{pmatrix} \dfrac{\alpha_1(1 - \alpha_1)}{n + 2} & \dfrac{\alpha_1(1 - \alpha_2)}{n + 2} \\ \dfrac{\alpha_1(1 - \alpha_2)}{n + 2} & \dfrac{\alpha_2(1 - \alpha_2)}{n + 2} \end{pmatrix}$$

where

$$\alpha_1 = \frac{i}{n + 1}, \qquad \alpha_2 = \frac{j}{n + 1}.$$

8. Consider Example 8.3.1. Define $Z_i = Y_i - Y_{i-1}$, $1 \leq i \leq n$ ($Y_0 = 0$). Show that $Z_1, \ldots, Z_n$ are mutually independent and that $(n - i + 1)Z_i$, $i = 1, \ldots, n$, has chi-square distribution with 2 degrees of freedom.

9. Let (X_1, X_2) follow a bivariate normal distribution with

$$E(X_1) = E(X_2) = 0, \qquad V(X_1) = V(X_2) = 1.$$

Let the coefficient of correlation between X_1 and X_2 be equal to ρ. Find the distribution of $Y_2 - Y_1$, where

$$Y_2 = \max(X_1, X_2), \qquad Y_1 = \min(X_1, X_2).$$

Calculate the probability $P\{Y_2 > 0\}$. Show that

$$E(Y_1) = -\left(\frac{1 - \rho}{\pi}\right)^{1/2}, \qquad E(Y_2) = \left(\frac{1 - \rho}{\pi}\right)^{1/2}.$$

10. (Glivenko–Cantelli theorem) Let $S_n(x)$ denote the empirical distribution function of a sample of size n from a random phenomenon specified by the distribution function $F_X(x)$. Show that $S_n(x)$, $-\infty < x < \infty$, converges to $F_X(x)$ as $n \to \infty$ with probability 1.

11. (Continuation of Exercise 10: Kolmogorov–Smirnov theorem) Let $D_n = \sup_{-\infty < x < \infty} |S_n(x) - F_X(x)|$. Show that

$$\lim_{n \to \infty} P(\sqrt{n} D_n < \lambda) = \begin{cases} \sum\limits_{i=-\infty}^{\infty} (-1)^i \exp(-2i^2\lambda^2), & \lambda > 0 \\ 0, & \lambda \le 0. \end{cases}$$

12. Let $X_1, \ldots, X_{n+1}$ be independently and identically distributed random variables with pdf $f_X(x) = \exp(-x)$, $x > 0$. Let

$$Y_i = \frac{\sum_{j=1}^{i} X_j}{\sum_{i=1}^{n+1} X_j}, \quad i = 1, \ldots, n.$$

Show that $Y_1, \ldots, Y_n$ form the first n order statistics from a uniform distribution over interval $(0,1)$.

9

Statistical Inference: Parametric Point Estimation

We now begin a systematic study of the problems of statistical inference. In this chapter we study the problems associated with parametric estimation when the sample size is fixed.

Consider a random phenomenon characterized by the stochastic model

$$\left(\Omega, \frac{\partial}{\partial x} F_X(x|\theta)\right), \qquad \theta \in \Omega,$$

where Ω represents the parametric space. The term $\partial F_X(x|\theta)/\partial x$ denotes the probability density or probability mass function, depending on whether the phenomenon is continuous or discrete. We know for certainty the functional form of f (or p), but we do not know the parameter θ and our problem is to make a good guess about θ or some function of θ. From this experiment we collect a sample of n (fixed) observations $(x_1,...,x_n)$. As pointed out in Section 5.5, $(x_1,...,x_n)$ is a value of the random sample $(X_1,...,X_n)$ associated with this experiment. What is generally needed for our purpose is a statistic, which is a function of $(X_1,...,X_n)$ such that its value at $(x_1,...,x_n)$ gives a good guess about the true variant θ. In this context the statistic is called an *estimator*, and its value at $(x_1,...,x_n)$ is called an *estimate* of the unknown parameter θ. For example, the sufficient statistic is an estimator that contains all relevant information present in the sample about θ. Various other estimators, depending on the purpose for which an estimate is obtained, are used for statistical inference. Unfortunately, no single estimator is appropriate for all situations. In this chapter we discuss several criteria for judging the relative merits and demerits of different estimators and lay down appropriate procedures for obtaining estimates.

9.1 CRITERIA FOR JUDGING ESTIMATORS

Let the stochastic model of a random phenomenon be specified by $(\Omega, \partial F_X(x|\theta)/\partial x)$, where for each $\theta \in \Omega$, $\partial F_X(x|\theta)/\partial x$ gives the law of randomness of the variant θ of the phenomenon, and let $(X_1,\ldots,X_n)$ be a random sample of size n associated with this model. Our problem is to find a statistic $T(X_1,\ldots,X_n)$ such that its value $T(x_1,\ldots,x_n)$ at the observed sample point $(x_1,\ldots,x_n)$ gives an estimate of θ or some function of it which is, in some sense, closest to its true value.

Definition 9.1.1: Unbiased estimator   A statistic $T(X_1,\ldots,X_n)$ is called an *unbiased estimator* of $g(\theta)$, where g is a known function, if $E_\theta[T(X_1,\ldots,X_n)] = g(\theta)$ for all θ. Otherwise, it is said to be *biased*.

Example 9.1.1   Let $(X_1,\ldots,X_n)$ be a random sample of size n from a discrete random phenomenon specified by the stochastic model $(\Omega, p_X(x|\theta))$, where

$$p_X(x|\theta) = \begin{cases} \dfrac{e^{-\theta}\theta^x}{x!}, & x = 0,1,2,\ldots \\ 0 & \text{otherwise.} \end{cases}$$

Let

$$T(X_1,\ldots,X_n) = \frac{1}{n}\sum_{j=1}^{n} X_j.$$

Obviously, $E(T)$ is equal to θ. (Also, from Section 5.5, it is sufficient.) It may be noted that

$$T_i(X_1,\ldots,X_n) = \frac{1}{i}\sum_{1}^{i} X_i, \qquad i = 1,2,\ldots,n$$

are also unbiased estimators of θ.

Example 9.1.2   Let $(X_1,\ldots,X_n)$ be a random sample of size n from the random phenomenon represented by the pdf

$$f_X(x|\theta) = \begin{cases} \dfrac{1}{\sqrt{2\pi}}\, e^{-(x-\theta)^2/2}, & -\infty < x < \infty, \quad -\infty < \theta < \infty \\ 0 & \text{otherwise.} \end{cases}$$

Here

$$T(X_1,\ldots,X_n) = \frac{1}{n} \sum_1^n X_i$$

is an unbiased estimator of θ and is also sufficient. As in Example 9.1.2,

$$\frac{1}{i} \sum_{j=1}^i X_j, \qquad i = 1,2,\ldots,n$$

are unbiased estimators of θ.

The property of unbiasedness, while generally desirable for an estimator, does not always lead to a single estimator. In all cases, we can find a large number of unbiased estimators of a parameter, and as a result it is necessary to hunt for some other criterion to determine a single unbiased estimator (if possible) that is closer to the true value in some sense. One such criterion follows.

Definition 9.1.2: Minimum-mean-square-error estimator Let $T(X_1,\ldots, X_n)$ be an estimator of the true variant θ (not necessarily unbiased). T is said to be a *minimum-mean-square-error estimator* of $g(\theta)$ if for any other estimator $T'(X_1,\ldots,X_n)$ of $g(\theta)$,

$$E_\theta(T - g(\theta))^2 \leq E(T' - g(\theta))^2 \tag{9.1}$$

for all θ.

Unfortunately, minimum-mean-square-error estimators do not generally exist. However, in some cases we can eliminate this difficulty by restricting our attention to the class of unbiased estimators only.

Definition 9.1.3: Minimum-variance estimator An unbiased estimator T of $g(\theta)$, a function of the true variant θ, is said to be a *minimum-variance estimator* if for any other unbiased estimator T' of $g(\theta)$,

$$\text{var}(T) \leq \text{var}(T') \qquad \text{for all } \theta.$$

It may be remarked that in some cases the restriction to unbiased estimators is very disturbing because there may exist a biased estimator with smaller variance.

Example 9.1.3 Let $(X_1,\ldots,X_n)$ be a random sample of size n from the random phenomenon specified by the probability density function

$$f_X(x|\mu,\sigma^2) = \begin{cases} \dfrac{1}{\sqrt{2\pi}\,\sigma} \exp\left[-\dfrac{1}{2}\left(\dfrac{x-\mu}{\sigma}\right)^2 \right], & -\infty < x < \infty \\ 0, & \text{otherwise.} \end{cases}$$

Let

$$S^2 = \sum_{1}^{n} \frac{(X_i - \overline{X})^2}{n - 1}$$

where $\overline{X} = \Sigma_1^n X_i/n$. From Chapter 6 we know that $(n - 1)S^2/\sigma^2$ has chi-square distribution with $n - 1$ degrees of freedom. Hence

$$E(S^2) = \sigma^2, \qquad V(S^2) = \frac{2\sigma^4}{n - 1}.$$

In other words, S^2 is an unbiased estimator of σ^2 with variance $2\sigma^4/(n - 1)$. Consider a class of estimators (not necessarily unbiased) given by CS^2, $C > 0$. The mean square error of CS^2 is

$$E(CS^2 - \sigma^2)^2 = C^2 \frac{2\sigma^4}{n - 1} + \sigma^4(C - 1)^2$$

$$= \sigma^4 \left[\frac{(n - 1)C^2}{n - 1} - 2C + 1 \right].$$

This attains its minimum value $2\sigma^4/(n + 1)$ when $C = (n - 1)/(n + 1)$. Since $1/(n + 1) < 1/(n - 1)$, the biased estimator $\Sigma_1^n (X_i - \overline{X})^2/n + 1$ has the smallest mean-square-error among the estimators given by CS^2, $C > 0$. It may be noted that if n is large, the difference of the variance of S^2 and $\Sigma_1^n (X_i - \overline{X})^2/(n + 1)$ is negligible.

The important property of a minimum-variance unbiased estimator of $g(\theta)$ lies in its ability to pool the information supplied by all independent estimators of the same parametric function $g(\theta)$.

Definition 9.1.4: Minimum-variance linear unbiased estimator Let $(X_1,\ldots,X_n)$ be a random sample from $\partial F_X(x|\theta)/\partial x$. A statistic $T(X_1,\ldots,X_n)$ is called the *minimum-variance linear unbiased estimator* of $g(\theta)$ if (a) T is a linear function of $X_1,\ldots,X_n$, (b) $E(T) = g(\theta)$, and (c) among all estimators T' satisfying (a) and (b), $\mathrm{var}(T) \leq \mathrm{var}(T')$.

Theorem 9.1.1 Let $T_1,\ldots,T_n$ be n independent linear unbiased estimators of the parametric function $g(\theta)$ with $\mathrm{var}(T_i) = \sigma^2 < \infty$. Then the statistic

$$T = \frac{1}{n} \sum_{1}^{n} T_i$$

is the minimum-variance linear unbiased estimator of $g(\theta)$.

Proof. Let

$$Z = \sum_{i=1}^{n} a_i T_i + b$$

with $a_1,...,a_n$, b real constants, be a linear unbiased estimator of $g(\theta)$. Then

$$E(Z) = g(\theta) \sum_{1}^{n} a_i + b = g(\theta).$$

This implies that $\sum_{1}^{n} a_i = 1$, $b = 0$. Now

$$\text{var}\left(\sum_{1}^{n} a_i T_i\right) = \sum_{1}^{n} a_i^2 \sigma^2.$$

We want to minimize $\sum_{1}^{n} a_i^2 \sigma^2$ subject to the restriction $\sum_{1}^{n} a_i = 1$. This is equivalent to unconditionally minimizing

$$\sum_{1}^{n} a_i^2 \sigma^2 - \lambda\left(\sum_{1}^{n} a_i - 1\right),$$

where λ is the Lagrange multiplier. Differentiating with respect to a_i we get (since $\sum_{1}^{n} a_i = 1$)

$$a_i = \frac{\lambda}{2}$$

$$= \frac{1}{n}. \tag{9.2}$$

Hence the linear estimator $T = (1/n) \sum_{1}^{n} T_i$ is the minimum-variance linear unbiased estimator of $g(\theta)$.   Q.E.D.

Remark.   Let $\text{var}(T_i) = \sigma_i^2 < \sigma^2$ for all i. Then

$$\text{var}(T) = \frac{1}{n^2} \sum_{1}^{n} \sigma^2 < \frac{\sigma^2}{n}. \tag{9.3}$$

Hence $\text{var}(T) \to 0$ as $n \to \infty$. In other words, T approaches the true value $g(\theta)$ with probability 1 as $n \to \infty$. If the T_i are biased estimators of $g(\theta)$, then T tends to the wrong value as $n \to \infty$.

We have seen in Section 5.5 that the sufficient estimator (statistic) of a parameter θ (which may be vector-valued) extracts all available information about the parameter θ present in the sample. No doubt sufficient estimators are not always unbiased. In some cases we can make them unbiased without losing sufficiency. In other words, if T is a sufficient estimator of θ, then $AT + b$, $a \neq 0$ is also a sufficient estimator of θ. A more general result along this line is given by the following theorem.

Theorem 9.1.2 Let $(X_1,\ldots,X_n)$ be a random sample from the random phenomenon specified by the probability density function $f_X(x|\theta)$, $\theta \in \Omega$ [or mass function $p_X(x|\theta)$ in the discrete case]. Let $T(X_1,\ldots,X_n)$ be a sufficient estimator of the parameter θ and let $Z = g(T(X_1,\ldots,X_n)) = h(X_1,\ldots,X_n)$ be a single-valued function of T not involving θ, with a single-valued inverse $T = W(Z)$. Then Z is also sufficient for θ.

Proof. We will prove this theorem for the case of a continuous random phenomenon. Note that

$$\prod_1^n f_{X_i}(x_i|\theta) = g_T(t(x_1,\ldots,x_n)|\theta)K(x_1,\ldots,x_n)$$

$$= g_W(w(h(x_1,\ldots,x_n))|\theta)K(x_1,\ldots,x_n). \tag{9.4}$$

Since the first factor in the right-hand side of Eq. (9.4) is a function of $Z = h(X_1,\ldots,X_n)$ and θ, and since the second factor does not depend on θ, Theorem 5.5.2 implies that Z is also sufficient for θ.

The proof for the case of a discrete random phenomenon is exactly similar.   Q.E.D.

In our search for a minimum-variance estimator of a parameter $g(\theta)$ the following theorem tells us that if a sufficient statistic exists we can restrict that search to functions of a sufficient statistic.

Theorem 9.1.3: Rao–Blackwell theorem Let $(X_1,\ldots,X_n)$ be a random sample of size n from the random phenomenon specified by the probability density function $f_X(x|\theta)$, $\theta \in \Omega$ [or mass function $p_X(x|\theta)$ in the discrete case]. Let $T(X_1,\ldots,X_n)$ be a sufficient estimator of θ and let $T_1(X_1,\ldots,X_n)$ (not necessarily a function of T alone) be an estimator of $g(\theta)$, where g is any function of θ. Then for all θ

$$E(T_1 - g(\theta))^2 \geq E(h(T)) - (g(\theta))^2, \tag{9.5}$$

where $h(t) = E(T_1|T = t)$ and $h(T)$ is an unbiased estimator of $g(\theta)$ if T_1 is an unbiased estimator of $g(\theta)$.

Proof. As usual, we deal with the continuous case. Let $f_{T,T_1}(t,t_1|\theta)$, $f_T(t|\theta)$, $f_{T_1}(t_1|\theta)$, $f_{T_1|T}(t_1|t,\theta)$ denote, respectively, the joint probability density function of T, T_1, the marginal probability density functions of T and T_1, and the conditional probability density function of T_1 given T. Now

$$h(t) = E(T_1|T = t)$$

$$= \int_{-\infty}^{\infty} t_1 f_{T_1|T}(t_1|t,\theta)\, dt_1$$

$$= \int_{-\infty}^{\infty} \frac{t_1 f_{T_1,T}(t_1,t|\theta)\, dt_1}{f_T(t|\theta)}. \tag{9.6}$$

Hence

$$h(t)f_T(t|\theta) = \int_{-\infty}^{\infty} t_1 f_{T_1,T}(t_1,t|\theta)\, dt_1. \tag{9.7}$$

Since T is sufficient for θ, $h(t)$ is independent of θ [i.e., $h(t)$ is a statistic] and

$$E(h(T)) = \int_{-\infty}^{\infty} h(t)f_T(t|\theta)\, dt$$

$$= \int_{-\infty}^{\infty}\int_{-\infty}^{\infty} t_1 f_{T_1,T}(t_1,t|\theta)\, dt_1\, dt$$

$$= E(T_1).$$

Now

$$E(T_1 - g(\theta))^2 = E(T_1 - h(T) + h(T) - g(\theta))^2$$

$$= E(T_1 - h(T))^2 + E(h(T)) - g(\theta))^2$$

$$\quad + 2E(T_1 - h(T))(h(T) - g(\theta))$$

$$= E(T_1 - h(T))^2 + E(h(T)) - g(\theta))^2$$

$$\geq E(h(T)) - g(\theta))^2$$

since

$$E(T_1 - h(T))(h(T) - g(\theta))$$

$$= \int_{-\infty}^{\infty}\int_{-\infty}^{\infty} [t_1 - h(t)][h(t) - g(\theta)]f_{T_1,T}(t_1,t|\theta)\, dt_1\, dt$$

$$= \int_{-\infty}^{\infty} [h(t) - g(\theta)]\left\{\int_{-\infty}^{\infty} [t_1 - h(t)]f_{T_1|T}(t_1|t)\, dt_1\right\}f_T(t|\theta)\, dt$$

$$= \int_{-\infty}^{\infty} [h(t) - g(\theta)](0)f_T(t|\theta)\, dt$$

$$= 0.$$

This proves the first part of the theorem. Further, if T_1 is an unbiased estimator of $g(\theta)$, then $E(h(T)) = g(\theta)$. In this case Eq. (9.5) becomes

$$\text{var}(T_1) \geq \text{var}(h(T)).$$

The proof for the discrete case can be handled in the same way.   Q.E.D.

This theorem was proved by Rao (1945) and later by Blackwell (1947). It tells us that given any estimator there exists an estimator that is a function of a sufficient estimator and is uniformly (in θ) better in the sense of mean square error.

Example 9.1.4 Let X be a binomial random variable with probability mass function, $0 < \theta < 1$,

$$p_X(x|\theta) = \begin{cases} \binom{N}{x}\theta^x(1 - \theta)^{N-x}, & x = 0,1,\dots,N \\ 0, & \text{otherwise.} \end{cases}$$

It can be verified that

$$\begin{aligned} E(X^2) &= N\theta(1 - \theta) + N^2\theta^2 \\ &= \theta^2 N(N - 1) + N\theta \\ &= \theta^2(N)(N - 1) + E(X). \end{aligned}$$

Hence $E(X^2 - X) = EX(X - 1) = N(N - 1)\theta^2$. This implies that $X(X - 1)/N(N - 1)$ is an unbiased estimator of θ^2. Let $(X_1,\dots,X_n)$ be a random sample of size n from $p_X(x|\theta)$. Obviously,

$$\sum_i^n \frac{X_i(X_i - 1)}{N(N - 1)n}$$

is an unbiased estimator of θ^2.

Example 9.1.5 Consider a random phenomenon specified by the normal distribution with known mean μ and unknown variance σ^2, $\sigma^2 > 0$. Let $X_1,\dots,X_n$ be a random sample from this distribution. Now, it is easy to check that

$$E\left(\sum_1^n (X_i - \mu)^2\right) = n\sigma^2.$$

Hence

$$S_1^2 = \sum_1^n \frac{(X_i - \mu)^2}{n}$$

is an unbiased estimator of σ^2. Consider S_1 an estimator of σ. Since nS_1^2/σ^2 is a chi-square random variable with n degrees of freedom, we get

$$E(S_1) = \frac{n^{n/2}}{(2\sigma^2)^{(n/2)\Gamma(n/2)}} \int_0^\infty \exp\left(-\frac{1}{2}\frac{ns_1^2}{\sigma^2}\right)(s_1^2)^{(n-1)/2}\,ds_1^2$$

$$= \frac{\sqrt{2}\,\Gamma[(n+1)/2]}{\sqrt{n}\,\Gamma(n/2)}\,\sigma$$

$$\neq \sigma.$$

Hence S_1 is not an unbiased estimator of σ.

Example 9.1.6 Consider a random phenomenon specified by the normal distribution with unknown mean μ and variance 1. Let $X_1,\dots,X_n$ be a random sample of size n from this distribution, and let

$$T(X_1,\dots,X_n) = \sum_{i=1}^n X_i, \qquad T_1(X_1,\dots,X_n) = X_1.$$

It is easy to see that T is a sufficient estimator of θ and T_1 is an unbiased estimator of θ. Furthermore, the conditional probability density function of T_1 given $T = t$ is given by

$$f_{T_1|T}(t_1|t) = \frac{1}{[2\pi(n-1)/n]^{1/2}} \exp\left[-\frac{n(t_1 - t/n)^2}{2(n-1)}\right].$$

From above,

$$E(T_1|T = t) = \frac{t}{n}.$$

Obviously, the estimator

$$\frac{T}{n} = \frac{1}{n}\sum_{i=1}^n X_i = \overline{X}$$

is unbiased, with smaller variance than that of $T_1 = X_1$.

EXERCISES

1. Let $(X_1,\dots,X_n)$ be a random sample of size n from a random phenomenon characterized by the uniform probability density function

$$f_X(x|0) = \begin{cases} 1 & \text{if } \theta - \frac{1}{2} < x < \theta + \frac{1}{2} \\ 0 & \text{otherwise.} \end{cases}$$

Show that (a) $\sum_1^n X_i/n$, (b) $(\max_{1\le i\le n} X_i + \min_{1\le i\le n} X_i)/2$ are unbiased estimators of θ.

2. In Exercise 1 show that the probability density functions of $X = \max_{1 \le i \le n} X_i$, $Y = \min_{1 \le i \le n} X_i$ are given by

$$f_X(x|\theta) = \begin{cases} n[x - (\theta - \frac{1}{2})]^{n-1} & \text{if } \theta - \frac{1}{2} \le x \le \theta + \frac{1}{2} \\ 0 & \text{otherwise} \end{cases}$$

$$f_Y(y|\theta) = \begin{cases} n(\theta + \frac{1}{2} - y)^{n-1} & \text{if } \theta - \frac{1}{2} \le y \le \theta + \frac{1}{2} \\ 0 & \text{otherwise} \end{cases}$$

Find the probability density function of $X + Y$.

3. Let $(x_1,\ldots,x_n)$ be a random sample from a Bernoulli distribution with $P(X_i = 1) = \theta = 1 - P(X_i = 0)$, where $0 < \theta < 1$. Find an unbiased estimator of θ^2. Compare this result with Example 9.1.4. Show that $1/\theta^r$ does not have an unbiased estimator for any integer $r > 0$.

4. Let $(x_1,\ldots,x_n)$ be a random sample from a random phenomenon specified by the probability mass function

$$p_X(x|\theta) = \begin{cases} \dfrac{e^{-\theta}\theta^x}{x!} & \text{if } x = 0,1,\ldots,n \\ 0 & \text{otherwise.} \end{cases}$$

Find an unbiased estimator of θ^2. Does θ^{-2} have an unbiased estimator?

5. Let $X_1,\ldots,X_n$ be n independent normally distributed random variables with $E(X_i) = \mu$ and $\text{var}(X_i) = 1/i$. Find the minimum-variance linear unbiased estimator of μ.

6. Show that for a normal population with mean μ and variance σ^2, sample mean is the minimum-variance unbiased linear estimator for μ.

9.2   COMPLETENESS AND THE BEST UNBIASED ESTIMATOR

We have seen in Section 9.1 that given any unbiased estimator ϕ of $g(\theta)$, there exists at least one unbiased estimator (a function of the sufficient estimator of θ only) whose variance is uniformly smaller than that of ϕ, and that this estimator is obtained by taking the conditional expectation of ϕ given the sufficient estimator. Now the question is: Is it possible to find a unique unbiased estimator of $g(\theta)$, based only on the sufficient estimator of θ, which has the minimum variance among all unbiased estimators (not functions of sufficient estimators only) of $g(\theta)$? In general, this is not possible. However, if the distribution of the sufficient estimator is complete in the sense defined below, and if there exists a function of the sufficient estimator that is unbiased for $g(\theta)$, that function will essentially be unique and will have minimum variance among all unbiased estimators.

Definition 9.2.1: Complete family of distributions Let T be a continuous (discrete) random variable with probability density (mass) function $f_T(t|\theta)(p_T(t|\theta))$, $\theta \in \Omega$, where Ω is the parametric space. Let $U(T)$ be a real-valued function of T. The family of density (mass) functions $\{f_T(t|\theta) : \theta \in \Omega\}$ (or $\{p_T(t|\theta) : \theta \in \Omega\}$) is said to be *complete* if

$$E_\theta(U(T)) = 0 \qquad \text{for every } \theta \in \Omega \tag{9.8}$$

implies that $U(T) = 0$ for all values of T at which $f_T(t|\theta)$ [or $p_T(t|\theta)$] is greater than zero for some $\theta \in \Omega$.

Example 9.2.1 Let $X_1,\ldots,X_n$ be a random sample of size n from a Poisson distribution with parameter $\lambda > 0$ and let $T(X_1,\ldots,X_n) = \Sigma_1^n X_i$. We know that T is a Poisson random variable with parameter $n\lambda = \theta$. For any integrable function $U(T)$,

$$E_\theta(U(T)) = \sum_{t=0}^{\infty} U(t) \frac{e^{-\theta}\theta^t}{t!}.$$

If $E_\theta(U(T))$ is identically equal to zero for all θ, $0 < \theta < \infty$, we get

$$\sum_{t=0}^{\infty} \frac{U(t)\theta^t}{t!} = 0 \qquad \text{for all } \theta, \quad 0 < \theta < \infty. \tag{9.9}$$

The left-hand side of Eq. (9.9) is a polynomial in θ; hence all the coefficients must be zero. So $U(t) = 0$ for $t = 0,1,2,\ldots,\infty$. Hence the family of distributions of T is complete.

Note 1: If a polynomial of degree n or, in general, a convergent power series

$$\sum_{n=0}^{\infty} a_n z^n$$

is zero for all z belonging to some open interval, then $a_n = 0$ for all n.

Note 2: If the family of distributions of a sufficient statistic is complete, we call it a complete sufficient statistic.

Example 9.2.2 Let $X_1,\ldots,X_n$ be a random sample of size n from a distribution specified by the probability density function

$$f_X(x|\theta) = \begin{cases} \dfrac{1}{\theta} & \text{if } 0 \le x \le \theta \\ 0 & \text{otherwise.} \end{cases}$$

The joint probability density function of $X_1,\ldots,X_n$ is given by

$$\prod_1^n f_{X_i}(x_i|\theta) = \frac{1}{\theta^n} h(\theta,U),$$

where $U = \max(x_1,\ldots,x_n)$ and h is defined by

$$h(\theta,U) = \begin{cases} 1 & \text{if } 0 \le U \le \theta \\ 0 & \text{otherwise.} \end{cases}$$

By the factorization theorem, U is sufficient for θ and the distribution of U is given by

$$F_U(u|\theta) = \begin{cases} 0 & \text{if } u \le 0 \\ \dfrac{u^n}{\theta^n} & \text{if } 0 < u < \theta \\ 1 & \text{if } u \ge \theta. \end{cases}$$

Hence the probability density function of U is given by

$$f_U(u|\theta) = \begin{cases} n\theta^n u^{n-1} & \text{if } 0 \le u < \theta \\ 0 & \text{otherwise.} \end{cases}$$

For any real-valued function $g(U)$, if

$$E(g(U)) = \int_0^\theta g(u)n\theta^{-n}u^{n-1}\, du$$

is identically equal to zero for all θ, $0 < \theta < \infty$, then $g(U) = 0$ for all values of U for which $f_U(u|\theta) > 0$. Thus the family of probability density functions $\{f_U(u|\theta)\}$ is complete, or, equivalently, U is a complete sufficient statistic.

An important class of distributions that permit complete sufficient statistic is the exponential family of distributions (see Section 5.6).

Theorem 9.2.1   Let $X_1,\ldots,X_n$ be a random sample of size n from the family of distributions specified by the probability density functions

$$f_X(x|\theta) = \begin{cases} C(\theta)h(x) \exp\left[\sum_1^m Q_j(\theta)T_j(x)\right] & \text{if } a < x < b \\ 0 & \text{otherwise,} \end{cases} \tag{9.10}$$

where a, b do not depend on θ, $\theta \in \Omega$, and C, h, Q_j, and T_j are defined as in Section 5.6 or by the probability mass function

$$p_X(x|\theta) = \begin{cases} C(\theta)h(x) \exp\left[\sum_1^m Q_j(\theta)T_j(x)\right] & \text{if } x = a_1,a_2,a_3,\ldots \\ 0 & \text{otherwise} \end{cases} \quad (9.11)$$

where the set $\{a_1,a_2,\ldots\}$ does not depend on θ. Let Ω contain an open set. Then if $n > m$,

$$\left(\sum_{j=1}^n T_1(X_j),\ldots,\sum_{j=1}^n T_m(X_j)\right)$$

is a complete sufficient statistic for θ.

[For a more general treatment of this theorem the reader is referred to Lehmann (1959, p. 132). The proof depends on the uniqueness of Laplace transforms. Since this is a problem in analysis, we will not try to demonstrate it here.]

Proof. We will prove the theorem for the case of continuous random variables only. The proof for the discrete case is exactly the same. Let

$$T = \left(\sum_1^n T_1(X_j),\ldots,\sum_{j=1}^n T_m(X_j)\right)$$

$$t = \left(\sum_1^n T_1(x_j),\ldots,\sum_1^n T_m(x_j)\right) = (t_1,\ldots,t_m). \quad (9.12)$$

From Eq. (9.10) the probability density function of T is given by

$$f_T(t|\theta) = C(\theta) \exp\left[\sum_{j=1}^k Q_j(\theta)t_j\right] \int \cdots \int \left[\prod_1^n h(x_i)\right] dx_1 \cdots dx_n$$

$$= C(\theta) \exp\left[\sum_{j=1}^m Q_j(\theta)t_j\right] K(t), \quad (9.13)$$

where $K(t)$ is positive and does not depend on θ, and ω is the set of $x_1,\ldots,x_n$ such that

$$\sum_{j=1}^n T_i(x_j) = t_i, \qquad j = 1,\ldots,m.$$

This follows from the fact that $h(x_i)$ is positive for all x_i and is independent of θ. Now for any real-valued function $g(T)$ for which $E_\theta(g(T)) = 0$ for all $\theta \in \Omega$, we get that

$$E_\theta(g(T)) = \int \cdots \int C(\theta)g(t)K(t) \exp\left[\sum_{j=1}^{m} Q_j(\theta)t_j\right] dt_1 \cdots dt_m \qquad (9.14)$$

is identically equal to zero for all $\theta \in \Omega$. Interpreting the integral above as the Laplace transform of $g(T)K(T)$, we conclude from Eq. (9.14) that $g(T)K(T) = 0$ for all values of T for which $f_T(t|\theta) > 0$ for some $\theta \in \Omega$.   Q.E.D.

Example 9.2.3 Let $X_1,\ldots,X_n$ be a random sample of size n from the normal probability density function

$$f_X(x|\theta) = \begin{cases} \dfrac{1}{\sqrt{2\pi}} \exp\left[-\dfrac{1}{2}(x^2 - 2\theta x + \theta^2)\right], \\ \qquad -\infty < x < \infty, \quad -\infty < \theta < \infty \\ 0 \qquad \text{otherwise.} \end{cases}$$

Obviously, $f_X(x|\theta)$ belongs to the exponential family, with

$$C(\theta) = \frac{1}{\sqrt{2\pi}} e^{-\theta^2/2}, \qquad h(x) = e^{-x^2/2},$$

$$Q_1(\theta) = \theta, \qquad T_1(X) = x, \qquad m = 1.$$

Hence $\Sigma_1^n X_i$ is a complete sufficient statistic for θ.

Example 9.2.4 Let $X_1,\ldots,X_n$ be a random sample from the normal distribution $[\theta = (\mu,\sigma^2), -\infty < \mu < \infty, \sigma^2 > 0]$

$$f_X(x|\theta) = \begin{cases} \dfrac{1}{\sqrt{2\pi}\,\sigma} \exp\left[-\dfrac{1}{2}\left(\dfrac{x-\mu}{\sigma}\right)^2\right] \\ \dfrac{1}{\sqrt{2\pi}\,\sigma} \exp\left(\dfrac{-\mu^2}{2\sigma^2}\right) \exp\left(\dfrac{-x^2}{2\sigma^2} + \dfrac{\mu x}{\sigma^2}\right), \quad -\infty < x < \infty \\ 0 \qquad\qquad\qquad\qquad\qquad\qquad\qquad\qquad \text{otherwise.} \end{cases}$$

Hence $f_X(x|\theta)$ belongs to the exponential family with $h(x) = 1$, $Q_1(\theta) = -1/2\sigma^2$, $Q_2(\theta) = \mu/\sigma^2$, $T_1(X) = X^2$, $T_2(X) = X$. So by Theorem 9.2.1,

$$\left(\sum_1^n X_i^2, \sum_1^n X_i\right), \qquad n \geq 2$$

is a complete sufficient statistic for θ.

Example 9.2.5 In Example 9.2.4 assume that $\sigma^2 = \mu^2$ and that $\Omega = \{\mu : -\infty < \mu < \infty\}$ is a one-dimensional Euclidian space. From the factorization theorem, $(\Sigma_1^n X_i^2, \Sigma_1^n X_i)$ is sufficient for μ. Consider the fact that for all μ,

$$E\left(2\left(\sum_1^n X_i\right)^2 - (n+1)\sum X_i^2\right)$$

$$= 2(n\mu^2 + n^2\mu^2) - n(n+1)(\mu^2 + \mu^2)$$

$$= 0.$$

But

$$2\left(\sum_1^n X_i\right)^2 - (n+1)\sum_1^n X_i^2 \neq 0$$

for all $(x_1,\ldots,x_n)$. So there does not exist a complete sufficient statistic for μ.

Note 3: This example tells us that a two-dimensional sufficient statistic for a one-dimensional parameter is not complete.

The notion of completeness of the sufficient estimator of the parameter θ can be used to find the best unbiased estimator of any function $g(\theta)$. If T is a complete sufficient estimator of θ and T_1 is any unbiased estimator (not necessarily a function of T only) of $g(\theta)$, then $\phi(T) = E(T_1|T)$ is the unique minimum-variance unbiased estimator of $g(\theta)$ among all unbiased estimators of $g(\theta)$. We will call it the best unbiased estimator.

Example 9.2.6 Let $(X_1,\ldots,X_n)$ be a random sample from the Poisson mass function

$$p_X(x|\theta) = \frac{e^{-\theta}\theta^x}{x!}$$

$$= \begin{cases} e^{-\theta}\dfrac{1}{x!}e^{x(\log\theta)} & \text{if } x = 1,2,\ldots\infty, \quad \theta > 0 \\ 0 & \text{otherwise.} \end{cases}$$

By Theorem 9.2.1, $T = \Sigma_1^n X_i$ is a complete sufficient statistic for θ. Let $\phi(T) = T/n$. Since the distribution of T is complete and $E(\phi(T)) = \theta$, $\phi(T)$ is the only function of T such that $E(\phi(T)) = \theta$. By Theorem 9.1.3 it has the minimum variance among all unbiased estimators of θ. Hence $\phi(T)$ is the best unbiased estimator of θ.

Example 9.2.7 Let $(X_1,\ldots,X_n)$ be a random sample from a normal distribution with mean θ, $-\infty < \theta < \infty$ and variance 1. It is easy to check that $T = \Sigma_1^n X_i$ is a complete sufficient statistic for θ. Now $\phi(T) = T/n$ is the only function of T such that $E(\phi(T)) = \theta$. By Theorem 9.1.3, $\phi(T)$ is the best unbiased estimator of θ.

EXERCISES

7. Let $(X_1,\ldots,X_n)$ be a random sample from the normal distribution with mean μ and variance σ^2. Let

$$n\overline{X} = \sum_1^n X_i, \qquad S^2 = \sum_1^n (X_i - \overline{X})^2.$$

   Show that $(\overline{X},S^2)$ is a complete sufficient statistic for $\theta = (\mu,\sigma^2)$, $-\infty < \mu < \infty$, $\sigma^2 > 0$.

8. Let $(X_1,\ldots,X_{n_1})$, $(Y_1,\ldots,Y_{n_2})$ be two random samples from two different normal distributions with the same mean μ and variances σ_1^2, σ_2^2, respectively. Let

$$n_1\overline{X} = \sum_1^{n_1} X_i, \quad n_2\overline{Y} = \sum_1^{n_2} Y_i, \quad T = \left(\overline{X},\overline{Y}, \sum_1^{n_1} X_i^2, \sum_1^{n_2} Y_i^2\right).$$

   Show that T is sufficient for $\theta = (\mu,\sigma_1^2,\sigma_2^2)$, but that T is not a complete sufficient statistic.

9. Let $X = (X_1,\ldots,X_p)'$ be a random p-vector, normally distributed with mean $\mu = (\mu_1,\ldots,\mu_p)'$ and covariance matrix I (identity matrix). Let $X_\alpha = (X_{\alpha 1},\ldots,X_{\alpha p})'$, $\alpha = 1,\ldots,N$, be a random sample of size N from this distribution. Show that $\overline{X} = \Sigma_1^N X_\alpha/N$ is a complete sufficient statistic for μ.

10. Let $X_1,\ldots,X_n$ be a random sample of size n from a normal distribution with mean μ and variance σ^2. Let $T_1 = \Sigma_1^n a_i X_i$, $T = \Sigma_1^n X_i$. Show that the complete sufficient statistic T is independent of T_1 if and only if $\Sigma_1^n a_i = 0$.

11. Let $X_1,\ldots,X_n$ be a random sample of size n from a normal distribution with mean μ and variance σ^2. Let c be a given constant such that

$$p = P(X > c).$$

    Find the best unbiased estimator of p. [*Hint:* Let

$$T(X_1,\ldots,X_n) = \begin{cases} 0 & \text{if } X_1 \le c \\ 1 & \text{if } X_1 > c. \end{cases}$$

    Then $E(T) = p$. Use Exercise 7 and the Rao–Blackwell theorem.]

12. Let $(X_1,\ldots,X_n)$ be a random sample of size n from a distribution with parameter θ, and let $T(X_1,\ldots,X_n)$ be a minimum-variance unbiased estimator of $g(\theta)$. Denote by $(x_a,x_b,\ldots)$ a permutation of $(x_1,\ldots,x_n)$. There are $n!$ such permutations of $x_1,\ldots,x_n$. Show that

$$\frac{\Sigma\, T(x_a,x_b,\ldots)}{n!},$$

where the summation is over all permutations of $x_1,\ldots,x_n$, is also an unbiased estimator of $g(\theta)$, and has variance not larger than that of T.

13. Let $X_1,\ldots,X_n$ be a random sample from a normal distribution with mean θ, $-\infty < \theta < \infty$, and variance 1. Let $n\overline{X} = \Sigma_1^n X_i$. Show that $\overline{X}^2 - (1/n)$ is a minimum-variance unbiased estimator of θ^2.

14. Let $X_1,\ldots,X_n$ be n independent, identically distributed random variables, distributed uniformly on the integers, $1,2,\ldots,N$, where N is unknown. Show that $T = \max(X_1,\ldots,X_n)$ is a sufficient estimator for N.

15. Suppose that T_1, T_2, T_3 are unbiased estimators of the parameter θ and T_1 is the minimum-variance unbiased estimator of θ. Assume that

$$V(T_2) = V(T_3) = \lambda V(T_1), \qquad \lambda > 1.$$

Show that the coefficient of correlation ρ satisfies
(a) $\rho(T_1,T_2) = \rho(T_1,T_3) = \lambda^{-1/2}$.
(b) $\rho(T_2,T_3) = (1/\lambda)(2 - \lambda)$.

9.3 MOST EFFICIENT ESTIMATOR AND CONSISTENT ESTIMATOR

In Section 9.2 we studied the criteria of unbiasedness and minimum variance to judge different estimators of the parametric function $g(\theta)$. We have studied these two criteria mainly to obtain what has been defined as the best estimator. Another important criterion for comparing different estimators is efficiency, which was introduced by Fisher (1921, 1925). It is obvious that the smaller the variance of an estimator, the more concentrated are its values (estimates) around the true value, hence the larger is the probability that an estimate will be close to the true value.

Definition 9.3.1: Most efficient estimator   A most efficient estimator of a parametric function is an estimator whose variance is as small as possible. The efficiency of an estimator is the inverse of the ratio of its variance to that of the most efficient estimator.

To give a precise definition of the most efficient estimator it is necessary to derive the following important inequality, called the Rao–Cramér inequality. We will consider here the case of a one-dimensional parameter only. For the case of a multidimensional parameter the reader is referred to Rao (1965). To derive this inequality we need the following definition.

Definition 9.3.2: Regular distribution   The joint probability density function $f_{X_1,\ldots,X_n}(x_1,\ldots,x_n|\theta)$, $\theta \in \Omega$, is said to be *regular* with respect to its first θ-derivative if

$$\frac{\partial}{\partial\theta} \int_{-\infty}^{\infty} \cdots \int_{-\infty}^{\infty} f_{X_1,\ldots,X_n}(x_1,\ldots,x_n|\theta)\ dx_1 \cdots dx_n$$

$$= \int_{-\infty}^{\infty} \cdots \int_{-\infty}^{\infty} \frac{\partial}{\partial\theta} \log[f_{X_1,\ldots,X_n}(x_1,\ldots,x_n|\theta)]$$

$$\times\ f_{X_1,\ldots,X_n}(x_1,\ldots,x_n|\theta)\ dx_1 \cdots dx_n. \tag{9.15}$$

In the case of discrete random variables a similar definition holds provided that we replace the integrals by summation and the probability density function by the probability mass function.

Rao–Cramér Inequality   Let $X_1,\ldots,X_n$ be a random sample of size n from a random phenomenon specified by the probability density function $f_X(x|\theta)$, $\theta \in \Omega$. Let $T(X_1,\ldots,X_n)$ be an unbiased estimator of the parametric function $g(\theta)$. Assume the following condition: $g(\theta)$, $f_X(x|\theta)$ have first-order derivatives with respect to θ and the joint probability density function of $X_1,\ldots,X_n$ is regular with respect to its first θ-derivative for θ belonging to an open interval in Ω. Then

$$V(T) \geq \frac{[g'(\theta)]^2}{n \int_{-\infty}^{\infty} \left[\dfrac{\partial}{\partial\theta} \log f_X(x|\theta)\right]^2 f_X(x|\theta)\ dx}. \tag{9.16}$$

For the case of discrete random phenomenon specified by the probability mass function $p_X(x_i|\theta)$, $i = 0,\ldots,k$, we have, instead,

$$V(T) \geq \frac{[g'(\theta)]^2}{n \sum_{0}^{k} \left[\dfrac{\partial}{\partial\theta} \log p_X(x|\theta)\right]^2 p_X(x_i|\theta)}. \tag{9.17}$$

Proof. Let the probability density function of T be $f_T(t|\theta)$. We are given that

$$\int_{-\infty}^{\infty} f_{X_i}(x_i|\theta) \, dx_i = 1, \qquad i = 1,\ldots,n$$

$$g(\theta) = \int_{-\infty}^{\infty} t f_T(t|\theta) \, dt$$

$$= \int_{-\infty}^{\infty} \cdots \int_{-\infty}^{\infty} T(x_1,\ldots,x_n) \left[\prod_1^n f_{X_i}(x_i|\theta) \, dx_i \right]. \tag{9.18}$$

If we differentiate both sides of each of these two equations with respect to θ, we get

$$0 = \int_{-\infty}^{\infty} \frac{\partial}{\partial\theta} f_{X_i}(x_i|\theta) \, dx_i$$

$$= \int_{-\infty}^{\infty} \frac{\partial}{\partial\theta} [\log f_{X_i}(x_i|\theta)] f_{X_i}(x_i|\theta) \, dx_i \tag{9.19}$$

$$g'(\theta) = \frac{\partial}{\partial\theta} \left[\int_{-\infty}^{\infty} \cdots \int_{-\infty}^{\infty} T(x_1,\ldots,x_n) \prod_1^n f_{X_i}(x_i|\theta) \, dx_i \right]$$

$$= \int_{-\infty}^{\infty} \cdots \int_{-\infty}^{\infty} T(x_1,\ldots,x_n) \left[\sum_1^n \frac{\partial}{\partial\theta} \log f_{X_i}(x_i|\theta) \right]$$

$$\times \prod_1^n f_{X_i}(x_i|\theta) \, dx_i$$

$$= E(TZ)$$

$$= E(T)E(Z) + \mathrm{cov}(TZ)$$

$$= E(T)E(Z) + \rho\sqrt{V(T)V(Z)}, \tag{9.20}$$

where

$$Z = \sum_1^n \frac{\partial}{\partial\theta} \log f_{X_i}(X_i|\theta)$$

and ρ is the coefficient of correlation between T and Z. From Eq. (9.19) we get

$$E(Z) = \sum_1^n E\left(\frac{\partial}{\partial\theta} \log f_{X_i}(X_i|\theta) \right)$$

$$= \sum_1^n \int_{-\infty}^{\infty} \frac{\partial}{\partial\theta} \log f_{X_i}(x_i|\theta) f_{X_i}(x_i|\theta) \, dx_i$$

$$= 0. \tag{9.21}$$

Furthermore, Z is the sum of n independent, identically distributed random variables each with mean zero; consequently,

$$V(Z) = nE\left[\frac{\partial}{\partial\theta} \log f_X(X|\theta)\right]^2$$

$$= n \int_{-\infty}^{\infty} \left[\frac{\partial}{\partial\theta} \log f_X(x|\theta)\right]^2 f_X(x|\theta)\, dx. \qquad (9.22)$$

From Eq. (9.20) we get

$$\rho^2 = \frac{[g'(\theta)]^2}{V(Z)V(T)}. \qquad (9.23)$$

Since $\rho^2 \leq 1$, from Eq. (9.23) we have

$$V(T) \geq \frac{[g'(\theta)]^2}{\text{var}(Z)}$$

$$= \frac{[g'(\theta)]^2}{nE\left[\dfrac{\partial}{\partial\theta} \log f_X(X|\theta)\right]^2}.$$

The proof of Eq. (9.17) is exactly similar.   Q.E.D.

According to this inequality, the variance of the unbiased estimator T of the parametric function $g(\theta)$ cannot be smaller than

$$\frac{[g'(\theta)]^2}{nE\left[\dfrac{\partial}{\partial\theta} \log f_X(X|\theta)\right]^2}$$

which is a constant for fixed n. We shall call this constant the Rao–Cramér lower bound of the variances of unbiased estimators of $g(\theta)$.

Note: An estimation problem is called regular if it satisfies the condition assumed for the Rao–Cramér inequality.

Definition 9.3.3 In a regular point-estimation case, an unbiased estimator is called the *most efficient estimator* of the parametric function $g(\theta)$ if and only if its variance attains the Rao–Cramér lower bound.

Theorem 9.3.1 The equality sign in Eq. (9.16) [similarly in Eq. (9.17)] holds if and only if:
(a) $(X_1,\ldots,X_n)$ is sufficient for θ.
(b) For $f_T(t|\theta) > 0$, the probability density function $f_T(t|\theta)$ satisfies the following equation:

$$\frac{\partial}{\partial\theta} \log f_T(t|\theta) = k[t - g(\theta)] \tag{9.24}$$

where the constant k is independent of t.

Proof. First we prove the sufficiency of these two conditions. Suppose that conditions (a) and (b) are satisfied. Since T is sufficient for θ, the conditional distribution of $X_1,\ldots,X_n$ given that T is independent of θ. Now using Theorem 5.5.2, after trivial simplification we get

$$\int_{-\infty}^{\infty} \cdots \int_{-\infty}^{\infty} \left\{ \frac{\partial}{\partial\theta} \left[\log \prod_1^n f_{X_i}(x_i|\theta) \right] \right\}^2 \prod_1^n [f_{X_i}(x_i|\theta) \, dx_i]$$

$$= k^2 \int_{-\infty}^{\infty} [t - g(\theta)]^2 f_T(t|\theta) \, dt$$

$$= k^2 V(T) \tag{9.25}$$

and

$$g'(\theta) = \int_{-\infty}^{\infty} t \frac{\partial}{\partial\theta} f_T(t|\theta) \, dt$$

$$= \int_{-\infty}^{\infty} [t - g(\theta)] \frac{\partial}{\partial\theta} f_T(t|\theta) \, dt$$

$$= \int_{-\infty}^{\infty} [t - g(\theta)] \left[\frac{\partial}{\partial\theta} \log f_T(t|\theta) \right] f_T(t|\theta) \, dt$$

$$= kV(T). \tag{9.26}$$

The second equality of Eq. (9.26) follows from the fact that

$$\int_{-\infty}^{\infty} f_T(t|\theta) \, dt = 1$$

and the fact that the integral can be differentiated under the integral sign. From Eqs. (9.25) and (9.26) we get

$$n \int_{-\infty}^{\infty} \left[\frac{\partial}{\partial\theta} \log f_X(x|\theta) \right]^2 f_X(x|\theta) \, dx = \frac{[g'(\theta)]^2}{V(T)}. \tag{9.27}$$

Conversely, when the equality in Eq. (9.16) is attained, the correlation between T and Z is

$$\frac{E(TZ)}{\sqrt{V(T)V(Z)}} = \frac{g'(\theta)}{V(T)V(Z)} = 1, \tag{9.28}$$

in which case $T - g(\theta) = CZ$, where C is a constant independent of T except possibly for a set of values of $x_1,\ldots,x_n$ of probability zero. Writing $C = 1/k$, we get

$$Z = \frac{\partial}{\partial\theta}\left[\log \prod_1^n f_X(x_i|\theta)\right] = k[t - g(\theta)]. \tag{9.29}$$

Now integrating Eq. (9.29) with respect to θ, we get

$$\prod_1^n f_{X_i}(x_i|\theta) = c(x_1,\ldots,x_n)K(\theta)\exp[Q_1(\theta)T(x_1,\ldots,x_n)], \tag{9.30}$$

where c, K, and Q_1 are known functions. From Theorem 5.5.1, Problem 11 of Chapter 5, and Eq. (9.30) we conclude that T is sufficient for θ. Since T is sufficient for θ, Eq. (9.29) reduces to (using Theorem 5.5.2)

$$\frac{\partial}{\partial\theta}\log f_T(t|\theta) = k[t - g(\theta)]. \qquad \text{Q.E.D.}$$

Example 9.3.1 Let $X_1,\ldots,X_n$ denote a random sample of size n from a random phenomenon specified by the probability mass function ($0 < \theta < 1$),

$$p_X(x|\theta) = \begin{cases} \dbinom{N}{x}\theta^x(1 - \theta)^{N-x} & \text{if } x = 0,1,\ldots,N \\ 0 & \text{otherwise} \end{cases}$$

and let $g(\theta) = \theta$. Here

$$n\sum_{x=0}^N\left[\frac{\partial}{\partial\theta}\log p_X(x|\theta)\right]^2 p_X(x|\theta)$$

$$= n\sum_{x=0}^N\left(\frac{x}{\theta} - \frac{N - x}{1 - \theta}\right)^2\binom{N}{x}\theta^x(1 - \theta)^{N-x}$$

$$= n\sum_{x=0}^N\left(\frac{x - N\theta}{\theta(1 - \theta)}\right)^2\binom{N}{x}\theta^x(1 - \theta)^{N-x}$$

$$= \frac{n}{\theta^2(1 - \theta)^2}V(X) = \frac{Nn}{\theta(1 - \theta)}.$$

From the Rao–Cramér inequality the variance of the most efficient estimator of $\hat{\theta}$ is equal to $\theta(1 - \theta)/nN$. Now consider the estimator $\overline{X} = \Sigma_i X_i/Nn$. From the expressions for the mean and variance of the binomial distribution, we get $E(\overline{X}) = \theta$, $V(\overline{X}) = \theta(1 - \theta)/Nn$. Thus $\overline{X}$ is the most efficient estimator of θ.

Example 9.3.2 Let $X_1,\ldots,X_n$ be a random sample of size n from a random phenomenon specified by the probability density function (which is known to be regular)

$$f_X(x|\theta) = \begin{cases} \dfrac{\theta^p}{\Gamma(p)}\, e^{-\theta x} x^{p-1}, & x > 0 \\ 0 & \text{otherwise} \end{cases}$$

and let

$$g(\theta) = \frac{1}{\theta}, \qquad \overline{X} = \frac{1}{n}\sum_1^n X_i.$$

Thus

$$E\!\left(\frac{\overline{X}}{p}\right) = g(\theta), \qquad V\!\left(\frac{\overline{X}}{p}\right) = \frac{[g(\theta)]^2}{np}.$$

Here

$$n\int_0^\infty \left[\frac{\partial}{\partial\theta}\log f_X(x|\theta)\right]^2 f_X(x|\theta)\,dx = n\int_0^\infty \left(\frac{p}{\theta} - x\right)^2 f_X(x|\theta)\,dx$$

$$= \frac{np}{\theta^2}.$$

By the Rao–Cramér inequality the variance of the most efficient unbiased estimator of $g(\theta)$ is $1/np\theta^2$. So $\overline{X}/p$ is the most efficient unbiased estimator of $g(\theta)$.

Example 9.3.3 Let $X_1,\ldots,X_n$ be a random sample of size n from a random phenomenon specified by the probability density function

$$f_X(x|\mu,\sigma^2) = \begin{cases} \dfrac{1}{\sqrt{2\pi}\,\sigma}\exp\left[-\frac{1}{2}\left(\dfrac{x-\mu}{\sigma}\right)^2\right], & -\infty < x < \infty \\ 0, & \text{otherwise,} \end{cases}$$

where $-\infty < \mu < \infty$, $\sigma^2 > 0$. Let

$$\overline{X} = \frac{1}{n}\sum_1^n X_i, \qquad S^2 = \sum_1^n \frac{(X_i - \overline{X})^2}{n-1}.$$

From Chapter 5, $(n - 1)S^2/\sigma^2$ has central chi-square distribution with $n - 1$ degrees of freedom. Hence $E(S^2) = \sigma^2$, $V(S^2) = 2\sigma^4/(n - 1)$. Now

$$\log f_X(x|\mu,\sigma^2) = -\frac{(x - \mu)^2}{2\sigma^2} - \log \sqrt{2\pi} \, \sigma^2$$

$$\frac{\partial}{\partial \sigma^2} \log f_X(x|\mu,\sigma^2) = -\frac{(x - \mu)^2}{2\sigma^4} - \frac{1}{2\sigma^2}$$

$$\frac{\partial^2}{\partial(\sigma^2)^2} \log f_X(x|\mu,\sigma^2) = -\frac{(x - \mu)^2}{\sigma^6} + \frac{1}{2\sigma^4}$$

$$-E\left(\frac{\partial^2}{\partial(\sigma^2)^2} \log f_X(x|\mu,\sigma^2)\right) = \frac{1}{2\sigma^4}.$$

From Problem 23 of this chapter, the Rao–Cramér lower bound of the variance of unbiased estimators of σ^2 is $2\sigma^4/n$. Hence the efficiency of the unbiased estimator S^2 of σ^2 is $(n - 1)/n$.

The theory of jointly most efficient estimators of several parameters has been developed with the associated concept of joint efficiency of several estimators. This is beyond the scope of this book, and the reader is referred to Rao (1965), Chapter 5 and Cramér (1946), Chapter 32.

Consider a random phenomenon specified by the distribution function $F_X(x|\theta)$, $\theta \in \Omega$. Let $X_1,\ldots,X_n$ be a random sample of size n from this phenomenon, and let $T_n = T(X_1,\ldots,X_n)$ be an estimator of $g(\theta)$. Now our aim is to examine the properties of T_n as the sample size n tends to infinity. It is obvious that, in general, the estimates obtained from T_n for fixed n will differ from its true value. However, it is desirable that as n increases to infinity, the probability of T_n being close to its true value $g(\theta)$ should be near unity.

Definition 9.3.4: Consistent estimator   An estimator T_n is said to be consistent for $g(\theta)$ if T_n converges to $g(\theta)$ in probability (i.e., for every $\epsilon > 0$, $\lim_{n\to\infty} P\{|T_n - g(\theta)| > \epsilon\} = 0$) or with probability 1 (i.e., $P\{\lim_{n\to\infty} T_n = g(\theta)\} = 1$). The former case is called *weak consistency* and the latter, *strong consistency*.

Consistency as defined here is a limiting property of the estimator as the sample size increases to infinity, and is not valid with reference to a particular sample size. Thus it is necessary to exercise a certain amount of caution in applying the criterion in practical situations. For example, suppose that T_n is a consistent estimator of $g(\theta)$. Define T'_n by

$$T'_n = \begin{cases} 0 & \text{if } n \le 10^5 \\ T & \text{if } n > 10^5. \end{cases} \tag{9.31}$$

Obviously, T_n' is also a consistent estimator of $g(\theta)$. But T_n' cannot be accepted as a reasonable estimator in all practical situations where the sample size is large but not sufficiently large.

Example 9.3.4 Let T_n be defined by

$$T_n = \frac{1}{n} \sum_1^n X_i, \qquad n = 1,2,\dots,$$

where the random variables X_i are independent, identically distributed normal random variables with mean θ and variance 1. By Chebychev's inequality for $\epsilon > 0$,

$$P\{|T_n - \theta| > \epsilon\} \le \frac{1}{n\epsilon^2} \longrightarrow 0 \qquad \text{as } n \longrightarrow \infty.$$

Hence T_n is a weakly consistent estimator of θ. From Theorem 4.1.6 it follows that

$$P\{\lim_{n\to\infty} T_n = \theta\} = 1.$$

Hence T_n is also a strongly consistent estimator of θ (s.c. $\Rightarrow$ w.c.).

EXERCISES

16. Let $X_1,\dots,X_n$ be a random sample of size n from a normal distribution with mean μ and variance σ^2. Let

$$nS_1^2 = \sum_1^n (X_i - \overline{X})^2, \qquad n\overline{X} = \sum_1^n X_i.$$

    Show that S_1^2 is a weakly consistent estimator of σ^2.
17. Let $(X_1,\dots,X_n)$ be a random sample of size n from a random phenomenon specified by binomial distribution with parameter (N,p). Find the most efficient estimator of p. Is it strongly consistent?
18. Let $(X_1,\dots,X_n)$ be a random sample of size n from a Poisson distribution with parameter λ. Find the most efficient estimator of λ. Is it strongly consistent?
19. Consider a random phenomenon specified by the exponential probability density function

$$f_X(x|\lambda) = \begin{cases} 0, & x \le 0 \\ \lambda e^{-\lambda x}, & x > 0. \end{cases}$$

    Find the most efficient estimator of $g(\lambda) = 1/\lambda$ on the basis of a random sample of size n. Is it consistent?

20. Let $(X_1,\ldots,X_n)$ be a random sample of size n from a random phenomenon specified by the Cauchy distribution with probability density function

$$f_X(x|\theta) = \begin{cases} \dfrac{1}{\pi(1 + (x - \theta)^2)}, & -\infty < x < \infty \\ 0, & \text{otherwise.} \end{cases}$$

Is $\overline{X}_n = (1/n) \sum_1^n X_i$ a consistent (weakly) estimator of θ? Find a consistent estimator of θ. (*Hint*: $\overline{X}_n$ has the same Cauchy distribution.)

21. Consider a random phenomenon specified by the probability density function (with p known)

$$f_X(x|\theta) = \begin{cases} \dfrac{\theta^p}{\Gamma(p)} e^{-\theta x} x^{p-1} & \text{if } x > 0 \\ 0 & \text{otherwise.} \end{cases}$$

For any random sample $(X_1,\ldots,X_n)$ from this let us define $\overline{X}_n$ by

$$n\overline{X}_n = \sum_1^n X_i.$$

(a) Show that

$$T = \frac{np - 1}{n} \overline{X}_n$$

is an unbiased, consistent, sufficient estimator of θ.

(b) Find the efficiency of the most efficient estimator of θ.

22. Let $(X_1,\ldots,X_n)$ be a random sample of size n from a random phenomenon specified by the uniform probability density function

$$f_X(x|\theta) = \begin{cases} \dfrac{1}{\theta} & \text{if } 0 < x \le \theta \\ 0 & \text{otherwise.} \end{cases}$$

(a) Show that $Y = \max(X_1,\ldots,X_n)$ is a consistent estimator of θ.

(b) Show that $(n + 1)Y$ is the minimum-variance unbiased estimator of θ.

23. Show that Rao–Cramér inequality can be written as

$$V(T) \ge \frac{[g'(\theta)]^2}{-nE\left(\dfrac{\partial^2}{\partial\theta^2} \log f_X(x|\theta)\right)}.$$

9.4   VARIOUS METHODS OF ESTIMATION

Our object in the present section is to discuss various methods for obtaining estimators. Sometimes it is difficult to single out the best method in a particular case: the merits of the various methods are to be judged in the light of the different criteria discussed earlier, keeping in mind the specified needs of each situation. In discussing these methods we will put greatest emphasis on the traditional method of maximum likelihood, with passing reference to the method of moments and the minimum-chi-squared method.

9.4.1   Method of Moments

The oldest method of obtaining estimators is the method of moments. Let $(X_1,...,X_n)$ be a random sample of size n from distribution function $F_X(x|\theta)$, with θ (real or vector) $\in \Omega$ (the parametric space), and θ unknown. Let $(x_1,...,x_n)$ denote a corresponding sample value. Assume that θ is a p-vector with components $\theta_1,...,\theta_p$. Let m_k denote the kth raw moment of order k of the distribution F. Obviously, m_k, $k = 1,...,p$ are functions of $\theta_1,...,\theta_p$. Define

$$A_k = \frac{1}{n} \sum_{1}^{n} X_i^k, \qquad k = 1,...,p. \tag{9.32}$$

A_k is called the *sample raw moment* of order k based on a random sample of size n. The method of moments consists of equating A_k with m_k for $k = 1,...,p$ and solving for $\theta_1,...,\theta_p$. The values we obtain are estimators of $\theta_1,...,\theta_p$. The estimates of $\theta_1,...,\theta_p$ corresponding to $(x_1,...,x_n)$ are obtained by replacing $X_1,...,X_n$ by $x_1,...,x_n$.

The estimator $(1/n) \sum_{1}^{n} X_i^k$ is the mean of n independent, identically distributed random variables X_i^k, $i = 1,...,n$. Thus if $E(X_i^k) < \infty$, then, by the strong law of large numbers, $(1/n) \sum_{1}^{n} X_i^k$ converges to m_k as $n \to \infty$ with probability 1, so that A_k is a strongly consistent estimator of m_k. It is now left to the reader to verify that if $m_1,...,m_p$ are one-to-one functions of $\theta_1,...,\theta_p$, and if the inverse functions

$$\theta_i = g_i(m_1,...,m_p), \qquad i = 1,...,p \tag{9.33}$$

are continuous in $m_1,...,m_p$, then [with $a_k = (1/n) \sum_{1}^{n} x_i^k$]

$$\hat{\theta}_i = g_i(a_1,...,a_p), \qquad i = 1,...,p \tag{9.34}$$

are the moment estimate of θ_i and $g_i(A_1,...,A_p)$, $i = 1,...,p$, are consistent (strongly) estimators of θ_i.

Comment: This method is not applicable when m_k, $k = 1,\ldots,p$ do not exist [e.g., if $F_X(x|\theta)$ is a Cauchy distribution].

Example 9.4.1.1   Let $X_1,\ldots,X_n$ be a random sample of size n from a normal distribution with μ and variance σ^2. Here

$$m_1 = \mu, \qquad m_2 = \sigma^2 + \mu^2,$$

$$A_1 = \frac{1}{n}\sum_1^n X_i = \overline{X}, \qquad A_2 = \frac{1}{n}\sum_1^n X_i^2.$$

Hence the moment estimator of σ^2 is given by

$$\frac{1}{n}\sum_1^n X_i^2 - \overline{X}^2 = \frac{1}{n}\sum_1^n (X_i - \overline{X})^2.$$

Example 9.4.1.2   Let $X_1,\ldots,X_n$ be a random sample of size n from a Poisson distribution with parameter λ. Here $m_1 = \lambda$, $A_1 = (1/n)\sum_1^n X_i = \overline{X}$. The moment estimator of λ is $\overline{X}$.

9.4.2 Method of Maximum Likelihood

Consider a random sample $(X_1,\ldots,X_n)$ of size n from a random phenomenon specified by the probability density function $f_X(x|\theta)$, $\theta = [(\theta_1,\ldots,\theta_k)$, $k > 1] \in \Omega$. For any given sample $(x_1,\ldots,x_n)$ of observations on $X_1,\ldots,X_n$,

$$L(x_1,\ldots,x_n;\theta) = \prod_1^n f_{X_i}(x_i|\theta), \qquad \theta \in \Omega \tag{9.35}$$

is called the *likelihood function* of the sample $(x_1,\ldots,x_n)$. For the discrete random phenomenon specified by the probability mass function $p_X(x|\theta)$, the likelihood function is given by

$$L(x_1,\ldots,x_n;\theta) = \prod_1^n p_{X_i}(x_i|\theta), \qquad \theta \in \Omega \tag{9.36}$$

Given $x_1,\ldots,x_n$, $L(x_1,\ldots,x_n;\theta)$ is a function of θ and is the probability of obtaining the observed sample $(x_1,\ldots,x_n)$ from the assumed random phenomenon. The maximum-likelihood estimate of the unknown parameter θ is that value of $\theta = \hat{\theta} = (\hat{\theta}_1,\ldots,\hat{\theta}_k)$ (provided that it exists) which maximizes $L(x_1,\ldots,x_n;\theta)$. If such a $\hat{\theta}$ exists, it is obtained (in most cases) by solving the following system of k equations:

$$\frac{\partial L}{\partial \theta_i} = 0, \qquad i = 1,\ldots,k. \tag{9.37}$$

Now, since L and $\log L$ assume their maximum for the same values of $\theta_1,...,\theta_k$, Eq. (9.37) can be replaced by

$$\frac{\partial}{\partial \theta_i} \log L = 0, \qquad i = 1,...,k. \tag{9.38}$$

The second set of equations is usually more convenient for practical computations. Obviously, $\hat{\theta}_i$ is a function of $(x_1,...,x_n)$ and will be written as $\hat{\theta}_i (x_1,...,x_n)$. The statistic $\hat{\theta}_i(X_1,...,X_n)$ will be called the maximum-likelihood estimator of θ_i. Before proceeding further, let us first give a few examples to clarify these concepts.

Example 9.4.2.1 Let $(x_1,...,x_n)$ be a sample of size n from a binomial distribution with probability mass function $(0 < \theta < 1)$

$$p_X(x|\theta) = \begin{cases} \binom{N}{x}\theta^x(1 - \theta)^{N-x} & \text{if } x = 0,1,...,N \\ 0 & \text{otherwise.} \end{cases}$$

$$L(x_1,...,x_n;\theta) = \prod_1^n \binom{N}{x_i}\theta^{\Sigma_1^n x_i}(1 - \theta)^{Nn - \Sigma_1^n x_i}.$$

Thus

$$\log L = \log \prod_1^n \binom{N}{x_i} + \sum_1^n x_i \log \theta + \left(Nn - \sum_1^n x_i\right)\log(1 - \theta)$$

$$\frac{\partial}{\partial \theta} \log L = \frac{1}{\theta}\sum_1^n x_i - \left(Nn - \sum_1^n x_i\right)(1 - \theta)^{-1}.$$

On putting this last expression equal to zero and solving for θ, we find that the maximum-likelihood estimate of θ is

$$\hat{\theta} = \frac{1}{Nn}\sum_1^n x_i.$$

It is easy to verify that $\theta = \hat{\theta}$ provides a maximum of the likelihood function L. Hence, for a random sample $(X_1,...,X_n)$ from this distribution the maximum-likelihood estimator of θ is $(1/Nn) \Sigma_1^n X_i$. We have seen earlier in this chapter that this estimator is unbiased, sufficient, strongly consistent, and most efficient.

Example 9.4.2.2 Let $(X_1,...,X_n)$ be a random sample of size n from a normal distribution with mean μ and variance σ^2. For any given sample

$(x_1,\ldots,x_n)$ from this population the log likelihood function is given by

$$\log L(x_1,\ldots,x_n;\mu,\sigma^2) = -\frac{n}{2}\log 2\pi - \frac{n}{2}\log \sigma^2 - \frac{1}{2\sigma^2}\sum_1^n (x_i - \mu)^2.$$

To find the maximum-likelihood estimates of μ, σ^2 we compute

$$\frac{\partial}{\partial \mu}\log L = \frac{1}{\sigma^2}\sum_1^n (x_i - \mu)$$

$$\frac{\partial}{\partial \sigma^2}\log L = -\frac{n}{2\sigma^2} + \frac{1}{2\sigma^4}\sum_1^n (x_i - \mu)^2.$$

Putting these partial derivatives equal to zero and solving the resulting equations, we find the maximum-likelihood estimates of μ, σ^2 to be

$$\hat{\mu} = \frac{1}{n}\sum_1^n x_i = \bar{x}, \qquad \hat{\sigma}^2 = \frac{1}{n}\sum_1^n (x_i - \bar{x})^2 = s_1^2.$$

where s_1^2 is an arbitrary designation. To show that $\bar{x}$, s_1^2, provide a maximum of the likelihood function we need to establish that

$$\log L = -\frac{n}{2}\log 2\pi - \frac{n}{2}\log \sigma^2 - \frac{1}{2\sigma^2}[n(\bar{x} - \mu)^2 + ns_1^2]$$

$$\leq -\frac{n}{2}\log 2\pi - \frac{n}{2}\log s_1^2 - \frac{n}{2} \tag{9.39}$$

for all μ, $-\infty < \mu < \infty$, $\sigma^2 > 0$, or, equivalently,

$$0 < \left[\left(\frac{s_1^2}{2\sigma^2} - \frac{1}{2}\right) - \log \frac{s_1}{\sigma}\right] + \frac{(\bar{x} - \mu)^2}{2\sigma^2}. \tag{9.40}$$

Now, for $x > 0$, $\log x < (x^2 - 1)/2$, so that

$$\frac{1}{2}\left(\frac{s_1^2}{\sigma^2} - 1\right) - \log \frac{s_1}{\sigma} > \frac{1}{2}\left(\frac{s_1^2}{\sigma^2} - 1\right) - \frac{1}{2}\left(\frac{s_1^2}{\sigma^2} - 1\right) = 0.$$

The second term in the right-hand side of Eq. (9.40) is nonnegative. Hence Eq. (9.39) is true.

The maximum-likelihood estimator of (μ, σ^2) is

$$\left(\frac{1}{n}\sum_1^n X_i = \bar{X},\, S_1^2 = \sum_1^n \frac{(X_i - \bar{X})^2}{n}\right).$$

We have seen previously (in Chapter 5) that $(\bar{X}, S_1^2)$ is sufficient for (μ, σ^2). Whereas $\bar{X}$ is an unbiased estimator of μ, the estimator S_1^2 is biased for σ^2. $\bar{X}$ is strongly consistent and most efficient for μ. The estimator S_1^2 is consistent for σ^2.

Example 9.4.2.3 For the distribution with probability density function

$$f_X(x|\alpha,\beta) = \begin{cases} \dfrac{1}{\beta - \alpha} & \text{if } \alpha < x < \beta \\ 0 & \text{otherwise.} \end{cases}$$

The likelihood function of the sample $(x_1,\ldots,x_n)$ is given by

$$L(x_1,\ldots,x_n|\alpha,\beta) = \begin{cases} \dfrac{1}{(\beta - \alpha)^n} & \text{if } \alpha < x_i < \beta, \quad i = 1,\ldots,n \\ 0 & \text{otherwise.} \end{cases}$$

Here

$$\frac{\partial}{\partial \alpha} \log L = -\frac{n}{\beta - \alpha}, \qquad \frac{\partial}{\partial \beta} \log L = +\frac{n}{\beta - \alpha}.$$

If we put these partial derivatives equal to zero and solve for α, β, we can conclude that one of α, β must be infinite, which is contrary to our expectation. To locate the maximum we proceed as follows. Let $y_1,\ldots,y_n$ be the ordered sample of $x_1,\ldots,x_n$. Obviously, given $x_1,\ldots,x_n$, α can be no larger than y_1 and β can be no smaller than y_n. Hence the maximum value of L lies at the minimum value of $\beta - \alpha$, which is $y_n - y_1$. On the basis of a random sample $X_1,\ldots,X_n$ of size n, the maximum-likelihood estimator of α is $Y_1 = \min(X_1,\ldots,X_n)$ and that of β is $Y_n = \max(X_1,\ldots,X_n)$.

We shall now consider some properties of maximum-likelihood estimators for a random sample of fixed size n, while in the next section asymptotic behavior of maximum-likelihood estimators for large values of n will be investigated.

Theorem 9.4.2.1 Let $X_1,\ldots,X_n$ denote a random sample of size n from a random phenomenon specified by the probability density function $f_X(x|\theta)$, θ (real or vector) $\in \Omega$. If there exists a sufficient statistic $T(X_1,\ldots,X_n)$ (real or vector) for θ and if a unique maximum-likelihood estimator $\hat{\theta}(X_1,\ldots,X_n)$ for θ exists, then $\hat{\theta}(X_1,\ldots,X_n)$ is a function of T.

Proof. Let $x_1,\ldots,x_n$ denote a sample of observations on $X_1,\ldots,X_n$. The likelihood function of $x_1,\ldots,x_n$ is given by Eq. (3.35):

$$L(x_1,\ldots,x_n;\theta) = \prod_{1}^{n} f_{X_i}(x_i|\theta).$$

Let $f_T(t|\theta)$ denote the probability density function of T. By Theorem 5.5.1,

$$\log L = \log f_T(t|\theta) + \log H(x_1,\ldots,x_n), \tag{9.41}$$

where $H(x_1,\ldots,x_n)$ does not depend on θ. Hence in maximizing log L with respect to θ, log $H(x_1,\ldots,x_n)$ behaves like a constant. Thus the maximum-likelihood estimate of θ is obtained by maximizing log $f_T(t|\theta)$ with respect to $\theta \in \Omega$, and thus is a function of t.   Q.E.D.

For the case of discrete random phenomenon, the same result holds and the proof is exactly the same. It is now left to the reader to prove the following theorem.

Theorem 9.4.2.2   Let $X_1,\ldots,X_n$ be a random sample from a random phenomenon specified by the probability density function $f_X(x|\theta)$ [or probability mass function $p_X(x|\theta)$], θ real or vector. If $T(X_1,\ldots,X_n)$ (real or vector) is the most efficient estimator of the parameter θ with probability density function $f_T(t|\theta)$ [or $p_T(t|\theta)$ in the discrete case], then almost everywhere where $f_T(t|\theta) > 0$ [or $p_T(t|\theta) > 0$] T is a solution of the equation

$$\frac{\partial}{\partial \theta_i} \log f_T(t|\theta) = 0, \qquad i = 1,\ldots,k \tag{9.42}$$

where $\theta = (\theta_1,\ldots,\theta_k)$.

We shall now investigate the asymptotic distribution of the maximum-likelihood estimator of θ when the sample size n tends to infinity. We shall deal with the case where θ is real. As before, it will be sufficient to deal with the case of continuous random phenomena. An analogous theorem for the discrete random phenomenon can be proved in the same way.

Theorem 9.4.2.3   Let $X_1,\ldots,X_n$ be a random sample of size n from a random phenomenon specified by the probability density function $f_X(x|\theta)$, $\theta \in \Omega$. Suppose that the following conditions are satisfied:
   (a)   The parameter θ is an interior point of some nondegenerate interval $I \subset \Omega$.
   (b)   The partial derivatives

$$\frac{\partial}{\partial \theta} \log f_X(x|\theta), \quad \frac{\partial^2}{\partial \theta^2} \log f_X(x|\theta), \quad \frac{\partial^3}{\partial \theta^3} \log f_X(x|\theta)$$

exist for every $\theta \in I$ and for almost all x.
   (c)   For every $\theta \in I$, we have

$$\left| \frac{\partial}{\partial \theta} \log f_X(x|\theta) \right| < H_1(x)$$

$$\left| \frac{\partial^2}{\partial \theta^2} \log f_X(x|\theta) \right| < H_2(x) \tag{9.43}$$

$$\left| \frac{\partial^3}{\partial \theta^3} \log f_X(x|\theta) \right| < H_3(x),$$

where H_1, H_2 are integrable functions over the real axis $(-\infty,\infty)$ and H_3 satisfies

$$\int_{-\infty}^{\infty} H_3(x)f_X(x|\theta)\ dx < M, \tag{9.44}$$

where M is independent of θ.

(d) For every $\theta \in I$,

$$0 < \int_{-\infty}^{\infty} \left[\frac{\partial}{\partial\theta} \log f_X(x|\theta)\right]^2 f_X(x|\theta)\ dx < \infty.$$

Then the equation

$$\sum_{1}^{n} \frac{\partial}{\partial\theta} \log f_{X_i}(X_i|\theta) = 0 \tag{9.45}$$

has a solution $\hat{\theta}_n(X_1,...,X_n)$ that converges in probability to θ as $n \to \infty$ and $\sqrt{n}(\hat{\theta}_n - \theta)$ is asymptotically normally distributed with mean 0 and variance

$$1/E\left(\frac{\partial}{\partial\theta} \log f_X(x|\theta)\right)^2$$

as the sample size n increases to infinity.

Remarks

1. If there are several solutions of Eq. (3.4.5), it will be difficult to find out, in particular, which solution does have these properties.
2. This theorem can be generalized to the maximum-likelihood estimator of the vector parameter $\theta = (\theta_1,...,\theta_k)$ as follows: The maximum-likelihood estimators $\hat{\theta}_1,...,\hat{\theta}_k$ from a random sample of size n have a joint normal distribution with means $\theta_1,...,\theta_k$, respectively, and with covariances

$$\operatorname{cov}(\hat{\theta}_i,\hat{\theta}_j) = -E\left(\frac{\partial^2}{\partial\theta_i\ \partial\theta_j} \log f_X(X|\theta)\right).$$

3. Remark 2 and the theorem in no way depend on the fact that we have used $X_1,...,X_n$ as a random sample from a univariate distribution. In all statements $X_1,...,X_n$ can be replaced by vectors.

Proof of the theorem. Let $(x_1,...,x_n)$ be a sample of observations on $(X_1,...,X_n)$. Let

$$L(\theta) = \sum_{1}^{n} \log f_X(x_i|\theta). \tag{9.46}$$

Let $\theta_0 \in I$ be the unknown true value of the parameter θ. Since $\log f_X(x|\theta)$ has three derivatives, expanding $\log f_X(x|\theta)$ by means of Taylor series in the neighborhood of $\theta_0 \in I$, we get

$$\frac{\partial}{\partial \theta} \log f_X(x|\theta)$$

$$= \left[\frac{\partial}{\partial \theta} \log f_X(x|\theta)\right]_{\theta=\theta_0} + (\theta - \theta_0)\left[\frac{\partial^2}{\partial \theta^2} \log f_X(x|\theta)\right]_{\theta=\theta_0}$$

$$+ \frac{(\theta - \theta_0)^2}{2}\left[\frac{\partial^3}{\partial \theta^3} \log f_X(x|\theta)\right]_{\theta=\theta_0+\xi(\theta-\theta_0)}$$

$$= \left[\frac{\partial}{\partial \theta} \log f_X(x|\theta)\right]_{\theta=\theta_0} + (\theta - \theta_0)\left[\frac{\partial^2}{\partial \theta^2} \log f_X(x|\theta)\right]_{\theta=\theta_0}$$

$$+ \frac{(\theta - \theta_0)^2}{2} v H_3(x), \tag{9.47}$$

where $0 < \xi < 1$, $|v| < 1$. Now write

$$B_0 = \frac{1}{n} \sum_1^n \left[\frac{\partial}{\partial \theta} \log f_X(x_i|\theta)\right]_{\theta=\theta_0}$$

$$B_1 = \frac{1}{n} \sum_1^n \left[\frac{\partial^2}{\partial \theta^2} \log f_X(x_i|\theta)\right]_{\theta=\theta_0} \tag{9.48}$$

$$B_2 = \frac{1}{n} \sum_1^n H_3(x_i).$$

From Eqs. (9.46) and (9.47), the likelihood equation can be written as

$$nB_0 + nB_1(\theta - \theta_0) + \tfrac{1}{2}(\theta - \theta_0)^2 \sum_1^n v_i H_3(x_i) = 0, \tag{9.49}$$

where $|v_i| < 1$, $i = 1,\ldots,n$. Since $|v_i| < 1$,

$$-nB_2 \le \sum_1^n v_i H_3(x_i) \le nB_2.$$

So we can write

$$\sum_1^n v_i H_3(x_i) = vnB_2, \qquad |v| < 1.$$

From Eq. (9.49) we get

$$nB_0 + nB_1(\theta - \theta_0) + \tfrac{1}{2}(\theta - \theta_0)^2 v B_2 = 0. \tag{9.50}$$

To prove the first part of the theorem we have to establish that with a probability tending to 1 as the sample size n increases indefinitely, Eq. (9.50) has a root between $(\theta_0 - \delta, \theta_0 + \delta)$, however small the $\delta > 0$ is chosen. By assumptions (b) and (c) we have

$$E\left(\frac{\partial}{\partial\theta}\log f_X(X|\theta)\right) = \int_{-\infty}^{\infty}\frac{\partial}{\partial\theta}f_X(x|\theta)\,dx = 0 \qquad (9.51)$$

$$E\left(\frac{\partial^2}{\partial\theta^2}\log f_X(X|\theta)\right) = \int_{-\infty}^{\infty}\left\{\frac{\frac{\partial^2}{\partial\theta^2}f_X(x|\theta)}{f_X(x|\theta)} - \frac{\left[\frac{\partial}{\partial\theta}f_X(x|\theta)\right]^2}{[f_X(x|\theta)]^2}\right\}$$

$$\times f_X(x|\theta)\,dx$$

$$= -E\left(\frac{\partial}{\partial\theta}\log f_X(x|\theta)\right)^2$$

$$= -k^2(\theta), \qquad (9.52)$$

where $k^2(\theta)$ is an arbitrary designation. By the weak law of large numbers:

1. B_0 converges in probability as $n \to \infty$ to $E_{\theta_0}(B_0) = 0$
2. B_1 converges in probability as $n \to \infty$ to $-k^2(\theta_0)$
3. B_2 converges in probability to $E_{\theta_0}(H_3(x)) \le M$ for all θ_0.

From (1) to (3) it follows that given $|\delta| > 0$, $\epsilon > 0$, there exists $N_0(\epsilon,\delta)$ such that for $n \ge N_0$,

$$P\{|B_0| \ge \delta^2\} \le \tfrac{1}{3}t$$

$$P\{B_1 \ge -\tfrac{1}{2}k^2(\theta_0)\} \le \tfrac{1}{3}t \qquad (9.53)$$

$$P\{|B_2| \ge 2M\} \le \tfrac{1}{3}t.$$

Define the event S:

$$S = \{|B_0| < \delta^2\} \cap \{B_1 < -\tfrac{1}{2}k^2(\theta_0)\} \cap \{|B_2| < 2M\}. \qquad (9.54)$$

Clearly, for all $n \ge N_0$, we have

$$P(S) \ge 1 - P\{|B_0| \ge \delta^2\} - P\{B_1 \ge -\tfrac{1}{2}k^2(\theta_0)\} - P\{|B_2| \ge 2M\}$$

$$\ge 1 - t. \qquad (9.55)$$

Let $\delta = \theta - \theta_0$. Assume that S has occurred and that

$$|\delta| < \frac{k^2(\theta_0)}{2(M + 1)}. \qquad (9.56)$$

If $\delta > 0$, the left-hand side of Eq. (9.50) is less than

$$\delta^2 - \frac{\delta}{2} k^2(\theta_0) + M\delta^2 = \delta[(M+1)\delta - \tfrac{1}{2}k^2(\theta_0)] < 0.$$

If $\delta < 0$, the left-hand side of Eq. (9.50) is greater than

$$-\delta^2 + \tfrac{1}{2}|\delta|k^2(\theta_0) - M\delta^2 = -|\delta|\{(M+1)|\delta| - \tfrac{1}{2}k^2(\theta_0)\} > 0.$$

Thus if S occurs,

$$\sum_1^n \frac{\partial}{\partial\theta} \log f_{X_i}(x_i|\theta)$$

a continuous function of θ, is positive when $\theta = \theta_0 - \delta$ and is negative when $\theta = \theta_0 + \delta$. Hence if S occurs, there exists a root $\hat\theta = \theta(x_1,\dots,x_n)$ of the likelihood equation

$$\sum_1^n \frac{\partial}{\partial\theta} \log f_{X_i}(x_i|\theta) = 0$$

in the interval $(\theta_0 - \delta, \theta_0 + \delta)$. Now, choosing δ very small subject to $|\delta| < k^2(\theta_0)/2(M+1)$, we can make $P(S)$ larger and larger. Hence there exists a consistent maximum-likelihood estimator. This completes the proof of the first part of the theorem.

Let $\hat\theta = \hat\theta(x_1,\dots,x_n)$ be a solution of the likelihood equation

$$\sum_1^n \frac{\partial}{\partial\theta} \log f_{X_i}(x_i|\theta) = 0,$$

the existence of which has just been established. From Eqs. (9.48) and (9.50), we obtain

$$\sqrt{n}(\theta - \theta_0) = \frac{-\sqrt{n}B_0}{B_1 + \tfrac{1}{2}B_2(\theta - \theta_0)}. \tag{9.57}$$

Further, $B_1 + \tfrac{1}{2}vB_2(\theta - \theta_0)$ converges in probability to $-k^2(\theta_0)$ as the sample size n increases to ∞. Now

$$-\sqrt{n}B_0 = \frac{-1}{\sqrt{n}} \sum_1^n \left[\frac{\partial}{\partial\theta} \log f_{X_i}(x_i|\theta) \right]_{\theta=\theta_0}. \tag{9.58}$$

By Eqs. (9.51) and (9.52) and the Lindeberg–Lévy theorem of Chapter 4, $-\sqrt{n}\, B_0$ is asymptotically normally distributed with mean 0 and variance $k^2(\theta_0)$ as $n \to \infty$. Hence $\sqrt{n}(\hat\theta - \theta_0)$ is asymptotically normally distributed with mean 0 and variance $(k^2(\theta_0))^{-1}$ as $n \to \infty$.   Q.E.D.

Example 9.4.2.4 Let $X_1,...,X_n$ be independently distributed normal random variables with $E(X_i) = d_i\theta$ and $V(X_i) = 1$ for all i, where $d_1,...,d_n$ are known; then we have

$$\log L(x_1,...,x_n;\theta) = \frac{n}{2}\log 2\pi - \frac{1}{2}\sum_1^n (x_i - d_i\theta)^2.$$

The maximum-likelihood estimate of θ is

$$\hat{\theta}_n = \hat{\theta}(x_i,...,x_n) = \frac{\sum_1^n d_i x_i}{\sum_1^n d_i^2}.$$

Obviously,

$$E(\hat{\theta}_n) = \theta, \quad V(\hat{\theta}_n) = \frac{1}{\sum_1^n d_i^2}$$

and $\hat{\theta}_n$ is normally distributed with mean θ and variance $(\sum_1^n d_i^2)^{-1}$. Suppose that $\sum_1^\infty d_i^2 < \infty$, so that as $n \to \infty$ the distribution of $\hat{\theta}_n$ tends to a particular normal distribution with mean θ and positive (nonzero) variance. Hence $\hat{\theta}_n$ is not (weakly) consistent.

9.4.3 Minimum-Chi-Square Method

This method is applicable to the case of a grouped discrete or continuous distribution. In either case we have a set of n observations divided into k groups. We denote the observed group frequencies by $n_1,...,n_k$ such that $\sum_1^k n_i = n$, while the corresponding hypothetical probabilities for these groups are $p_i(\theta)$, $i = 1,...,k$ such that p_i are functions of the parameter $\theta = (\theta_1,...,\theta_q)$ with $\sum_1^k p_i(\theta) = 1$. In actual practice, the values of the parameter θ are unknown and must be estimated from the sample. They are obtained by minimizing a suitably defined measure of deviations between the observed frequencies $n_1,...,n_k$ and the hypothetical expected frequencies $np_1(\theta),...,np_k(\theta)$. Some such measures used to estimate θ are:

(a) Chi-square: $\chi^2(\theta) = \sum_1^k \dfrac{[n_i - np_i(\theta)]^2}{np_i(\theta)}.$

(b) Modified chi-square: $\chi_M^2(\theta) = \sum_1^k \dfrac{[n_i - np_i(\theta)]^2}{n_i}.$

The minimum-chi-square estimate of the parameter θ is obtained by solving the following equations:

$$-\frac{1}{2}\frac{\partial}{\partial\theta_j}\chi^2(\theta) = \sum_1^k \left\{\frac{n_i - np_i(\theta)}{p_i(\theta)} + \frac{[n_i - np_i(\theta)]^2}{2np_i^2(\theta)}\right\}\frac{\partial}{\partial\theta_j}p_i(\theta) = 0$$

$$(9.59)$$

for $j = 1,\ldots,q$, with respect to the unknown parameters $\theta_1,\ldots,\theta_q$. Similarly, the minimum-modified-chi-square estimate of θ is obtained by solving the following equations with respect to the parameters $\theta_1,\ldots,\theta_q$:

$$-\frac{1}{2}\frac{\partial}{\partial\theta_j}\chi_M^2(\theta) = \sum_1^k \frac{n_i - np_i(\theta)}{n_i}\frac{\partial}{\partial\theta_j}p_i(\theta) = 0, \qquad j = 1,\ldots,q. \qquad (9.60)$$

Even in simple cases the system of equations (9.59) is often very difficult to solve. For large n the influence of the second term inside the braces is negligible, and thus (9.60) reduces to

$$\sum_1^k \frac{n_i - np_i(\theta)}{p_i(\theta)}\frac{\partial}{\partial\theta_j}p_i(\theta) = 0, \qquad\qquad\qquad (9.61)$$

which is easier to solve.

EXERCISES

24. Find the maximum-likelihood estimator of the variance of the Poisson distribution.
25. Find the maximum-likelihood estimator of the parameter θ of the chi-square distribution with $\theta/2$ degrees of freedom.
26. Let $X_1,\ldots,X_n$ be n independent normal random variables with the same mean μ and $V(X_i) = \sigma_i^2$, $i = 1,\ldots,n$.
    (a) Is it possible to estimate all the parameters?
    (b) Assuming that the σ_i^2 are known, find the maximum-likelihood estimator of μ.
    (c) Is the maximum-likelihood estimator of μ unbiased?
27. Find the maximum-likelihood estimators of θ for the following probability density functions:

(a) $f_X(x) = \begin{cases} e^{-|x-\theta|}, & x > \theta \\ 0, & x < \theta. \end{cases}$

(b) $f_X(x) = \begin{cases} \dfrac{1}{\theta}, & 0 \le x \le \theta, \quad 0 < \theta < \infty \\ 0, & \text{otherwise.} \end{cases}$

(c) $f_X(x) = \begin{cases} \frac{1}{2}e^{-|x-\theta|}, & -\infty < x < \infty, \quad 0 < \theta < \infty \\ 0, & \text{otherwise.} \end{cases}$

28. Let $(x_1,y_1),\ldots,(x_n,y_n)$ be a sample of size n from a bivariate normal population with probability density function

$$f_{X,Y}(x,y) = \frac{1}{2\pi\sigma_1\sigma_2(1 - \rho^2)^{1/2}} \exp\left\{ -\frac{1}{2(1 - \rho^2)} \left[\left(\frac{x - \mu_1}{\sigma_1}\right)^2 \right. \right.$$
$$\left. \left. + \left(\frac{y - \mu_2}{\sigma_2}\right)^2 - 2\rho\left(\frac{x - \mu_1}{\sigma_1}\right)\left(\frac{y - \mu_2}{\sigma_2}\right) \right] \right\}.$$

(a) Find the maximum-likelihood estimators for μ_1, μ_2, ρ, σ_1^2, and σ_2^2.

(b) Show that the maximum-likelihood estimators of σ_1^2, σ_2^2 are biased.

29. Find the maximum-likelihood estimators of the parameters of a multinomial distribution.

9.5 ESTIMATION OF PARAMETERS IN $N_p(\mu,\Sigma)$

In this section we consider only the estimation of parameters of a multivariate normal population by the method of maximum likelihood and study some properties of these estimates. We first explain the method of maximum likelihood estimation for the bivariate normal population without using the vector and matrix notations. In the general case of a p-variate normal population with mean μ and covariance Σ we will use techniques involving vectors and matrices to find the maximum likelihood estimates of μ and Σ.

Example 9.5.1 Let $X = (X_1,X_2)$ be a bivariate normal random variable with pdf

$$f_X(x|\theta) = \frac{1}{2\pi\sigma_1\sigma_2(1 - \rho^2)^{1/2}}$$
$$\times \exp\left\{\frac{1}{2(1 - \rho^2)} \left[\left(\frac{x_1 - \mu_1}{\sigma_1}\right)^2 + \left(\frac{x_2 - \mu_2}{\sigma_2}\right)\right.\right.$$
$$\left.\left. - 2\rho\left(\frac{x_1 - \mu_1}{\sigma_1}\right)\left(\frac{x_2 - \mu_2}{\sigma_2}\right)\right]\right\},$$

where $\theta = (\mu_1,\mu_2,\sigma_1,\sigma_2,\rho)$. We want to find the maximum-likelihood estimators of these parameters on the basis of observations

$$x_\alpha = (x_{\alpha 1},x_{\alpha 2})', \qquad \alpha = 1,\ldots,N.$$

We shall assume that $N > 2$; the reason for this assumption will be made clear later in this section.

The likelihood of these observations is given by

$$L(x_1,\ldots,x_N|\theta) = \frac{1}{(2\pi)^N(\sigma_1^2\sigma_2^2)^{N/2}(1 - \rho^2)^{N/2}}$$

$$\times \exp\left\{-\frac{1}{2(1 - \rho^2)}\left[\sum_{\alpha=1}^N \left(\frac{x_{\alpha 1} - \mu_1}{\sigma_1}\right)^2\right.\right.$$

$$+ \sum_{\alpha=1}^N \left(\frac{x_{\alpha 2} - \mu_2}{\sigma_2}\right)^2$$

$$\left.\left.- 2\rho \sum_{\alpha=1}^N \left(\frac{x_{\alpha 1} - \mu_1}{\sigma_1}\right)\left(\frac{x_{\alpha 2} - \mu_2}{\sigma_2}\right)\right]\right\}.$$

Hence, writing $L(x_1,\ldots,x_N|\theta)$ as L,

$$\log L = -N \log 2\pi - \frac{N}{2}[\log \sigma_1^2 + \log \sigma_2^2 + \log(1 - \rho^2)]$$

$$- \frac{1}{2(1 - \rho^2)}\left[\sum_{\alpha=1}^N \left(\frac{x_{\alpha 1} - \mu_1}{\sigma_1}\right)^2 + \sum_{\alpha=1}^N \left(\frac{x_{\alpha 2} - \mu_2}{\sigma_2}\right)^2\right.$$

$$\left.- 2\rho \sum_{\alpha=1}^N \left(\frac{x_{\alpha 1} - \mu_1}{\sigma_1}\right)\left(\frac{x_{\alpha 2} - \mu_2}{\sigma_2}\right)\right],$$

from which we get

$$\frac{\partial \log L}{\partial \mu_1} = \frac{N}{\sigma_1(1 - \rho^2)}\left\{\frac{\bar{x}_1 - \mu_1}{\sigma_1} - \rho\left(\frac{\bar{x}_2 - \mu_2}{\sigma_2}\right)\right\}$$

$$\frac{\partial \log L}{\partial \mu_2} = \frac{N}{\sigma_2(1 - \rho^2)}\left\{\frac{\bar{x}_2 - \mu_2}{\sigma_2} - \rho\left(\frac{\bar{x}_1 - \mu_1}{\sigma_1}\right)\right\}$$

$$\frac{\partial \log L}{\partial \sigma_1^2} = -\frac{1}{2\sigma_1^2(1 - \rho^2)}\left\{N(1 - \rho^2) - \sum_{\alpha=1}^N \left(\frac{x_{\alpha 1} - \mu_1}{\sigma_1}\right)^2\right.$$

$$\left.+ \rho \sum_{\alpha=1}^N \left(\frac{x_{\alpha 1} - \mu_1}{\sigma_1}\right)\left(\frac{x_{\alpha 2} - \mu_2}{\sigma_2}\right)\right\}$$

$$\frac{\partial \log L}{\partial \sigma_2^2} = -\frac{1}{2\sigma_2^2(1 - \rho^2)}\left\{N(1 - \rho^2) - \sum_{\alpha=1}^N \left(\frac{x_{\alpha 2} - \mu_2}{\sigma_2}\right)^2\right.$$

$$\left.+ \rho \sum_{\alpha=1}^N \left(\frac{x_{\alpha 1} - \mu_1}{\sigma_1}\right)\left(\frac{x_{\alpha 2} - \mu_2}{\sigma_2}\right)\right\}$$

$$\frac{\partial \log L}{\partial \rho} = -\frac{1}{(1 - \rho^2)} \left\{ N\rho - \left(\frac{1}{1 - \rho^2}\right) \left[\rho \sum_{\alpha=1}^{N} \left(\frac{x_{\alpha 1} - \mu_1}{\sigma_1}\right)^2 \right. \right.$$

$$+ \rho \sum_{\alpha=1}^{N} \left(\frac{x_{\alpha 2} - \mu_2}{\sigma_2}\right)^2 - (1 + \rho^2)$$

$$\left. \left. \times \sum_{\alpha=1}^{N} \left(\frac{x_{\alpha 1} - \mu_1}{\sigma_1}\right)\left(\frac{x_{\alpha 2} - \mu_2}{\sigma_2}\right) \right] \right\}, \tag{9.62}$$

where

$$\bar{x}_1 = \frac{1}{N} \sum_{\alpha=1}^{N} x_{\alpha 1}, \qquad \bar{x}_2 = \frac{1}{N} \sum_{\alpha=1}^{N} x_{\alpha 2}.$$

Putting $\partial \log L/\partial \mu_1 = 0$, $\partial \log L/\partial \mu_2 = 0$, we get

$$\frac{\bar{x}_1 - \mu_1}{\sigma_1} = \rho\left(\frac{\bar{x}_2 - \mu_2}{\sigma_2}\right)$$

$$\frac{\bar{x}_2 - \mu_2}{\sigma_2} = \rho\left(\frac{\bar{x}_1 - \mu_1}{\sigma_1}\right). \tag{9.63}$$

The only solution of (9.63) is

$$\mu_1 = \bar{x}_1, \qquad \mu_2 = \bar{x}_2 \tag{9.64}$$

Substituting $\mu_1 = \bar{x}_1$, $\mu_2 = \bar{x}_2$ in

$$\frac{\partial \log L}{\partial \sigma_1^2} = 0, \quad \frac{\partial \log L}{\partial \sigma_2^2} = 0, \quad \text{and} \quad \frac{\partial \log L}{\partial \rho} = 0,$$

we get

$$N(1 - \rho^2) = \frac{s_{11}}{\sigma_1^2} - \rho \frac{s_{12}}{\sigma_1 \sigma_2}$$

$$N(1 - \rho^2) = \frac{s_{22}}{\sigma_2^2} - \rho \frac{s_{12}}{\sigma_1 \sigma_2} \tag{9.65}$$

$$\text{and} \quad N(1 - \rho^2) = \frac{s_{11}}{\sigma_1^2} + \frac{s_{22}}{\sigma_2^2} - \frac{1 + \rho^2}{\rho} \frac{s_{12}}{\sigma_1 \sigma_2},$$

where

$$s_{11} = \sum_{\alpha=1}^{N} (x_{\alpha 1} - \bar{x}_1)^2, \qquad s_{22} = \sum_{\alpha=1}^{N} (x_{\alpha 2} - \bar{x}_2)^2$$

$$s_{12} = \sum_{\alpha=1}^{N} (x_{\alpha 1} - \bar{x}_1)((x_{\alpha 2} - \bar{x}_2)).$$

From (9.65) we get

$$N(1 - \rho^2) = \frac{1 - \rho^2}{\rho} \frac{s_{12}}{\sigma_1 \sigma_2} \quad \text{or} \quad N\rho = \frac{s_{12}}{\sigma_1 \sigma_2},$$

$$\sigma_1^2 = \frac{s_{11}}{N}, \qquad \sigma_2^2 = \frac{s_{22}}{N}.$$

Thus the maximum-likelihood estimators of $\hat{\mu}_1$, $\hat{\mu}_2$, $\hat{\sigma}_1^2$, $\hat{\sigma}_2^2$, $\hat{\rho}$ of μ_1, μ_2, σ_1^2, σ_2^2, ρ, respectively, are given by

$$\hat{\mu}_1 = \bar{x}_1, \qquad \hat{\mu}_2 = \bar{x}_2$$

$$\hat{\sigma}_1^2 = \frac{s_{11}}{N}, \qquad \hat{\sigma}_2^2 = \frac{s_{22}}{N}, \qquad \hat{\rho} = \frac{s_{12}}{\sqrt{s_{11} s_{22}}}.$$

Let

$$\mu = (\mu_1, \mu_2)', \qquad \Sigma = \begin{pmatrix} \sigma_1^2 & \rho\sigma_1\sigma_2 \\ \rho\sigma_1\sigma_2 & \sigma_2^2 \end{pmatrix}.$$

From Section 7.2 we can write

$$f_X(x|\theta) = \frac{1}{(2\pi)|\Sigma|^{1/2}} \exp\left[-\frac{1}{2} (x - \mu)'\Sigma^{-1}(x - \mu) \right]$$

On the basis $x_{\alpha 1}$, $\alpha = 1,\ldots,N$ the maximum-likelihood estimators

$$\hat{\mu} = (\hat{\mu}_1, \hat{\mu}_2)', \qquad \hat{\Sigma} = \begin{pmatrix} \hat{\sigma}_1^2, & \hat{\rho}\hat{\sigma}_1\hat{\sigma}_2, \\ \hat{\rho}\hat{\sigma}_1\hat{\sigma}_2, & \hat{\sigma}_2^2, \end{pmatrix}$$

of μ, Σ, respectively, are

$$\hat{\mu} = \bar{x} = (\bar{x}_1, \bar{x}_2)'$$

$$\hat{\Sigma} = \frac{S}{N} = \frac{1}{N} \begin{pmatrix} s_{11} & s_{12} \\ s_{21} & s_{22} \end{pmatrix} = \frac{1}{N} \sum_{\alpha=1}^{N} (x_\alpha - \bar{x})(x_\alpha - \bar{x})'.$$

We now consider the general case of a p-variate normal with mean $\mu = (\mu_1,\ldots,\mu_p)'$ and covariance matrix $\Sigma = (\sigma_{ij})$ with $\sigma_{ii} = \sigma_i^2$. Let

$$x_\alpha = (x_{\alpha 1},\ldots,x_{\alpha p})', \qquad \alpha = 1,\ldots,N$$

be a sample of size N from this population with the pdf given in (7.8). The likelihood of x_α, $\alpha = 1,\ldots,N$ is

$$L(x_1,\ldots,x_N|\theta) = \prod_{\alpha=1}^{N} \frac{1}{(2\pi)^{p/2}|\Sigma|^{1/2}} \exp\left[-\frac{1}{2} (x_\alpha - \mu)'\Sigma^{-1}(x_\alpha - \mu) \right]$$

$$= \frac{1}{(2\pi)^{Np/2}|\Sigma|^{N/2}} \exp\left[-\frac{1}{2} \sum_{\alpha=1}^{N} (x_\alpha - \mu)'\Sigma^{-1}(x_\alpha - \mu) \right],$$

$$(9.66)$$

where $\theta = (\mu,\Sigma)$. To derive the maximum-likelihood estimators $\hat{\mu}$, $\hat{\Sigma}$ or μ, Σ, respectively, we need the following lemmas.

Lemma 9.5.1

$$\sum_{\alpha=1}^{N} (x_\alpha - \mu)'\Sigma^{-1}(x_\alpha - \mu) = \text{tr } \Sigma^{-1} \sum_{\alpha=1}^{N} (x_\alpha - \mu)(x_\alpha - \mu)'.$$

Proof. Since

$$\sum_{\alpha=1}^{N} (x_\alpha - \mu)'\Sigma^{-1}(x_\alpha - \mu)$$

is a scalar quantity, by using Theorem A.2.4.4, we get

$$\sum_{\alpha=1}^{N} (x_\alpha - \mu)'\Sigma^{-1}(x_\alpha - \mu)$$

$$= \text{tr}\left(\sum_{\alpha=1}^{N} (x_\alpha - \mu)'\Sigma^{-1}(x^\alpha - \mu) \right)$$

$$= \sum_{\alpha=1}^{N} \text{tr}(x_\alpha - \mu)'\Sigma^{-1}(x_\alpha - \mu)$$

$$= \sum_{\alpha=1}^{N} \text{tr } \Sigma^{-1}(x_\alpha - \mu)(x_\alpha - \mu)'$$

$$= \text{tr } \Sigma^{-1}\left[\sum_{\alpha=1}^{N} (x_\alpha - \mu)(x_\alpha - \mu)' \right]. \qquad \text{Q.E.D.}$$

Lemma 9.5.2   Let A be a $p \times p$ symmetric matrix with characteristic roots $\lambda_1,\ldots,\lambda_p$. Then
(a) $\text{tr } A = \Sigma_{i=1}^{p} \lambda_i$.
(b) $|A| = \Pi_{i=1}^{p} \lambda_i$.

Proof. (a) Since A is symmetric, by Theorem A.2.6.5 there exists an orthogonal matrix P such that PAP' is diagonal matrix of diagonal elements $\lambda_1,\ldots,\lambda_p$, where the λ_i are the characteristic roots of A. Hence

$$\text{tr } A = \text{tr } AP'P = \text{tr } PAP' = \sum_{i=1}^{p} \lambda_i$$

$$|A| = |AP'P| = |AP'||P| = |P||AP'|$$

$$= |PAP'| = \prod_{i=1}^{p} \lambda_i. \qquad \text{Q.E.D.}$$

Let

$$\overline{X} = \frac{1}{N} \sum_{\alpha=1}^{N} \sum_{\alpha=1}^{N} X_\alpha, \qquad S = \sum_{\alpha=1}^{N} (X_\alpha - \overline{X})(X_\alpha - \overline{X})'$$

It is easy to check that S is positive definite with probability one if $N > p$, but it is beyond the scope of this book to prove this result. However, we shall assume that $N > p$. Now

$$\sum_{\alpha=1}^{N} (X_\alpha - \mu)(X_\alpha - \mu)' = \sum_{\alpha=1}^{N} (X_\alpha - \overline{X} + \overline{X} - \mu)$$

$$\times (X_\alpha - \overline{X} + \overline{X} - \mu)'$$

$$= \sum_{\alpha=1}^{N} (X_\alpha - \overline{X})(X_\alpha - \overline{X})'$$

$$+ \sum_{\alpha=1}^{N} (X_\alpha - \overline{X})(\overline{X} - \mu)'$$

$$+ \sum_{\alpha=1}^{N} (\overline{X} - \mu)(X_\alpha - \overline{X})'$$

$$+ N(\overline{X} - \mu)(\overline{X} - \mu)'$$

$$= S + N(\overline{X} - \mu)(\overline{X} - \mu)'. \qquad (9.67)$$

Since

$$\sum_{\alpha=1}^{N} (X_\alpha - \overline{X})(\overline{X} - \mu)' = \left(\sum_{\alpha=1}^{N} (X_\alpha - \overline{X}) \right)(\overline{X} - \mu)' = 0$$

$$\sum_{\alpha=1}^{N} (\overline{X} - \mu)(X_\alpha - \overline{X})' = (\overline{X} - \mu)\left(\sum_{\alpha=1}^{N} (X_\alpha - \overline{X}) \right)' = 0.$$

Note: We shall write

$$s = \sum_{\alpha=1}^{N} (x_\alpha - \bar{x})(x_\alpha - \bar{x})'$$

as a value of S and $\bar{x} = (1/N) \sum_{\alpha=1}^{N} x_\alpha$ as a value of $\bar{X}$.

Using Lemma 9.5.1, we get, from (9.66),

$$L(x_1,\ldots,x_N|\theta) = \frac{1}{(2\pi)^{Np/2}|\Sigma|^{N/2}} \exp\left[-\frac{1}{2} \operatorname{tr} \Sigma^{-1}(s + N(\bar{x} - \mu)(\bar{x} - \mu)')\right]$$

$$= \frac{1}{(2\pi)^{Np/2}|\Sigma|^{N/2}} \exp\left(-\frac{1}{2} \operatorname{tr} \Sigma^{-1}s\right)$$

$$\times \exp\left[-\frac{N}{2} (\bar{x} - \mu)'\Sigma^{-1}(\bar{x} - \mu)\right]. \tag{9.68}$$

Since Σ^{-1} is positive definite, $(\bar{x} - \mu)'\Sigma^{-1}(\bar{x} - \mu)$, being the distance between $\sqrt{N}\,\bar{x}$ and $\sqrt{N}\,\mu$, is always nonnegative and is 0 when $\mu = \bar{x}$. Hence the maximum likelihood estimator $\hat{\mu}$ of μ is $\hat{\mu} = \bar{x}$. Thus, from (9.68),

$$\max_{\mu,\Sigma} L(x_1,\ldots,x_N|\theta) = \max_{\Sigma} \frac{1}{(2\pi)^{Np/2}|\Sigma|^{N/2}} \exp\left(-\frac{1}{2} \operatorname{tr} \Sigma^{-1}s\right) \tag{9.69}$$

Theorem 9.5.1

$$\frac{1}{(2\pi)^{Np/2}|\Sigma|^{N/2}} \left(-\frac{1}{2} \operatorname{tr} \Sigma^{-1}s\right)$$

is maximum when $\Sigma = s/N$.

Proof. Since s is a $p \times p$ symmetric positive definite matrix, by Theorem A.2.6.6, there exists a $p \times p$ nonsingular matrix C such that $CsC' = I$, which implies that

$$s^{-1} = C'C.$$

Hence

$$\frac{1}{(2\pi)^{Np/2}|\Sigma|^{N/2}} \left(-\frac{1}{2} \operatorname{tr} \Sigma^{-1}s\right)$$

$$= \frac{1}{(2\pi)^{Np/2}|s|^{N/2}} |(C\Sigma C')^{-1}|^{N/2} \exp\left[-\frac{1}{2} \operatorname{tr}(C\Sigma C')^{-1}\right]. \tag{9.70}$$

Since C is nonsingular and Σ is positive definite, $(C\Sigma C')^{-1}$ is positive definite. Let $\lambda_1,\ldots,\lambda_p$ be the characteristic roots of $(C\Sigma C')^{-1}$. In terms of $\lambda_1,\ldots,\lambda_p$, using Lemma 9.5.2, we can write (9.70) as

$$\frac{1}{(2\pi)^{Np/2}|s|^{N/2}}\left(\prod_{i=1}^{p}(\lambda_i)^{N/2}\right)\exp\left(-\frac{1}{2}\sum_{i=1}^{p}\lambda_i\right)$$

$$\frac{1}{(2\pi)^{Np/2}|s|^{N/2}}\left(\prod_{i=1}^{p}\lambda_i^{N/2}\right)\exp\left(-\frac{1}{2}\lambda_i\right).$$

The function

$$(\lambda_i)^{N/2}\exp\left(-\frac{1}{2}\lambda_i\right)$$

has a maximum with respect to λ_i occurring at $\lambda_i = N$. Since the maximum-likelihood estimates possess the invariant property; that is, if $\hat{\theta}$ is a maximum-likelihood estimator of θ, then the maximum-likelihood estimator of $h(\theta)$, which is a function of θ, is $h(\hat{\theta})$. Denoting $\hat{\Sigma}$ as the maximum-likelihood estimator of Σ, we get

$$(C\hat{\Sigma}C')^{-1} = NI \quad \text{or} \quad \hat{\Sigma} = \frac{1}{N}(C'C)^{-1} = \frac{s}{N}. \qquad \text{Q.E.D.}$$

Let

$$\mu = (\mu_1,\ldots,\mu_p)', \qquad \bar{x} = (\bar{x}_1,\ldots,\bar{x}_p)'$$
$$\Sigma = (\sigma_{ij}), \qquad s = (s_{ij}).$$

Then by the invariance property of the maximum-likelihood estimator (mentioned in Theorem 9.5.1) the maximum-likelihood estimator by μ_i, σ_{ij}, and $\rho_{ij} = \sigma_{ij}/\sigma_{ij}\sigma_{ij}$ are given by

$$\hat{\mu}_i = \bar{x}_i = \frac{1}{N}\sum_{\alpha=1}^{N}x_{\alpha i}, \qquad i = 1,\ldots,p$$

$$\hat{\sigma}_{ij} = \frac{1}{N}\sum_{\alpha=1}^{N}(x_{\alpha i} - \bar{x}_i)(x_{\alpha j} - \bar{x}_j) = \frac{1}{N}s_{ij} \qquad \text{if } i \neq j$$

$$\hat{\sigma}_{ii} = \frac{1}{N}\sum_{\alpha=1}^{N}(x_{\alpha i} - \bar{x}_i)^2 = s_{ii}$$

$$\hat{\rho}_{ij} = \frac{\sum_{\alpha=1}^{N}(x_{\alpha i} - \bar{x}_i)(x_{\alpha j} - \bar{x}_j)}{\{[\sum_{\alpha=1}^{N}(x_{\alpha i} - \bar{x}_i)^2][\sum_{\alpha=1}^{N}(x_{\alpha j} - \bar{x}_j)^2]\}^{1/2}}$$

$$= \frac{s_{ij}}{\sqrt{s_{ii}s_{ij}}}. \qquad (9.71)$$

Definition 9.5.1: Sample covariance matrix s/N is called the *sample covariance matrix* based on N observations $x_1,\ldots,x_N$ from $N_p(\mu,\Sigma)$.

Definition 9.5.2: Sample correlation coefficient   The $r_{ij} = \hat{\rho}_{ij}$, $i,j = 1,\ldots,p$ are called the *sample correlation coefficients* between the ith and the jth components of the normal p-vector.

The $p \times p$ matrix R

$$R = (r_{ij})$$

is called the *sample correlation matrix*.

From Theorem 7.3.4 the regression surface of $X_{(2)}$ on $X_{(1)} = (X_1,\ldots, X_q)' = x_{(1)} = (x_1,\ldots,x_q)$ is given by

$$E(X_{(2)}|X_{(1)} = x_{(1)}) = \mu_{(2)} + \beta(x_{(1)} - \mu_{(1)}),$$

where

$$\beta = \Sigma_{21}\Sigma_{11}^{-1}$$

is called the *matrix of regression coefficients* of $X_{(2)}$ on $X_{(1)} = x_{(1)}$. By the invariance property of the maximum likelihood estimates, the maximum likelihood estimate of β is obtained by replacing Σ_{11}, Σ_{12} by their corresponding maximum likelihood estimate. Arguing in the same way, we get the following theorem.

Theorem 9.5.2   On the basis of N observations $x_\alpha = (x_{\alpha 1},\ldots,x_{\alpha p})'$, $\alpha = 1,\ldots,N$ from $N_p(\mu,\Sigma)$ the maximum-likelihood estimate of

$$\beta, \Sigma_{11} = \text{cov}(X_{(1)}), \qquad \Sigma_{22.1} = \Sigma_{22} - \Sigma_{21}\Sigma_{11}^{-1}\Sigma_{12} = \text{cov}(X_{(2)}|X_{(1)} = x_{(1)})$$

are given by

$$\hat{\beta} = s_{(21)}s_{(11)}^{-1}, \quad \hat{\Sigma}_{11} = \frac{s_{(11)}}{N}, \quad \hat{\Sigma}_{22.1} = \frac{1}{N}\left(s_{(22)} - s_{(21)}s_{(11)}^{-1}s_{(12)}\right)$$

where s is partitioned as

$$s = \begin{pmatrix} s_{(11)} & s_{(12)} \\ s_{(21)} & s_{(22)} \end{pmatrix}$$

and $s_{(11)}$ is the upper left-hand corner submatrix of s.

Definition 9.5.3: Matrix of sample regression coefficients   The matrix $\hat{\beta}$ is called the *matrix of sample regression coefficients* of $X_{(2)}$ on $X_{(1)} = x_{(1)}$.

From Theorem 7.3.4,

$$\text{cov}(X_{(2)}|X_{(1)} = x_{(1)}) = \Sigma_{22.1} = \Sigma_{22} - \Sigma_{21}\Sigma_{11}^{-1}\Sigma_{12}.$$

Hence the maximum likelihood estimate of $\Sigma_{22.1}$ is given by

$$\hat{\Sigma}_{22.1} = \frac{1}{N}\left(s_{(22)} - s_{(21)}s_{(11)}^{-1}s_{(12)}\right). \tag{9.72}$$

Let

$$s_{ij.1,\ldots,q}, \qquad i,j = q + 1,\ldots,p$$

denote the (i,j)th element of the $(p - q) \times (p - q)$ matrix $(1/N)$ $(s_{(22)} - s_{(21)}s_{(11)}^{-1}s_{(12)})$. From (7.12) the partial correlation coefficient between the components

$$X_i, X_j, \qquad i,j = q + 1,\ldots,p$$

when the components $X_1,\ldots,X_q$ are held fixed, is given by

$$\rho_{ij.1,\ldots,q} = \frac{\sigma_{ij.1,\ldots,q}}{\sqrt{\sigma_{ii.1,\ldots,q}\sigma_{ij.1,\ldots,q}}}.$$

Hence its maximum-likelihood estimate is

$$\hat{\rho}_{ij.1,\ldots,q} = \frac{s_{ij.1,\ldots,q}}{\sqrt{s_{ii.1,\ldots,q}s_{ij.1,\ldots,q}}}$$

$$= r_{ij.1,\ldots,q}.$$

Definition 9.5.4: Sample partial correlation coefficient The $r_{ij.1,\ldots,q}$ is called the *sample partial correlation coefficient* between the components X_i, X_j, $i, j = q + 1,\ldots,p$ when the components $X_1,\ldots,X_q$ are held fixed.

From (7.16) the multiple correlation coefficient ρ of the last component X_p on the remaining components $X_1,\ldots,X_{p-1}$ of $X = (X_1,\ldots,X_p)'$ is given by

$$\rho = \sqrt{\frac{\Sigma_{21}\Sigma_{11}^{-1}\Sigma_{12}}{\Sigma_{22}}}, \tag{9.73}$$

where the covariance matrix Σ of X is partitioned as

$$\Sigma = \begin{pmatrix} \Sigma_{11} & \Sigma_{12} \\ \Sigma_{21} & \Sigma_{22} \end{pmatrix}$$

with Σ_{11} is the upper left-hand corner submatrix of Σ of dimension $(p - 1) \times (p - 1)$. By replacing $\Sigma_{ij}, i,j = 1,2$ by their corresponding maximum likelihood estimates, the maximum likelihood estimates of ρ^2 is given by

$$\hat{\rho}^2 = \frac{s_{(21)}s_{(11)}^{-1}s_{(12)}}{s_{(22)}} = R^2, \tag{9.74}$$

where s is partitioned as

$$\begin{pmatrix} s_{(11)} & s_{(12)} \\ s_{(21)} & s_{(22)} \end{pmatrix}$$

with $s_{(11)}$ the upper left-hand corner $(p - 1) \times (p - 1)$ submatrix of s.

Definition 9.5.5: Sample multiple correlation coefficient The square root of R^2 in (9.74) is called the *sample multiple correlation coefficient* of X_p on $X_1,\ldots,X_{p-1}$.

Example 9.5.2 Table 9.1 gives observations made in the Indian Agricultural Research Institute, New Delhi, India on six different characters:

X_1 = plant height at harvesting (cm)

X_2 = number of effective tillers

X_3 = length of ear (cm)

X_4 = number of fertile spikelets per 10 ears

X_5 = number of grains per 10 ears

X_6 = weight of grains per 10 ears (gm)

for 27 randomly selected plants of Sonalika, a late-sown variety of wheat in 1972.

Let us assume that the data constitute a sample from a six-variate normal population with mean μ and covariance matrix Σ. The maximum likelihood estimates of μ,Σ are

$$\hat{\mu} = \begin{pmatrix} 77.1444 \\ 167.1852 \\ 10.4585 \\ 13.0963 \\ 361.55553 \\ 14.7630 \end{pmatrix}$$

Table 9.1

Plant	X_1	X_2	X_3	X_4	X_5	X_6
1	74.35	162	9.76	12.2	337	13.7
2	66.05	145	10.10	12.5	351	13.9
3	80.30	156	10.71	13.8	424	17.7
4	77.60	148	10.75	13.0	379	17.3
5	80.45	142	9.56	12.4	327	13.8
6	81.00	200	10.48	13.9	378	15.7
7	85.05	163	10.90	13.3	367	16.3
8	80.75	170	10.65	13.0	172	15.1
9	80.95	165	10.57	13.8	357	14.6
10	64.40	142	10.21	12.2	352	14.8
11	75.90	157	10.79	13.6	357	13.8
12	69.00	170	8.61	9.8	258	9.4
13	82.25	156	11.06	13.8	404	17.5
14	80.75	156	11.14	14.7	412	17.1
15	82.25	164	10.30	13.3	390	17.2
16	79.55	174	10.75	13.4	400	16.9
17	81.90	163	10.75	13.4	355	16.5
18	83.55	182	11.43	14.3	406	15.0
19	65.45	147	9.55	11.3	300	12.0
20	68.00	156	9.88	11.7	330	13.2
21	66.85	194	9.56	11.9	304	9.3
22	81.45	192	11.12	14.2	384	17.8
23	75.65	191	10.93	13.7	380	13.0
24	77.30	170	11.09	14.1	404	16.3
25	81.35	186	10.41	13.3	340	12.5
26	79.45	165	10.79	13.6	384	14.4
27	81.35	198	10.53	13.4	310	13.8

$$\hat{\Sigma} = \frac{s}{27} = \begin{pmatrix} 1496 & 73.83 & 136.50 & -11.33 & 20.67 & 34.26 \\ 73.83 & 5.122 & 8.695 & -6.438 & 1.050 & 1.714 \\ 136.50 & 8.695 & 37.39 & 31.65 & 2.508 & 4.720 \\ -11.33 & -6.438 & 31.65 & 288.7 & 2.026 & 4.817 \\ 20.67 & 1.050 & 2.508 & 2.026 & 0.390 & 0.609 \\ 34.26 & 1.714 & 4.720 & 4.817 & 0.609 & 1.097 \end{pmatrix}.$$

The matrix of sample correlation coefficients $R = (r_{ij})$ is

$$R = \begin{pmatrix} 1.00000 & 0.84339 & 0.57699 & -0.01729 & 0.85654 & 0.84636 \\ 0.84339 & 1.000000 & 0.62820 & -0.16753 & 0.74304 & 0.72306 \\ 0.57699 & 0.62820 & 1.000000 & 0.30469 & 0.65693 & 0.73694 \\ -0.01729 & -0.16753 & 0.30469 & 1.00000 & 0.19090 & 0.27053 \\ 0.85654 & 0.74304 & 0.65693 & 0.19090 & 1.00000 & 0.93191 \\ 0.84636 & 0.72306 & 0.73694 & 0.27053 & 0.93191 & 1.00000 \end{pmatrix}$$

The maximum-likelihood estimate of the regression of X_6 on $X_1 = x_1,\ldots,$ $X_5 = x_5$ is

$$\hat{E}(X_6|x_1,\ldots,X_5 = x_5) = -4.82662 + 0.12636x_1 - 0.03436x_2$$
$$+ 0.61897x_3 - -0.28526x_4 + 0.03553x_5.$$

The maximum-likelihood estimate R^2 of the square of the multiple correlation coefficient of X_6 on $(X_1,\ldots,X_5)$ is $R^2 = 0.80141$. The maximum-likelihood estimates of some of the partial correlation coefficients are

$r_{23.4} = 0.7234$ $r_{23.45} = 0.4576$ $r_{23.15} = 0.3194$ $r_{23.146} = 0.4439$

$r_{23.5} = 0.2776$ $r_{23.46} = 0.3539$ $r_{23.16} = 0.3709$ $r_{23.156} = 0.3795$

$r_{23.6} = 0.2042$ $r_{23.14} = 0.4887$ $r_{23.456} = 0.3861$ $r_{23.1456} = 0.4532$

$r_{23.1} = 0.3226$ $r_{23.56} = 0.2494$ $r_{23.145} = 0.4578$

9.5.1 Some Properties of $\hat{\mu}, \hat{\Sigma}$

Let $X_\alpha = (X_{\alpha 1},\ldots,X_{\alpha p})'$, $\alpha = 1,\ldots,N$ be independently and identically distributed $N_p(\mu,\Sigma)$. In Section 7.6 we have shown that $\hat{\mu} = \bar{X}$, $N\hat{\Sigma} = S$ are independent and $\sqrt{N}\,\bar{X}$ is $N_p(\sqrt{N}\,\mu,\Sigma)$,

$$S \text{ is distributed as } \sum_{\alpha=2}^{N} Y_\alpha Y_\alpha', \tag{9.75}$$

where $Y_\alpha = (Y_{\alpha 1},\ldots,Y_{\alpha p})'$, $\alpha = 2,\ldots,N$ are independently distributed p-variate normals with $E(Y_\alpha) = 0$, $\text{cov}(Y_\alpha) = \Sigma$, $\alpha = 2,\ldots,N$. Hence

$$E(\bar{X}) = \mu$$

$$E(S) = E\left(\sum_{\alpha=2}^{N} E(Y_\alpha Y_\alpha')\right)$$

$$= \sum_{\alpha=2}^{N} E(Y_\alpha Y_\alpha')$$

$$= (N-1)\Sigma.$$

Thus the maximum-likelihood estimator $\hat{\mu} = \bar{X}$ is unbiased for μ. But $S/(N-1)$ is an unbiased estimator of Σ. We now state, without proof, a few properties of these estimators. For a proof we refer to Giri (1974).

(a) $(\bar{X},S)$ is a complete sufficient statistic for (μ,Σ).

(b) $\bar{X}$ is an efficient and consistent estimator of Σ.

(c) For the estimation of μ with Σ unknown, the estimator

$$\left(1 - \frac{p-2}{(N-p+2)\bar{X}'S^{-1}\bar{X}}\right)\bar{X} \tag{9.76}$$

dominates $\overline{X}$ with $p > 2$ under the quadratic loss function. This is known as *Stein's estimator* of μ.

9.6 ESTIMATION OF PARAMETERS IN MULTINOMIAL POPULATION

In Section 7.5 we have defined the multinomial distribution. Let $X = (X_1,\ldots,X_k)'$ be a k-variate random vector having the multinomial distribution with parameters $\theta = (\theta_1,\ldots,\theta_q)$. Its pdf is given by

$$p(X_1 = x_1,\ldots,X_k = x_k) = \frac{n!}{\prod_{i=1}^{k} x_i!} \prod_{i=1}^{k} \theta_i^{x_i} \tag{9.77}$$

with $\Sigma_{i=1}^{k} x_i = n$ and $\Sigma_{i=1}^{k} \theta_i = 1$, $\theta_i \geq 0$. Since $\Sigma_{i=1}^{k} \theta_i = 1$, to obtain the maximum likelihood estimates $\hat{\theta}_i$ of θ_i, $i = 1,\ldots,k$; we treat $p(X_1 = x_1,\ldots, X_k = x_k)$, given $x_1,\ldots,x_k$ as a function of $\theta_1,\ldots,\theta_{k-1}$ and the parameters θ_i, $i = 1,\ldots,k$ vary subject only to the restriction that $\theta_i \geq 0$, $i = 1,\ldots,k - 1$, with $\Sigma_{i-1}^{k-1} \theta_i \leq 1$. From (9.77) we get

$$\log p(X_1 = x_1,\ldots,X_k = x_k) = \log n! - \log\left(\prod_{i=1}^{k} (x_i)!\right) - \sum_{i=1}^{k} x_i \log \theta_i. \tag{9.78}$$

Hence $\hat{\theta} = (\hat{\theta}_1,\ldots,\hat{\theta}_k)$ satisfies

$$\sum_{j=1}^{k} \frac{\partial}{\sigma\theta_i} (x_j \log \theta_j) = \sum_{j=1}^{k} \frac{x_j}{\theta_j} \frac{\partial\theta_j}{\partial\theta_i} = 0, \qquad i = 1,\ldots,k - 1. \tag{9.79}$$

Since $\theta_k = 1 - \Sigma_{i=1}^{k-1} \theta_i$, we get $\partial\theta_k/\partial\theta_i = -1$. Thus any solution of (9.79) satisfies

$$\frac{\hat{\theta}_i}{\hat{\theta}_k} = \frac{x_i}{x_k}, \qquad i = 1,\ldots,k - 1.$$

Using $\hat{\theta}_k = 1 - \Sigma_{i=1}^{k-1} \hat{\theta}_i$, we get the unique solution of θ_i as

$$\hat{\theta}_i = \frac{x_i}{n}, \qquad i = 1,\ldots,k.$$

From Section 7.5,

$$E(\hat{\theta}_i) = E\frac{X_i}{n} = \frac{n\theta_i}{n} = \theta_i$$

$$\text{var}(\hat{\theta}_i) = \frac{\theta_i(1 - \theta_i)}{n}$$

$$\text{cov}(\hat{\theta}_i, \hat{\theta}_j) = \frac{\theta_i\theta_j}{n}, \qquad i \neq j.$$

EXERCISES

30. Let $X_\alpha = (X_{\alpha 1}, \ldots, X_{\alpha p})'$, $\alpha = 1, \ldots, N$ be a random sample of size N from $N_p(\mu, \Sigma)$.
    (a) Find the maximum likelihood estimator of μ when Σ is known.
    (b) Find the maximum likelihood estimator of Σ when μ is known. Is it unbiased?

Table 9.2

Farm	X_1	X_2	X_3
1	75	15	6.0
2	156	6	2.5
3	145	60	0.5
4	175	24	3.0
5	70	5	2.0
6	179	8	1.5
7	165	14	4.0
8	134	13	4.0
9	137	7	1.5
10	175	19	1.5
11	135	13	0.5
12	175	12	2.5
13	240	7	2.0
14	175	27	4.0
15	197	16	6.0
16	125	6	5.0
17	227	13	5.0
18	172	13	11.0
19	170	34	2.0
20	172	19	6.5

31. The variability in the price of farmland per acre is to be studied in relation to three factors which are assumed to have a major influence in determining the selling price. For 20 randomly selected firms, the price (in dollars) per acre (X_1), the depreciated cost (in dollars) of building per acre (X_2), and the distance to the nearest shopping mall (in miles) are as given in Table 9.2. Assume that $X = (X_1, X_2, X_3)$ is $N_3(\mu, \Sigma)$.

(a) Find the maximum likelihood estimates of μ, Σ.

(b) Find the sample correlation matrix.

(c) Find the maximum likelihood estimator of

$$E(X_1 | X_2 = x_2, X_3 = x_3).$$

(d) Find the sample multiple correlation coefficient of X_1 on (X_2, X_2).

(e) Find the sample partial correlation coefficients $r_{12.3}$ and $r_{13.2}$.

BIBLIOGRAPHY

Blackwell, D. (1947). Conditional expectation and unbiased sequential estimation. *Ann. Math. Stat.*, Vol. 18, p. 105.

Cramér, H. (1946). *Mathematical Methods of Statistics*, Princeton University Press, Princeton, N.J.

Fisher, R. A. (1921). On the mathematical foundation of theoretical statistics. *Phil. Trans. Royal Soc.* p. 309.

Fisher, R. A. (1925). Theory of statistical estimation. *Proc. Cambridge Philos. Soc.*, Vol. 22.

Giri, N. (1974). *Introduction to Probability and Statistics*, Part I, Marcel Dekker, New York.

Giri, N. (1977). *Multivariate Statistical Inference*, Academic Press, N.Y.

Lehmann, E. L. (1959). *Testing Statistical Hypotheses*, Wiley, New York.

Rao, C. R. (1945). Information and accuracy obtainable in an estimator. *Bull. Calcutta Math. Soc.*, Vol. 37, p. 81.

Rao, C. R. (1965). *Linear Statistical Inference and Its Applications*, Wiley, New York.

Zacks, S. (1971). *The Theory of Statistical Inference*, Wiley, New York.

10

Testing of Statistical Hypotheses

The two major areas of statistical inference that have been most thoroughly explored are (a) point and interval estimation of parameters and (b) the testing of statistical hypotheses. We have studied the problem of point estimation of parameters in Chapter 9. We study the problem of the testing of statistical hypotheses in this chapter. Our main aim is to develop a general theory of the testing of different kinds of statistical hypotheses and to apply this theory to some problems of common occurrence.

In contrast to point estimation, in which we try to find the value of a parameter on the basis of information contained in a sample of observations from the random phenomenon under study, the testing of statistical hypotheses deals with situations where we are forced to choose between two possible actions regarding the parameter. As the term suggests, these two possible actions are acceptance or rejection of the hypothesis under consideration.

Consider a random phenomenon characterized by the probability distribution function $F_X(x|\theta)$, where θ (scalar or vector) belongs to the parametric space Ω. A hypothesis is simply a statement about $F_X(x|\theta)$ characterizing the random phenomenon. Obviously, there are two distinct types of statements that can be made about $F_X(x|\theta)$. The first of these is to assume that the functional form of $F_X(x|\theta)$ is completely known except for the parameter θ, and the hypothesis is merely a statement about the parameter θ. The second is a statement regarding the functional form of $F_X(x|\theta)$.

A test of a hypothesis amounts to getting a sample of size n from the random phenomenon under study and deciding on the basis of the sample observations to accept or reject the hypothesis. In this chapter we study the case where the sample size n is fixed in advance. In the terminology of Chapter 5, we call the test a fixed-sample-size test. The case where the sample size is not fixed in advance and the observations are taken sequentially is studied in Chapter 13.

10.1 BASIC DEFINITIONS

Let ω be a proper nonempty subset of Ω. A hypothesis is a statement "$\theta \in \omega$." An alternative hypothesis is the statement "$\theta \in (\Omega - \omega)$" (complement of ω). To be consistent in our use of the term *alternative hypothesis*, we will call "$\theta \in \omega$" the *null hypothesis*. A standard notation for the null hypothesis is H_0 and that of the alternative hypothesis is H_1. If ω consists of one point only, we say that H_0 is simple. Otherwise, it is composite. Similarly, H_1 is simple or composite, depending on whether $\Omega - \omega$ contains one or more points.

Example 10.1.1 Suppose that the average life of light bulbs made under a standard manufacturing process is 1200 hours. A new process is designed to increase the lifetime of the light bulbs. Naturally, before replacing the standard process, the manufacturer wants to be sure scientifically whether the light bulbs manufactured under the new process are more durable (on the average) than those made under the standard process. The statistical model is that we are dealing with two populations of light bulbs—those made by the standard process and those made by the new one. The average life for the first one is known and is equal to 1200 hours. To answer the manufacturer's question we first set it as a problem of testing of hypotheses. The null hypothesis, in this case, is: "The mean life of the light bulbs manufactured by the new process is equal to 1200 hours." The alternative hypothesis is: "The mean life is greater than 1200 hours." Obviously, the null hypothesis is simple and the alternative hypothesis is composite. Two possible actions are: (1) we accept the null hypothesis, in which case we retain the existing process, or (2) we reject the null hypothesis, in which case we recommend installing the new process to manufacture light bulbs.

As mentioned earlier, our decision to accept or to reject the null hypothesis depends entirely on the information contained in the sample

$x_1,\ldots,x_n$ of size n from the distribution. We may recall that a sample $x_1,\ldots,x_n$ of size n has been defined as a value of a random sample $X_1,\ldots,X_n$ of size n from the random phenomenon, specified by the distribution function under consideration. Let S denote the sample space of $X_1,\ldots,X_n$ (i.e., the set of all possible values of $X_1,\ldots,X_n$).

The choice between the two possible actions is made by means of a rule that associates with each point in the sample space one of the two possible actions and thereby divides the sample space S into two complementary regions S_0 and S_1, with $S_0 \cup S_1 = S$, such that if the observed sample point $(x_1,\ldots,x_n) \in S_1$, we accept the null hypothesis, and if $(x_1,\ldots,x_n) \in S_0$, we reject the null hypothesis (i.e., we accept the alternative hypothesis). Such a rule (which is a function of $X_1,\ldots,X_n$) is called a test and the corresponding S_1, S_0 are called the acceptance region and the rejection region (critical region), respectively, of the test. In what follows we will use S_0, S_1 as the generic notations for the critical region and the acceptance region, respectively, of a test. It should be clear that the regions S_0, S_1 will differ from test to test. When it is necessary to specify these regions for more than one test, we will use the symbols $S_0, S_0', S_0'', \ldots, S_1, S_1', S_1'', \ldots$ to denote the critical regions and the acceptance regions, respectively.

Example 10.1.2 Consider Example 10.1.1. To test the null hypothesis we make a number (say, n) of light bulbs by the new process and measure their lives $x_1,\ldots,x_n$. Let us assume that $x_1,\ldots,x_n$ constitute a sample of size n from the distributions of all lifetimes of light bulbs made by the new process. Let

$$\bar{x} = \frac{1}{n} \sum_1^n x_i.$$

A rule that rejects the null hypothesis if the observed sample is

$$(x_1,\ldots,x_n) \in S_0 = \{(x_1,\ldots,x_n) | \bar{x} \geq 1225\}$$

and that accepts the null hypothesis if the observed point is

$$(x_1,\ldots,x_n) \in S_1 = \{(x_1,\ldots,x_n) | \bar{x} < 1225\}$$

is a test. Obviously, unless a specific criterion is laid out for choosing the regions S_0, S_1, one can define various other tests for the same problem. Thus one would naturally be interested in finding a criterion for choosing a test that is the best, in some sense, of all available tests.

In performing a test one may commit one of the following two kinds of errors: (1) rejecting the null hypothesis when it is really true, and (2) accepting the null hypothesis when the alternative hypothesis is true. We shall call the first one *type I error* and the second one *type II error*. The relative importance of these two types of errors depends on the individual problem under study. In Example 10.1.2, if it is expensive to replace the standard manufacturing process, one should be very careful about type I error.

Whatever may be the relative importance of these errors, it is preferable to choose a test for which the probabilities of both kinds of error are as small as possible. Unfortunately, when the sample size n is fixed in advance, it is not mathematically possible to control, simultaneously, both kinds of error. What is possible is to choose a test that keeps the probability of one type of error to a minimum when the probability of the other type of error is fixed. It is customary to fix an upper bound to type I error for tests, and to choose a test that minimizes the probability of type II error. In other words, we choose a number α, $0 < \alpha < 1$, called the *level of significance*, and impose a condition on tests so that the probability of type I error of each will be less than or equal to α; that is, the probability $P_\theta\{(X_1,\ldots,X_n) \in S_0\}$ that $X_1,\ldots,X_n$ belongs to S_0 when θ is the true value of the parameter satisifes the inequality

$$P_\theta\{(X_1,\ldots,X_n) \in S_0\} = P_\theta(S_0) \le \alpha, \qquad \theta \in \omega. \tag{10.1}$$

We are interested in choosing, among all tests satisfying the inequality (10.1), a test for which the probability of type II error $P_\theta(S_1)$ is minimum at every $\theta \in (\Omega - \omega)$. Since $S_0 \cup S_1 = S$, we know that $P_\theta(S_1) = 1 - P_\theta(S_0)$ for all $\theta \in \Omega$. The condition that $P_\theta(S_1)$ be minimum is equivalent to the condition that $P_\theta(S_0)$ be maximum at every point $\theta \in (\Omega - \omega)$.

Definition 10.1.1: Size of a test $\text{Sup}_{\theta \in \omega}\, P_\theta(S_0)$ is called the *size of the test* with critical region S_0.

Definition 10.1.2: Power of a test $\beta(\theta) = P_\theta(S_0)$, $\theta \in (\Omega - \omega)$, is called the *power of the test* with critical region S_0 at θ. $\beta(\theta)$, as a function of $\theta \in \Omega$, is called the *power function* of the test.

In terms of the power function, the size of a test is given by $\text{Sup}_{\theta \in \omega}\, \beta(\theta)$. Thus in our search for a suitable criterion to choose a better test among all available tests, we consider tests of size α only and find among them (if it exists) a test whose power $\beta(\theta)$, $\theta \in (\Omega - \omega)$, is maximum at every θ. The existence of such a test depends in general, on the nature of the distribution functions and, in particular, on the nature of the null and alternative hypotheses. This is discussed for different cases in this chapter.

There is no hard and fast rule for the choice of α. Most statistical tables assign values 0.05, 0.01, 0.005 to α. As a result, it has become customary to take one of these values for α. However, a more realistic choice of α must take into account the degree of confidence in the null hypothesis. If one strongly believes that the null hypothesis is true, naturally one will not be willing to reject the null hypothesis unless strong evidence against it is present in the data. One will then be tempted to assign α a very small value, so that there will be very little chance of the null hypothesis being rejected. On the other hand, if one rejects the null hypothesis in favor of the alternative hypothesis, the power of the test should be reasonably high for different values of $\theta \in (\Omega - \omega)$. If the power is too low for a particular choice of α, one should look for a larger value of α, which will yield a larger power.

10.2 TESTS OF SIMPLE H_0 AGAINST SIMPLE H_1

In this section we require that both H_0 and H_1 be simple. Thus in all cases the parametric space consists of two distinct points only. Consider a continuous random phenomenon specified by the distribution function $F_X(x|\theta)$, $\theta \in \Omega$. Let $f_X(x|\theta)$ be the corresponding probability density function. Assume that Ω contains only two distinct points, θ_0 and θ_1. It should be clear that θ is not necessarily a scalar quantity. The probability density function $f_X(x|\theta)$ may depend on a finite number of parameters, in which case θ is a vector and the symbol θ_0 or θ_1 simply means that all components of θ are specified. We are now interested in testing H_0: $\theta = \theta_0$ against H_1: $\theta = \theta_1$.

Let $X_1,...,X_n$ denote a random sample of size n (n a fixed positive integer) from this random phenomenon. We now define a most powerful test of size α for testing H_0 against H_1.

Definition 10.2.1: Most powerful test of size α   For testing H_0 against H_1, a test with critical region S_0 is called a *most powerful test of size* α if:
  (a)   It is of size α [i.e., $P_{\theta_0}(S_0) = \alpha$].
  (b)   For any other test of the same size α with critical region S'_0,

$$P_{\theta_1}(S_0) \geq P_{\theta_1}(S'_0). \tag{10.2}$$

The critical region associated with the most powerful test is called the *best critical region*.

The following lemma gives the sufficient conditions for the existence of a most powerful test.

Neyman–Pearson lemma   Let $X_1,...,X_n$ be a random sample of size n from a continuous random phenomenon, specified by the probability density function $f_X(x|\theta)$, $\theta \in \Omega$ and let Ω contain two distinct points θ_0, θ_1 only. If there exists a test with regions S_0, S_1, defined by

$$S_0 = \left\{ (x_1,\ldots,x_n) : \prod_1^n f_{X_i}(x_i|\theta_i) \geq k \prod_{i=1}^n f_{X_i}(x_i|\theta_0) \right\}$$

$$S_1 = \left\{ (x_1,\ldots,x_n) : \prod_1^n f_{X_i}(x_i|\theta_i) < k \prod_1^n f_{X_i}(x_i|\theta_0) \right\} \qquad (10.3)$$

for some positive number k such that

$$P_{\theta_0}(S_0) = \alpha, \qquad (10.4)$$

then it is most powerful of size α for testing $H_0 \colon \theta = \theta_0$ against the alternative $H_1 \colon \theta = \theta_1$.

Proof. If S_0 is the only critical region of size α in the sample space of $X_1,\ldots,X_n$, the lemma is obvious. Let S_0' be any other critical region of size α. Write

$$S_0 = (S_0 \cap S_0') \cup (S_0 \cap S_1')$$

$$S_0' = (S_0' \cap S_0) \cup (S_0' \cap S_1), \qquad (10.5)$$

where $S_0 \cap S_0'$, $S_0 \cap S_1'$ and $S_0' \cap S_0$, $S_0' \cap S_1$ are pairs of disjoint sets. For convenience we shall write the n-fold integral over a region R as

$$\int \cdots \int_R dx_1 \, dx_2 \cdots dx_n = \int_R dx_1 \, dx_2 \cdots dx_n.$$

Now, it is clear that

$$P_{\theta_1}(S_0) - P_{\theta_1}(S_0') = \int_{S_0} \prod_{i=1}^n \{f_{X_i}(x_i|\theta_1) \, dx_i\} - \int_{S_0'} \prod_{i=1}^n \{f_{X_i}(x_i|\theta_1) \, dx_i\}$$

$$= \int_{S_0 \cap S_1'} \prod_{i=1}^n \{f_{X_i}(x_i|\theta_1) \, dx_i\} - \int_{S_0' \cap S_1} \prod_{i=1}^n \{f_{X_i}(x_i|\theta_1) \, dx_i\}.$$

$$(10.6)$$

However, by Eqs. (10.3), for each point $(x_1,\ldots,x_n) \in S_0$, we have

$$\prod_{i=1}^n f_{X_i}(x_i|\theta_1) \geq k \prod_{i=1}^n f_{X_i}(x_i|\theta_0).$$

Hence

$$\int_{S_0 \cap S_1'} \prod_{i=1}^n \{f_{X_i}(x_i|\theta_1) \, dx_i\} \geq k \int_{S_0 \cap S_1'} \prod_{i=1}^n \{f_{X_i}(x_i|\theta_0) \, dx_i\}. \qquad (10.7)$$

Similarly, we get

$$\int_{S_0' \cap S_1} \prod_{i=1}^{n} \{f_{X_i}(x_i|\theta_1) \, dx_i\} < k \int_{S_0' \cap S_1} \prod_{i=1}^{n} \{f_{X_i}(x_i|\theta_0) \, dx_i\}. \tag{10.8}$$

From Eqs. (10.6) to (10.8), we get

$$P_{\theta_1}(S_0) - P_{\theta_1}(S_0') \geq k \left[\int_{S_0 \cap S_1'} \prod_{i=1}^{n} \{f_{X_i}(x_i|\theta_0) \, dx_i\} \right.$$

$$\left. - \int_{S_0' \cap S_1} \prod_{i=1}^{n} \{f_{X_i}(x_i|\theta_0) \, dx_i\} \right]$$

$$= k \left[\int_{S_0} \prod_{i=1}^{n} \{f_{X_i}(x_i|\theta_0) \, dx_i\} - \int_{S_0'} \prod_{i=1}^{n} \{f_{X_i}(x_i|\theta_0) \, dx_i\} \right]$$

$$= k(\alpha - \alpha)$$

$$= 0. \tag{10.9}$$

Thus the power of the test with critical region S_0 is greater than or equal to the power of any other test of the same size. In other words, the test with critical region S_0 is most powerful for testing H_0 against H_1. Q.E.D.

Note: If $k = \infty$, then

$$S_0 = \{(x_1,\ldots,x_n) : \prod_{i=1}^{n} f_{X_i}(x_i|\theta_0) = 0\}$$

$$S_1 = \{(x_1,\ldots,x_n) : \prod_{i=1}^{n} f_{X_i}(x_i|\theta_0) > 0\}.$$

For the case of a random phenomenon specified by the probability mass function $p_X(x|\theta)$, $\theta \in \Omega$, the Neyman–Pearson lemma holds good with the probability density function replaced by a probability mass function. The proof is the same, with integration replaced by summation. However, a difficulty is seen to arise here. It may not be possible to achieve

$$P_{\theta_0}(S_0) = \sum_{x \in S_0} p_X(x|\theta_0) = \alpha \tag{10.10}$$

because sometimes if a certain point is included in S_0 the value α is not achieved, whereas α is exceeded if the next point is included in S_0. This difficulty can be eliminated by divising a test that permits splitting the next point and including the required portion of it in S_0 to make $P_{\theta_0}(S_0) = \alpha$. Such a test will be termed a *randomized test*. Before we discuss such tests in detail, and before we examine whether conditions (10.3) and (10.4) are also necessary for the existence of a most powerful test, we shall first consider some applications of the Neyman–Pearson lemma.

Example 10.2.1 Let $X_1,\ldots,X_n$ be a random of size n from a random phenomenon specified by the normal probability density function with unknown mean θ and unit variance, and let $x_1,\ldots,x_n$ be the observed sample. Here we have

$$\prod_{i=1}^{n} f_{X_i}(x_i|\theta) = \frac{1}{(2\pi)^{n/2}} \exp\left[-\frac{1}{2} \sum_{i=1}^{n} (x_i - \theta)^2 \right]$$

$$= \frac{1}{(2\pi)^{n/2}} \exp\left\{ -\frac{1}{2} \left[\sum_{i=1}^{n} (x_i - \bar{x})^2 + n(\bar{x} - \theta)^2 \right] \right\},$$

where

$$\bar{x} = \frac{1}{n} \sum_{1}^{n} x_i.$$

For testing H_0: $\theta = \theta_0$ against H_1: $\theta = \theta_1$, we obtain

$$\frac{\prod_{i=1}^{n} f_{X_i}(x_i|\theta_1)}{\prod_{i=1}^{n} f_{X_i}(x_i|\theta_0)} = \exp\left\{ -\frac{n}{2} (\theta_0 - \theta_1)[2\bar{x} - (\theta_0 + \theta_1)] \right\}. \tag{10.11}$$

Case 1: $\theta_1 > \theta_0$. To find the best critical region, we find the set

$$S_0 = (x_1,\ldots,x_n) : \exp\left[\frac{n}{2} (\theta_1 - \theta_0)(2\bar{x} - \theta_0 - \theta_1) \right] \geq k$$

$$= \left\{ (x_1,\ldots,x_n) : \bar{x} \geq \frac{1}{2} \left[\frac{2 \log k}{n(\theta_1 - \theta_0)} + \theta_1 + \theta_0 \right] = A \right\}.$$

Now

$$\bar{X} = \frac{1}{n} \sum_{1}^{n} X_i$$

is normally distributed with mean θ and variance $1/n$. For a given level of significance (size) α, if S_0 is of size α, then A must satisfy

$$\int_{A}^{\infty} \left(\frac{n}{2\pi} \right)^{1/2} \exp\left[-\frac{n}{2} (\bar{x} - \theta_0)^2 \right] d\bar{x} = \alpha.$$

From the table of the normal distribution in Appendix B we can conclude that such an A (and hence such a k) always exists. For example, if $\alpha = 0.05$, $\theta_0 = 0$, $\theta_1 = 1$, then $A = 0.329$, $k = e^{-4.276}$. Hence we conclude that for testing H_0 against H_1, there exists a most powerful test of size α which is given by the critical region

$$S_0 = \{(x_1,\ldots,x_n) | \bar{x} \geq A\},$$

where A is chosen in such a way that S_0 is of size α.

Case 2: $\theta_1 < \theta_0$. The set S_0, in this case, is the set of points $(x_1,\ldots,x_n)$ such that

$$S_0 = \left\{(x_1,\ldots,x_n) : \bar{x} \leq \frac{1}{2}\left[\frac{2\log k^{-1}}{n(\theta_0 - \theta_1)} + \theta_1 + \theta_0\right] = B\right\},$$

where B satisfies the equation

$$\int_{-\infty}^{B} \left(\frac{n}{2\pi}\right)^{1/2} \exp\left[-\frac{n}{2}(\bar{x} - \theta_0)^2\right] d\bar{x} = \alpha.$$

From the table of the normal distribution we can conclude that such a B (and hence such a k) always exists.

Example 10.2.2 Let $X_1,\ldots,X_n$ be a random sample of size n from a random phenomenon specified by the Bernoulli distribution with parameter θ, and let $H_0: \theta = \theta_0$, $H_1: \theta = \theta_1 > \theta_0$. Here

$$\prod_{i=1}^{n} p_{X_i}(x_i|\theta) = \prod_{1}^{n} \binom{1}{x_i} \theta^{\sum_{i=1}^{n} x_i} (1 - \theta)^{n - \sum_{i=1}^{n} x_i}$$

and

$$\frac{\prod_{i=1}^{n} p_{X_i}(x_i|\theta_1)}{\prod_{i=1}^{n} p_{X_i}(x_i|\theta_0)} = \left(\frac{\theta_1}{\theta_0}\right)^{\sum_{i=1}^{n} x_i} \left(\frac{1 - \theta_1}{1 - \theta_0}\right)^{n - \sum_{i=1}^{n} x_i}$$

$$= \left(\frac{1 - \theta_1}{1 - \theta_0}\right)^{n} \left[\frac{\theta_1(1 - \theta_0)}{\theta_0(1 - \theta_1)}\right]^{\sum_{i=1}^{n} x_i}.$$

Since $\theta_1 > \theta_0$, we get $\theta_1(1 - \theta_0) > (1 - \theta_1)\theta_0$ and hence

$$\left(\frac{1 - \theta_1}{1 - \theta_0}\right)^{n} \left[\frac{(1 - \theta_0)\theta_1}{(1 - \theta_1)\theta_0}\right]^{\sum_{i=1}^{n} x_i} \geq k$$

is equivalent to

$$\sum_{i=1}^{n} x_i \geq \frac{\log k[(1 - \theta_0)/(1 - \theta_1)]^n}{\log[\theta_1(1 - \theta_0)/\theta_0(1 - \theta_1)]} = B,$$

where B is merely a label. If there exists a k such that the critical region

$$S_0 = \left\{ (x_1,...,x_n) : \frac{\Pi_{i=1}^n \, p_{X_i}(x_i|\theta_1)}{\Pi_{i=1}^n \, p_{X_i}(x_i|\theta_0)} \geq k \right\}$$

satisfies

$$P_{\theta_0}(S_0) = \alpha,$$

then there exists a B satisfying

$$P_{\theta_0}\left(\sum_{i=1}^n X_i \geq B \right) = \alpha.$$

Now

$$Y = \sum_{i=1}^n X_i$$

is a binomial random variable with parameter (n,θ), that is,

$$p_Y(j) = \begin{cases} \binom{n}{j} \theta^j (1 - \theta)^{n-j}, & j = 0,1,...,n \\ 0 & \text{otherwise.} \end{cases} \tag{10.12}$$

Let us now consider a particular case $n = 5$, $\alpha = 0.05$, $\theta_0 = 0.5$, $\theta_1 = 0.6$. From Eq. (4.12), with $\theta = 0.5$, we get

$$p_Y(5) = \tfrac{1}{32}, \qquad p_Y(4) = \tfrac{5}{32}.$$

Thus

$$P_{\theta_0}(Y \geq 5) = \tfrac{1}{32} < 0.05, \qquad P_{\theta_0}(Y \geq 4) = \tfrac{6}{32} > 0.05.$$

Hence in general there does not exist a B (and hence a k) such that $P_{\theta_0}(Y \geq B) = \alpha$. It may be noted that if $\alpha = \tfrac{1}{32}$, the test with critical region S_0 is most powerful for testing H_0: $\theta = 0.5$ against H_1: $\theta = 0.6$ at size α.

Example 10.2.3 Let $X_1,...,X_n$ be a random sample of size n from a normal distribution with mean μ and variance σ^2 and let H_0: $\mu = 0$, $\sigma^2 = 1$; H_1: $\mu = \mu_1$, $\sigma^2 = \sigma_1^2$, where μ_1, σ_1^2 are specified constants. Now,

$$\frac{\Pi_{i=1}^n \, f_{X_i}(x_i|\mu_1, \sigma_1^2)}{\Pi_{i=1}^n \, f_{X_i}(x_i|0,1)} = \frac{1}{\sigma_1^n} \exp\left\{ \frac{1}{2\sigma_1^2} [(\sigma_1^2 - 1)(ns^2 + n\bar{x}^2) \right.$$

$$\left. + 2n\mu_1\bar{x} - n\mu_1^2] \right\},$$

where

$$\bar{x} = \frac{1}{n} \sum_{i=1}^{n} x_i, \qquad s^2 = \frac{1}{n} \sum_{i=1}^{n} (x_i - \bar{x})^2.$$

To find the best critical region of size α, we find the set

$$S_0 = \{(x_1,\ldots,x_n) : (\sigma_1^2 - 1)(s^2 + \bar{x}^2) + 2\mu_1 \bar{x} \geq C\},$$

where

$$C = \frac{2\sigma_1^2}{n} \log k\sigma_1^n + \mu_1^2$$

$$P_{\mu=0,\sigma^2=1}(S_0) = \alpha. \tag{10.13}$$

If $\sigma_1^2 - 1 > 0$, then

$$(\sigma_1^2 - 1)(s^2 + \bar{x}^2) + 2\mu_1 \bar{x} = C$$

represents a cricle in the plane of $\bar{x}$, s and the best critical region in this case is the boundary and the exterior of this circle. If $\sigma_1^2 - 1 < 0$, the best critical region is, similarly, the interior and the boundary of a certain circle in the plane of $\bar{x}$, s. Since

$$\bar{X} = \frac{1}{n} \sum_{i=1}^{n} X_i, \qquad S^2 = \sum_{i=1}^{n} (X_i - \bar{X})^2$$

are continuous random variables, the existence of a k, satisfying Eq. (10.13), is evident.

In the examples above we have observed the following: Given the level of significance α, in the case of a continuous random phenomenon there exists a nonegative constant k satisfying conditions (10.3) and (10.4), but in the case of a discrete phenomenon a nonnegative constant k satisfying conditions (10.3) and (10.4) may not always exist. In general, however, if such a constant does not exist, the Neyman–Pearson lemma can be modified to yield a most powerful test of size α. Such a test will be called a *randomized test* and will permit the inclusion of a fraction of a point or a subset of a set into the critical region to attain the required size α. We will consider only the case of a continuous random phenomenon. The argument for the discrete case is exactly the same.

Let us assume that the sample space of $X_1,\ldots,X_n$ consists only of points that have positive probabilities under both H_0 and H_1. Let $\alpha(k)$ denote the probability that the random variable

$$\frac{\Pi_{i=1}^{n} f_{X_i}(X_i|\theta_1)}{\Pi_{i=1}^{n} f_{X_i}(X_i|\theta_0)}$$

(for given θ_0, θ_1) is greater than or equal to k when H_0 is true, that is,

$$\alpha(k) = P_{\theta_0}\{(X_1,\ldots,X_n) \in S_0(k)\}$$

$$= P_{\theta_0}\left\{(x_1,\ldots,x_n) : \prod_{i=1}^{n} f_{X_i}(x_i|\theta_1) \geq k \prod_{i=1}^{n} f_{X_i}(x_i|\theta_0)\right\}. \qquad (10.14)$$

Obviously, $\alpha(k)$ is a nonincreasing function of k. For $k_1 < k_2$, any point $(x_1,\ldots,x_n)$ that satisfies

$$\frac{\prod_{i=1}^{n} f_{X_i}(x_i|\theta_1)}{\prod_{i=1}^{n} f_{X_i}(x_i|\theta_0)} \geq k_2$$

also satisfies

$$\frac{\prod_{i=1}^{n} f_{X_i}(x_i|\theta_1)}{\prod_{i=1}^{n} f_{X_i}(x_i|\theta_0)} \geq k_1$$

Thus we have

$$\left\{(x_1,\ldots,x_n) : \frac{\prod_{i=1}^{n} f_{X_i}(x_i|\theta_1)}{\prod_{i=1}^{n} f_{X_i}(x_i|\theta_0)} \geq k_2\right\} \subseteq \left\{(x_1,\ldots,x_n) : \frac{\prod_{i=1}^{n} f_{X_i}(x_i|\theta_1)}{\prod_{i=1}^{n} f_{X_i}(x_i|\theta_0)} \geq k_1\right\}$$

so that $\alpha(k_2) \leq \alpha(k_1)$. Again

$$\alpha(0) = P_{\theta_0}\left\{(x_1,\ldots,x_n) : \prod_{i=1}^{n} f_{X_i}(x_i|\theta_1) \geq 0\right\} = 1.$$

Furthermore, we also have

$$\alpha(k) \leq \frac{1}{k}, \qquad k > 0. \qquad (10.15)$$

To prove this, assume that the inequality (10.15) does not hold: Suppose that $\alpha(k) > 1/k$, $k > 0$. Then for $k > 0$,

$$\int_{S_0(k)} \prod_{i=1}^{n} \{f_{X_i}(x_i|\theta_1)\,dx_i\} \geq k \int_{S_0(k)} \prod_{i=1}^{n} \{f_{X_i}(x_i|\theta_0)\,dx_i\}$$

$$= k\alpha(k)$$

$$> 1. \qquad (10.16)$$

But the left-hand side represents the probability of the set $S_0(k)$ under H_1, so it cannot be greater than 1. This contradicts the assumption $\alpha(k) > 1/k$. Hence we have $\alpha(k) < 1/k$ for $k \geq 0$. From this we get

$$\lim_{k \to \infty} \alpha(k) = 0.$$

If there exists a k such that $\alpha(k) = \alpha$, there exists a most powerful test of size α with critical region S_0 [also $S_0(k)$ in the notation above] given in (10.3).

Now if there does not exist a constant k such that $\alpha(k) = \alpha$, there does exist a $k = k'$ (say) such that

$$\alpha \leq \alpha(k' - 0), \qquad \alpha(k' + 0) \leq \alpha. \tag{10.17}$$

Then we have

$$\alpha(k' - 0) - \alpha(k' + 0) = \int_A \prod_{i=1}^{n} \{f_{X_i}(x_i|\theta_0) \, dx_i\}, \tag{10.18}$$

where

$$A = \left\{ (x_1,\ldots,x_n) : \prod_{i=1}^{n} f_{X_i}(x_i|\theta_1) = k' \prod_{i=1}^{n} f_{X_i}(x_i|\theta_0) \right\}.$$

The next theorem will show that a most powerful test of size α has critical region

$$S_0 = S_0(k' + 0) \cup B, \tag{10.19}$$

where B is a subset of A and is chosen in such a way that $P_{\theta_0}(S_0) = \alpha$.

Let $\phi(x_1,\ldots,x_n)$ denote the conditional probability of rejecting H_0: $\theta = \theta_0$ when $x_1,\ldots,x_n$ is observed. For the test defined in the lemma above, we have

$$\phi(x_1,\ldots,x_n) = \begin{cases} 1 & \text{if } (x_1,\ldots,x_n) \in S_0 \\ 0 & \text{if } (x_1,\ldots,x_n) \in S_1. \end{cases}$$

In a situation where $\alpha(k)$ satisfies (10.17) it is necessary to include only a part of the boundary set A in the region

$$C = \left\{ (x_1,\ldots,x_n) : \prod_{i=1}^{n} f_{X_i}(x_i|\theta_1) > k' \prod_{i=1}^{n} f_{X_i}(x_i|\theta_0) \right\}$$

in order to get a critical region of size α. Obviously, then, it is necessary to assign to any point $(x_1,\ldots,x_n) \in A$ a probability $\phi(x_1,\ldots,x_n) = \gamma$ ($0 < \gamma < 1$), so that we get a critical region of required size α. In other words, we need to select randomly a fraction γ of the totality of points of the boundary set A for inclusion in the set C. A test ϕ that permits randomization on the boundary set is called a *randomized test*. $E_{\theta_0}\phi(X_1,\ldots,X_n)$ is called the *size* and $E_{\theta_1}\phi(X_1,\ldots,X_n)$ is called the *power* of the randomized test. A most powerful randomized test of size α is a randomized test of size α having maximum power among all tests of the same size α. Tests that do not permit randomization on the boundary will be called *nonran-domized tests*. Randomized and nonrandomized tests differ only on the boundary. However, in actual practice one would prefer to adopt a different size rather than randomizing the boundary.

Theorem 10.2.1 Let $X_1,\ldots,X_n$ be a random sample of size n from a continuous random phenomenon specified by the probability density function $f_X(x|\theta)$, $\theta \in \Omega$, and let Ω contain only two distinct points, θ_0 and θ_1.

(a) For testing H_0: $\theta = \theta_0$ against the alternative H_1: $\theta = \theta_1$ there exists a randomized test ϕ given by

$$\phi(x_1,\ldots,x_n) = \begin{cases} 1 & \text{if } \prod_{i=1}^{n} f_{X_i}(x_i|\theta_1) > k \prod_{i=1}^{n} f_{X_i}(x_i|\theta_0) \\ \gamma & \text{if } \prod_{i=1}^{n} f_{X_i}(x_i|\theta_1) = k \prod_{i=1}^{n} f_{X_i}(x_i|\theta_0) \\ 0 & \text{if } \prod_{i=1}^{n} f_{X_i}(x_i|\theta_1) < k \prod_{i=1}^{n} f_{X_i}(x_i|\theta_0), \end{cases} \tag{10.20}$$

where k, γ are nonnegative constants such that

$$E_{\theta_0}(X_1,\ldots,X_n) = \alpha. \tag{10.21}$$

(b) If a test $\phi(X_1,\ldots,X_n)$ satisfies Eqs. (10.20) and (10.21) for some k and γ, it is most powerful of size α for testing H_0 against H_1.

Proof of (a). Let

$$\gamma = \frac{\alpha - \alpha(k + 0)}{\alpha(k - 0) - \alpha(k + 0)} \qquad \text{if } \alpha(k - 0) \neq \alpha(k + 0).$$

Then we have

$$E_{\theta_0}\phi(X_1,\ldots,X_n)$$

$$= P_{\theta_0}\left\{\frac{\prod_{i=1}^{n} f_{X_i}(X_i|\theta_1)}{\prod_{i=1}^{n} f_{X_i}(X_i|\theta_0)} > k\right\}$$

$$+ \frac{\alpha - \alpha(k + 0)}{\alpha(k - 0) - \alpha(k + 0)} P_{\theta_0}\left\{\frac{\prod_{i=1}^{n} f_{X_i}(X_i|\theta_1)}{\prod_{i=1}^{n} f_{X_i}(x_i|\theta_0)} = k\right\}$$

$$= \alpha(k + 0) + \frac{\alpha - \alpha(k + 0)}{\alpha(k - 0) - \alpha(k + 0)} [\alpha(k - 0) - \alpha(k + 0)]$$

$$= \alpha(k + 0) + \alpha - \alpha(k + 0)$$

$$= \alpha.$$

If $\alpha(k - 0) = \alpha(k + 0)$, then

$$P_{\theta_0}\left\{\frac{\Pi_{i=1}^n f_{X_i}(x_i|\theta_1)}{\Pi_{i=1}^n f_{X_i}(X_i|\theta_0)} = k\right\} = 0,$$

and hence the entire boundary set can be added to the critical region without any randomization.

Proof of (b). Let $\phi^*(X_1,...,X_n)$ be any other randomized test of size α for testing H_0: $\theta = \theta_0$ and let

$$S^+ = \{(x_1,...,x_n) : \phi(x_1,...,x_n) - \phi^*(x_1,...,x_n) > 0\}$$

$$S^- = \{(x_1,...,x_n) : \phi(x_1,...,x_n) - \phi^*(x_1,...,x_n) < 0\}$$

be defined in the sample space S of $X_1,...,X_n$. It is easy to conclude that for any point $(x_1,...,x_n) \in S^+$, $\phi(x_1,...,x_n) = 1$ or γ and for any point $(x_1,...,x_n) \in S^-$, $\phi(x_1,...,x_n) = 0$ or γ. Thus we have

$$E_{\theta_1}\phi(X_1,...,X_n) - E_{\theta_1}\phi^*(X_1,...,X_n)$$

$$= \int_S (\phi - \phi^*) \prod_{i=1}^n \{f_{X_i}(x_i|\theta_1)\,dx_i\}$$

$$= \int_{S^+} (\phi - \phi^*) \prod_{i=1}^n \{f_{X_i}(x_i|\theta_1)\,dx_i\}$$

$$+ \int_{S^-} (\phi - \phi^*) \prod_{i=1}^n \{f_{X_i}(x_i|\theta_1)\,dx_i\}$$

$$\geq k\left[\int_{S^+} (\phi - \phi^*) \prod_{i=1}^n \{f_{X_i}(x_i|\theta_0)\,dx_i\}\right.$$

$$\left. + \int_{S^-} (\phi - \phi^*) \prod_{i=1}^n \{f_{X_i}(x_i|\theta_0)\,dx_i\}\right]$$

$$= k\left[\int_S \phi(x_1,...,x_n) \prod_{i=1}^n \{f_{X_i}(x_i|\theta_0)\,dx_i\}\right.$$

$$\left. - \int_S \phi^*(x_1,...,x_n) \prod_{i=1}^n \{f_{X_i}(x_i|\theta_0)\,dx_i\}\right]$$

$$= k(\alpha - \alpha)$$

$$= 0. \qquad \text{Q.E.D.}$$

Now the question arises: Are the conditions (10.20) and (10.21) also necessary? The reader is referred to Lehmann (1959) for a detailed discussion.

Example 10.2.4   Consider Example (10.2.2). Let $\alpha = 0.05$, $n = 5$, $\theta_0 = 0.5$, $\theta_1 = 0.6$. By Theorem 10.2.1, the randomized test $\phi(X_1,...,X_n)$ given by

$$\phi(x_1,...,x_n) = \begin{cases} 1 & \text{if } \bar{x} > 4 \\ \gamma & \text{if } \bar{x} = 4 \\ 0 & \text{if } \bar{x} < 4, \end{cases}$$

where γ is chosen such that

$$\tfrac{1}{32} + \gamma \tfrac{5}{32} = 0.05 \quad \text{or} \quad \gamma = 0.12$$

is most powerful for testing H_0: $\theta = 0.5$ against H_1: $\theta = 0.6$ at $\alpha = 0.05$.

10.3   TESTS OF SIMPLE H_0 AGAINST COMPOSITE H_1

For practical purposes, the problem of testing a simple null hypothesis against a simple alternative is not very interesting. In this section we consider a random phenomenon whose probability density function $f_X(x|\theta)$ [or probability mass function $p_X(x|\theta)$] depends on a single continuous parameter $\theta \in \Omega$. Consider the problem of testing the simple null hypothesis H_0: $\theta = \theta_0$ against the composite alternatives H_1: $\theta \in \omega \subseteq (\Omega - \{\theta_0\})$. Sometimes it may happen that the most powerful test of size α (randomized or nonrandomized) for testing H_0: $\theta = \theta_0$ against the simple alternative $\theta = \theta_1 \in \omega$ is also most powerful against any simple alternative $\theta = \theta' \in \omega$. Such a test is called a *uniformly most powerful test of size α* for testing H_0: $\theta = \theta_0$ against the composite alternatives H_1: $\theta \in \omega$. As will be seen, such a test does not always exist. However, when it exists, the Neyman–Pearson lemma or its randomized counterpart provides a technique for finding it.

Example 10.3.1   Let $X_1,...,X_n$ be a random sample of size n from a continuous random phenomenon specified by the normal probability density function with unknown mean θ and variance 1. In Example 10.2.1 we saw that for testing the simple null hypothesis H_0: $\theta = \theta_0$ against the simple alternative H_1: $\theta = \theta_1$, the most powerful test of size α, when $\theta_1 > \theta_0$, is given by

$$\phi(x_1,...,x_n) = \begin{cases} 1, & \bar{x} \geq A \\ 0 & \text{otherwise,} \end{cases} \tag{10.22}$$

where the constant A is chosen such that $E_{\theta_0}\phi(X_1,...,X_n) = \alpha$. Furthermore, from Eq. (10.11) it follows that the test ϕ is most powerful against any simple alternative $\theta = \theta' > \theta_0$. Thus, for testing H_0: $\theta = \theta_0$ against the composite alternatives H_1: $\theta > \theta_0$, the test ϕ is uniformly most powerful of size α.

Similarly, when $\theta_1 < \theta_0$, we can conclude that the test $\phi^*(X_1,...,X_n)$, given by

$$\phi^*(x_1,...,x_n) = \begin{cases} 1 & \text{if } \overline{x} \leq B \\ 0 & \text{otherwise,} \end{cases} \tag{10.23}$$

where the constant B is chosen such that $E_{\theta_0}\phi^*(X_1,...,X_n) = \alpha$ is uniformly most powerful of size α for testing H_0: $\theta = \theta_0$ against the composite alternatives H_1: $\theta < \theta_0$.

Obviously, these two tests ϕ and ϕ^* are different: The first one rejects the null hypothesis for large values of

$$\overline{X} = \frac{1}{n} \sum_{i=1}^n X_i,$$

while the second one rejects the null hypothesis for small values of $\overline{X}$. Hence we may conclude that there does not exist a uniformly most powerful test of size α for testing H_0: $\theta = \theta_0$ against the two-sided alternatives H_1: $\theta \neq \theta_0$.

Example 10.3.2   Let $X_1,...,X_n$ be a random sample of size n from a random phenomenon specified by the probability mass function (Poisson)

$$p_X(x|\theta) = \begin{cases} \dfrac{e^{-\theta}\theta^x}{x!}, & x = 0,1,...,\infty \\ 0 & \text{otherwise.} \end{cases}$$

Here we have

$$\prod_{i=1}^n p_{X_i}(x_i|\theta) = \frac{e^{-n\theta}\theta^{\Sigma_{i=1}^n x_i}}{\Pi_{i=1}^n (x_i!)},$$

so that for any $\theta_1 \neq \theta_0$,

$$\frac{\Pi_{i=1}^n p_{X_i}(x_i|\theta_1)}{\Pi_{i=1}^n p_{X_i}(x_i|\theta_0)} = e^{-n(\theta_1-\theta_0)}\left(\frac{\theta_1}{\theta_0}\right)^{\Sigma_{i=1}^n x_i}. \tag{10.24}$$

Obviously,

$$\left(\frac{\theta_1}{\theta_0}\right)^{\Sigma_{i=1}^n x_i}$$

increases or decreases monotonically in $\Sigma_{i=1}^{n} x_i$ according as $\theta_1 > \theta_0$ or $\theta_1 < \theta_0$. For any constant k,

$$e^{-n(\theta_1 - \theta_0)} \left(\frac{\theta_1}{\theta_0}\right)^{\Sigma_{i=1}^{n} x_i} > k$$

implies that $\Sigma_{i=1}^{n} x_i > A$ or $< B$ according as $\theta_1 > \theta_0$ or $\theta_1 < \theta_0$, respectively, where

$$A = \frac{\log(k e^{n(\theta_1 - \theta_0)})}{\log(\theta_1/\theta_0)}, \qquad B = \frac{\log[(1/k)e^{n(\theta_0 - \theta_1)}]}{\log(\theta_1/\theta_0)}.$$

Thus we conclude the following:

(a)   For testing H_0: $\theta = \theta_0$ against the simple alternative $\theta = \theta_1 > \theta_0$, the most powerful test $\phi(X_1,\ldots,X_n)$ of size α is given by

$$\phi(x_1,\ldots,x_n) = \begin{cases} 1 & \text{if } \sum_{i=1}^{n} x_i > A \\[2mm] \gamma & \text{if } \sum_{i=1}^{n} x_i = A \\[2mm] 0 & \text{if } \sum_{i=1}^{n} x_i < A, \end{cases}$$

where A (or k), γ are positive constants such that $E_{\theta_0}\phi(X_1,\ldots,X_n) = \alpha$. Now $\Sigma_{i=1}^{n} X_i$ is the Poisson random variable with parameter $n\theta$. For $\theta_0 = 0.3$, $\theta_1 = 0.4$, $\alpha = 0.05$, $n = 10$, we have

$$P_{\theta_0}\left(\sum_{i=1}^{10} X_i > 4\right) = 1 - \sum_{i=0}^{4} \frac{e^{-3} \cdot 3^i}{i!} = 1 - 0.815 = 0.185$$

$$P_{\theta_0}\left(\sum_{i=1}^{10} X_i > 5\right) = 1 - \sum_{i=0}^{5} \frac{e^{-3} \cdot 3^i}{i!} = 1 - 0.961 = 0.039$$

$$P_{\theta_0}\left(\sum_{i=1}^{10} X_i = 5\right) = 0.961 - 0.815 = 0.146$$

$$A = 5, \qquad \gamma = 0.075.$$

From Eq. (10.24) it is clear that the test is uniformly most powerful of size α for testing H_0: $\theta = \theta_0$ against the alternatives H_1: $\theta > \theta_0$.

(b)   Similarly, for testing H_0: $\theta = \theta_0$ against the alternatives H_1: $0 < \theta_0$ the uniformly most powerful test of size α is given by

$$
\phi^*(x_1,\ldots,x_n) =
\begin{cases}
1 & \text{if } \sum_{i=1}^{n} x_i < B \\[2mm]
\gamma & \text{if } \sum_{i=1}^{n} x_i = B \\[2mm]
0 & \text{if } \sum_{i=1}^{n} x_i > B,
\end{cases}
$$

where the positive constants B and γ are chosen such that $E_{\theta_0}\phi^*(X_1,\ldots,X_n) = \alpha$. For $n = 10$, $\theta_0 = 0.3$, $\alpha = 0.05$, we have

$$
P_{\theta_0}\!\left(\sum_{i=1}^{10} X_i = 0\right) = 0.05,
$$

which tells us to take $\gamma = 1$, $B = 0$.

Obviously, then, for testing $\theta = \theta_0$ against two-sided alternatives $\theta \neq \theta_0$ there does not exist any uniformly most powerful test of size α.

10.4   TEST OF COMPOSITE H_0 AGAINST COMPOSITE H_1 (ONE-PARAMETER CASE)

We consider here testing problems where both the null and alternative hypotheses are composite. Because of our limitation of the scope of this book we consider here only the family of distributions that depend continuously on a single parameter $\theta \in \Omega$ and that possess a monotone likelihood ratio in the following sense.

Definition 10.4.1: Monotone likelihood ratio   Let $X_1,\ldots,X_n$ be a random sample of size n from a random phenomenon specified by the probability density function $f_X(x|\theta)$ in the continuous case [or $p_X(x|\theta)$ in the discrete case], $\theta \in \Omega$. The family of densities

$$
\prod_{i=1}^{n} f_{X_i}(x_i|\theta) \qquad \left[\text{or} \quad \prod_{i=1}^{n} p_{X_i}(x_i|\theta) \right]
$$

is said to have a *monotone likelihood ratio* in $T(X_1,\ldots,X_n)$ if there exists a real-valued function $T(X_1,\ldots,X_n)$ such that for $\theta_1 < \theta_2$:

(a)   The distribution functions corresponding to θ_1, θ_2 are distinct;

(b) $\dfrac{\Pi_{i=1}^n f_{X_i}(x_i|\theta_2)}{\Pi_{i=1}^n f_{X_i}(x_i|\theta_1)}$ $\left[\text{or}\ \dfrac{\Pi_{i=1}^n p_{X_i}(x_i|\theta_2)}{\Pi_{i=1}^n p_{X_i}(x_i|\theta_1)}\right]$

is a nondecreasing function of $T(x_1,\ldots,x_n)$.

Example 10.4.1   Let $X_1,\ldots,X_n$ be a random sample of size n from a binomial distribution with parameter (m,θ). Then

$$\prod_{i=1}^n p_{X_i}(x_i|\theta) = \prod_{i=1}^n \binom{m}{x_i}\theta^{\Sigma_{i=1}^n x_i}(1-\theta)^{mn-\Sigma_{i=1}^n x_i}.$$

For $\theta_1 < \theta_2$ [or $\theta_2(1-\theta_1) > \theta_1(1-\theta_2)$]

$$\frac{\Pi_{i=1}^n p_{X_i}(x_i|\theta_2)}{\Pi_{i=1}^n p_{X_i}(x_i|\theta_1)} = \left(\frac{1-\theta_2}{1-\theta_1}\right)^{mn}\left[\frac{\theta_2(1-\theta_1)}{\theta_1(1-\theta_2)}\right]^{\Sigma_{i=1}^n x_i}$$

is a nondecreasing function of $T(x_1,\ldots,x_n) = \Sigma_{i=1}^n x_i$. Hence $\Pi_{i=1}^n p_{X_i}(x_i|\theta)$ has monotone likelihood ratio in $T(x_1,\ldots,x_n) = \Sigma_{i=1}^n x_i$.

Example 10.4.2   Let $X_1,\ldots,X_n$ be a random sample of size n from a random phenomenon specified by the normal distribution with mean θ and unit variance. For $\theta_1 < \theta_2$,

$$\frac{\Pi_{i=1}^n f_{X_i}(x_i|\theta_2)}{\Pi_{i=1}^n f_{X_i}(x_i|\theta_1)} = \exp\left[(\theta_2-\theta_1)\sum_{i=1}^n x_i + \frac{n}{2}(\theta_1^2 - \theta_2^2)\right]$$

is a nondecreasing function of $\Sigma_{i=1}^n x_i$. Hence $\Pi_{i=1}^n f_{X_i}(x_i|\theta)$ has a monotone likelihood ratio in $T(x_1,\ldots,x_n) = \Sigma_{i=1}^n x_i$.

Theorem 10.4.1   Let $X_1,\ldots,X_n$ be a random sample of size n from a random phenomenon specified by the probability density function $f_X(x|\theta)$ [(or mass function $p_X(x|\theta)$] which depends continuously on a single parameter $\theta \in \Omega$ and let $\Pi_{i=1}^n f_{X_i}(x_i|\theta)$ [or $\Pi_{i=1}^n p_{X_i}(x_i|\theta)$] have monotone likelihood ratio in $T(x_1,\ldots,x_n)$. Then for testing H_0: $\theta \le \theta_0$ against the alternatives H_1: $\theta > \theta_0$ there exists a uniformly most powerful test $\phi(X_1,\ldots,X_n)$ of size α, given by

$$\phi(x_1,\ldots,x_n) = \begin{cases} 1 & \text{if } T(x_1,\ldots,x_n) > k \\ \gamma & \text{if } T(x_1,\ldots,x_n) = k \\ 0 & \text{if } T(x_1,\ldots,x_n) < k, \end{cases} \tag{10.25}$$

where k, γ are nonnegative constants satisfying $E_{\theta_0}\phi(X_1,\ldots,X_n) = \alpha$.

Proof. We give the proof for the continuous case. The proof for the discrete case is exactly the same. Since $\Pi_{i=1}^{n} f_{X_i}(x_i|\theta)$ has monotone likelihood ratio in $T(x_1,\ldots,x_n)$, for any $\theta_1 > \theta_0$ and constant k,

$$\frac{\Pi_{i=1}^{n} f_{X_i}(x_i|\theta_1)}{\Pi_{i=1}^{n} f_{X_i}(x_i|\theta_0)} \begin{Bmatrix} > \\ = \\ < \end{Bmatrix} k \tag{10.26}$$

is equivalent to

$$T(x_1,\ldots,x_n) \begin{Bmatrix} > \\ = \\ < \end{Bmatrix} c$$

for some constant c. Hence, by Theorem 10.2.1, there exists a test $\phi(X_1,\ldots,X_n)$ defined in Eq. (10.2.5) which is most powerful of size α for testing $H_0: \theta = \theta_0$ against any simple alternative $\theta = \theta_1$ provided that $\theta_1 > \theta_0$. Furthermore, for any pair (θ',θ'') with $\theta' < \theta''$, the test ϕ is most powerful for testing the simple hypothesis $\theta = \theta'$ against the simple alternative $\theta = \theta''$ for size $E_{\theta'}\phi(X_1,\ldots,X_n)$.

For any test ϕ let us define

$$\beta_\phi(\theta) = E_\theta \phi(X_1,\ldots,X_n).$$

From Exercises 8 of this chapter it follows that

$$\beta_\phi(\theta') < \beta_\phi(\theta'')$$

unless $\beta_\phi(\theta') = 1$. Thus $\beta_\phi(\theta)$ is nondecreasing in θ for all θ satisfying $\beta_\phi(\theta) < 1$.

Let $\{\phi^*\}$ be a class of tests $\phi^*(X_1,\ldots,X_n)$ that satisfy

$$E_\theta \phi^*(X_1,\ldots,X_n) \leq \alpha \quad \text{for } \theta \leq \theta_0.$$

Obviously, $\{\phi^*\}$ forms a subclass of the class of tests $\{\phi^{**}\}$ that satisfy $E_{\theta_0}\phi^{**}(X_1,\ldots,X_n) = \alpha$. Since the test ϕ maximizes $\beta_{\phi^{**}}(\theta_1)$ among all tests in the class $\{\phi^{**}\}$, it also maximizes $\beta_{\phi^*}(\theta_1)$ among all tests in the subclass $\{\phi^*\}$. From (10.26) it follows that ϕ is independent of any particular $\theta_1 > \theta_0$, and hence ϕ is uniformly most powerful for testing $H_0: \theta \leq \theta_0$ against the alternatives $H_1: \theta > \theta_0$. Q.E.D.

Corollary 10.4.1.1 For testing $H_0: \theta \geq \theta_0$ against $H_1: \theta < \theta_0$, the test ϕ given by

$$\phi(x_1,\ldots,x_n) = \begin{cases} 1 & \text{if } T(x_1,\ldots,x_n) < c \\ \gamma & \text{if } T(x_1,\ldots,x_n) = c \\ 0 & \text{if } T(x_1,\ldots,x_n) > c, \end{cases} \tag{10.27}$$

where γ, c are constants satisfying $E_{\theta_0}\phi(X_1,\ldots,X_n) = \alpha$ is uniformly most powerful of size α.

Proof. For $\theta_1 < \theta_0$,

$$\frac{\Pi_{i=1}^n f_{X_i}(x_i|\theta_1)}{\Pi_{i=1}^n f_{X_i}(x_i|\theta_0)}$$

is a nonincreasing function of $T(x_1,\ldots,x_n)$. Now the proof follows trivially from that of the theorem.   Q.E.D.

Example 10.4.3   Let $X_1,\ldots,X_n$ be a random sample of size n from a random phenomenon specified by the normal probability density function with mean 0 and unknown variance σ^2. Find the uniformly most powerful test of size α of H_0: $\sigma^2 \le \sigma_0^2$ against the alternatives H_1: $\sigma^2 > \sigma_0^2$, where σ_0^2 is specified. For $\sigma_1^2 < \sigma_2^2$,

$$\frac{\Pi_{i=1}^n f_{X_i}(x_i|\sigma_2^2)}{\Pi_{i=1}^n f_{X_i}(x_i|\sigma_1^2)} = \left(\frac{\sigma_1^2}{\sigma_2^2}\right)^{n/2} \exp\left[\frac{1}{2}\sum_{i=1}^n x_i^2\left(\frac{1}{\sigma_1^2} - \frac{1}{\sigma_2^2}\right)\right] \tag{10.28}$$

is a nondecreasing function of $\sum_{i=1}^n x_i^2$. Thus $\Pi_{i=1}^n f_{X_i}(x_i|\sigma^2)$ has a monotone likelihood ratio in $\sum_{i=1}^n x_i^2$. Hence, by Theorem 10.4.1, the $\phi(X_1,\ldots,X_n)$ given by

$$\phi(x_1,\ldots,x_n) = \begin{cases} 1 & \text{if } \sum_{i=1}^n x_i^2 > c \\ \gamma & \text{if } \sum_{i=1}^n x_i^2 = c \\ 0 & \text{if } \sum_{i=1}^n x_i^2 < c, \end{cases}$$

where γ, c are nonnegative constants such that $E_{\sigma_0^2}\phi(X_1,\ldots,X_n) = \alpha$, is uniformly most powerful for testing H_0 against H_1. Since, under H_0: $\sum_{i=1}^n X_i^2/\sigma_0^2$ has central chi-square distribution with n degrees of freedom, we have

$$P_{\sigma_0^2}\left(\sum_{i=1}^n \frac{X_i^2}{\sigma_0^2} = c\right) = 0.$$

Thus ϕ can be redefined as

$$\phi(x_1,\ldots,x_n) = \begin{cases} 1 & \text{if } \sum_{i=1}^n x_i^2 \ge c \\ 0 & \text{if } \sum_{i=1}^n x_i^2 < c, \end{cases}$$

where the nonnegative constant c is determined by $E_{\sigma_0^2}\phi(X_1,...,X_n) = \alpha$. For example, if $n = 10$, $\alpha = 0.05$, $\sigma_0^2 = 2$, then $c = 4.456$.

Example 10.4.4 Let X be a Cauchy random variable with pdf

$$f_X(x|\theta) = \frac{1}{\pi[1 + x - \theta)^2]}, \qquad -\infty < x < \infty.$$

For $\theta' > \theta$,

$$\frac{f_X(x|\theta')}{f_X(x|\theta)} = \frac{1 + (x - \theta)^2}{1 + (x - \theta')^2} \longrightarrow 1 \qquad \text{if } x \to \infty \text{ or } x \to -\infty.$$

Thus the Cauchy distribution does not have a monotone likelihood ratio.

10.5 UNBIASED TESTS

In Examples 10.3.1 and 10.3.2 we have observed that for testing the simple hypothesis H_0: $\theta = \theta_0$ against two-sided alternatives H_1: $\theta \neq \theta_0$, no uniformly most powerful test exists. For any test $\phi(X_1,...,X_n)$, let $\beta_\phi(\theta) = E_\theta\phi(X_1,...,X_n)$. It may be verified that the $\beta_\phi(\theta)$ (for $\theta < \theta_0$) of the uniformly most powerful test ϕ of H_0: $\theta = \theta_0$ against the alternatives H_1: $\theta > \theta_0$, is less than the size of the test ϕ. Similarly, $\beta_{\phi^*}(\theta)$ (for $\theta > \theta_0$) of the uniformly most powerful test ϕ^* of H_0: $\theta = \theta_0$ against the alternatives H_1: $\theta < \theta_0$, is less than the size of the test ϕ^*. In cases where a uniformly most powerful test among the class of all available tests does not exist, we may sometimes find a test that is uniformly most powerful among a subclass of the class of all available tests. Such a subclass is obtained by imposing certain extra restrictions on tests. Obviously, one such restriction that one may wish to impose on tests of H_0 against alternatives H_1: $\theta \neq \theta_0$ is that for any test ϕ in the subclass

$$E_{\theta_0}\phi(X_1,...,X_n) = \alpha$$

$$E_\theta\phi(X_1,...,X_n) = \beta_\phi(\theta) \geq \alpha$$

(10.29)

for $\theta \neq \theta_0$. This condition tells us that $\beta_\phi(\theta)$ has its minimum at $\theta = \theta_0$ only. A test $\phi(X_1,...,X_n)$ that satisfies conditon (10.29) is called an unbiased test of size α for testing H_0 against H_1: $\theta \neq \theta_0$. Obviously, the class of all unbiased tests of size α forms a subclass of the class of all tests of size α. In this section we attempt to find a test (if it exists) which is uniformly most powerful in the class of all unbiased tests of size α for testing H_0 against H_1: $\theta \neq \theta_0$. Such a test will be called a *uniformly most powerful unbiased test of size α for testing H_0 against H_1: $\theta \neq \theta_0$*. Because of our limitations on mathematical content, we shall consider the one-parameter exponential family of distributions only.

Theorem 10.5.1 Let $X_1,\ldots,X_n$ be a random sample of size n from a random phenomenon with probability density function $f_X(x|\theta)$ [or probability mass function $p_X(x|\theta)$], which depends continuously on a single parameter θ. Suppose that

$$\left.\begin{array}{l} \prod_{i=1}^{n} f_{X_i}(x|\theta) \\ \text{or} \\ \prod_{i=1}^{n} p_{X_i}(x|\theta) \end{array}\right\} = K(\theta)C(x_1,\ldots,x_n)\,\exp[Q(\theta)T(x_1,\ldots,x_n)] \qquad (10.30)$$

where $Q(\theta)$ is strictly monotone. For testing $H_0: \theta = \theta_0$ against the alternatives $H_1: \theta \neq \theta_0$ there exists a uniformly most powerful unbiased size α test. If $Q(\theta)$ is strictly increasing, then $\phi(X_1,\ldots,X_n)$ is given by

$$\phi(x_1,\ldots,x_n) = \begin{cases} 1 & \text{if } T(x_1,\ldots,x_n) < k_1 \text{ or } > k_2 \\ \gamma_i & \text{if } T(x_1,\ldots,x_n) = k_i, i = 1,2 \\ 0 & \text{if } k_1 < T(x_1,\ldots,x_n) < k_2, \end{cases} \qquad (10.31)$$

where k_1, k_2, γ_1, γ_2 are constants such that

$$E_{\theta_0}\phi(X_1,\ldots,X_n) = \alpha$$
$$(10.32)$$
$$E_{\theta_0}(\phi(X_1,\ldots,X_n)T(X_1,\ldots,X_n)) = \alpha E_{\theta_0}T(X_1,\ldots,X_n).$$

We shall not be able to prove this theorem here because of the advanced mathematical character of its proof. The reader is referred to Lehmann (1959) for a proof of this theorem.

Example 10.5.1 Let $X_1,\ldots,X_n$ be a random sample of size n from a random phenomenon specified by the normal distribution with unknown mean θ and variance 1. Consider the problem of testing $H_0: \theta = \theta_0$ against H_1: $\theta \neq \theta_0$. Now

$$\prod_{i=1}^{n} f_X(x_i|\theta) = \frac{1}{(2\pi)^{n/2}} \exp\left[-\frac{1}{2}\sum_{i=1}^{n}(x_i - \theta)^2\right]$$

$$= \frac{1}{(2\pi)^{n/2}} \exp\left(-\frac{n}{2}\theta^2\right) \exp\left(-\frac{1}{2}\sum_{i=1}^{n}x_i^2\right) \exp\left(\theta\sum_{i=1}^{n}x_i\right).$$

Hence by Theorem 10.5.1 the uniformly most powerful unbiased test of size α of H_0 against H_1 is given by

$$\phi(x_1,\ldots,x_n) = \begin{cases} 1 & \text{if } \sum_{i=1}^{n} x_i < k_1 \text{ or } > k_2 \\ \gamma_i & \text{if } \sum_{i=1}^{n} x_i = k_i, \quad i = 1,2 \\ 0 & \text{if } k_1 < \sum_{i=1}^{n} x_i < k_2, \end{cases} \tag{10.33}$$

where (with $T = \Sigma_{i=1}^{n} X_i$)

$$P_{\theta_0}\left(\sum_{i=1}^{n} X_i < k_1\right) + P_{\theta_0}\left(\sum_{i=1}^{n} X_i > k_2\right) + \gamma_1 P_{\theta_0}\left(\sum_{i=1}^{n} X_i = k_1\right)$$

$$+ \gamma_2 P_{\theta_0}\left(\sum_{i=1}^{n} X_i = k_2\right) = \alpha \tag{10.34}$$

$$\int_{-\infty}^{k_1} t f_T(t|\theta_0)\, dt + \int_{k_2}^{\infty} t f_T(t|\theta_0)\, dt + \gamma_1 k_1 P_{\theta_0}(T = k_1)$$

$$+ \gamma_2 k_2 P_{\theta_0}(T = k_2) = \alpha E_{\theta_0}(T). \tag{10.35}$$

When H_0 is true, T is normally distributed with mean $n\theta_0$ and variance n. Hence $P_{\theta_0}(T = k_i) = 0$, $i = 1,2$, and

$$P_{\theta_0}(T - n\theta_0 < -u) = P_{\theta_0}(T - n\theta_0 > u)$$

for all real u. Thus the Eq. (10.34) reduces to

$$\int_{-\infty}^{k_1} t f_T(t|\theta_0)\, dt + \int_{k_2}^{\infty} t f_T(t|\theta_0)\, dt = \alpha. \tag{10.36}$$

Now, using (10.36), we get

$$\int_{-\infty}^{k_1} f_T(t|\theta_0)\, dt + \int_{k_2}^{\infty} f_T(t|\theta_0)\, dt = \int_{-\infty}^{k_1} (t - n\theta_0) f_T(t|\theta_0)\, dt$$

$$+ \int_{k_2}^{\infty} (t - n\theta_0) f_T(t|\theta_0)\, dt = n\theta_0\alpha.$$

Let $Y = (T - n\theta_0)/\sqrt{n}$. Then Eq. (10.35) reduces to

$$\int_{-\infty}^{k_1} (t - n\theta_0) f_T(t|\theta_0)\, dt + \int_{k_2}^{\infty} (t - n\theta_0) f_T(t|\theta_0)\, dt$$

$$= \int_{-\infty}^{(k_1 - n\theta_0)/\sqrt{n}} y\, \frac{1}{\sqrt{2\pi}}\, e^{-(1/2)y^2}\, dy + \int_{(k_2 - n\theta_0)/\sqrt{n}}^{\infty} y\, \frac{1}{\sqrt{2\pi}}\, e^{-(1/2)y^2}\, dy.$$

$$= 0. \tag{10.37}$$

From Eq. (10.3.7) we conclude that

$$-\frac{k_1 - n\theta_0}{\sqrt{n}} = \frac{k_2 - n\theta_0}{\sqrt{n}}. \tag{10.38}$$

Now, from Eq. (10.3.6), we get

$$\alpha = \int_{-\infty}^{(k_1 - n\theta_0)/\sqrt{n}} \frac{1}{\sqrt{2\pi}}\, e^{-y^2/2}\, dy + \int_{(k_2 - n\theta_0)/\sqrt{n}}^{\infty} \frac{1}{\sqrt{2\pi}}\, e^{-y^2/2}\, dy$$

$$= \int_{-\infty}^{(k_2 - n\theta_0)/\sqrt{n}} \frac{1}{\sqrt{2\pi}}\, e^{-y^2/2}\, dy + \int_{(k_2 - n\theta_0)/\sqrt{n}}^{\infty} \frac{1}{\sqrt{2\pi}}\, e^{-y^2/2}\, dy$$

$$= 2 \int_{(k_2 - n\theta_0)/\sqrt{n}}^{\infty} \frac{1}{\sqrt{2\pi}}\, e^{-y^2/2}\, dy.$$

So the equal-tailed test, given by

$$\phi(x_1,\ldots,x_n) = \begin{cases} 1 & \text{if } |t - n\theta_0| \geq k_2 \\ 0 & \text{if } |t - n\theta_0| < k_2, \end{cases}$$

where k_2 is such that $E_{\theta_0}\phi(X_1,\ldots,X_n) = \alpha$ is uniformly most powerful unbiased of size α for testing H_0: $\theta = \theta_0$ against H_1: $\theta \neq \theta_0$.

Example 10.5.2 Let $X_1,\ldots,X_n$ be a random sample of size n from a normal distribution with known mean m_0 and unknown variance σ^2. Let us consider the problem of testing H_0: $\sigma^2 = \sigma_0^2$ against the alternatives H_1: $\sigma^2 \neq \sigma_0^2$. The joint probability density function of $X_1,\ldots,X_n$ is given by

$$\prod_{i=1}^{n} f_{X_i}(x_i|\sigma^2) = \frac{1}{(2\pi)^{n/2}(\sigma^2)^{n/2}} \exp\left[-\frac{1}{2\sigma^2} \sum_{i=1}^{n} (x_i - m_0)^2 \right].$$

Thus $\prod_{i=1}^{n} f_{X_i}(x_i|\sigma^2)$ with variance $\sigma^2 > 0$ is an exponential distribution with

$$C(x_1,\ldots,x_n) = (2\pi)^{-n/2}, \qquad K(\theta) = \left(\frac{1}{\sigma^2}\right)^{n/2}, \qquad Q(\theta) = -\frac{1}{2\sigma^2}$$

and

$$T(x_1,\ldots,x_n) = \sum_{i=1}^{n} (x_i - m_0)^2.$$

Let $Y = \sum_{i=1}^{n} (X_i - m_0)^2$. By Theorem 10.5.1, the uniformly most powerful unbiased test of size α of H_0 against H_1 is given by

$$\phi(x_1,\ldots,x_n) = \begin{cases} 1 & \text{if } y = \sum_{i=1}^{n} (x_i - m_0)^2 < k_1 \text{ or } > k_2 \\ \gamma_i & \text{if } y = k_i, \, i = 1,2 \\ 0 & \text{if } k_1 < y < k_2 \end{cases}$$

where the constants k_1, k_2, γ_1, γ_2 are determined by the condition

$$\begin{aligned} E_{\sigma_0^2}(\phi(X_1,\ldots,X_n)) &= \alpha \\ E_{\sigma_0^2}(Y\phi(X_1,\ldots,X_n)) &= \alpha E_{\sigma_0^2}(Y). \end{aligned} \tag{10.39}$$

Since $X_1,\ldots,X_n$ are independently normally distributed with mean m_0 and variance σ^2, Y/σ_0^2 has chi-square distribution with n degrees of freedom when H_0 is true. Hence

$$\begin{aligned} P_{\sigma^2}(Y = k_i) &= 0, \quad i = 1,2 \\ E_{\sigma_0^2}(Y) &= n\sigma_0^2 \end{aligned}$$

and the probability density function of Y under H_0 is given by

$$f_Y(y|\sigma_0^2) = \begin{cases} \dfrac{1}{2^{n/2}\Gamma(n/2)(\sigma_0^2)^{n/2}} \, y^{n/2-1} \exp\left[-\dfrac{y}{2\sigma_0^2}\right] & \text{if } y \geq 0 \\ 0 & \text{otherwise.} \end{cases} \tag{10.40}$$

The condition (10.39) reduces to

$$\int_{k_1}^{k_2} f_Y(y|\sigma_0^2) \, dy = 1 - \alpha \tag{10.41}$$

and

$$n\sigma_0^2 - \int_{k_1}^{k_2} y f_Y(y|\sigma_0^2) \, dy = \alpha n\sigma_0^2$$

or

$$\int_{k_1}^{k_2} \frac{1}{2^{(n+2)/2}\Gamma[(n+2)/2](\sigma_0^2)^{(n+2)/2}} \, y^{[(n+2)/2]-1} \, \exp\left(\frac{-y}{2\sigma_0^2}\right) \, dy$$

$$= \int_{k_1}^{k_2} f_Z(z|\sigma_0^2) \, dz$$

$$= 1 - \alpha, \tag{10.42}$$

where the probability density function $f_Z(z|\sigma_0^2)$ of the random variable Z is such that Z/σ_0^2 has chi-square distribution with $n+2$ degrees of freedom. Integrating by parts, we get, from Eq. (10.42),

$$\frac{1}{2^{n/2}\Gamma[(n+2)/2](\sigma_0^2)^{n/2}} \, e^{-y/2\sigma_0^2} y^{[(n+2)/2]-1} \Big|_{k_1}^{k_2}$$

$$+ \int_{k_1}^{k_2} \frac{1}{2^{n/2}\Gamma(n/2)(\sigma_0^2)^{n/2}} \, e^{-y/2\sigma_0^2} y^{(n/2)-1} = 1 - \alpha. \tag{10.43}$$

Thus, from Eqs. (10.41) and (10.43), we get

$$k_1^{n/2} \exp\left(-\frac{k_1}{2\sigma_0^2}\right) = k_2^{n/2} \exp\left(-\frac{k_2}{2\sigma_0^2}\right). \tag{10.44}$$

Hence the uniformly most powerful unbiased test of size α of H_0 against H_1 is given by

$$\phi(x_1,\ldots,x_n) = \begin{cases} 1 & \text{if } y \leq k_1 \text{ or } \geq k_2 \\ 0 & \text{if } k_1 < y < k_2, \end{cases} \tag{10.45}$$

where the constants k_1, k_2 are determined by Eqs. (10.41) and (10.44).

Let us now consider the case where n is sufficiently large that

$$U = \frac{Y - n\sigma_0^2}{\sqrt{2n\sigma_0^4}}$$

has approximately the normal distribution with mean 0 and variance 1 when H_0 is true. Since the normal distribution is symmetric about the mean, we conclude that

$$P_{\sigma_0^2}(Y - n\sigma_0^2 < -u) = P_{\sigma_0^2}(Y - n\sigma_0^2 > u)$$

for all real u. As in the example above, the equal-tailed test given by (10.45), where k_1, k_2 are such that

$$\int_0^{k_1} f_Y(y|\sigma_0^2) \, dy = \int_{k_2}^{\infty} f_Y(y|\sigma_0^2) \, dy = \frac{\alpha}{2}$$

is uniformly most powerful unbiased of size α for testing H_0 against H_1.

Example 10.5.3 Let $X_1,\ldots,X_n$ be a random sample of size n from a Poisson distribution with parameter θ. The joint probability mass function of $X_1,\ldots,X_n$ is given by

$$\prod_{i=1}^{n} p_{X_i}(x_i|\theta) = \frac{e^{-n\theta}\theta^{\sum_{i=1}^{n} x_i}}{\prod_{i=1}^{n} x_i}.$$

Hence the conditions of Theorem 10.5.1 are satisfied with

$$Q(\theta) = \log\theta, \qquad K(\theta) = e^{-n\theta}$$

$$C(x_1,\ldots,x_n) = \left(\prod_{i=1}^{n} x_i!\right)^{-1}$$

$$T(x_1,\ldots,x_n) = \sum_{i=1}^{n} x_i.$$

By Theorem 10.5.1 it follows that the uniformly most powerful unbiased test of size α for testing $H_0\colon \theta = \theta_0$ against $H_1\colon \theta \neq \theta_0$ is given by

$$\phi(x_1,\ldots,x_n) = \begin{cases} 1 & \text{if } \sum_{i=1}^{n} x_i < k_1 \text{ or } > k_2 \\ \gamma_i & \text{if } \sum_{i=1}^{n} x_i = k_i, \quad i = 1,2 \\ 0 & \text{if } k_1 < \sum_{i=1}^{n} x_i < k_2, \end{cases}$$

where $k_1, k_2, \gamma_1, \gamma_2$ are determined by

$$E_{\theta_0}\phi(X_1,\ldots,X_n) = \alpha \tag{10.46}$$

$$E_{\theta_0}(T(X_1,\ldots,X_n)\phi(X_1,\ldots,X_n)) = \alpha E_{\theta_0}(T(X_1,\ldots,X_n)).$$

Since $\sum_{i=1}^{n} X_i$ has Poisson distribution with mean $n\theta_0$ under H_0, Eq. (10.46) becomes

$$\sum_{j=0}^{k_1-1} \frac{e^{-n\theta_0}(n\theta_0)^j}{j!} + \sum_{j=k_2+1}^{\infty} \frac{e^{-n\theta_0}(n\theta_0)^j}{j!} + \frac{\gamma_1 e^{-n\theta_0}(n\theta_0)^{k_1}}{k_1!}$$

$$+ \frac{\gamma_2 e^{-n\theta_0}(n\theta_0)^{k_2}}{k_2!} = \alpha \tag{10.47a}$$

$$\sum_{j=0}^{k_1-1} j\frac{e^{-n\theta_0}(n\theta_0)^j}{j!} + \sum_{j=k_2+1}^{\infty} j\frac{e^{-n\theta_0}(n\theta_0)^j}{j!} + \frac{\gamma_1 k_1 e^{-n\theta_0}(n\theta_0)^{k_1}}{k_1!}$$

$$+ \frac{\gamma_2 k_2 e^{-n\theta_0}(n\theta_0)^{k_2}}{k_2!} = n\theta_0\alpha \tag{10.47b}$$

In general, these two equations are not very easy to solve. However, in cases where n is not very small and θ_0 is not close to zero,

$$U = \frac{\sum_{i=1}^{n} X_i - n\theta_0}{\sqrt{n\theta_0}}$$

is approximately normally distributed with mean 0 and variance 1 when H_0 is true and the distribution of U is approximately symmetric about zero. As we have seen in the examples above, the equal-tail test where k_1, k_2, γ_1, γ_2 are such that

$$\sum_{j=0}^{k_1-1} \frac{e^{-n\theta_0}(n\theta_0)^j}{j!} + \gamma_1 \frac{e^{-n\theta_0}(n\theta_0)^{k_1}}{k_1!} = \sum_{j=k_2+1}^{\infty} \frac{e^{-n\theta_0}(n\theta_0)^j}{j!} + \gamma_2 \frac{e^{-n\theta_0}(n\theta_0)^{k_2}}{k_2!}$$

$$= \frac{\alpha}{2}$$

is approximately uniformly most powerful unbiased of size α for testing H_0 against H_1.

If $n = 10$, $\theta_0 = 1$, and $\alpha = 0.05$, we get

$$\sum_{j=0}^{3} \frac{e^{-10}(10)^j}{j!} = 0.010, \qquad \sum_{j=0}^{4} \frac{e^{-10}(10)^j}{j!} = 0.029,$$

$$\sum_{j=0}^{17} \frac{e^{-10}(10)^j}{j!} = 0.986, \qquad \sum_{j=0}^{16} \frac{e^{-10}(10)^j}{j!} = 0.973.$$

Thus $k_1 = 4$, $k_2 = 16$, $\gamma_1 = 0.789$, and $\gamma_2 = 0.846$.

10.6 TESTS OF COMPOSITE H_0 AGAINST COMPOSITE H_1 (MULTIPARAMETER CASE): LIKELIHOOD-RATIO TESTS

In this section we consider random phenomena the distribution of which depends on more than one parameter, where the null hypothesis or the alternative hypothesis or both are composite. Let $X_1,...,X_n$ be a random sample of size n from the random phenomenon specified by the probability density function $f_X(x|\theta)$ [or the probability mass function $p_X(x|\theta)$ in the discrete case], where $\theta = (\theta_1,...,\theta_p) \in \Omega$. The likelihood of the observed sample $(x_1,...,x_n)$ on $(X_1,...,X_n)$ is

$$L(x_1,...,x_n|\theta) = \prod_{i=1}^{n} f_{X_i}(x_i|\theta) \quad \left[\text{or} \quad \prod_{i=1}^{n} p_{X_i}(x_i|\theta) \right]. \tag{10.48}$$

Given $x_1,\ldots,x_n$, $L(x_1,\ldots,x_n|\theta)$ is a function of θ only and we shall denote it simply by $L(\theta)$. Consider the problem of testing H_0: $\theta \in \omega \subset \Omega$ against the alternatives H_1: $\theta \in (\Omega - \omega)$. $L(\theta)$ will ordinarily have a maximum as θ is allowed to vary over the subspace ω and let $L(\hat{\omega})$ denote the maximum value of $L(\theta)$ when θ is restricted to ω. Similarly, let $L(\hat{\Omega})$ be the maximum value of $L(\theta)$ when θ may take any value in Ω (union of H_0 and H_1). The ratio

$$\lambda = \frac{L(\hat{\omega})}{L(\hat{\Omega})} \tag{10.49}$$

is called the *likelihood ratio*. Since $\omega \subset \Omega$, for any given sample $(x_1,\ldots,x_n)$, $L(\hat{\omega}) \leq L(\hat{\Omega})$. Since $L(\theta)$, $\theta \in \Omega$, is the product of probability density functions, $\lambda \geq 0$. The ratio λ is a function of sample observations $x_1,\ldots,x_n$ and it does not depend on θ. So $\lambda(X_1,\ldots,X_n)$ can be used as a test statistic. Let λ_0 be a positive constant. The test that rejects H_0 whenever

$$\lambda \leq \lambda_0, \tag{10.50}$$

where λ_0 is chosen in such a way that the test has size α, is called the *likelihood-ratio test* (of size α) of H_0.

The likelihood-ratio test principle is purely an intuitive one. In cases where both H_0 and H_1 are simple it leads to the best test as given by Neyman–Pearson theory. Furthermore, for many classical problems, likelihood-ratio tests have been shown to possess some optimum properties when the sample size n tends to infinity. For scalar random variables, the sample sizes generally encountered in practice are usually large enough for these results to hold true. But for a vector random variable with dimension p the situation is not quite the same. There are cases where the sample size n must be such that n/p^3 is required to be large enough for this result to hold.

To determine the critical region of the likelihood-ratio test we need to know the distribution of $\lambda(X_1,\ldots,X_n)$ under the null hypothesis H_0. This, in turn, will depend on the nature of H_0. If the null hypothesis is composite the distribution of λ may not always be unique: It may be different for different points in ω. Thus, to determine λ_0 uniquely in such cases, we must impose extra conditions on tests (invariance, similarity, etc.). We shall not be concerned with these problems here; instead, we shall consider only problems where λ_0 can be obtained uniquely.

It may also be added that in cases where λ_0 cannot be determined uniquely, a very satisfactory solution of the testing problem can be obtained by taking the sample size n sufficiently large. For large n, under the null hypothesis, $-2 \log \lambda(X_1,\ldots,X_n)$ has approximately chi-squre distribution with $k - r$ degrees of freedom, k and r being the dimensions of Ω and ω, respectively. Since $-2 \log \lambda$ increases as λ decreases, the critical region of the null hypothesis H_0 in terms of $-2 \log \lambda$ will be given by

$$-2 \log \lambda(x_1,\ldots,x_n) \geq C \tag{10.51}$$

where the constant C is determined by

$$P(\chi^2_{k-r} \geq C) = \alpha.$$

10.6.1 Student's *t* Test

Let $X_1,\ldots,X_n$ be a random sample of size n from a random phenomenon specified by the normal distribution with unknown mean μ and unknown variance σ^2. Consider the problem of testing $H_0: \mu = \mu_0$ against $H_1: \mu \neq \mu_0$. The likelihood of the sample of observations $x_1,\ldots,x_n$ on $X_1,\ldots,X_n$ is given by

$$L(\theta) = L(x_1,\ldots,x_n|\theta)$$
$$= \frac{1}{(2\pi)^{n/2}(\sigma^2)^{n/2}} \exp\left[-\frac{1}{2\sigma^2} \sum_{i=1}^{n} (x_i - \mu)^2\right],$$

where $\theta = (\mu,\sigma^2)$. Here

$$\omega = \{\theta = (\mu_0,\sigma^2) : \sigma^2 > 0\}$$
$$\Omega = \{\theta = (\mu,\sigma^2) : -\infty < \mu < \infty, \sigma^2 > 0\}.$$

For $\theta \in \omega$ we have

$$L(\theta) = \frac{1}{(2\pi)^{n/2}(\sigma^2)^{n/2}} \exp\left[-\frac{1}{2\sigma^2} \sum_{i=1}^{n} (x_i - \mu_0)^2\right].$$

It can easily be verified that $L(\theta)$, $\theta \in \omega$, is maximum if

$$\sigma^2 = \sum_{i=1}^{n} \frac{(x_i - \mu_0)^2}{n}.$$

Thus

$$L(\hat{\omega}) = \frac{1}{(2\pi)^{n/2}} \left[\frac{n}{\sum_{i=1}^{n} (x_i - \mu_0)^2}\right]^{n/2} e^{-n/2}.$$

Similarly, for $\theta \in \Omega$ we have

$$L(\theta) = \frac{1}{(2\pi)^{n/2}(\sigma^2)^{n/2}} \exp\left[-\frac{1}{2\sigma^2} \sum_{i=1}^{n} (x_i - m)^2\right]$$

and $L(\theta)$ is maximum when

$$m = \frac{1}{n} \sum_{i=1}^{n} x_i = \bar{x}, \qquad \sigma^2 = \frac{1}{n} \sum_{i=1}^{n} (x_i - \bar{x})^2 = \frac{s^2}{n}.$$

Thus we have

$$L(\hat{\Omega}) = \frac{1}{(2\pi)^{n/2}} \left(\frac{n}{s^2}\right)^{n/2} e^{-n/2},$$

so that

$$\lambda = \frac{L(\hat{\omega})}{L(\hat{\Omega})}$$

$$= \left[\frac{1}{1 + [y^2/(n-1)]}\right]^{n/2},$$

where

$$y = \frac{\sqrt{n}\,(\bar{x} - \mu_0)}{\sqrt{s^2/(n-1)}}.$$

We may now recall (Corollary 6.2.1.1) that when H_0 is true,

$$Y = \frac{\sqrt{n}\,(\bar{X} - \mu_0)}{\sqrt{S^2/(n-1)}},$$

where

$$\bar{X} = \frac{1}{n} \sum_{i=1}^{n} X_i, \qquad S^2 = \sum_{i=1}^{n} (X_i - \bar{X})^2,$$

has central t distribution with $n - 1$ degrees of freedom, that is,

$$f_Y(y) = \begin{cases} \dfrac{1}{\sqrt{(n-1)\pi}} \dfrac{\Gamma(n/2)}{\Gamma[(n-1)/2]} \left(1 + \dfrac{y^2}{n-1}\right)^{-n/2} \\ \qquad \text{if } -\infty < y < \infty \\ 0 \qquad \text{otherwise.} \end{cases} \tag{10.52}$$

Since λ is a monotonically decreasing function of $y^2, \lambda < \lambda_0$ implies that $y^2 \geq C$ for some constant C. For testing H_0 against H_1, the critical region of the likelihood-ratio test of size α is given by

$$\phi(x_1,\ldots,x_n) = \begin{cases} 1 & \text{if } y \leq -C \text{ or } y \geq C \\ 0 & \text{if } -C < y < C \end{cases}$$

where C is chosen such that $P(Y \le -C) + P(Y \ge C) = \alpha$. From Eq. (10.52) it may be verified that $P(Y \le -C) = 1 - P(Y \le C) = P(Y \ge C)$. Thus the constant C can be determined by $P(Y \ge C) = \alpha/2$. For example, if $n = 25$, $\alpha = 0.05$, then $C = 2.064$.

10.6.2 Tests of Equality of Two Means

Let $X_1,\ldots,X_m$ be a random sample of size m from a normal distribution with unknown mean μ_1 and unknown variance σ_1^2 and let $Y_1,\ldots,Y_n$ be a random sample of size n (independent of $X_1,\ldots,X_m$) from another normal distribution with unknown mean μ_2 and unknown variance σ_2^2. Consider the problem of testing $H_0\colon \mu_1 = \mu_2 = \mu$ against $H_1\colon \mu_1 \ne \mu_2$. Here

$$\Omega = \{\Omega = (\mu_1,\mu_2,\sigma_1^2,\sigma_2^2) : -\infty < \mu_i < \infty, \ \sigma_i^2 > 0, \ i = 1,2\}$$

$$\omega = \{\Omega = (\mu,\sigma_1^2,\sigma_2^2) : -\infty < \mu < \infty, \ \sigma_i^2 > 0, \ i = 1,2\}.$$

The likelihood of the sample observations $x_1,\ldots,x_m,\ y_1,\ldots,y_n$ on $X_1,\ldots,X_m,\ Y_1,\ldots,Y_n$ is given by

$$L(x_1,\ldots,x_m,\ y_1,\ldots,y_n|\theta) = L(\theta)$$

$$= \frac{1}{(2\pi)^{(m+n)/2}(\sigma_1^2)^{m/2}(\sigma_2^2)^{n/2}} \exp\left\{-\frac{1}{2} \times \left[\sum_{i=1}^{m} \frac{(x_i - \mu_1)^2}{\sigma_1^2} + \sum_{i=1}^{n} \frac{(y_i - \mu_2)^2}{\sigma_2^2}\right]\right\}.$$

Let

$$\bar{x} = \frac{1}{m}\sum_{i=1}^{m} x_i, \qquad \bar{y} = \frac{1}{n}\sum_{i=1}^{n} y_i$$

$$s_1^2 = \sum_{i=1}^{m} (x_i - \bar{x})^2, \qquad s_2^2 = \sum_{i=1}^{n} (y_i - \bar{y})^2$$

$$\bar{X} = \frac{1}{m}\sum_{i=1}^{m} X_i, \qquad \bar{Y} = \frac{1}{n}\sum_{i=1}^{n} Y_i$$

$$S_1^2 = \sum_{i=1}^{m} (X_i - \bar{X})^2, \qquad S_2^2 = \frac{1}{n}\sum_{i=1}^{n} (Y_i - \bar{Y})^2.$$

The maximum of $L(\theta)$, $\theta \in \Omega$ can be easily seen to be

$$L(\hat{\Omega}) = \frac{1}{(2\pi)^{(m+n)/2}} \left(\frac{m}{s_1^2}\right)^{m/2} \left(\frac{n}{s_2^2}\right)^{n/2} e^{-(m+n)/2}.$$

For $\theta \in \omega$, we have

$$L(\theta) = \frac{1}{(2\pi)^{(m+n)/2}(\sigma_1^2)^{m/2}(\sigma_2^2)^{n/2}}$$

$$\times \exp\left\{-\frac{1}{2}\left[\sum_{i=1}^{m}\frac{(x_i - \mu)^2}{\sigma_1^2} + \sum_{i=1}^{n}\frac{(y_i - \mu)^2}{\sigma_2^2}\right]\right\}.$$

Differentiating $L(\theta)$ with respect to μ, σ_1^2, σ_2^2 we conclude that $L(\theta)$ is maximum when μ, σ_1^2, σ_2^2 satisfy

$$\bar{x} - \mu + \frac{n\sigma_1^2}{m\sigma_2^2}(\bar{y} - \mu) = 0$$

$$\sigma_1^2 = \frac{1}{m}\sum_{i=1}^{m}(x_i - \mu)^2, \qquad \sigma_2^2 = \frac{1}{n}\sum_{i=1}^{n}(y_i - \mu)^2. \tag{10.53a}$$

From above we get

$$m^2(\bar{x} - \mu)[s_2^2 + n(\bar{y} - \mu)^2] + n^2(\bar{y} - \mu)[s_1^2 + m(\bar{x} - \mu)^2] = 0, \tag{10.53b}$$

which is a cubic equation in μ. Thus the maximum-likelihood estimate of μ is given by the roots of a cubic equation and is a complex function of the observations. Obviously, in this case the likelihood-ratio test is not very suitable for practical use. We shall now consider some special cases that are amenable to simpler solutions.

Case 1: $\sigma_1^2 = \sigma_2^2 = \sigma^2$

Let us now redefine Ω, ω as follows:

$$\Omega = \{\theta = (\mu_1, \mu_2, \sigma^2) : -\infty < \mu_i < \infty, i = 1,2; \sigma^2 > 0\}$$

$$\omega = \{\theta = (\mu, \sigma^2) : -\infty < \mu < \infty, \sigma^2 > 0\}.$$

For $\theta \in \Omega$, we have

$$L(\theta) = \frac{1}{(2\pi)^{(m+n)/2}(\sigma^2)^{(m+n)/2}}$$

$$\times \exp\left\{-\frac{1}{2\sigma^2}\left[\sum_{i=1}^{m}(x_i - \mu_1)^2 + \sum_{i=1}^{n}(y_i - \mu_2)^2\right]\right\}.$$

The maximum of $L(\theta)$ when θ varies in Ω can easily be seen to be

$$L(\hat{\Omega}) = \frac{1}{(2\pi)^{m+n/2}}\left(\frac{m + n}{s_1^2 + s_2^2}\right)e^{-(m+n)/2}.$$

For $\theta \in \omega$, we have

$$L(\theta) = \frac{1}{(2\pi)^{(m+n)/2}(\sigma^2)^{(m+n)/2}}$$

$$\times \exp\left\{-\frac{1}{2\sigma^2}\left[\sum_{i=1}^{m}(x_i - \mu)^2 + \sum_{i=1}^{n}(y_i - \mu)^2\right]\right\},$$

so the maximum-likelihood estimates of μ and σ^2 are given by

$$\mu = \hat{\mu} = \frac{m\bar{x} + n\bar{y}}{m + n}$$

$$\sigma^2 = \hat{\sigma}^2 = \frac{\sum_{i=1}^{m}(x_i - \hat{\mu})^2 + \sum_{i=1}^{n}(y_i - \hat{\mu})^2}{m + n}$$

$$= \frac{1}{m + n}\left[s_1^2 + s_2^2 + \frac{mn}{m + n}(\bar{x} - \bar{y})^2\right].$$

Hence we get

$$L(\hat{\omega}) = \frac{1}{(2\pi)^{(m+n)/2}}\left[\frac{m + n}{s_1^2 + s_2^2 + [mn/(m + n)](\bar{x} - \bar{y})^2}\right]^{(m+n)/2} e^{-(m+n)/2}.$$

Thus the likelihood ratio

$$\lambda = \left(\frac{1}{1 + \dfrac{[mn/(m + n)](\bar{x} - \bar{y})^2}{s_1^2 + s_2^2}}\right)^{(m+n)/2}$$

$$= \left(\frac{1}{1 + y^2/(m + n - 2)}\right)^{(m+n)/2}, \tag{10.54}$$

where

$$y^2 = \frac{(\bar{x} - \bar{y})^2/(1/m + 1/n)}{(s_1^2 + s_2^2)/(m + n - 2)}.$$

Let

$$Y = \frac{(\bar{X} - \bar{Y})/(1/m + 1/n)^{1/2}\sigma}{[(S_1^2 + S_2^2)/(m + n - 2)\sigma^2]^{1/2}}$$

$$= \frac{(\bar{X} - \bar{Y})/(1/m + 1/n)^{1/2}}{[(S_1^2 + S_2^2)/(m + n - 2)]^{1/2}}.$$

When the null hypothesis is true, $\bar{X} - \bar{Y}$ is normally distributed with mean 0 and variance $\sigma^2[(1/m) + (1/n)]$. Furthermore, S_1^2/σ^2, S_2^2/σ^2 are independently distributed chi-square random variables with $m - 1$, $n - 1$

degrees of freedom and $\overline{X} - \overline{Y}$ is independent of S_1^2, S_2^2. Thus, by Corollary 6.2.1.1.1, Y has a t distribution with $m + n - 2$ degrees of freedom. The probability density function of Y (under H_0) is given by

$$
f_Y(y) = \begin{cases} \dfrac{1}{\sqrt{(m + n - 2)}} \dfrac{\Gamma[(m + n - 1)/2]}{\pi\Gamma[(m + n - 2)/2]} \\ \quad \times \left(1 + \dfrac{y^2}{m + n - 2}\right)^{-(m+n-1)/2} & \text{if } -\infty < y < \infty \\ 0 & \text{otherwise.} \end{cases}
$$

$$(10.55)$$

From Eq. (10.54) it follows that the likelihood-ratio test of H_0: $\mu_1 = \mu_2$ against H_1: $\mu_1 \neq \mu_2$ is given by

$$
\phi(x_1,\ldots,x_n) = \begin{cases} 1 & \text{if } y \leq -C \text{ or } \geq C \\ 0 & \text{if } -C < y < C, \end{cases}
$$

where C is chosen in such a way that under H_0, $P(Y \leq -C) + P(Y \geq C) = \alpha$. Since from Eq. (10.55) we have $(P \leq -C) = 1 - P(Y \leq C) = P(Y \geq C)$ we can choose C such that $P(Y \geq C) = \alpha/2$.

Case II: *Paired t Test:* $\sigma_1^2 \neq \sigma_2^2$ but $m = n$

When $\sigma_1^2 \neq \sigma_2^2$ but $m = n$, a suitable solution is reached by using the following pairing device. Define

$$Z_i = X_i - Y_i, \qquad i = 1,\ldots,m.$$

Obviously, $Z_1,\ldots,Z_m$ constitute a random sample of size m from a normal distribution with mean $\mu_1 - \mu_2 = \nu$ and variance $\sigma_1^2 + \sigma_2^2 = \sigma^2$. In terms of ν and σ^2, the testing problem reduces to that of testing H_0: $\nu = 0$ against H_1: $\nu \neq 0$ when σ^2 is unknown. From Section 10.6.1 we can conclude the following: On the basis of a sample of observations $z_1,\ldots,z_m$ on $Z_1,\ldots,Z_m$, the likelihood-ratio test of size α of H_0: $\nu = 0$ against H_1: $\nu \neq 0$ when σ^2 is unknown is given by

$$
\phi(z_1,\ldots,z_m) = \begin{cases} 1 & \text{if } y \leq -C \text{ or } \geq C \\ 0 & \text{if } -C < y < C \end{cases} \qquad (10.56)
$$

where

$$
y = \frac{\sqrt{m}\,\overline{z}}{\{\Sigma_{i=1}^m\,[(z_i - \overline{z})^2/(m - 1)]\}^{1/2}}, \qquad \overline{z} = \frac{1}{m}\sum_{i=1}^m z_i
$$

and under H_0,

$$Y = \frac{\sqrt{m}\,\overline{Z}}{\{\sum_{i=1}^{m} [(Z_i - \overline{Z})^2/(m-1)]\}^{1/2}}, \qquad \overline{Z} = \frac{1}{m} \sum_{i=1}^{m} Z_i$$

has central Student's t distribution with $m - 1$ degrees of freedom, and the constant C is chosen such that $P(Y \geq C) = \alpha/2$ when H_0 is true.

Case III: *Scheffé's Solution for* $\sigma_1^2 \neq \sigma_2^2$, $m \neq n$

Assume that $m < n$. Define

$$Z_i = Z_i - \left(\frac{m}{n}\right)^{1/2} Y_i + \frac{1}{\sqrt{mn}} \sum_{j=1}^{m} Y_j - \frac{1}{n} \sum_{j=1}^{n} Y_j, \qquad i = 1,\ldots,m.$$

Then

$$E(Z_i) = \mu_1 - \left(\frac{m}{n}\right)^{1/2} \mu_2 + \frac{1}{\sqrt{mn}}\, m\mu_2 - \frac{n\mu_2}{n}$$

$$= \mu_1 - \mu_2$$

$$\text{var}(Z_i) = E\left[X_i - \mu_1 - \left(\frac{m}{n}\right)^{1/2} (Y_i - \mu_2) + \frac{1}{\sqrt{mn}} \sum_{i=1}^{m} (Y_i - \mu_2) \right.$$

$$\left. - \frac{1}{n} \sum_{i=1}^{n} (Y_i - \mu_2) \right]^2$$

$$= \sigma_1^2 + \frac{m}{n}\,\sigma_2^2 + \frac{2\sigma_2^2}{n} - \frac{2}{n}\,\sigma_2^2 + \frac{2\sqrt{m}}{n^{3/2}}\,\sigma_2^2 - \frac{2\sqrt{m}}{n^{3/2}}\,\sigma_2^2$$

$$= \sigma_1^2 + \frac{m}{n}\,\sigma_2^2,$$

and similarly,

$$\text{cov}(Z_i, Z_j) = 0, \qquad i \neq j.$$

Thus $Z_1,\ldots,Z_m$ constitute a random sample of size m from a random phenomenon specified by normal distribution with mean $\nu = \mu_1 - \mu_2$ and variance $\sigma_1^2 + (m/n)\sigma_2^2$. From Section 10.5.1 it follows that on the basis of sample observations $z_1,\ldots,z_m$, the likelihood-ratio test of $H_0: \nu = 0$ against $H_1: \nu \neq 0$ is given by Eq. (10.56).

10.6.3 Chi-Square Test

Let $X_1,\ldots,X_n$ be a random sample of size n from a normal distribution with unknown mean μ and unknown variance σ^2. Consider the problem of testing H_0: $\sigma^2 = \sigma_0^2$ (specified) against H_1: $\sigma^2 \neq \sigma_0^2$. Here

$$\Omega = \{\theta = (\mu,\sigma^2) : -\infty < \mu < \infty; \sigma^2 > 0\}$$

$$\omega = \{\theta = (\mu,\sigma_0^2) : -\infty < \mu < \infty\}.$$

The likelihood of the sample observations $x_1,\ldots,x_n$ on $X_1,\ldots,X_n$ is given by

$$L(x_1,\ldots,x_n|\theta) = L(\theta) = \frac{1}{(2\pi)^{n/2}(\sigma^2)^{n/2}} \exp\left[-\frac{1}{2\sigma^2}\sum_{i=1}^{n}(x_i - \mu)^2\right].$$

It is easy to verify that

$$L(\hat{\Omega}) = \frac{1}{(2\pi)^{n/2}} \left[\frac{n}{\sum_{i=1}^{n}(x_i - \bar{x})^2}\right]^{n/2} e^{-n/2}$$

$$L(\hat{\omega}) = \frac{1}{(2\pi)^{n/2}(\sigma_0^2)^{n/2}} \exp\left[-\frac{1}{2\sigma_0^2}\sum_{i=1}^{n}(x_i - \bar{x})^2\right]$$

where $\bar{x} = (1/n)\sum_{i=1}^{n} x_i$. Let

$$y = \sum_{i=1}^{n}\frac{(x_i - \bar{x})^2}{\sigma_0^2}, \qquad Y = \sum_{i=1}^{n}\frac{(X_i - \bar{X})^2}{\sigma_0^2}, \qquad \bar{X} = \frac{1}{n}\sum_{i=1}^{n} X_i.$$

Then the likelihood ratio for testing H_0 against H_1 is given by

$$\lambda = \frac{L(\hat{\omega})}{L(\hat{\Omega})}$$

$$= \left(\frac{y}{n}\right)^{n/2} e^{-(1/2)(y-n)}.$$

On plotting λ as a function of y, it is apparent (Figure 10.7) that the inequality $\lambda \leq \lambda_0$ is equivalent to

$$y \leq k_1 \quad \text{or} \quad y \geq k_2$$

where k_1, k_2 are values of y such that $\lambda = \lambda_0$, that is,

$$k_1^{n/2}e^{-k_1/2} = k_2^{n/2}e^{-k_2/2}. \tag{10.57}$$

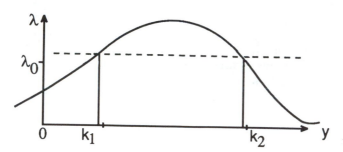

Figure 10.7

By Theorem 6.3.1.1,

$$Y = \sum_{i=1}^{n} \frac{(X_i - \overline{X})^2}{\sigma_0^2},$$

where $\overline{X} = (1/n) \sum_{i=1}^{n} X_i$ has central chi-square distribution with $n - 1$ degrees of freedom when H_0 is true.

Thus the likelihood-ratio test of size α of H_0 against H_1 is equivalent to

$$\phi(x_1,...,x_n) = \begin{cases} 1 & \text{if } y \le k_1 \text{ or } \ge k_2 \\ 0 & \text{if } k_1 < y < k_2, \end{cases}$$

where k_1, k_2 are constants satisfying Eq. (10.57) and

$$\int_{k_1}^{k_2} \frac{1}{2[(n-1)/2]\Gamma[(n-1)/2]} y^{(n-1)/2-1}e^{-y/2} \, dy = 1 - \alpha.$$

From Example 10.5.2 it follows that if n is large, we can choose k_1, k_2 such that

$$\int_0^{k_1} \frac{1}{2[(n-1)/2]\Gamma[(n-1)/2]}$$
$$\times \, y^{(n-1)/2-1}e^{-y/2} \, dy = \int_{k_2}^{\infty} \frac{1}{2[(n-1)/2]\Gamma[(n-1)/2]} y^{(n-1)/2-1}e^{-y/2}$$
$$= \frac{\alpha}{2}.$$

10.6.4 F Test

Let $X_1,...,X_m$ be a random sample of size m from a random phenomenon specified by the normal distribution with unknown mean μ_1 and unknown variance σ_1^2 and let $Y_1,...,Y_n$ be a random sample of size n (independent of $X_1,...,X_m$) from another random phenomenon specified by the normal distribution with unknown mean μ_2 and unknown variance σ_2^2. Consider the problem of testing H_0: $\sigma_1^2 = \sigma_2^2 = \sigma^2$ against the alternative H_1: $\sigma_1^2 \neq \sigma_2^2$. Here

$$\Omega = \{\theta = (\mu_1,\mu_2,\sigma_1^2,\sigma_2^2) : -\infty < \mu_i < \infty, \ \sigma_i^2 > 0, \ i = 1,2\};$$

$$\omega = \{\theta = (\mu_1,\mu_2,\sigma^2) : -\infty < \mu_i < \infty, \ i = 1,2; \ \sigma^2 > 0\}.$$

The likelihood of the sample observations $x_1,\ldots,x_m$; $y_1,\ldots,y_n$ is given by

$$L(\theta) = \frac{1}{(2\pi)^{(m+n)/2}(\sigma_1^2)^{m/2}(\sigma_2^2)^{n/2}}$$

$$\times \exp\left\{-\frac{1}{2}\left[\sum_{i=1}^{n}\frac{(x_i - \mu_1)^2}{\sigma_1^2} + \sum_{i=1}^{n}\frac{(y_i - \mu_2)^2}{\sigma_2^2}\right]\right\}.$$

By Section 10.6.2 we have

$$L(\hat\Omega) = \frac{1}{(2\pi)^{(m+n)/2}}\left(\frac{m}{s_1^2}\right)^{m/2}\left(\frac{n}{s_2^2}\right)^{n/2}e^{-(m+n)/2}.$$

Furthermore, $L(\theta)$, $\theta \in \omega$, is maximum when

$$\mu_1 = \bar{x}, \qquad \mu_2 = \bar{y}, \qquad \sigma^2 = \frac{s_1^2 + s_2^2}{m + n},$$

so that

$$L(\hat\omega) = \frac{1}{(2\pi)^{(m+n)/2}}\left(\frac{m + n}{s_1^2 + s_2^2}\right)^{(m+n)/2}e^{-(m+n)/2}.$$

Hence the likelihood ratio for testing H_0 against H_1 is given by

$$\lambda = \frac{(m + n)^{(m+n)/2}}{m^{m/2}n^{n/2}}\frac{\{[(m - 1)/(n - 1)]f\}^{m/2}}{\{1 + [(m - 1)/(n - 1)]f\}^{(m+n)/2}}$$

where

$$f = \frac{(n - 1)s_1^2}{(m - 1)s_2^2}.$$

Let us define

$$F = \frac{(n - 1)S_1^2}{(m - 1)S_2^2}.$$

Under H_0 we know that S_1^2/σ^2, S_2^2/σ^2 are independently distributed central chi-square random variables with $m - 1$ and $n - 1$ degrees of freedom, respectively. Hence by Example 3.10.2, the random variable F has central F distribution with $m - 1$, $n - 1$ degrees of freedom. Its probability density function is given by

$$f_F(f) = \begin{cases} \dfrac{\Gamma\left(\dfrac{m + n - 2}{2}\right)}{\Gamma\left(\dfrac{m - 1}{2}\right)\Gamma\left(\dfrac{n - 1}{2}\right)} \left(\dfrac{m - 1}{n - 1}\right)^{(m-1)/2} \\ \quad \times \dfrac{(f)^{[(m-1)/2]-1}}{\left[1 + \dfrac{(m - 1)f}{n - 1}\right]^{(m+n-2)/2}} \quad \text{if } 0 < f < \infty \\ 0 \qquad\qquad\qquad\qquad\qquad\qquad \text{otherwise.} \end{cases} \qquad (10.58)$$

Now, plotting λ as a function of f it is apparent that the inequality $\lambda \le \lambda_0$ is equivalent to $f \le k_1$ or $f \ge k_2$, where k_1, k_2 are values of f such that $\lambda = \lambda_0$, that is,

$$\left(\frac{m - 1}{n - 1} k_1\right)^{m/2}\left(1 + \frac{m - 1}{n - 1} k_2\right)^{(m+n)/2}$$

$$= \left(\frac{m - 1}{n - 1} k_2\right)^{n/2}\left(1 + \frac{m - 1}{n - 1} k_1\right)^{(m+n)/2} \qquad (10.59)$$

Hence the likelihood-ratio test for testing H_0 against H_1 is given by

$$\phi(x_1,\ldots,x_m;y_1,\ldots,y_n) = \begin{cases} 1 & \text{if } f \le k_1 \text{ or } f \ge k_2 \\ 0 & \text{if } k_1 < f < k_2, \end{cases}$$

where k_1, k_2 are constants satisfying Eq. (10.59) and

$$\int_0^{k_1} f_F(f)\, df + \int_{k_2}^{\infty} f_F(f)\, df = \alpha.$$

It is customary to choose k_1, k_2 such that

$$\int_0^{k_1} f_F(f)\, df = \int_{k_2}^{\infty} f_F(f)\, df = \frac{\alpha}{2}. \qquad (10.60)$$

This is mainly because the existing statistical tables of the F distribution make it easy to determine k_1, k_2 satisfying Eq. (10.60). No doubt this choice of k_1, k_2 does not necessarily lead to the best test.

10.7 TEST OF ZERO CORRELATION

Let $(X_1, Y_1), \ldots, (X_n, Y_n)$ be a random sample of size n from a random phenomenon specified by the bivariate normal probability density function

$$f_{X,Y}(x,y|\theta) = \frac{1}{(2\pi)\sigma_1\sigma_2(1-\rho^2)^{1/2}} \exp\left\{-\frac{1}{2(1-\rho)^2}\left[\left(\frac{x-\mu_1}{\sigma_1}\right)^2\right.\right.$$
$$\left.\left. + \left(\frac{y-\mu_2}{\sigma_2}\right)^2 - 2\rho\left(\frac{x-\mu_1}{\sigma_1}\right)\left(\frac{y-\mu_2}{\sigma_2}\right)\right]\right\}$$

for $-\infty < x < \infty$, $-\infty < y < \infty$, (10.61)

where $\theta = (\mu_1, \mu_2, \sigma_1^2, \sigma_2^2, \rho)$. Consider the problem of testing H_0: $\rho = 0$ against the alternatives H_1: $\rho \neq 0$. Here

$$\Omega = \{\theta = (\mu_1, \mu_2, \sigma_1^2, \sigma_2^2, \rho) : -\infty < \mu_i < \infty, \ \sigma_i^2 > 0, \ -1 < \rho < 1;$$
$$i = 1,2\}$$
$$\omega = \{\theta = (\mu_1, \mu_2, \sigma_2^2, \sigma_1^2, 0) : -\infty < \mu_i < \infty, \ \sigma_i^2 > 0; \ i = 1,2\}.$$

The likelihood of the observed sample $(x_1, y_1), \ldots, (x_n, y_n)$ is given by

$$L(\theta) = \frac{1}{(2\pi)^n[\sigma_1^2\sigma_2^2(1-\rho^2)]^{n/2}} \exp\left\{-\frac{1}{2(1-\rho^2)}\left[\sum_{i=1}^{n}\frac{(x_i-\mu_1)^2}{\sigma_1^2}\right.\right.$$
$$\left.\left. + \sum_{i=1}^{n}\frac{(y_i-\mu_2)^2}{\sigma_2^2} - 2\rho\sum_{i=1}^{n}\frac{(x_i-\mu_1)(y_i-\mu_2)}{\sigma_1\sigma_2}\right]\right\}$$
$$= \frac{1}{(2\pi)^n(\sigma_1^2\sigma_2^2(1-\rho^2))^{n/2}} \exp\left[-\frac{1}{2(1-\rho^2)}\right.$$
$$\times \left.\left(\frac{s_1^2}{\sigma_1^2} + \frac{s_2^2}{\sigma_2^2} - 2\rho\frac{s_{12}}{\sigma_1\sigma_2}\right)\right]\exp\left[-\frac{1}{2(1-\rho^2)}\left(\frac{n(\bar{x}-\mu_1)^2}{\sigma_1^2}\right.\right.$$
$$\left.\left. + \frac{n(\bar{y}-\mu_2)^2}{\sigma_2^2} - 2\rho\frac{n(\bar{x}-\mu_1)(\bar{y}-\mu_2)}{\sigma_1\sigma_2}\right)\right],$$

where

$$s_1^2 = \sum_{i=1}^{n}(x_i-\bar{x})^2, \qquad s_2^2 = \sum_{i=1}^{n}(y_i-\bar{y})^2,$$

$$s_{12} = \sum_{i=1}^{n}(x_i-\bar{x})(y_i-\bar{y}), \qquad \bar{x} = \frac{1}{n}\sum_{i=1}^{n}x_i, \qquad \bar{y} = \frac{1}{n}\sum_{i=1}^{n}y_i.$$

For $\theta \in \Omega$, $L(\theta)$ is maximum when (see Example 9.5.1)

$$\sigma_1^2 = \frac{s_1^2}{n}, \qquad \sigma_2^2 = \frac{s_2^2}{n}, \qquad \rho = \frac{s_{12}}{\sqrt{s_1^2 s_2^2}} = r$$

$$\mu_1 = \bar{x}, \qquad \mu_2 = \bar{y}.$$

Hence, from Example 9.5.1 we get

$$L(\hat{\Omega}) = \frac{1}{(2\pi)^n} \left(\frac{n}{s_1^2}\right)^{n/2} \left(\frac{n}{s_2^2}\right)^{n/2} \frac{e^{-n}}{(1-r^2)^{n/2}}.$$

For $\theta \in \omega$,

$$L(\theta) = \frac{1}{(2\pi)^n (\sigma_1^2 \sigma_2^2)^{n/2}} \exp\left\{ -\frac{1}{2}\left[\sum_{i=1}^{n} \frac{(x_i - \mu_1)^2}{\sigma_1^2} + \sum_{i=1}^{n} \frac{(y_i - \mu_2)^2}{\sigma_2^2} \right] \right\}$$

is maximum when

$$\sigma_1^2 = \frac{s_1^2}{n}, \qquad \sigma_2^2 = \frac{s_2^2}{n}, \qquad \mu_1 = \bar{x}, \qquad \mu_2 = \bar{y}.$$

Hence we have

$$L(\hat{\omega}) = \frac{1}{(2\pi)^n} \left(\frac{n}{s_1^2}\right)^{n/2} \left(\frac{n}{s_2^2}\right)^{n/2} e^{-n},$$

and thus

$$\lambda = (1 - r^2)^{n/2}.$$

The likelihood-ratio test is thus equivalent to

$$\phi(x_1, y_1, \ldots, x_n, y_n) = \begin{cases} 1 & \text{if } r \le -k \text{ or } \ge k \\ 0 & \text{if } -K < r < k, \end{cases}$$

where k is such that the test ϕ has size α. To determine k, we need to find the probability density function of

$$R = \frac{S_{12}}{\sqrt{S_1^2 S_2^2}}$$

where

$$S_{12} = \sum_{i=1}^{n} (X_i - \bar{X})(Y_i - \bar{Y}), \qquad S_1^2 = \sum_{i=1}^{n} (X_i - \bar{X})^2,$$

$$S_2^2 = \sum_{i=1}^{n} (Y_i - \bar{Y})^2, \qquad \bar{X} = \frac{1}{n}\sum_{i=1}^{n} X_i, \qquad \bar{Y} = \frac{1}{n}\sum_{i=1}^{n} Y_i$$

when $\rho = 0$.

We will show below that the probability density function of R when $\rho = 0$ is given by

$$f_R(r) = \begin{cases} \dfrac{\Gamma[(n-1)/2]}{\Gamma(\frac{1}{2})\Gamma[(n-2/2]} (1 - r^2)^{(n-4)/2} & \text{if } -1 < r < 1 \\ 0 & \text{otherwise.} \end{cases} \tag{10.62}$$

The cumulative distribution of R has been tabulated by F. N. David (1938); these tables can be used to find k for different values of α. The values of k for different values of α can also be obtained from Table IV of Fisher and Yates (1938). From Eq. (10.62) it is easy to verify that the random variable $\sqrt{n-2}\, R/\sqrt{1-R^2}$ has t distribution with $n-2$ degrees of freedom. Hence the Student's t table can also be used for the values of k for different values of α.

Let us now prove (10.62). The distribution of R^2 when $\rho = 0$ can easily be obtained from the Wishart distribution given in Eq. (7.33). Since we have merely stated the result there, we will instead give a straightforward proof of Eq. (10.62). From (10.61) it is easy to conclude that when $\rho = 0$, X_i is independent of Y_i for each i and Y_i is normally distributed with mean μ_2 and variance σ_2^2. Since $(X_1, Y_1), \ldots, (X_n, Y_n)$ is a random sample of size n, $\rho = 0$ implies that $(X_1, \ldots, X_n)$, $(Y_1, \ldots, Y_n)$ are independent sequences of mutually independent random variables. Thus, given $X_1 = x_1, \ldots, X_n = x_n$, the conditional distribution of S_2^2/σ_2^2 is of central chi-square type with $n-1$ degrees of freedom and the conditional distribution of

$$U = \frac{\sum_{i=1}^{n} (x_i - \bar{x})(Y_i - \bar{Y})}{\sqrt{s_1^2}} = \frac{\sum_{i=1}^{n} (x_i - \bar{x}) Y_i}{\sqrt{s_1^2}}$$

is normal with mean 0 and variance σ_2^2. Now

$$S_2^2 = U^2 + \sum_{i=1}^{n} \left[Y_i - \bar{Y} - \frac{\sum_{i=1}^{n} (x_i - \bar{x})(Y_i - \bar{Y})}{s_1^2} (x_i - \bar{x}) \right]^2$$

$$+ \frac{2[\sum_{i=1}^{n} (x_i - \bar{x})(Y_i - \bar{Y})]^2}{s_1^2} - \frac{2[\sum_{i=1}^{n} (x_i - \bar{x})(Y_i - \bar{Y})]^2}{(s_1^2)^2} s_1^2$$

$$= U^2 + V^2,$$

where

$$V^2 = \sum_{i=1}^{n} \left[Y_i - \bar{Y} - \frac{\sum_{i=1}^{n} (x_i - \bar{x})(Y_i - \bar{Y})^2}{s_1^2} (x_i - \bar{x}) \right]^2.$$

In accordance with Theorem 7.6.3. (Cochran's theorem), given $X_1 = x_1, \ldots, X_n = x_n$, the conditional distribution of V/σ_2^2 is a central chi-square distribution with $n-2$ degrees of freedom. Note that conditionally

U^2/σ_2^2 is a central chi-square distribution with 1 degree of freedom and is independent of V^2/σ_2^2. Thus given $X_1 = x_1,\ldots,X_n = x_n$, the ratio (when $\rho = 0$)

$$t = \frac{U}{\sqrt{V^2/n-2}} = \sqrt{n-2}\,\frac{R_1}{\sqrt{1-R_1^2}},$$

where

$$R_1 = \frac{\sum_{i=1}^{n}(x_i - \bar{x})(Y_i - \bar{Y})}{\sqrt{s_1^2 S_2^2}}$$

has t distribution with $n - 2$ degrees of freedom. Hence the conditional probability density function of this ratio is given by

$$f_t(y) = \begin{cases} \dfrac{1}{\sqrt{(n-2)\pi}}\dfrac{\Gamma[(n-1)/2]}{\Gamma[(n-2)/2]}\left(1 + \dfrac{y^2}{n-2}\right)^{-(n-1)/2}, \\ \qquad -\infty < y < \infty \\ 0, \qquad \text{otherwise.} \end{cases}$$

From this the conditional probability density function of R_1 when $\rho = 0$ is given by

$$f_{R_1}(r_1) = \begin{cases} \dfrac{\Gamma[(n-1)/2]}{\Gamma(\frac{1}{2})\Gamma[(n-2)/2]}(1 - r_1^2)^{(n-4)/2}, \\ \qquad -1 < r_1 < 1 \\ 0 \qquad \text{otherwise.} \end{cases} \qquad (10.63)$$

Since $f_{R_1}(r_1)$ does not depend on $x_1,\ldots,x_n$, it follows that the probability density function of R when $\rho = 0$ is given by (10.62).

10.8 CONFIDENCE INTERVALS

In many statistical problems, instead of specifying a definite estimate for a certain unknown parameter θ, it may be sufficient to provide an interval within which the same may be expected to lie. In Chapter 9 we described different methods of obtaining estimators that are functions of a random sample of fixed size n; when evaluated at a definite sample point, these provided an estimate of the parameter involved. Here we discuss the problem of setting up lower and upper bounds for θ in terms of some functions of a random sample. The word "bounds" is in quotation marks because the parameter θ is always assumed to be fixed, and the bounds, as described, are merely random variables. The following discussion will clarify the sense in which the term should be interpreted.

Let $X_1,\ldots,X_n$ be a random sample of size n from a random phenomenon specified by a distribution function that involves an unknown parameter θ (scalar). Let $\theta_1 = \theta_1 (X_1,\ldots,X_n)$ and $\theta_2 = \theta_2 (X_1,\ldots,X_n)$ be two functions (random variables) of $X_1,\ldots,X_n$, and let $\theta_1 < \theta_2$ for all values of $X_1,\ldots,X_n$.

Definition 10.8.1: Confidence interval The interval (θ_1,θ_2) is called a $100(1 - \alpha)\%$ confidence interval for θ if for a given α,

$$P\{\theta_1(X_1,\ldots,X_n) < \theta < \theta_2(X_1,\ldots,X_n)|\theta\} = 1 - \alpha \qquad (10.64)$$

for all θ. The functions θ_1 and θ_2 are called the *lower* and the *upper confidence limits* (or bounds) of θ, respectively, and $1 - \alpha$ is called the *confidence coefficient* of the interval (θ_1,θ_2).

A statement such as Eq. (10.64) must not be misunderstood. Recall that θ is an unknown constant, not a random variable, while θ_1, θ_2 are random variables. Equation (10.64) merely says that in repeated experiments the random interval (θ_1,θ_2) will include θ in $100(1 - \alpha)\%$ cases. Sometimes it may be enough to know only either an upper or a lower bound for θ, in which case notions similar to those just described apply. Also, the definitions above can easily be generalized in cases where θ is vector valued.

The choice of an appropriate confidence interval obviously needs to be based on a certain "consistency" and other considerations also. For example, in estimating the mean μ of a certain normal distribution, the trivial confidence interval $(-\infty,\infty)$ with 100% confidence coefficient hardly throws any further light on our knowledge about μ. If there is more than one confidence interval satisfying Eq. (10.64) and all other things being equal, a general rule might be to choose the one that has the minimum range. But this is a gross simplification of the situation in most cases. We shall not discuss the sophisticated techniques needed to handle such complex cases; interested students are referred to Lehmann (1959).

Usually, in an effort to set up a suitable confidence interval for θ, one looks for a random variable Z dependent on the X's and θ, whose distribution is independent of θ. Given α, it may be possible to find two quantities z_1 and z_2 such that

$$P\{z_1 < Z < z_2|\theta\} = 1 - \alpha \qquad (10.65)$$

for all θ. This will provide confidence bounds for θ if the inequality is translatable in terms of X's and θ's.

Example 10.8.1 Suppose that $X_1,X_2,\ldots,X_n$ is a random sample of size n from a normal population having mean μ (unknown) and variance 1. We want to provide a 95% confidence interval for μ.

Observe first that $\overline{X} = (1/n) \sum_{i=1}^{n} X_i$ is also normally distributed with mean μ and variance $1/n$. Thus

$$Z = \sqrt{n}(\overline{X} - \mu)$$

is a normal deviate. From tables it can easily be found that

$$P\{|Z| \leq 1.96|\mu\} = 0.95.$$

But this statement is equivalent to

$$P\left\{\overline{X} - \frac{1.96}{\sqrt{n}} < \mu < \overline{X} + \frac{1.96}{\sqrt{n}}\middle|\mu\right\} = 0.95.$$

Thus $(\overline{X} - (1.96/\sqrt{n}), \overline{X} + (1.96/\sqrt{n})$ provides a 95% confidence interval for μ. This means that in the long run, the interval above will include μ in 95% of the cases.

Example 10.8.2   Suppose that $X_1, X_2, \ldots, X_n$ is a random sample of size n from a normal population with mean 0 and (unknown) variance σ^2. Write $S_n^2 = (1/n) \sum_{i=1}^{n} X_i^2$. Then nS_n^2/σ^2 is distributed as a χ^2 with n degrees of freedom for all $\sigma^2 > 0$. Let α_1 and α_2 be such that $\alpha = \alpha_1 + \alpha_2$. Also suppose that $\chi^2_{\alpha_1}$ and $\chi^2_{1-\alpha_2}$ are the two percentage points of the χ^2 distribution with n degrees of freedom such that

$$P\{\chi_n^2 \geq \chi_{\alpha_1}^2\} = \alpha_1$$

$$P\{\chi_n^2 \geq \chi_{1-\alpha_2}^2\} = 1 - \alpha_2.$$

Then

$$P\left\{\chi_{1-\alpha_2}^2 \leq \frac{nS_n^2}{\sigma^2} \leq \chi_{\alpha_1}^2\middle|\sigma\right\} = 1 - \alpha.$$

Thus $(nS_n^2\chi_{\alpha_1}^{-2}, nS_n^2\chi_{1-\alpha_2}^{-2}$ provides a $100(1 - \alpha)\%$ confidence interval for σ^2. Note that varying α_1 and α_2 subject to $\alpha = \alpha_1 + \alpha_2$, one may obtain an uncountable number of $100(1 - \alpha)\%$ confidence intervals for σ^2. A one-sided $100(1 - \alpha)\%$ confidence interval for σ^2 can be obtained from the relation

$$P\left\{\frac{nS_n^2}{\sigma^2} \leq \chi_{\alpha}^2\middle|\sigma\right\} = 1 - \alpha$$

for all σ where χ_{α}^2 is such that

$$P\{\chi_n^2 \geq \chi_{\alpha}^2\} = \alpha.$$

The theory of uniformly most powerful tests of size α can be applied to obtain a uniformly most accurate $100(1 - \alpha)\%$ confidence interval for a scalar parameter θ. We will not attempt to discuss this here and the reader is referred to Lehmann (1959).

10.9 TESTS AND CONFIDENCE INTERVAL FOR MEAN IN $N_p(\mu, \Sigma)$

This section deals with testing problems concerning the mean and covariance matrix of a multivariate population. Using this development we will find the confidence region for the mean vector. In most problems the covariance matrix Σ is rarely known; tests and confidence regions concerning mean vector must be based on the appropriate estimate of Σ (e.g., s/N in Theorem 9.5.1). However, in cases of long experience with the same experimental variables, we can sometimes assume that Σ to be known.

In the univariate normal case ($p = 1$) the difference between the sample mean and the population mean is normally distributed with mean 0 and known variance. So one can use the table of easily available standard normal distribution to get the confidence limits or the significant point (in tests of hypothesis). In the multivariate case the difference between the sample mean and the population mean has a multivariate normal distribution with mean 0 and known covariance matrix. So one can set the confidence interval or prescribe the test for each component of p-dimensional vector μ as in the univariate normal case.

Such a procedure has several drawbacks and in many cases it leads to procedures where performance are not quite satisfactory. We shall consider here only single population problems because of the limitations of the scope of this book. We refer to Giri (1977) for further results in this context.

Let $x_\alpha = (x_{\alpha 1}, \ldots, x_{\alpha p})'$, $\alpha = 1, \ldots, N$ ($N > p$) be a sample of size N from $N_p(\mu, \Sigma)$. Consider the problem of testing H_0: $\mu = \mu_0$ (specified against the alternatives H_1, $\mu \neq \mu_0$ when Σ is known.

Let $\bar{x} = (1/N) \Sigma_{\alpha=1}^N x_\alpha$. From (9.75),

$$N(\overline{X} - \mu_0)' \Sigma^{-1} (\overline{X} - \mu_0)$$

has the central chi-square distribution under H_0 with p degrees of freedom, and has the noncentral chi-square distribution under H_1 with p degrees of freedom with the noncentrality parameter δ^2

$$\delta^2 = N(\mu - \mu_0)' \Sigma^{-1}(\mu - \mu_0).$$

Under H_0, $\delta^2 = 0$.

For testing H_0: $\mu = \mu_0$ against the alternatives H_1: $\mu \neq \mu_0$ a commonly used test procedure is given by

$$\begin{cases} 1, & \text{if } N(\bar{x} - \mu_0)'\Sigma^{-1}(\bar{x} - \mu_0) \geq \chi^2_{p,\alpha} \\ 0, & \text{otherwise,} \end{cases}$$

where $\chi^2_{p,\alpha}$ is a value of the χ^2_p such that

$$P(\chi^2_p \geq \chi^2_{p,\alpha}) = \alpha.$$

From the properties of the noncentral chi-square distribution (6.7) the power function of ϕ $[=E_{\delta^2}(\phi)]$ is a monotonically increasing function of δ^2 and it attains its minimum value α at $\delta^2 = 0$ (i.e., $\mu = \mu_0$).

We now proceed to find the confidence region of μ with confidence coefficient $1 - \alpha$ when Σ is known. Since

$$P(N(\bar{X} - \mu)'\Sigma^{-1}(\bar{X} - \mu) \leq \chi^2_{p,\alpha}) \geq 1 - \alpha,$$

we conclude that the probability is $1 - \alpha$ that the sample mean $\bar{x}$, based on N observations from $N_p(\mu,\Sigma)$ with known Σ, satisfies

$$N(\bar{x} - \mu)'\Sigma^{-1}(\bar{x} - \mu) \leq \chi^2_{p,\alpha}.$$

Hence the set

$$\{\mu : N(\bar{x} - \mu)'\Sigma^{-1}(\bar{x} - \mu) \leq \chi^2_{p,\alpha}\} \tag{10.66}$$

gives the confidence region for μ with confidence coefficient $1 - \alpha$. This region includes the interior and the boundary of an ellipsoid with center $\bar{x}$. Its shape depends on Σ and its size depends on Σ and $\chi^2_{p,\alpha}$.

10.9.1 Hotelling's T^2 Test

We now assume that Σ is unknown. We want to test the null hypothesis H_0: $\mu = \mu_0$ against the alternatives H_1: $\mu \neq \mu_0$ on the basis of observations $x_\alpha = (x_{\alpha 1},...,x_{\alpha p})'$, $\alpha = 1,...,N$ $(N > p)$ from $N_p(\mu,\Sigma)$ with Σ unknown. In Section 10.6.1 we have considered the univariate $(p = 1)$ problem of testing the null hypothesis H_0: $\mu = \mu_0$ against H_1: $\mu \neq \mu_0$ when the variance σ^2 is unknown. The likelihood ratio test for this problem has been found to be the two-sided Student's t test. We now treat the analogous multivariate problem and show that the likelihood-ratio test in this case is the analog of Student's t test, which is called Hotelling's T^2 test, since Hotelling (1931) first proposed the extension of Student's t test to the multivariate case and derived its distribution under the null hypothesis as given in (7.34) with

$\delta^2 = 0$. Using (9.67) and Theorem 9.5.1 the likelihood of the observations x_α, $\alpha = 1,\ldots,N$ can be written as

$$L(x_1,\ldots,x_N|\mu,\Sigma) = \prod_{\alpha=1}^{N} \frac{1}{(2\pi)^{p/2}|\Sigma|^{1/2}} \exp\left[-\frac{1}{2}(x_\alpha - \mu)'\Sigma^{-1}(x_\alpha - \mu)\right]$$

$$= \frac{1}{(2\pi)^{Np/2}|\Sigma|^{N/2}} \exp\left[-\frac{1}{2}\sum_{\alpha=1}^{N}(x_\alpha - \mu)'\Sigma^{-1}(x_\alpha - \mu)\right]$$

$$= \frac{1}{(2\pi)^{Np/2}|\Sigma|^{N/2}}$$
$$\times \exp\left\{-\frac{1}{2}\operatorname{tr}\left[\sum_{\alpha=1}^{N}(x_\alpha - \mu)'\Sigma^{-1}(x_\alpha - \mu)\right]\right\}$$

$$= \frac{1}{(2\pi)^{Np/2}|\Sigma|^{N/2}}$$
$$\times \exp\left\{-\frac{1}{2}\operatorname{tr}\Sigma^{-1}\left[\sum_{\alpha=1}^{N}(x_\alpha - \mu)(x_\alpha - \mu)'\right]\right\}$$

$$= \frac{1}{(2\pi)^{Np/2}|\Sigma|^{N/2}} \exp\left(-\frac{1}{2}\operatorname{tr}\Sigma^{-1}s\right)$$
$$\times \exp\left[-\frac{N}{2}(\bar{x} - \mu)'\Sigma^{-1}(\bar{x} - \mu)\right],$$

where

$$\bar{x} = \frac{1}{N}\sum_{\alpha=1}^{N}x^\alpha, \qquad s = \sum_{\alpha=1}^{N}(x_\alpha - \bar{x})(x_\alpha - \bar{x})'.$$

By Theorem 9.5.1, $L(x_1,\ldots,x_N|\mu,\Sigma)$ is maximum when $\mu = \hat{\mu} = \bar{x}$, $\Sigma = \hat{\Sigma} = s/N$. Hence

$$\max_{\Omega} L(x_1,\ldots,x_N|\mu,\Sigma) = \frac{1}{(2\pi)^{Np/2}|s/N|^{N/2}} \exp\left(-\frac{Np}{2}\right), \qquad (10.67)$$

where Ω is the parametric space of μ, Σ. Under the hypothesis H_0, Ω is reduced to $\omega = \{(\mu_0,\Sigma) : \Sigma \text{ positive definite}\}$ with $\omega \subset \Omega$.

Under H_0 the likelihood of x_α, $\alpha = 1,\ldots,N$ is given by

$$L(x_1,\ldots,x_N|\mu_0,\Sigma) = \frac{1}{(2\pi)^{Np/2}|\Sigma|^{N/2}}$$

$$\times \exp\left\{-\frac{1}{2}\operatorname{tr}\Sigma^{-1}\left[\sum_{\alpha=1}^{N}(x_\alpha - \mu_0)(x_\alpha - \mu_0)'\right]\right\}$$

$$= \frac{1}{(2\pi)^{Np/2}|\Sigma|^{N/2}}\exp\left(-\frac{1}{2}\operatorname{tr}\Sigma^{-1}A\right),$$

where

$$A = \sum_{\alpha=1}^{N}(x_\alpha - \mu_0)(x_\alpha - \mu_0)'$$

$$= \sum_{\alpha=1}^{N}(x_\alpha - \bar{x})(x_\alpha - \bar{x})' + N(\bar{x} - \mu_0)(\bar{x} - \mu_0)'$$

$$= s + N(\bar{x} - \mu_0)(\bar{x} - \mu_0)'.$$

Since $N > p$, s is positive definite with probability 1, A is also positive definite with probability 1. From Theorem 9.5.1, $L(x_1,\ldots,x_N|\mu,\Sigma)$ is maximum when

$$\Sigma = \frac{A}{N}.$$

Hence

$$\max_{\omega} L(x_1,\ldots,x_N|\mu,\Sigma) = \frac{1}{(2\pi)^{Np/2}|A/N|^{N/2}}\exp\left(-\frac{Np}{2}\right). \qquad (10.68)$$

From (10.67)–(10.68),

$$\lambda = \frac{\max_{\omega}L(x_1,\ldots,x_N|\mu_0,\Sigma)}{\max_{\Omega}L(x_1,\ldots,x_N|\mu,\Sigma)}$$

$$= \frac{|s|^{N/2}}{|s + N(\bar{x} - \mu_0)(\bar{x} - \mu_0)'|^{N/2}}. \qquad (10.69)$$

Lemma 10.9.1.1

$$|s + N(\bar{x} - \mu_0)(\bar{x} - \mu_0)'| = |s|(1 + N(\bar{x} - \mu_0)'s^{-1}(\bar{x} - \mu_0)).$$

Proof. Using Theorem A.2.7,

$$|s + N(\bar{x} - \mu_0)(\bar{x} - \mu_0)'| = \left|\begin{array}{cc} \sqrt{N}\,\frac{1}{\sqrt{N}}(\bar{x} - \mu_0) & -\sqrt{N}\,(\bar{x} - \mu_0) \\ & s \end{array}\right|.$$

Q.E.D.

Hence, by Lemma 10.9.1, we get, from (10.69),

$$\lambda = (1 + N(\bar{x} - \mu_0)'s^{-1}(\bar{x} - \mu_0))^{-N/2}$$

$$= \left(1 + \frac{T^2}{N-1}\right)^{-N/2}, \tag{10.70}$$

where

$$T^2 = N(N-1)(\bar{x} - \mu_0)'s^{-1}(\bar{x} - \mu_0) \tag{10.71}$$

and T^2 is called Hotelling's T^2 statistic.

From (10.70), the likelihood ratio test of H_0: $\mu = \mu_0$ rejects H_0 if

$$T^2 \geq c,$$

where c is a constant such that $P(T^2 \geq c | H_0) = \alpha$.

EXERCISES

1. Let $X_1,\ldots,X_n$ be a random sample of size n from a random phenomenon specified by a normal distribution with unknown mean μ and variance 4.
   (a) For testing H_0: $\mu = 0$, show that each of the following critical regions is of the same size $\alpha = 0.05$.
      1. $2\bar{X} \geq 1.645$
      2. $2\bar{X} \leq -1.645$
      3. $2\bar{X} \leq -1.960$ and $2\bar{X} \geq 1.960$
      4. $1.50 \leq 2\bar{X} \leq 2.125$
      where $\bar{X} = (1/n) \sum_{i=1}^n X_i$.
   (b) Find the one of the four regions above that is most appropriate for testing (a) H_0: $\mu = 0$ against H_1: $\mu < 0$; (b) H_0: $\mu = 0$ against H_1: $\mu > 0$; (c) H_0: $\mu = 0$ against H_1: $\mu \neq 0$.
2. Let $X_1,\ldots,X_n$ be a random sample of size n from a random phenomenon specified by the probability density function

$$f_X(x|\theta) = \begin{cases} \dfrac{1}{\theta} e^{-x/\theta}, & 0 < x < \infty \\ 0 & \text{otherwise.} \end{cases}$$

Find the most powerful test of (a) H_0: $\theta = 2$ against H_1: $\theta = 4$; (b) H_0: $\theta = 2$ against H_1: $\theta = 1$.

3. Let $X_1,\ldots,X_n$ (n sufficiently large) be a random sample of size n from a random phenomenon specified by the Bernoulli distribution with parameter θ, that is,

$$p_X(x) = \begin{cases} \binom{1}{x}\theta^x(1-\theta)^{1-x}, & x = 0,1 \\ 0 & \text{otherwise.} \end{cases}$$

Find the most powerful test of size 0.05 of (a) H_0: $\theta = \frac{1}{2}$ against H_1: $\theta = \frac{1}{3}$; (b) H_0: $\theta = \frac{1}{3}$ against H_0: $\theta = \frac{1}{2}$.

4. Let $X_1,\ldots,X_n$ be a random sample of size n from a random phenomenon specified by the probability density function

$$f_X(x|\delta,\beta) = \begin{cases} \beta e^{-\beta(x-\delta)} & \text{if } \delta \le x < \infty \\ 0 & \text{otherwise.} \end{cases}$$

Find the most powerful test of size 0.05 of H_0: $\delta = \delta_0$, $\beta = \beta_0$ against H_1: $\delta = \delta_1 < \delta_0$, $\beta = \beta_1 > \beta_0$. Is it uniformly most powerful for testing H_0 against the composite alternatives $\delta < \delta_0$, $\beta > \beta_0$?

5. Let $X_1,\ldots,X_n$ be a random sample of size n from a random phenomenon specified by probability density function

$$f_X(x|\theta) = \begin{cases} \dfrac{1}{\theta} & \text{if } 0 < x \le \theta \\ 0 & \text{otherwise.} \end{cases}$$

Find the uniformly most powerful test of

(a) H_0: $\theta = \theta_0$ against H_1: $\theta > \theta_0$.
(b) H_0: $\theta = \theta_0$ against H_1: $\theta < \theta_0$.

6. Let $X_1,\ldots,X_{10}$ be a random sample of size 10 from a random phenomenon characterized by the probability mass function

$$p_X(x|\theta) = \begin{cases} \dfrac{e^{-\theta}\theta^x}{x!} & \text{if } x = 0,1,\ldots,\infty \\ 0 & \text{otherwise.} \end{cases}$$

Find the most powerful (randomized) test of H_0: $\theta = 0.01$ against H_1: $\theta = 1$ at level of significance $\alpha = 0.05$.

7. Let $X_1,\ldots,X_n$ be a random sample of size n from a normal distribution with mean 0 and unknown variance σ^2. Find the uniformly most powerful test of H_0: $\sigma^2 = \sigma_0^2$ against H_1: $\sigma^2 > \sigma_0^2$.

8. Show that if $\phi(X_1,\ldots,X_n)$ is the most powerful test of size α for testing the simple hypothesis H_0: $\theta = \theta_0$ against the simple alternative H_1: $\theta = \theta_1$, then $E_{\theta_0}(\phi) < E_{\theta_1}(\phi)$.

9. On the basis of the results of the tossing of a coin 25 times it is to be tested whether or not the coin is biased. Give the suitable unbiased test procedure with a level of significance not exceeding 0.05. Draw the corresponding power curve to verify the unbiasedness of the test. Also consider the case where we set H_0: $p = 0.4, p$ being the probability of heads.

10. With the inception of a new system of traffic control it is hypothesized that the number of daily road accidents will be reduced. For the last few years the average number of road accidents per day in a certain city has been 12. On the basis of current-year data, how do you propose to test the hypothesis? Draw the corresponding power curve of the test with a level of significance not exceeding 0.075. (Use the Poisson approximation for road accidents.)

11. Suppose that we have two univariate normal distributions with means μ_1, μ_2 and variances 4, 9, respectively. Two samples of sizes 10 and 15, respectively, are taken from these two populations in order to test the equality of the means at the level of significance 0.025. Draw the power curve of the hypothesis above against (i) $\mu_1 - \mu_2 > 0$, (ii) $\mu_1 - \mu_2 < 0$.

12. A sample of size 11 is drawn from a normal distribution with an unknown mean μ and unknown variance σ^2. Construct the power curves for the two one-sided uniformly most powerful tests of
    (a) H_0: $\sigma^2 = 25$ against H_1: $\sigma^2 < 25$.
    (b) H_0: $\sigma^2 = 25$ against H_1: $\sigma^2 > 25$.
    Take $\alpha = 0.05$.

13. A random sample of size 12 is drawn from a random phenomenon specified by the probability density function

$$f_X(x|\theta_1) = \begin{cases} \dfrac{1}{\theta_1^p \Gamma(p)} e^{-x/\theta_1} x^{p-1} & \text{if } x > 0 \\ 0 & \text{otherwise} \end{cases}$$

with $p = 16$ and $\theta_1 > 0$ and a second random sample of size 15 is drawn from another random phenomenon specified by the probability density function

$$f_Y(y|\theta_2) = \begin{cases} \dfrac{1}{\theta_2^q \Gamma(q)} e^{-y/\theta_2} y^{q-1} & \text{if } y > 0 \\ 0 & \text{otherwise} \end{cases}$$

with $q = 18$ and $\theta_2 > 0$. Find a suitable test for the equality of $\theta_1 = \theta_2$ against the alternatives $\theta_1 > \theta_2$. Draw a power curve of this test.

14. Two random samples of sizes 12 and 14, respectively, are drawn from two normal populations with unknown means and a common but unknown variance σ^2. Find an appropriate test for the equality of the two means. Draw a power curve of the test against
    (a) One-sided alternatives with $\alpha = 0.05$.
    (b) Two-sided alternatives with $\alpha = 0.05$.

15. A cigarette manufacturer sent each of two testing laboratories presumably identical samples of tobacco. Each laboratory made six determinations of nicotine content in milligrams as follows:

Lab 1	24	26	20	19	17	25
Lab 2	28	25	19	18	21	23

Were the laboratories testing the same product? (Assume normality with the same variance.)

16. Find an alternative 95% confidence interval for μ in Example 10.8.1

17. Give a 95% confidence interval for μ in Example 10.8.1 supposing that the variance σ^2 of the underlying normal distribution is unknown.

18. How would you modify the confidence interval given in Example 10.8.2? if the mean μ of the underlying normal distribution is unknown?

19. Let $(X_1, Y_1), \ldots, (X_n, Y_n)$ be a random sample of size n from a bivariate normal distribution with parameters μ_1, μ_2, σ_1^2, σ_2^2, ρ. Give a 90% confidence interval for the difference $\mu_1 - \mu_2$ of the means.

20. Let $X_1, \ldots, X_n$, be a random sample of size n_1 from a random phenomenon specified by the normal distribution with mean μ_1 and variance σ_1^2; and let $Y_1, \ldots, Y_{n_2}$ be a random sample of size n_2 from another random phenomenon specified by the normal distribution with mean μ_2 and variance σ_2^2.
    (a) Set up a $100(1 - \alpha)\%$ confidence interval for σ_1^2/σ_2^2.
    (b) Assuming that $\sigma_1^2 = \sigma_2^2$, give a $100(1 - \alpha)\%$ confidence interval for $\mu_1 - \mu_2$.

21. Let $X_1, \ldots, X_n$ be a random sample of size n from a random phenomenon specified by the exponential probability density function

$$f_X(x) = \begin{cases} \theta e^{-\theta x}, & x \geq 0 \\ 0, & \text{otherwise}, \end{cases}$$

where $\theta > 0$. Find an upper $100(1 - \alpha)\%$ confidence bound for θ.

BIBLIOGRAPHY

David, F. N. (1938). *Tables of the Correlation Coefficient*, Cambridge University Press, London.

Fisher, R. A. (1915). The frequency distribution of the values of the correlation coefficient in samples from an indefinitely large population, *Biometrika*, Vol. 10, pp. 507–521.

Fisher, R. A., and Yates, F. (1938). *Statistical Tables*, Oliver & Boyd, London.

Fisz, M. (1963). *Probability Theory and Mathematical Statistics*, Wiley, New York.

Giri, N. (1979). *Multivariate Statistical Inference*, Academic Press, New York.

Hotelling, H. (1931). The generalization of Student's ratio, *Ann. Math. Statist.*, Vol. 2, pp. 217–234.

Lehmann, E. L. 91959). *Testing Statistical Hypotheses*, Wiley, New York.

Neyman, J., and Pearson, E. S. (1928). On the use and interpretation of certain test criteria for purposes of statistical inference, *Biometrika*, Vol. 20A, pp. 175–240.

Pearson, K. (1900). On a criterion that a given system of deviations from the probable in the case of a correlated system of variables is such that it can reasonably be supposed to have arisen in random sampling, *Philos. Mag.*, Vol. 50, pp. 157–175.

Student (1908). The probable error of a mean. *Biometrika*, Vol. 6, pp. 351–360.

11
Large-Sample Methods

11.1 INTRODUCTION

The basis of most large-sample approximations are based on the central limit theorem (Theorem 4.2.4), which states that if $\{X_n\}$ be a sequence of independent, identically distributed random variables with mean θ and finite variance σ^2, then

$$Y_n = \frac{\sum_{i=1}^{n} (X_i - \theta)}{\sigma\sqrt{n}} = \frac{\sqrt{n}(\overline{X}_n - \theta)}{\sigma},$$

where $\overline{X}_n = (1/n) \sum_{i=1}^{n} X_i$, is approximately normal with mean 0 and unit variance as $n \to \infty$. In other words,

$$\operatorname*{\mathfrak{L}t}_{n \to \infty} F_{Yn}(y) = \int_{-\infty}^{y} \frac{1}{\sqrt{2\pi}} e^{-1/2x^2} \, dx = \phi(y), \tag{11.1}$$

or we say that $\overline{X}_n$ is asymptotically normal with mean θ and variance σ^2/n. This approximation depends on the first two moments of $\overline{X}_n$.

Theorem 11.1.1 Let $g(\overline{X}_n)$ be any function of $\overline{X}_n$ with nonzero derivative $g'(\theta)$ at $\overline{X}_n = \theta$. Then

$$\frac{\sqrt{n}(g(\overline{X}_n) - g(\theta))}{\sigma g'(\theta)}$$

is asymptotically normal with zero mean and unit variance.

Proof. Expand $g(\overline{x}_n)$ by Tayler series as

$$g(\overline{x}_n) = g(\theta) + (\overline{x}_n - \theta)(g'(\theta) + \epsilon_n)$$

with $\epsilon_n \to 0$ as $g(\overline{x}_n) \to g(\theta)$ when $n \to \infty$. This implies that for every small k there exists a δ such that

$$|\epsilon_n| < k \quad \text{whenever} \quad |\overline{x}_n - \theta| < \delta.$$

Hence

$$P\{|\epsilon_n| < k\} \geq P\{|\overline{X}_n - \theta| < \delta\} \longrightarrow 1 \qquad \text{as } n \longrightarrow \infty.$$

Since k is arbitrary, $\{\epsilon_n\}$ converges in probability to zero. Now since $\sqrt{n}(\overline{X}_n - \theta)$ is asymptotically normal with mean zero and variance σ^2, and $\epsilon_n \to 0$ in probability,

$$\sqrt{n}(g(\overline{X}_n) - g(\theta)) - \sqrt{n}(\overline{X}_n - \theta)g'(\theta) = \sqrt{n}(\overline{X}_n - \theta)\epsilon_n \longrightarrow 0$$

in probability. Hence

$$\sqrt{n}(g(\overline{X}_n) - g(\theta)) \quad \text{and} \quad \sqrt{n}(\overline{X}_n - \theta)g'(\theta)$$

have the same asymptotic distribution which is normal with mean 0 and variance $\sigma^2(g'(\theta))^2$. Q.E.D.

Example 11.1.1 Let W be distributed as the central chi-square with n degrees of freedom. Then

$$W = \sum_{i=1}^{n} X_i^2,$$

where $X_1,\ldots,X_n$ are independently, identically distributed normal random variables with mean 0 and variance 1. From (6.8),

$$E(X_i^2) = 1, \qquad \text{var}(X_i^2) = 2,$$

which implies that $E(W) = n$, $\text{var}(W) = 2n$. Hence the asymptotic distribution of

$$\frac{W - n}{\sqrt{2n}}$$

is $N(0,1)$.

Example 11.1.2   Let $X_1,\ldots,X_n$ be independent, identically distributed Bernoulli random variables with parameter θ and let

$$\overline{X}_n = \frac{1}{n}\sum_{i=1}^{n} X_i.$$

Then

$$E(\overline{X}_n) = \theta, \qquad \text{var}(\overline{X}_n) = \frac{\theta(1 - \theta)}{n}.$$

Define

$$g(\overline{X}_n) = \overline{X}_n(1 - \overline{X}_n).$$

With $g(\theta) = \theta(1 - \theta)$,

$$\sqrt{n}(g(\overline{X}_n) - g(\theta))$$

is asymptotically normally distributed with mean zero and variance $\theta(1 - \theta)[g'(\theta)]^2$ as $n \to \infty$.
   Since $(g'(\theta))^2 = (1 - 2\theta)^2$,

$$\theta(1 - \theta)[g'(\theta)]^2 = \theta(1 - \theta)(1 - 2\theta)^2.$$

Example 11.1.3   Let $X_1,\ldots,X_n$ be independent, identically distributed normal random variables with mean 0 and variance 1, and let

$$F = \frac{(1/k) \sum_{i=1}^{k} X_i^2}{[1/(n - k)] \sum_{j=k+1}^{n} X_j^2}.$$

From Section 6.1.17 F is distributed as central F with k, $n - k$ degrees of freedom. Let us assume k is fixed and $n \to \infty$. This implies that $n - k \to \infty$. By Theorem 4.1.3 (the weak law of large numbers)

$$\frac{1}{(n - k)} \sum_{j=k+1}^{n} X_j^2$$

tends to 1 in probability as $n \to \infty$.
   By Slutsky's theorem (Exercise 4 in Chapter 4), F is asymptotically distributed as $(1/k) \chi_k^2$ as $n \to \infty$.

11.2 EDGEWORTH APPROXIMATION

Let

$$F_n(y) = P(Y_n \leq y).$$

The coefficient of skewness γ_{1n} and kurtosis γ_{2n} of Y_n are given by

$$\gamma_{1n}^2 = \frac{[E(Y_n^3)]^2}{[E(Y_n^2)]^3}, \qquad \gamma_{2n} = \frac{E(Y_n^4)}{[E(Y_n^2)]^2} - 3. \tag{11.2}$$

The Edgeworth approximation of $F_n(y)$ is given by (using 11.1)

$$F_n(y) = \phi(y) - \phi(y)[\tfrac{1}{6}\gamma_{1n}H_2(y) + \tfrac{1}{24}\gamma_{2n}H_3(y) + \tfrac{1}{72}\gamma_{1n}^2 H_5(y)] + \epsilon_n \tag{11.3}$$

as $n \to \infty$ where $\epsilon_n \to 0$ at a faster rate than $1/n$ and

$$H_2(y) = y^2 - 1, \qquad H_3(y) = y^3 - 3y, \qquad H_5(y) = y^5 - 10y^3 + 15y$$

are Hermit polynomials. The Edgeworth approximation uses the first, second, third, and the fourth moments of Y_n.

Example 11.2.1 From Example 11.1.1,

$$\frac{W - n}{\sqrt{2n}}$$

is asymptotically normal with mean zero and variance unity, where W is the χ_n^2 random variable. Since $E(W - n)^3 = 8n$, $E(W - n)^4 = 4n(12 + 3n)$, we get

$$\gamma_{1n} = \frac{2\sqrt{2}}{\sqrt{n}}, \qquad \gamma_{2n} = \frac{12}{n}.$$

Hence the Edgeworth approximation of $Y_n = (W - n)/\sqrt{2n}$ is given by

$$F_n(y) = \phi(y) - \phi(y)\left[\frac{\sqrt{2}}{3\sqrt{n}}(y^2 - 1) + \frac{1}{2n}(y^3 - 3y)\right.$$

$$\left. + \frac{1}{9n}(y^5 - 10y^3 + 15y)\right] + \epsilon_n.$$

11.3 VARIANCE STABILIZING TRANSFORMATIONS

Let $X_1,\ldots,X_n$ be a sample of size n from a population with mean θ and finite variance σ^2. We have observed that $\overline{X}_n = (1/n) \Sigma_{i=1}^n X_i$ is asymptotically normal with mean θ and variance σ^2/n. Furthermore, any function $g(\overline{X}_n)$, with nonzero derivative $g'(\theta)$, is also asymptotically normal with mean $g(\theta)$ and variance $(\sigma^2/n) (g'(\theta))^2$.

In many applications it is useful to obtain smooth transformation $g(\overline{X}_n)$ such that its variance for large n is independent of the parameters. Such transformations are called *variance stabilizing transformations*. In other words, we are looking for transformation g such that

$$\sigma g'(\theta) = c(\text{constant})$$

or

$$g(\theta) = \int \frac{c}{\sigma}\, d\theta. \tag{11.4}$$

Example 11.3.1 Consider Example 11.1.2. Here

$$g(\theta) = \int \frac{c\, d\sigma}{\theta(1 - \theta)} = \frac{c}{2} \sin^{-1}\sqrt{\theta} + d,$$

where d is an arbitrary constant. Choosing $c = 2$, $d = 0$, we get

$$g(\overline{X}_n) = \sin^{-1}\sqrt{\overline{X}_n}$$

is asymptotically normal with mean $\sin^{-1}\sqrt{\theta}$ and variance

$$\frac{\theta(1 - \theta)}{n} [g'(\theta)]^2 = \frac{\theta(1 - \theta)}{4n\, \theta(1 - \theta)} = \frac{1}{4n}.$$

Example 11.3.2 Let $X_1,\ldots,X_n$ be a random sample of size n from the Poisson distribution with parameter λ and let $\overline{X}_n = (1/n) \Sigma_{i=1}^n X_i$. Since $E(\overline{X}_n) = \lambda$, $\text{var}(\overline{X}_n) = \lambda/n$, the variance stabilizing transformation g is given by

$$g(\lambda) = \int \frac{c}{\sigma}\, d\lambda$$

$$= c(2\sqrt{\lambda} + d)$$

$$= 2c\sqrt{\lambda} + dc,$$

where d is an arbitrary constant. Now choosing $d = 0$, $c = \frac{1}{2}$, the variance stabilizing transformation

$$g(\overline{X}_n) = \sqrt{\overline{X}_n}$$

is asymptotically normal with mean $\sqrt{\lambda}$ and variance $1/4n$.

11.4 TESTS OF GOODNESS OF FIT

In Section 7.5 we defined the multinomial distribution as a generalization of the binomial distribution when the outcome of a single random experiment admits of $p > 2$ classifications. Let p_i, $i = 1,...,p$, denote the probability that the outcome of a single random experiment belong to the ith class such that $\sum_{i=1}^{p} p_i = 1$. Let us define $X_i = 1$ or 0 according as the outcome does or does not belong to the ith class. Then $(X_1,...,X_p)$ with $\sum_{i=1}^{p} X_i = 1$ is a multinomial random variable with probability mass function

$$p_{X_1,...,X_p}(x_1,...,x_p) = \prod_{i=1}^{p} p_i^{x_i}$$

$$x_i = 1 \quad \text{or} \quad 0, \qquad \sum_{i=1}^{p} x_i = 1.$$

Let $(x_{1i},...,x_{pi})$, $i = 1,...,n$, be a random sample of size n (results of n independent repetitions) from this distribution. The likelihood of an observed sample $(x_{1i},...,x_{pi})$, $i = 1,...,n$, is given by

$$L(\theta) = \prod_{i=1}^{p} p_i^{\sum_{j=1}^{n} x_{ij}} = \prod_{i=1}^{p} p_i^{n_i},$$

where

$$\theta = \left(p_1,...,p_p; \sum_{i=1}^{p} p_i = 1 \right) \quad \text{and} \quad \sum_{j=1}^{n} x_{ij} = n_i, \; i = 1,...,p.$$

Consider the problem of testing

$$H_0: p_i = p_i^0, \quad i = 1,...,p; \qquad \sum_{i=1}^{p} p_i^0 = 1,$$

against the alternatives

$$H_1: p_i \neq p_i^0, \qquad i = 1,...,p,$$

where

$$\Omega = \left\{ \theta = (p_1,...,p_p) : 0 \leq p_i \leq 1, \sum_{i=1}^{p} p_i = 1 \right\}$$

is of dimension $p - 1$ and

$$\omega = \left\{ (p_1^0,...,p_p^0) : 0 \leq p_i^0 \leq 1, \sum_{i=1}^{p} p_i^0 = 1 \right\}$$

is a single point. For $\theta \in \Omega$, it is easy to see that $L(\theta)$ is maximum if $p_i = n_i/n$, $i = 1,\ldots,p$. Hence we have

$$L(\hat{\Omega}) = \prod_{i=1}^{p} \left(\frac{n_i}{n}\right)^{n_i}.$$

For $\theta \in \omega$, we have

$$L(\hat{\omega}) = \prod_{i=1}^{p} (p_i^0)^{n_i},$$

so that

$$\lambda = \frac{L(\hat{\omega})}{L(\hat{\Omega})}$$

$$= n^n \prod_{i=1}^{n} \left(\frac{p_i^0}{n_i}\right)^{n_i},$$

and the likelihood-ratio test of size α for testing H_0 against H_1 is given by

reject H_0 if $\lambda \geq \lambda_0$

where λ_0 depends on the size α of the test. The small sample distribution of λ involves lengthy computations, and no ready-made table is available. For large n, $-2 \log \lambda$ has central chi-square distribution with $p - 1$ degrees of freedom. Hence one can use a chi-square table to evaluate λ_0 for large n.

Another test (goodness of fit) that is commonly used for this problem is due to Karl Pearson (1900). Let

$$Y = \sum_{i=1}^{p} \frac{(N_i - np_i^0)^2}{np_i^0},$$

where

$$N_i = \sum_{j=1}^{n} X_{ij}, \qquad n = \sum_{i=1}^{p} N_i.$$

Obviously,

$$y = \sum_{i=1}^{p} \frac{(n_i - np_i^0)^2}{np_i^0}$$

is an observed value of Y. It is proved in more advanced courses in statistics that as $n \to \infty$, Y has approximately chi-square distribution with $p - 1$ degrees of freedom when the null hypothesis H_0 is true. Since under H_0,

$E(N_i) = np_i^0$, it is expected that Y should not be too large when H_0 is true. The goodness-of-fit test suggested by Pearson is to reject H_0 if the observed $y \geq c$, where the constant c is such that $P(Y \geq c) = \alpha$ when H_0 is true.

Example 11.4.1 A genetic law states that the children having one parent of blood group M and the other parent of blood group N will always be one of the three blood groups M, MN, N, and that the expected number of children in these three groups will be in the ratio $1:2:1$. Of 300 children tested, each having one M parent and one N parent, 30% were found to be of type M, 45% of type MN, and 25% of type N. Do the data conform to the genetic law?

Here $p = 3$ and the null hypothesis is

$$H_0: p_1 = \tfrac{1}{4}, \; p_2 = \tfrac{1}{2}, \; p_3 = \tfrac{1}{4}.$$

Also, $n_1 = 90$, $n_2 = 135$, $n_3 = 75$, $n = 300$, and under H_0, $E(N_1) = 75$, $E(N_2) = 150$, and $E(N_3) = 75$. Thus we have

$$y = \frac{(90 - 75)^2}{75} + \frac{(135 - 150)^2}{150} + \frac{(75 - 75)^2}{75}$$

$$= 4.5.$$

With $\alpha = 0.05$, from Table 4B in Appendix B we get $c = 5.99$. Hence we conclude that the data observed conform to the law.

11.5 TEST OF INDEPENDENCE IN A CONTINGENCY TABLE

Let us assume that the outcome of a random experiment is classified according to the characteristics X and Y, and that the domain of values of the characteristics X and Y can be divided into r and s subclasses, respectively. For any n independent repetitions of the experiment, denote by n_{ij}, $i = 1,\dots,r$; $j = 1,\dots,s$ the number of outcomes that belong to the ith subclass of X and the jth subclass of Y. Let

$$\sum_{j=1}^{s} n_{ij} = n_{i\cdot}, \qquad \sum_{j=1}^{r} n_{ij} = n_{\cdot j}, \qquad \sum_{j=1}^{r} n_{i\cdot} = \sum_{j=1}^{s} n_{\cdot k} = n.$$

The element n_{ij} obtained in this way can be represented in the form of Table 11.1, which is commonly called the *contingency table*. The test of independence in a contingency table amounts to testing the independence of the characteristics X and Y. Let p_{ij} denote the probability that the

Table 11.1

	Y_1	Y_2	$\cdots$	Y_s	Total
X_1	n_{11}	n_{12}	$\cdots$	n_{1s}	$n_{1\cdot}$
X_2	n_{21}	n_{22}	$\cdots$	n_{2s}	$n_{2\cdot}$
.	.	.	$\cdots$	.	.
.	.	.	$\cdots$	.	.
.	.	.	$\cdots$	.	.
X_r	n_{r1}	n_{r2}	$\cdots$	n_{rs}	$n_{r\cdot}$
Total	$n_{\cdot 1}$	$n_{\cdot 2}$	$\cdots$	$n_{\cdot s}$	n

outcome of a random experiment will belong to the ith subclass of X and the jth subclass of Y and let

$$\sum_{j=1}^{s} p_{ij} = p_{i\cdot}, \qquad \sum_{i=1}^{r} p_{ij} = p_{\cdot j}.$$

Obviously,

$$\sum_{i=1}^{r} p_{i\cdot} = \sum_{j=1}^{s} p_{\cdot j} = 1.$$

The hypothesis of independence amounts to testing

$$H_0: p_{ij} = p_{i\cdot}p_{\cdot j}, \qquad i = 1,\ldots,r; \quad j = 1,\ldots,s.$$

Under H_0 the likelihood of the observed sample n_{ij} is

$$\prod_{i=1}^{r} \prod_{j=1}^{s} p_{ij}^{n_{ij}} = \prod_{i=1}^{r} \prod_{j=1}^{s} (p_{i\cdot}p_{\cdot j})^{n_{ij}}.$$

This is maximum when

$$p_{i\cdot} = \frac{n_{i\cdot}}{n}, \qquad p_{\cdot j} = \frac{n_{\cdot j}}{n}.$$

Denoting by N_{ij} the random variable with values n_{ij}, it is proved in advanced courses in statistics that for large n,

$$Y = n \sum_{i=1}^{r} \sum_{j=1}^{s} \frac{[N_{ij} - N_{i\cdot}N_{\cdot j}/n]^2}{N_{i\cdot}N_{\cdot j}}$$

Table 11.2

	Rural	Urban	Total
A	620	550	1170
B	380	450	830
Total	1000	1000	2000

has a central chi-square distribution with $v = (r - 1)(s - 1)$ degrees of freedom. The test of independence in the contingency table amounts to rejecting H_0 is

$$y = n \sum_{i=1}^{r} \sum_{j=1}^{s} \frac{[n_{ij} - (n_{i.}n_{.j}/n)]^2}{n_{i.}n_{.j}} \geq c,$$

where the constant c is such that $P(\chi_v^2 \geq c) = \alpha$. This was also proposed by Karl Pearson.

Example 11.5.1 Two sample polls of voters for two candidates A and B for a public office are taken, one for the residents of rural areas and the other for the residents of urban areas. The results are given in Table 11.2. Examine whether the voting preference for the candidates is dependent on the nature of the area (rural or urban). Take $\alpha = 0.05$. Here $y = 10.089$, $v = 1$, $c = 3.841$. Thus we conclude that the voting preference is not independent of areas.

EXERCISES

1. A die was thrown 200 times and the following results were obtained.

Face up	1	2	3	4	5	6
Frequency	27	32	39	28	49	25

Are the results consistent with the hypothesis that the die is fair? (Take $\alpha = 0.05$).

2. The following data show the distribution of digits 0,1,…,9 in telephone numbers chosen at random from the telephone numbers from the telephone directory of the Montreal area:

Digit	0	1	2	3	4	5	6	7	8	9
Frequency	1026	1107	997	966	1075	933	1107	972	964	853

Test whether the digits may be taken to occur equally frequently in the directory.

3. Test for independence between the two characteristics, health and working capacity, tabulated below.

	Working capacity	
Health	Good	Not good
Very good	20	10
Good	25	15
Fair	15	15

4. In experiments on immunization of cattle against tuberculosis (TB) the following results were obtained:

	Died of TB or very seriously affected	Unaffected or very slightly affected
Inoculated	6	13
Not inoculated	8	3

Test the hypothesis that inoculation and susceptibility to tuberculosis are statistically independent.

BIBLIOGRAPHY

Pearson, K. (1900). On the criterion that a given system of deviations from the probable in the case of a correlated system of variables is such that it can be reasonably supposed to have arisen from random sampling, *Philosophical Magazine*, Vol. 50, pp. 157–175.

Rao, C. R. *Linear Statistical Inference and Its Application*. Wiley, New York.

12

Statistical Decision Theory

12.1 BASIC CONCEPTS

Statistics is the science of decision making in the face of uncertainties about some characteristic properties of a population on the basis of sample observations $X_1,...,X_n$ from it. These uncertainties are measured in numerical quantities that could be a scalar or a vector or a matrix. We shall denote them by θ. The space of all possible values of θ is called the *state space* and is denoted by Ω. The job of a statistician is to take an action among the set of all possible actions available for the problem at hand on the basis of decision rules. These decision rules depend only on the sample observations $X_1,...,X_n$. The set of all actions will be denoted by $\mathfrak{a}$. A decision rule is denoted by $\delta(X_1,...,X_n)$. If we observe $x_1,...,x_n$ and use the decision δ, we take action $\delta(x_1,...,x_n)$. The space of all values $(x_1,...,x_n)$ is denoted by χ, the sample space. The set of all decision rules will be denoted by $\mathfrak{D}$.

In classical statistical analysis we use the sample observations $x_1,...,x_n$ for taking action (or making inference) about θ without any concern about the consequences of choosing a particular decision function that leads to take such an action. The statistical decision theory, on the other hand, is concerned with both of these two aspects, namely, taking a particular action and also evaluating the consequence of choosing a particular decision rule in order to take the best action possible.

We define by

$$L(\theta, \delta(X_1, \ldots, X_n)), \tag{12.1}$$

the loss incurred in choosing the decision rule $\delta(X_1, \ldots, X_n)$ when θ is the true value of the parameter. It is generally a nonnegative real-valued function of θ, δ. It is a random variable and we measure the performance of a particular δ by its average, denoted by

$$R(\theta, \delta) = E(L(\theta, \delta(X_1, \ldots, X_n))). \tag{12.2}$$

$R(\theta, \delta)$ is called the *risk function* of δ when θ is the true parameter. The expectation is taken with respect to the joint probability density function

$$f_{X_1 \cdots X_n}(x_1, \ldots, x_n | \theta) = f_X(x | \theta) \tag{12.3}$$

of $X_1, \ldots, X_n$, where $X = (X_1, \ldots, X_n)'$, $\theta \in \Omega$, $x = (x_1, \ldots, x_n)' \in \chi$. In the case of discrete random variable, we write it as $p_X(x | \theta)$.

Example 12.1.1   Let $X_1, \ldots, X_n$ be a random sample of size n from a normal population with mean θ ($\in R^1$) and variance 1 with values $(x_1, \ldots, x_n) \in \chi$. We are interested in estimating θ under the loss function

$$L(\theta, \delta(x_1, \ldots, x_n)) = [\delta(x_1, \ldots, x_n) - \theta]^2.$$

This is called the *squared-error loss*. Here $A = R^1$, the real line. Let

$$\delta(X_1, \ldots, X_n) = n^{-1} \sum_{i=1}^{n} X_i.$$

Then

$$R(\theta, \delta) = E(\overline{X} - \theta)^2 = \frac{1}{n}.$$

Example 12.1.2   In the setting of Example 12.1.1 consider the problem of testing the simple hypothesis H_0: $\theta = \theta_0$ against the simple alternative H_1: $\theta = \theta_1$, where θ_0, θ_1 are known. Here A has two elements 0 and 1, where

0 implies acceptance of H_0,

1 implies acceptance of H_1.

The loss function $L(\theta,\delta(x_1,\ldots,x_n))$ is given by

$$L(\theta_0,\delta(x_1,\ldots,x_n)) = \begin{cases} 0 & \text{if } \delta(x_1,\ldots,x_n) = 0 \\ 1 & \text{if } \delta(x_1,\ldots,x_n) = 1 \end{cases}$$

$$L(\theta_1,\delta(x_1,\ldots,x_n)) = \begin{cases} 0 & \text{if } \delta(x_1,\ldots,x_n) = 1 \\ 1 & \text{if } \delta(x_1,\ldots,x_n) = 0. \end{cases}$$

Here

$$R(\theta,\delta) = E(L(\theta,\delta(X_1,\ldots,X_n)))$$

$$= L(\theta,0)P(\delta(X_1,\ldots,X_n) = 0) + L(\theta,1)P(\delta(X_1,\ldots,X_n) = 1)$$

$$= \begin{cases} P(\delta(X_1,\ldots,X_n) = 1) & \text{if } \theta = \theta_0 \\ P(\delta(X_1,\ldots,X_n) = 0) & \text{if } \theta = \theta_1. \end{cases}$$

Thus $R(\theta,\delta)$ gives the level of significance of the test procedure δ when $\theta = \theta_0$ and it gives the probability of type II error of the test procedure δ when $\theta = \theta_1$.

12.2 ADMISSIBLE DECISION RULES

For most statistical problems $\mathscr{D}$ contains more than one decision rule. We need to compare them in terms of their risk functions to select the best whenever it is possible to do so. Given two decision rules δ_1 and δ_2 in $\mathscr{D}$, we say that δ_1 is better than δ_2 if and only if

$$R(\theta,\delta_1) \le R(\theta,\delta_2)$$

for all θ with strict inequality for at least one θ in Ω. If

$$R(\theta,\delta_1) = R(\theta,\delta_2)$$

for all θ, then δ_1 and δ_2 are equivalent. It may happen that

$$R(\theta,\delta_1) \le R(\theta,\delta_2) \qquad \text{for some } \theta\text{'s in } \Omega;$$

$$R(\theta,\delta_1) > R(\theta,\delta_2) \qquad \text{for the remaining } \theta\text{'s in } \Omega.$$

In other words, none is uniformly better than the other. In many statistical problems the pointwise comparisons may not lead to an unique best decision rule. A better approach has been to compare the risk functions by global criteria such as the average risk or the maximum risk.

Definition 12.1.1: Admissible decision rule   A decision rule $\delta \in \mathcal{D}$ is *admissible* if there exists no other better decision rule than δ in $\mathcal{D}$.

A decision rule δ is inadmissible if there exists a rule better than δ. Obviously, one should not use inadmissible rules. Since one can find a better one with smaller risk. Thus choosing admissible decision rules one eliminates bad decision rules from $\mathcal{D}$. Generally, there are many admissible decision rules in a given problem. To obtain unique best rules we need some global criteria such as Bayes and minimax decision rules.

12.3   BAYES' DECISION RULE

Assume that θ is a particular realization of a random variable or vector or matrix whose range is contained in Ω with probability density function $\Pi(\theta)$. This $\Pi(\theta)$ can be viewed as the information on θ prior to the experimentation, so it is called the *prior probability density function* of θ. The quantity

$$r(\delta) = E(R(\theta,\delta)) = \begin{cases} \displaystyle\int R(\theta,\delta)\Pi(\theta)\,d\theta & \text{if } \theta \text{ is continuous} \\ \displaystyle\sum_{\Omega} R(\theta,\delta)\Pi(\theta) & \text{if } \theta \text{ is discrete} \end{cases}$$

$$(12.4)$$

is called the *Bayes risk* of δ.

Definition 12.3.1: Bayes' decision rule   A decision rule δ_0 is *Bayes* with respect to the prior density $\Pi(\theta)$ if

$$r(\delta_0) = \min_{\delta \in \mathcal{D}} r(\delta).$$

Bayes decision rule is not necessarily unique. However, if $L(\theta,\delta)$ is a convex function of δ for a given θ, then δ_0 is essentially unique.

Let $\Pi(\theta|X = x)$ denote the conditional probability density function of θ given $X = x$. Then

$$P(\theta|X = x) = \begin{cases} \dfrac{\Pi(\theta)f_X(x|\theta)}{\int \Pi(\theta)f_X(x|\theta)\,d\theta} & \text{if } \theta \text{ is continuous} \\[2ex] \dfrac{\Pi(\theta)p_X(x|\theta)}{\sum_{\theta} \Pi(\theta)p_X(x|\theta)} & \text{if } \theta \text{ is discrete} \end{cases}$$

$$(12.5)$$

$\Pi(\theta|X = x)$ is called the *posterior probability density function* of θ given $X = x$. The quantity (in the continuous case)

$$E(L(\theta,\delta)|X = x) = \int L(\theta,\delta(X))\Pi(\theta|X = x)\, d\theta$$

is called the *posterior risk* of δ given $X = x$. In the discrete case the integral has to be replaced by summation. The following theorem is very helpful in calculating the Bayes decision rule.

Theorem 12.3.1 The Bayes decision rule minimizes the posterior risk of δ for all $\delta \in \mathcal{D}$ and for each x.

Proof. Let

$$f_X(x) = \int f_X(x|\theta)\Pi(\theta)\, d\theta$$

be the marginal probability density function of X (with obvious modification in the discrete case). Then

$$
\begin{aligned}
r(\delta) &= \int_\Omega R(\theta,\delta)\Pi(\theta)\, d\theta \\
&= \int_\Omega \left[\int_X L(\theta,\delta)f_X(x|\theta)\Pi(\theta)\, d\theta \right] dx \\
&= \int_X E(L(\theta,\delta)|X = x)f_X(x)\, dx.
\end{aligned}
\tag{12.6}
$$

Thus δ_0 which minimizes $r(\delta)$ for all δ in $\mathcal{D}$ also minimizes $E(L(\theta,\delta)|X = x)$ for each x.   Q.E.D.

The following theorem gives the Bayes estimator of θ under some loss functions that are commonly used in statistics.

Theorem 12.3.2

(a) Let $L(\theta,\delta) = (\theta - \delta)^2$ be the squared-error loss function for estimating θ. The Bayes estimator of θ is given by $\delta_0 = E(\theta|X = x)$.
(b) Let $L(\theta,\delta) = w(\theta)(\theta - \delta)^2$ be the weighted squared-error loss function for estimating θ, where $w(\theta)$ is the weight function. The Bayes estimator is given by

$$\delta_0 = \frac{E(W(\theta)\theta|X = x)}{E(W(\theta)|X = x)}.
\tag{12.7}$$

(c)   Let $L(\theta,\delta) = |\theta - \delta|$ be the absolute-value loss for estimating θ. The Bayes estimator is given by any median of $\Pi(\theta|X = x)$.

Proof

(a) $E(L(\theta,\delta)|X = x) = E((\theta^2 - 2\theta\delta(X) + \delta^2(X))|X = x))$.

Differentiating with respect to δ the value δ_0 that minimize the conditional expectation is $\delta_0(x) = E(\theta|X = x)$ (assuming that all integrals exist).

(b)   Proceeding in the same way as in (a), we get the result.
(c)   Let m denote a median of $\Pi(\theta|X = x)$. Then

$$P(\theta < m|X = x) \leq \tfrac{1}{2}, \qquad p(\theta \geq m|X = x) \geq \tfrac{1}{2}.$$

Let δ be any other decision rule such that $\delta < m$. Then

$$L(\theta,m) - L(\theta,\delta) = \begin{cases} m - \delta & \text{if } \theta \leq \delta \\ m + \delta - 2\theta & \text{if } \delta < \theta < m \\ \delta - m & \text{if } m \leq \theta. \end{cases}$$

Since

$$m + \delta - 2\theta < m - \delta \qquad \text{for } \delta < \theta < m,$$

we get

$$E(L(\theta,m) - L(\theta,\delta)|X = x)) \leq (m - \delta)P(\theta < m|X = x)$$
$$+ (\delta - m)P(\theta \geq m|X = x)$$
$$\leq 0.$$

Thus the posterior risk of m is at least as small as that of any other $\delta < m$. Similar results holds for $\delta > m$.   Q.E.D.

For a particular type of loss function the Bayes estimator of $Q(\theta)$, a function of θ, is obtained by replacing θ by $Q(\theta)$ in the expressions for δ_0 above.

In calculating the Bayes estimator, one can without any loss of generality replace the sample $X_1,\ldots,X_n$ by the sufficient statistic $T(X_1,\ldots,X_n)$ for θ. In most cases such a replacement will facilitate the computation.

Theorem 12.3.3   Let $T(X) = T(X_1,\ldots,X_n)$ be a sufficient statistic for θ with probability density function $g_T(t|\theta)$. Then

$$E(\theta|X = x) = E((\theta|T = t).$$

Proof.   By Fisher–Neyman factorization theorem we can write

$$f_X(x|\theta) = g_T(t|\theta)H(x),$$

where $H(x)$ does not depend on θ. Hence

$$
\begin{aligned}
E(\theta|X = x) &= \frac{\int \theta f_X(x|\theta)\Pi(\theta)\, d\theta}{\int f_X(x|\theta)\Pi(\theta)\, d\theta} \\
&= \frac{\int \theta g_T(t|\theta)H(x)\Pi(\theta)\, d\theta}{\int g_T(t|\theta)H(x)\Pi(\theta)\, d\theta} \\
&= \frac{\int \theta g_T(t|\theta)\Pi(\theta)\, d\theta}{\int g_T(t|\theta)\Pi(\theta)\, d\theta} \\
&= E(\theta|T = t). \qquad \text{Q.E.D.}
\end{aligned}
$$

Example 12.3.1 Let $X_1,\dots,X_n$ be a random sample of size n from $N(\theta,\sigma^2)$ with σ^2 known. Let $\overline{X} = (1/n)\sum_{i=1}^n X_i$. Then $\overline{X}$ is sufficient for θ. Suppose that we are interested to estimate θ with loss function $L(\theta,\delta) = (\theta - \delta)^2$. Let $\Pi(\theta)$ be $N(\mu,\tau^2)$. Then

$$
\begin{aligned}
\Pi(\theta|X = x) &= \frac{\rho}{\sqrt{2\pi}}\exp\left\{\frac{-\rho^2}{2}\left[\theta - \frac{1}{\rho^2}\left(\frac{n\overline{x}}{\sigma^2} + \frac{\mu}{\tau^2}\right)\right]^2\right\} \\
&= N\left(\left(\frac{n\overline{x}}{\sigma^2} + \frac{\mu}{\tau^2}\right)\frac{1}{\rho^2}, \frac{1}{\rho^2}\right),
\end{aligned}
$$

where $\rho^2 = n/\sigma^2 + 1/\tau^2$. Hence

$$
\begin{aligned}
E(\theta|X = x) &= \left(\frac{n\overline{x}}{\sigma^2} + \frac{\mu}{\tau^2}\right)\left(\frac{\sigma^2\tau^2}{n\tau^2 + \sigma^2}\right) \\
&= \mu\left(\frac{\sigma^2}{n\tau^2 + \sigma^2}\right) + \overline{x}\left(\frac{n\tau^2}{n\tau^2 + \sigma^2}\right),
\end{aligned}
$$

$$
\operatorname{var}(\theta|X = x) = \frac{\sigma^2\tau^2}{n\tau^2 + \sigma^2}. \tag{12.8}
$$

The same computation can be carried out with respect to the probability density function of $\overline{X}$. Since $\overline{X}$ is normal with mean θ and variance σ^2/n,

$$
\Pi(\theta|\overline{X} = \overline{x}) = \frac{\begin{array}{l}(\sqrt{n}/\sqrt{2\pi}\sigma)\exp[-(n/2\sigma^2)(\overline{x} - \theta)^2]\\ \times\,(1/\sqrt{2\pi}\tau^2)\exp[(-1/2\tau^2)(\theta - \mu)^2]\end{array}}{\int \begin{array}{l}(\sqrt{n}/\sqrt{2\pi}\sigma)\exp[-(n/2\sigma^2)(\overline{x} - \theta)^2]\\ \times\,(1/\sqrt{2\pi}\tau^2)\exp[-(1/2\tau^2)(\theta - \mu)^2]\, d\theta\end{array}}
$$

$$
= N\left(\frac{n\overline{x}}{\sigma^2} + \frac{\mu}{\tau^2}, \frac{1}{\rho^2}\right).
$$

Hence the Bayes estimator of θ with the squared-error loss is given by (12.8). It is not $\overline{X}$, but it tends to $\overline{X}$ as $\tau^2 \to \infty$.

Example 12.3.2 Let $X = X_1 + \cdots + X_n$, where $X_1, \ldots, X_n$ are independently and identically distributed Bernoulli random variables with

$$P(X_i = 1) = \theta, \qquad P(X_i = 0) = 1 - \theta, \quad 0 < \theta < 1.$$

The sufficient statistic for θ is X and the probability density function of X is binomial $B(n,\theta)$, with

$$p_X(x|\theta) = \begin{cases} \binom{n}{x} \theta^x (1 - \theta)^{n-x}, & x = 0,1,\ldots,n, \\ 0, & \text{otherwise.} \end{cases}$$

We wish to estimate θ with loss

$$L(\theta,\delta) = W(\theta)(\theta - \delta)^2, \qquad W(\theta) = \frac{1}{\theta(1 - \theta)}.$$

Let $\Pi(\theta)$ be given by

$$\Pi(\theta) = \begin{cases} 1, & 0 < \theta < 1 \\ 0, & \text{otherwise.} \end{cases}$$

Then the Bayes estimator of θ is

$$\delta_0(x) = \frac{E(\theta/\theta(1 - \theta)|X = x)}{E(1/\theta(1 - \theta)|X = x)}.$$

The marginal probability density function of X is given by

$$\begin{aligned} p_X(x) &= \binom{n}{x} \int_0^1 \theta^x (1 - \theta)^{n-x} \, d\theta \\ &= \binom{n}{x} \frac{\Gamma(x + 1)\Gamma(n - x + 1)}{\Gamma(x + 2)} = \frac{1}{n + 1}. \end{aligned}$$

Hence

$$\Pi(\theta|X = x) = \frac{1}{n + 1} \theta^x (1 - \theta)^{n-x}, \qquad 0 < \theta < 1.$$

Thus

$$E\left(\frac{1}{(1 - \theta)} \middle| X = x\right) = \frac{1}{n + 1} \frac{\Gamma(x + 1)\Gamma(n - x)}{\Gamma(n + 1)}$$

$$E\left(\frac{1}{\theta(1 - \theta)} \middle| X = x\right) = \frac{1}{n + 1} \frac{\Gamma(x)\Gamma(n - x)}{\Gamma(n)}.$$

So

$$\delta_0 = \frac{X}{n}. \tag{12.9}$$

This is also the maximum likelihood estimator of θ. Hence the maximum likelihood estimator is Bayes with respect to the uniform prior. Uniform priors are noninformative priors, they treat all θ equally likely.

Let us take $\Pi(\theta)$ as Beta(r,s), that is,

$$\Pi(\theta) = \frac{\Gamma(r + s)}{\Gamma(r)\Gamma(s)} \theta^{r-1}(1 - \theta)^{s-1}, \qquad 0 < \theta < 1. \tag{12.10}$$

Proceeding in the same way as above, the posterior probability density function of θ given $X = x$ can be obtained as

$$\Pi(\theta|X = x) = \frac{\Gamma(n + r + s)}{\Gamma(x + r)\Gamma(n + s - x)} \theta^{x+r-1}(1 - \theta)^{n+s-x-1}. \tag{12.11}$$

Hence the Bayes estimator in this case is

$$\delta_0(x) = \frac{E(1/(1 - \theta)|X = x)}{E(1/\theta(1 - \theta)|X = x)} = \frac{x + r - 1}{n + s + r - 2}. \tag{12.12}$$

This reduces to x/n if $s = r = 1$.

Example 12.3.3 In Example 12.3.2, let us consider the problem of estimating θ with the squared-error loss

$$L(\theta,\delta) = (\theta - \delta)^2$$

and $\Pi(\theta) = \beta(r,s)$. From (12.11) the Bayes estimator is given by

$$\delta_0 = E(\theta|X = x)$$

$$= \frac{x + r}{n + s + r}.$$

We have so far considered problems of estimation only. In the example below we calculate Bayes decision rule for testing a simple hypothesis against a simple alternative.

Example 12.3.4 Consider the problem of testing a simple hypothesis H_0: $\theta = \theta_0$ against the simple alternative H_1: $\theta = \theta_1$ when the probability density function of X is $f_X(x|\theta)$ [or $p_X(x|\theta)$ in the discrete case]. Let the loss function be 0–1 with two actions 0 and 1, where 0 indicates accepting H_0 and 1 indicates accepting H_1. Let

$$\Pi(\theta) = \begin{cases} \Pi_0 & \text{when } \theta = \theta_0 \\ 1 - \Pi_0 & \text{when } \theta = \theta_1. \end{cases}$$

Then

$$\Pi(\theta|X = x) = \begin{cases} \dfrac{\Pi_0 f_X(x|\theta_0)}{\Pi_0 f_X(x|\theta_0) + (1 - \Pi_0) f_X(x|\theta_1)} & \text{if } \theta = \theta_0 \\[3mm] \dfrac{(1 - \Pi_0) f_X(x|\theta_1)}{\Pi_0 f_X(x|\theta_0) + (1 - \Pi_0) f_X(x|\theta_1)} & \text{if } \theta = \theta_1. \end{cases}$$

Hence the posterior risk $r(i|X = x)$ of taking action i is given by

$$r(i|X = x) = \begin{cases} \dfrac{(1 - \Pi_0) f_X(x|\theta_1)}{\Pi_0 f_X(x|\theta_0) + (1 - \Pi_0) f_X(x|\theta_1)} & \text{if } i = 0 \\[3mm] \dfrac{\Pi_0 f_X(x|\theta_0)}{\Pi_0 f_X(x|\theta_0) + (1 - \Pi_0) f_X(x|\theta_1)} & \text{if } i = 1. \end{cases}$$

Thus the Bayes decision rule with $0 - 1$ loss is given by

accept H_0 if $r(0|X = x) < r(1|X = x)$

accept H_1 if $r(1|X = x) < r(0|X = x)$

(reject H_0)

which implies

accept H_0 if $\dfrac{f_X(x|\theta_1)}{f_X(x|\theta_0)} < \dfrac{\Pi_0}{1 - \Pi_0}$

accept H_1 if $\dfrac{f_X(x|\theta_1)}{f_X(x|\theta_0)} > \dfrac{\Pi_0}{1 - \Pi_0}$.

This is the most powerful test as given by the lemma of Neyman–Pearson for testing H_0 against H_1 with the critical value $k = \Pi_0/(1 - \Pi_0)$.

12.4   MINIMAX DECISION RULE

In calculating a Bayes decision rule we need to know the prior probability density $\Pi(\theta)$. It sometimes creates an additional problem to choose the proper prior $\Pi(\theta)$.

To avoid this, we look for another criterion to choose a decision rule that minimizes the worst possible risk of all δ in $\mathcal{D}$.

Definition 12.4.1: Minimax decision rule A decision rule δ_0 is *minimax* if

$$\sup_{\theta \in \Omega} R(\theta, \delta_0) = \inf_{\delta \in \mathscr{D}} \sup_{\theta \in \Omega} R(\theta, \delta). \qquad (12.13)$$

A popular approach to find the minimax decision rule is to find a prior probability density function $\Pi(\theta)$ for which the Bayes rule has a constant risk (does not depend on θ). This $\Pi(\theta)$ need not be the most appropriate one. In fact, this prior is the least favorable one in the sense that it maximizes among all available priors the Bayes risks for all Bayes decision rules.

In what follows we denote by $E_\Pi(\cdot)$, $r_\Pi(\cdot)$ the expectation and the Bayes risk with respect to the given prior $\Pi(\theta)$.

Theorem 12.4.1

(a) If δ_0 is a Bayes rule with respect to the prior $\Pi(\theta)$ such that $R(\theta, \delta_0)$ = constant, then δ_0 is minimax.
(b) $\Pi(\theta)$ is least favorable.

Proof. (a) Since $R(\theta, \delta_0)$ is constant for all θ,

$$r_\Pi(\delta_0) = E_\Pi(R(\theta, \delta_0)) = \sup_{\theta \in \Omega} R(\theta, \delta_0).$$

Let δ be any other decision rule, $\delta \neq \delta_0$. Then

$$\sup_{\theta \in \Omega} R(\theta, \delta) \geq E_\Pi(R(\theta, \delta_0)) = r_\Pi(\delta).$$

Since δ_0 is Bayes with respect to $\Pi(\theta)$, we get

$$r_\Pi(\delta_0) \leq r_\Pi(\delta).$$

Thus

$$\sup_{\theta \in \Omega} R(\theta, \delta) \geq r_\Pi(\delta) \geq r_\Pi(\delta_0) = \sup_{\theta \in \Omega} R(\theta, \delta).$$

Hence

$$\inf_{\delta \in \mathscr{D}} \sup_{\theta \in \Omega} R(\theta, \delta) \geq \sup_{\theta \in \Omega} R(\theta, \delta_0).$$

Thus δ_0 is minimax.

(b)   Let $\eta(\theta)$ be any other prior probability density function of θ. Then

$$\inf_{\delta \in \mathscr{D}} r_\Pi(\delta) = \inf_{\delta \in \mathscr{D}} E_\Pi(R(\theta,\delta)) = E_\Pi(R(\theta,\delta_0))$$

$$= \sup_{\theta \in \Omega} R(\theta,\delta_0) \geq E_\eta R(\theta,\delta_0)$$

$$\geq \inf_{\delta \in \mathscr{D}} E_\eta(R(\theta,\delta)) = \inf_{\delta \in \mathscr{D}} r_\eta(\delta).$$

Hence $\Pi(\theta)$ is least favorable.   Q.E.D.

Definition 12.4.2: Equalizer decision rule   A decision rule δ_0 is an *equalizer decision rule* if $R(\theta,\delta_0)$ is a constant for all $\theta \in \Omega$.

Example 12.4.1   Consider Example 12.3.2 with $r = s = 1$. Since

$$R\left(\theta, \frac{X}{n}\right) = E\left\{ \frac{(X/n - \theta)^2}{\theta(1 - \theta)} \right\}$$

$$= \frac{1}{n^2\theta(1 - \theta)} n\theta(1 - \theta) = \frac{1}{n},$$

the Bayes estimator $\delta_0 = X/n$ has constant risk for all θ. Hence X/n is the minimax estimator of θ.

Example 12.4.2   In Example 12.3.3 let the loss function be the squared-error loss and let $\Pi(\theta)$ be Beta(r,s). Then the Bayes estimator of θ is

$$\delta_0(x) = \frac{x + r}{n + s + r}$$

$$= \frac{x}{n + s + r} + \frac{r}{n + s + r}.$$

It is of the form $cX + d$, where c, d are reals.
      We now find an estimator of the form $cX + d$ for θ such that $R(\theta, cX + d)$ is constant for all θ. Since

$$R(\theta, cX + d) = E(cX + d - \theta)^2$$
$$= E(c(X - n\theta) + (cn - 1)\theta + d)^2,$$
$$= c^2n\theta(1 - \theta) + (d + (cn - 1)\theta)^2$$
$$= \theta^2[(cn - 1)^2 - c^2n] + \theta[2d(n - 1) + c^2n] + d^2,$$

$R(\theta, cX + d)$ is constant for all θ if

$$(cn - 1)^2 - c^2n = 0 \quad \text{and} \quad 2d(n - 1) + c^2n = 0.$$

Solving these two equations for c, d, we get

$$c = (n + \sqrt{n})^{-1}, \qquad d = \sqrt{n}(2(n + \sqrt{n}))^{-1}.$$

Thus the estimator

$$\delta_0(X) = \frac{X}{n + \sqrt{n}} + \frac{\sqrt{n}}{2(n + \sqrt{n})}$$

has constant risk for all θ and it is Bayes with respect to $\Pi(\theta) = \text{Beta}(r,s)$ with $r = s = \frac{1}{2}\sqrt{n}$. Hence it is minimax.

In many applications the Bayes estimators are not equalizer rules but they tend to equalize decision rules in the limit. As an example, consider Example 12.3.1. Here the Bayes estimator tends to $\overline{X}$ when $\tau^2 \to \infty$ and with the squared-error loss

$$R(\theta,\overline{X}) = E(\overline{X} - \theta)^2 = \frac{\sigma^2}{n}$$

is constant for all θ.

Such estimators are often called *extended Bayes estimators* and Theorem 12.4.3 will prove that they are minimax.

Definition 12.4.3: Extended Bayes decision   A decision rule $\delta_0 \in \mathcal{D}$ is *extended Bayes* for $\theta \in \Omega$ if it is ϵ-Bayes; that is, for every $\epsilon > 0$ there exists a prior probability density $\Pi_\epsilon(\theta)$ on Ω such that (taking the expectation with respect to Π_ϵ)

$$E(R(\theta,\delta_0)) \leq \inf_{\delta \in \mathcal{D}} E(R(\theta,\delta)) + \epsilon$$

Theorem 12.4.2   An extended Bayes decision rule with constant risk is minimax.

Proof.   Let δ_0^* be an extended Bayes rule with constant risk $R(\theta,\delta_0^*) = c$ for all $\theta \in \Omega$. Since it is extended Bayes, there exists a prior density function $\Pi_\epsilon(\theta)$ such that

$$E(R(\theta,\delta_0^*)) \geq \inf_{\delta \in \mathcal{D}} E(R(\theta,\delta)) + \epsilon, \tag{12.14}$$

where the expection is taken with respect to Π_ϵ. From (2.14) we get

$$c - \epsilon \leq \inf_{\delta \in \mathcal{D}} E(R(\theta,\delta)) + \epsilon.$$

Now suppose that δ_0^* is not minimax; then there exists δ^* in $\mathscr{D}$ such that

$$\sup_{\theta \in \Omega} R(\theta,\delta^*) < \sup_{\theta \in \Omega} R(\theta,\delta_0^*) = c.$$

Hence

$$\sup_{\theta \in \Omega} R(\theta,\delta^*) \leq c - \epsilon_0 \qquad \text{for some } \epsilon_0 > 0.$$

This implies that

$$R(\theta,\delta^*) \leq c - \epsilon_0$$

for all $\theta \in \Omega$, or

$$E(R(\theta,\delta^*)) \leq c - \epsilon_0, \tag{12.15}$$

where the expectation is taken with respect to any prior density function on Ω. From (12.14)–(12.15) we conclude that

$$c - \epsilon \leq \inf_{\theta \in \Omega} E(R(\theta,\delta^*)) \leq E(R(\theta,\delta^*)) \leq c - \epsilon_0.$$

which is a contradiction for $0 < \epsilon < \epsilon_0$.   Q.E.D.

Example 12.4.3   Consider Example 12.3.1. The Bayes estimator

$$\delta_0^* = \frac{n\tau^2}{n\tau^2 + \sigma^2} \overline{X} + \frac{\sigma^2}{n\tau^2 + \sigma^2} \mu$$

has the risk

$$R(\theta,\delta_0^*) = \frac{\sigma^2\tau^2}{n\tau^2 + \sigma^2}.$$

Hence taking the expection with respect to $\Pi(\theta)$, we get

$$\inf_{\delta \in \mathscr{D}} E(R(\theta,\delta)) = E(R(\theta,\delta_0^*)) = \frac{\sigma^2\tau^2}{n\tau^2 + \sigma^2}.$$

To show that $\overline{X}$ is extended Bayes with respect to $\Pi(\theta)$, we first observe that

$$E(R(\theta,\overline{X})) = \frac{\sigma^2}{n}$$

$$E(R(\theta,\overline{X}) - \inf_{\delta \in \mathscr{D}} E(R(\theta,\delta)) = \frac{\sigma^2}{n} - \frac{\sigma^2\tau^2}{n\tau^2 + \sigma^2}$$

$$= \frac{\sigma^4}{n(n\tau^2 + \sigma^2)} = \epsilon,$$

where $\epsilon = \sigma^4/n(n\tau^2 + \sigma^2)$ and the expection is taken with respect to $\Pi(\theta)$. Hence $\overline{X}$ is ϵ-Bayes for every $\epsilon > 0$. Since $\overline{X}$ has constant risk, by Theorem 12.4.2 $\overline{X}$ is minimax for estimating θ with respect to the squared-error loss when σ^2 is known.

Example 12.4.4 In the setting of Example 12.3.1, let us assume that both θ, σ^2 are unknown. We want to estimate θ with the loss function

$$L(\theta,\delta) = R(\theta - \delta)^2, \tag{12.16}$$

where $R = 1/\sigma^2$. Let the joint prior probability density of (θ,R) be given by

$$\Pi(\theta,R) = \Pi(\theta|R = r)\Pi(r)$$

$$= \sqrt{\frac{rb}{2\pi}} \exp\left[-\frac{br}{2}(\theta - \mu)^2\right] \frac{\beta^\alpha}{\Gamma(\alpha)} \exp(-\beta r)r^{\alpha-1},$$

where $\alpha > 0$, $\beta > 0$, $b > 0$ and $\Pi(\theta|R = r)$ is $N(\mu,b^{-1}r^{-1})$, $\Pi(r)$ is gamma with parameters α, β. The posterior joint density of (θ,R) given $X = x$ can be obtained as

$$\Pi(\theta,R|X = x) = \Pi(\theta|R = r,X = x)\Pi(R|X = x)$$

$$= N\left(\frac{b\mu + n\overline{x}}{b + n}, (b + n)^{-1}r^{-1}\right) G(\alpha',\beta'), \tag{12.17}$$

where $G(\alpha',\beta')$ is gamma with parameter

$$\alpha' = \alpha + \frac{n}{2}$$

$$\beta' = \beta + \frac{1}{2}\sum_{i=1}^{n}(x_i - \overline{x})^2 + \frac{bn(\overline{x} - \mu)^2}{2(b + n)}.$$

From this it follows that the Bayes estimator of θ is given by

$$\delta^* = \frac{b\mu + b\overline{x}}{b + n}$$

and its risk is

$$E\left(\frac{1}{\sigma^2}\left(n\frac{(\overline{x} - \theta)}{b + n} - \frac{b(\theta - \mu)}{b + n}\right)^2\right) = \frac{1}{b + n}.$$

Hence taking the expectation with respect to $\Pi(\theta)$,

$$E(R(\theta,\overline{X})) - \inf_{\delta \in \mathcal{D}} E(R(\theta,\delta)) = \frac{1}{n} - \frac{1}{b + n} = \frac{b}{(b + n)n} = \epsilon > 0.$$

Since $\overline{X}$ is extended Bayes with constant risk, we conclude that $\overline{X}$ is minimax.

12.5 ADMISSIBILITY OF BAYES' RULES

In Section 12.2 we have introduced the concept of admissible decision rules. We prove here several theorems that enable us to prove the admissibility property of Bayes decision rules.

Theorem 12.5.1 Let Ω be a finite space with elements θ_i, $i = 1,...,n$, and let the prior density $\Pi(\theta)$ be given by $\Pi(\theta_i) = \Pi_i > 0$, $i = 1,...,n$, such that $\Sigma_1^n \Pi_i = 1$. Then the Bayes rule δ_0 with respect $\Pi(\theta)$ is admissible.

Proof. Assume that θ_0 is not admissible. Then there exists a decision rule δ such that

$$R(\theta_i,\delta) \leq R(\theta_i,\delta_0)$$

for $i = 1,...,n$ with strict inequality for at least one i. Then

$$r(\delta) = \sum_{i=1}^{n} \Pi_i R(\theta_i,\delta)$$

$$< \sum_{i=1}^{n} \Pi_i R(\theta_i,\delta_0) = r(\delta_0).$$

This contradicts the fact that θ_0 is Bayes.   Q.E.D.

Theorem 12.5.2 A unique Bayes rule is admissible.

Proof. The proof is similar to that of Theorem 12.5.1.

Theorem 12.5.3 Let $\Pi_i(\theta)$, $i = 1,2,...$ be a sequence of prior density function on Ω. Let $r_i(\delta)$ denote the Bayes risk of δ with respect Π_i. Denote

$$r_i = \inf_{\delta \in \mathcal{D}} r_i(\delta). \tag{12.18}$$

Suppose that δ^* is a decision rule such that

$$\frac{r_i(\delta^*) - r_i}{\int_a^b \Pi_i(\theta)\, d\theta} \longrightarrow 0 \tag{12.19}$$

as $i \to \infty$ for every fixed $a < b$. If $R(\theta,\delta)$ is a continuous function of θ for each δ, then δ^* is admissible.

Proof. Suppose that δ^* is not admissible. Then there exists a δ in $\mathcal{D}$ such that

$$R(\theta,\delta) \leq R(\theta,\delta^*)$$

for all θ and strict inequality for some θ. Hence

$$R(\theta,\delta^*) - R(\theta,\delta) \geq 0$$

for all θ and strict inequality for some θ. Since $R(\theta,\delta^*) - R(\theta,\delta)$ is a continuous function of θ, there must exist an interval (a,b), $a < b$ and an $\epsilon > 0$ such that

$$\frac{\int \{R(\theta,\delta^*) - R(\theta,\delta)\}\Pi_i(\theta)\ d\theta}{\int_a^b \Pi_i(\theta)\ d\theta} = \frac{r_i(\delta^*) - r_i(\delta)}{\int_a^b \Pi_i(\theta)\ d\theta} \geq \epsilon.$$

But $r_i(\delta^*) - r_i(\delta) \leq r_i(\delta^*) - r_i$. Hence we conclude that for all i,

$$\frac{r_i(\delta^*) - r_i}{\int_a^b \Pi_i(\theta)\ d\theta} \geq \epsilon,$$

which contradicts (12.19).   Q.E.D.

Example 12.5.1   Consider Example 12.3.1 with $\Pi_i(\theta) = N(\mu, \tau^2 = i)$. It may be verified that $R(\theta,\delta)$ is a continuous function of θ for each δ. Taking the expectation with respect to $\Pi_i(\theta)$,

$$r_i(\overline{X}) = E\left(\frac{\sigma^2}{n}\right) = \frac{\sigma^2}{n}$$

$$r_i = \inf r_i(\delta) = \frac{\sigma^2}{n}\frac{i}{i + \sigma^2/n}.$$

Hence

$$r_i(\overline{X}) - r_i = \frac{\sigma^2}{n}\left(\frac{1}{1 + in/\sigma^2}\right).$$

Since

$$\int_a^b \Pi_i(\theta)\ d\theta = \phi\left(\frac{b - \mu}{\sqrt{i}}\right) - \phi\left(\frac{a - \mu}{\sqrt{i}}\right)$$

$$= \phi(\xi)\left(\frac{b - a}{\sqrt{i}}\right)$$

for some ξ with

$$\frac{a - \mu}{\sqrt{i}} \leq \xi \leq \frac{b - \mu}{\sqrt{i}} \quad \text{and} \quad \phi(z) = \int_{-\infty}^z \frac{1}{\sqrt{2\pi}}\exp\left(\frac{-1}{2}x^2\right)\ dx.$$

We conclude that as $i \to \infty$

$$\frac{r_i(\overline{X}) - r_i}{\int_a^b \Pi_i(\theta) \, d\theta} \to 0.$$

Hence $\overline{X}$ is an admissible estimator of θ.

EXERCISES

1. Let X be a Poisson random variable with parameter θ, that is, with probability mass function

$$p_x(x|\theta) = \begin{cases} \dfrac{e^{-\theta}\theta^x}{x!}, & x = 0,1,\dots \\ 0, & \text{otherwise,} \end{cases}$$

where $\theta > 0$, and let the loss function $L(\theta,\delta) = (\theta - \delta)^2/\theta$.
   (a) Show that the maximum-likelihood estimator $\delta_0 = X$ of θ is an equalizer decision rule.
   (b) Find the Bayes decision rule with respect to the prior density $\Pi(\theta)$, where

$$\Pi(\theta) = \begin{cases} \dfrac{\beta^\alpha}{\Gamma(\alpha)} \theta^{\alpha-1}e^{-\beta\theta}, & \theta \geq 0 \\ 0, & \text{otherwise,} \end{cases}$$

which is a gamma $G(\alpha,\beta)$ with parameters (α,β).
   (c) Show that δ_0 is extended Bayes and hence minimax.
   (d) Show that δ_0 is admissible.
2. Let $X_1,\dots,X_n$ be independent, identically distributed Poisson random variables with parameter θ and let $\Pi(\theta)$ be Gamma(α,β).
   (a) Show that the posterior density of θ given $X_1 = x_1,\dots,X_n = x_n$ is Gamma$(\alpha + \sum_{i=1}^n x_i, \beta + n)$.
   (b) Find the Bayes estimator of θ with the squared-error loss.
3. Let X be a random variable distributed uniformly over the interval $(0,\theta)$, where θ is the unknown parameter, and let

$$\Pi(\theta) = \begin{cases} \theta e^{-\theta}, & \theta > 0 \\ 0, & \text{otherwise.} \end{cases}$$

Show that $X + 1$ is the Bayes estimator of θ with the squared-error loss.
4. In Exercise 1 assume that $L(\theta,\delta) = (\theta - \delta)^2$. Show that the Bayes estimator of θ with $\Pi(\theta) = $ Gamma(α,β), is $(x + \alpha)/(1 + \beta)$.

5. Suppose that X is a exponential random variable with parameter θ; that is, its probability density function is given by

$$f_X(x|\theta) = \begin{cases} \theta e^{-\theta x}, & x > 0 \\ 0, & \text{otherwise} \end{cases}$$

and let $\Pi(\theta)$ be Gamma(α, β).

(a) Find the Bayes estimator of θ with the squared-error loss.

(b) Let $X_1, \ldots, X_n$ be independent, identically distributed exponential random variables with parameter θ. Show that the posterior density of θ given $X_1 = x_1, \ldots, X_n = x_n$ is Gamma$(\alpha + n, \beta + \Sigma_{i=1}^n x_i)$.

6. In Example 12.4.2 assume that $L(\theta, \delta) = (\theta - \delta)^2$. Show that the minimax estimator of θ is admissible.

7. Prove the results stated in Eq. (12.17).

8. Show that for any two random variables X, Y; $\text{var}(X) = E(\text{var}(X|Y)) + \text{var}(E(X|Y))$.

9. Let $X_1, \ldots, X_n$ be independently and identically distributed random variables with pdf $f_X(x|\theta)$. Consider the following three decision rules:

$$\delta_1(X_1, \ldots, X_n) = \frac{1}{n} \sum_{i=1}^n X_i$$

$$\delta_2(X_1, \ldots, X_n) = \frac{1}{n}(X_1 + \cdots + X_{n-1} - nX_n)$$

$$\delta_3(X_1, \ldots, X_n) = \frac{X_1 + X_2 + X_3}{3}$$

and two loss functions,

$$L_1(\theta, \delta) = (\theta - \delta)^2$$
$$L_2(\theta, \delta) = |\theta - \delta|.$$

Find $R_i(\theta, \delta_j) = E(L_i(\theta, \delta_j))$, $i = 1, 2$, $j = 1, 2, 3$, when

(a) $f_X(x|\theta)$ is $N(\theta, 1)$.

(b) $f_X(x|\theta)$ is $N(\theta, \sigma^2)$, $\sigma^2 > 0$ known.

(c) $f_X(x|\theta)$ is binomial $B(N, \theta)$.

(d) $f_X(x|\theta)$ is Poisson (θ).

10. Let $X_1, \ldots, X_n$ be independent, identically distributed $N(0, \theta)$. Find $R(\theta, \delta)$ when $L(\theta, \delta) = (\theta - \delta)^2$ and $\delta(X_1, \ldots, X_n) = n^{-1} \Sigma_{i=1}^n X_i^2$.

11. Let Z be a random variable such that $E(Z) < \infty$. Show that $(Z - b)^2$ is minimum when $b = E(Z)$.

12. Let X be distributed as $N(\theta, 1)$, $\Pi(\theta)$ be $N(0, 1)$, and $L(\theta, \delta) = \exp(3\theta^2/4)(\theta - \delta)^2$. Show that the Bayes rule is not necessarily admissible. (*Hint*: The Bayes risk is ∞ for all δ.)

13. Let X be a random variable with probability density function

$$f_X(x|\theta) = \frac{e^{-\theta}\theta^x}{x!}, \qquad x = 0,1,\ldots$$

and let

$$\Pi(\theta) = \frac{1}{\beta^\alpha\Gamma(\alpha)} \exp\left(-\frac{\theta}{\beta}\right)\theta^{\alpha-1}, \qquad \theta > 0, \quad \alpha > 0, \quad \beta > 0.$$

(a) Find the posterior probability density $\Pi(\theta|X = x)$.
(b) Find the Bayes decision rule.
(c) Show that the maximum-likelihood estimator is not a Bayes rule.

BIBLIOGRAPHY

Berger, J. O. (1985). *Statistical Decision Theory*, Springer-Verlag, New York.

DeGroot, M. H. (1970). *Optimal Statistical Decisions*, McGraw-Hill, New York.

Ferguson, T. S. (1967). *Mathematical Statistics: A Decision-Theoretic Approach*, Academic Press, New York.

Lindley, D. V. (1965). *Introduction to Probability and Statistics from a Bayesian Viewpoint*, Part 1, *Probability*; Part 2, *Inference*, Cambridge University Press, London.

13
Sequential Analysis

So far, while dealing with a problem of statistical inference we have always worked with a sample of a predetermined size. The sample size in such a fixed-sample procedure is determined by various factors guiding the experiment. These may include, among others, the cost of experimentation, knowledge of some previous experiments of a similar nature, and so on. There may be situations where a fixed-sample method may not be very advantageous to apply. For example, while determining the effectiveness of a certain fatal drug, an experiment with a predetermined sample size may prove to be extremely costly, in the long run, in terms of the lives of the subjects, money, time, and effort, particularly when little or no knowledge is available on its effectiveness. In such cases it may be desirable to determine the sample size from the knowledge of the incoming samples. The theory of sequential analysis of statistical problems is based on this basic idea. Here the statistician studies a sequence of observations one at a time, and on the basis of the data obtained thus far decides at each stage either to stop sampling with a terminal decision or to continue with one more observation. Thus in this case the sample size is itself a random variable.

To justify the use of such a procedure even more, consider the following example (see Example 10.5.2): Let $X_1, X_2, \ldots, X_n$ be a random sample of size n from a normal population with mean 0 and variance σ^2, and let $x_1, \ldots, x_n$ be observations on $X_1, \ldots, X_n$. The most powerful test for testing the null hypothesis H_0: $\sigma^2 = \sigma_0^2$ against the alternative H_1: $\sigma^2 = \sigma_1^2$ ($\sigma_1^2 > \sigma_0^2$) rejects H_0 if and only if $\sum_{j=1}^{n} x_j^2 > C$, where C is determined by the level of significance α. Now, if there exists an N, $1 \le N \le n$ for which $\sum_{j=1}^{N} x_j^2 > C$, the latter $n - N$ observations do not influence the conclusion of rejecting H_0, and thus there is really no reason to continue sampling beyond the Nth stage. (It may be remarked at this point that despite such apparent advantages of the sequential procedures, fixed-sample methods often offer simpler and better solutions in many situations.)

Although the idea of sequential analysis dates as far back as 1929 with the introduction of the double-sampling schemes of Dodge and Romig, the pioneering work with full mathematical generality is due to Wald (1947). In this chapter we discuss sequential probability ratio tests (SPRT) only to appreciate some basic aspects of the theory; more sophisticated results and analysis are beyond the scope of the book.

13.1 SEQUENTIAL PROBABILITY RATIO TESTS

Consider a random phenomenon characterized by the stochastic model

$$\left(\Omega, \frac{\partial}{\partial x} F_X(x|\theta) \right), \qquad \theta \in \Omega \tag{13.1}$$

where Ω represents the parametric space. $\partial F_X(x|\theta)/\partial x$ is the probability density or mass function, as the case may be, and in this chapter we always write

$$f_X(x|\theta) = \frac{\partial}{\partial x} F_X(x|\theta).$$

The functional form of $f_X(x|\theta)$ is assumed known. Suppose that for θ_0, $\theta_1 \in \Omega$, $\theta_0 \ne \theta_1$, $f_X(x|\theta_0)$ and $f_X(x|\theta_1)$ are distinct for all x. We are interested in testing the null hypothesis H_0: $\theta = \theta_0$ against the alternative H_1: $\theta = \theta_1$.

Suppose that we observe the mutually independent outcomes $X_1, X_2, \ldots$ in sequence, say $\{x_i\}$, $i = 1, 2, \ldots$, where for any fixed n, $X_1, X_2, \ldots, X_n$ constitute a random sample of size n from the random phenomenon characterized by Eq. (13.1). At each step we compute the likelihood ratio

$$L_n = \frac{\Pi_{j=1}^{n} f_{X_j}(x_j|\theta_1)}{\Pi_{j=1}^{n} f_{X_j}(x_j|\theta_0)}. \tag{13.2}$$

The sequential probability ratio test with boundaries A and B, denoted by $S(A,B;\theta_0,\theta_1)$, where $0 < A < B < \infty$ is defined as follows. At state $n \geq 1$:

1. Accept H_0 if $L_n \leq A$.
2. Reject H_0 if $L_n \geq B$.
3. Continue sampling by taking one more observation if $A < L_n < B$.

To fix ideas, one may at this point make a comparison between the structure of such a test and that of one described by the Neyman–Pearson lemma (Chapter 10).

Let us define

$$z_j = \log \frac{f_{X_j}(x_j|\theta_1)}{f_{X_j}(x_j|\theta_0)}, \qquad Z_j = \log \frac{f_{X_j}(X_j|\theta_1)}{f_{X_j}(X_j|\theta_0)} \tag{13.3}$$

and

$$s_n = \sum_{j=1}^{n} z_j, \qquad S_n = \sum_{j=1}^{n} Z_j. \tag{13.4}$$

Obviously, z_j is an observed value of Z_j; $S(A,B;\theta_0,\theta_1)$ may then be equivalently defined as follows. At stage $n \geq 1$:

1'. Accept H_0 if $s_n \leq \log A$.
2'. Reject H_0 if $s_n \geq \log B$.
3'. Continue sampling by taking one more observation if $\log A < s_n < \log B$.

Since $s_n = \sum_{j=1}^{n} z_j$, $S(A,B;\theta_0,\theta_1)$ in the latter form (1' to 3') is computationally more convenient to handle. Moreover, its probabilistic properties can be obtained by appealing to the rich theory of sums of independent, identically distributed random variables.

It may be noted that the sample size N, here, is an integer-valued random variable that can be defined as

$$N = \begin{cases} \text{the first } n \geq 1 \text{ for which either} \\ \quad \lambda_n \leq \log A \text{ or } \lambda_n \geq \log B, \\ \infty \quad \text{if no such } n \text{ exist.} \end{cases} \tag{13.5}$$

Plainly, N is the value of the sample size when the sampling is terminated. Technically, it is called the *stopping variable* of $S(A,B;\theta_0,\theta_1)$.

Example 13.1.1 Let $\{x_j\}$, $j = 1,2,...$ be a sequence of independent observations from $N(\mu,1)$. We are interested in testing the null hypothesis

H_0: $\mu = \mu_0$ against the alternative H_1: $\mu = \mu_1$, $\mu_1 > \mu_0$. According to the procedure described above, we observe the x_j one by one and compute

$$
\begin{aligned}
z_j &= \log f_{X_j}(x_j|\mu_1) - \log f_{X_j}(x_j|\mu_0) \\
&= \tfrac{1}{2}[(x_j - \mu_0)^2 - (x_j - \mu_1)^2] \\
&= \tfrac{1}{2}[2x_j(\mu_1 - \mu_0) + \mu_1^2 - \mu_0^2] \\
&= (\mu_1 - \mu_0)[x_j + \tfrac{1}{2}(\mu_1 + \mu_0)].
\end{aligned}
$$

Therefore, we have

$$
S_n = \sum_{j=1}^{n} z_j = (\mu_1 - \mu_0)\left[\sum_{j=1}^{n} x_j + \frac{n}{2}(\mu_1 + \mu_0)\right].
$$

The SPRT $S(A,B;\mu_0,\mu_1)$ is then described as follows (recall that $\mu_1 > \mu_0$). At stage $n \geq 1$:

1'. Accept H_0 if

$$
\sum_{j=1}^{n} x_j \leq \frac{\log A}{\mu_1 - \mu_0} - \frac{n}{2}(\mu_1 + \mu_0).
$$

2'. Reject H_0 if

$$
\sum_{j=1}^{n} x_j \geq \frac{\log B}{\mu_1 - \mu_0} - \frac{n}{2}(\mu_1 + \mu_0).
$$

3'. Continue sampling by taking one more observation if

$$
\frac{\log A}{\mu_1 - \mu_0} - \frac{n}{2}(\mu_1 + \mu_0) < \sum_{j=1}^{n} x_j < \frac{\log B}{\mu_1 - \mu_0} - \frac{n}{2}(\mu_1 + \mu_0).
$$

Example 13.1.2 Let $\{x_j\}$, $j = 1,2,\ldots$ be a sequence of independent observations from a population characterized by the probability mass function

$$
p_X(x|p) = \begin{cases} p^x(1 - p)^{1-x}, & x = 0,1 \\ 0, & \text{otherwise,} \end{cases}
$$

where $0 < p < 1$. We are interested in testing the null hypothesis H_0: $p = p_0$ against the alternative H_1: $p = p_1$, $0 < p_0 < p_1 < 1$. As before, the x_j are observed one by one and we compute

$$
\begin{aligned}
z_j &= \log f_X(x_j|p_1) - \log f_X(x_j|p_0) \\
&= x_j \log \frac{p_1}{p_0} + (1 - x_j) \log \frac{1 - p_1}{1 - p_0}.
\end{aligned}
$$

Therefore, we have

$$
S_n = \sum_{j=1}^{n} z_j
$$

$$
= \left(\sum_{j=1}^{n} x_j \right) \log \frac{p_1}{p_0} + \left(n - \sum_{j=1}^{n} x_j \right) \log \frac{1 - p_1}{1 - p_0}
$$

$$
= \left(\sum_{j=1}^{n} x_j \right) \left[\log \frac{p_1}{p_0} - \log \frac{1 - p_1}{1 - p_0} \right] + n \log \frac{1 - p_1}{1 - p_0}.
$$

Since $p_1 > p_0$, the coefficient of $\sum_{j=1}^{n} x_j$ is positive in the expression above. Thus the SPRT $S(A,B;p_0,p_1)$ is described as follows. At stage $n \geq 1$:

1'. Accept H_0 if

$$
\sum_{j=1}^{n} x_j \leq \frac{\log A - n \log[(1 - p_1)/(1 - p_0)]}{\log(p_1/p_0) - \log[(1 - p_1)/(1 - p_0)]}.
$$

2'. Reject H_0 if

$$
\sum_{j=1}^{n} x_j \geq \frac{\log B - n \log[(1 - p_1)/(1 - p_0)]}{\log(p_1/p_0) - \log[(1 - p_1)/(1 - p_0)]}.
$$

3'. Continue sampling by taking one more observation if $\sum_{j=1}^{n} x_j$ falls in between the two limits above.

13.2 FUNDAMENTAL RELATIONSHIP BETWEEN *A, B* AND THE ERROR PROBABILITIES α, β

Let α and β denote the probabilities of errors of first and second kinds, respectively, for testing the simple null hypothesis $H_0: \theta = \theta_0$ against the simple alternative $H_1: \theta = \theta_1$. That is,

$$
\alpha = P\{\text{reject } H_0 | \theta_0\} = P_{\theta_0}\{\text{reject } H_0\} \tag{13.6}
$$

and

$$
\beta = P\{\text{accept } H_0 | \theta_1\} = P_{\theta_1}\{\text{accept } H_0\}. \tag{13.7}
$$

Let $\mathcal{R}_n^0$ be the set of all n-dimensional vectors $(x_1, x_2, \ldots, x_n)$ that lead to the acceptance of H_0 at the nth stage, and let $\mathcal{R}_n^1$ be the set of all n-dimensional vectors $(x_1, x_2, \ldots, x_n)$ that lead to the rejection of H_0 (i.e., the acceptance of H_1) at the nth stage. Let $S(A,B;\theta_0,\theta_1)$ be as before. Then the following theorem holds.

Theorem 13.2.1 Subject to notations described above,

$$A \geq \frac{\beta}{1 - \alpha}, \qquad B \leq \frac{1 - \beta}{\alpha}. \tag{13.8}$$

Proof. Let E_n, $n = 1, 2, \ldots$ denote the event of terminating the procedure at the nth stage with the acceptance of H_0, that is,

$$E_n = \{(\text{accept } H_0) \cap (N = n)\}.$$

Thus we have

$$P\{E_n | \theta\} = \int \int \cdots \int_{\mathfrak{R}_n^0} \prod_{j=1}^{n} f_{X_j}(x_j | \theta) \, dx_j.$$

Clearly, E_i and E_j are disjoint if $i \neq j$; then we have by the theorem of total probability,

$$1 - \alpha = P_{\theta_0}\{\text{accept } H_0\}$$

$$= P\left\{ \bigcup_{n=1}^{\infty} E_n | \theta_0 \right\}$$

$$= \sum_{n=1}^{\infty} P\{E_n | \theta_0\}$$

$$= \sum_{n=1}^{\infty} \int \int \cdots \int_{\mathfrak{R}_n^0} \prod_{j=1}^{n} f_X(x_j | \theta_0) \, dx_j. \tag{13.9}$$

Similarly,

$$\beta = P_{\theta_1}\{\text{accept } H_0\} = \sum_{n=1}^{\infty} \int \int \cdots \int_{\mathfrak{R}_n^0} \prod_{j=1}^{n} f_{X_j}(x_j | \theta_1) \, dx_j. \tag{13.10}$$

Now, by conditions 1 to 3 of Section 13.1, we accept H_0 if and only if $L_n \leq A$; that is, $(x_1, x_2, \ldots, x_n) \in \mathfrak{R}_n^0$ if and only if

$$\prod_{j=1}^{n} f_{X_j}(x_j | \theta_1) \leq A \prod_{j=1}^{n} f_{X_j}(x_j | \theta_0).$$

Integrating both sides of the inequality above over $\mathfrak{R}_n^0$ and summing over $1 \leq n < \infty$, we find by Eqs. (13.9) and (13.10) that

$$\beta \leq A(1 - \alpha),$$

which is the first inequality of the theorem. The other can be proved in a similar way. (The reader is advised to carry out the proof for practice.) Q.E.D.

In practice, α and β are given and A and B are chosen to meet the specification set by α and β. Let us now see what happens in terms of the error probabilities α and β if one chooses $A = \beta/(1 - \alpha)$ and $B = (1 - \beta)/\alpha$. Let α' and β' be the actual probabilities of errors of the first and second kinds with such a choice of A and B. In conformity with this, α and β may be termed the desired error probabilities. Then, from Theorem 13.2.1 we must have

$$\frac{\beta'}{1 - \alpha'} \leq \frac{\beta}{1 - \alpha}, \qquad \frac{1 - \beta}{\alpha} \leq \frac{1 - \beta'}{\alpha'}. \tag{13.11}$$

These imply

$$\beta' \leq \frac{\beta}{1 - \alpha}, \qquad \alpha' \leq \frac{\alpha}{1 - \beta} \tag{13.12}$$

and

$$\beta' - \alpha\beta' \leq \beta - \alpha'\beta \tag{13.13}$$
$$\alpha' - \alpha'\beta \leq \alpha - \alpha\beta'.$$

Adding the two inequalities in (13.13), we find that

$$\alpha' + \beta' \leq \alpha + \beta. \tag{13.14}$$

This inequality implies that the sum of the actual probabilities of errors do not exceed the sum of their desired values. Moreover, at most one of the actual error probabilities can exceed its desired value. In actual practice, the desired error probabilities α and β are quite small, so that the right-hand sides of the inequalities (13.12) are also small. This, in turn, implies that the differences $\alpha - \alpha'$ and $\beta - \beta'$ cannot be large, so that for all practical purposes, we may choose $A = \beta/(1 - \alpha)$ and $B = (1 - \beta)/\alpha$.

13.3 PROPERTIES OF THE STOPPING RULE N IN THE SPRT

In this section we derive various probabilistic and moment properties of the stopping rule N as defined in Eq. (13.5). Recall that the random variables Z_j defined by Eq. (13.3) are all independent and identically distributed. In all the theorems given below, except for the last one, we shall assume that

$$P_\theta\{Z_1 = 0\} < 1, \qquad \theta \in \Omega. \tag{13.15}$$

It may be noted that this condition is not much of a restriction. It merely reaffirms that $f_X(x|\theta_0)$ and $f_X(x|\theta_1)$ are not identically equal with probability 1, when $\theta_0 \neq \theta_1$; otherwise, H_0 and H_1 become indistinguishable.

Theorem 13.3.1 If condition (13.15) holds, the SPRT terminates with probability 1, that is,

$$P_\theta\{N < \infty\} = 1, \quad \theta \in \Omega. \tag{13.16}$$

Proof. Let us write $a = \log A$, $b = \log B$, and $c = |a| + |b|$. The proof will consist in showing that under condition (13.15), there exists an integer r such that

$$P_\theta\{|Z_1 + Z_2 + \cdots + Z_r| < c\} = p < 1 \tag{13.17}$$

and

$$P_\theta\{N > mr\} < p^m \quad \text{for any integer } m. \tag{13.18}$$

The theorem will then follow by taking the limit of both sides of (13.18) as $m \to \infty$.

Note that (13.17) is trivially true with $r = 1$ if

$$P_\theta\{|Z_1| < c\} < 1.$$

Thus it will be sufficient to prove (13.17) under the assumption (13.15) and

$$P_\theta\{|Z_1| < c\} = 1. \tag{13.19}$$

In this case, there exists an $\epsilon > 0$ such that

$$\begin{aligned} P_\theta\{Z_1 > c - \epsilon\} &> 0 \\ P_\theta\{Z_1 < -(c - \epsilon)\} &> 0. \end{aligned} \tag{13.20}$$

[Otherwise, the left-hand sides are zero for every $\epsilon > 0$, so that

$$1 = \lim_{\epsilon \to c} P_\theta\{|Z_1| < c - \epsilon\} = P_\theta\{Z_1 = 0\},$$

which contradicts (13.15).] Choose $r = [1 + c/c - \epsilon]$, where $[y]$ is the integer part of y. Then there exists a δ, $0 \le \delta < 1$, such that

$$r = 1 + \frac{c}{c - \epsilon} - \delta,$$

that is,

$$r(c - \epsilon) = c + (1 - \delta)(c - \epsilon) \ge c.$$

Let us write

$$E_j^- \equiv \text{event } \{Z_j \le -(c - \epsilon)\}$$
$$E_j^+ \equiv \text{event } \{Z_j \ge (c - \epsilon)\},$$

$j = 1,2,...,r$. Then, clearly, we have

$$P_\theta\{|Z_1 + Z_2 + \cdots + Z_r| \geq c\} \geq P_\theta\{|Z_1 + Z_2 + \cdots + Z_r| \geq r(c - \epsilon)\}$$

$$\geq P_\theta\left\{\left(\bigcap_{j=1}^{r} E_j^-\right) \cup \left(\bigcap_{j=1}^{r} E_j^+\right)\right\}$$

$$= P_\theta\left\{\bigcap_{j=1}^{r} E_j^-\right\} + P_\theta\left\{\bigcap_{j=1}^{r} E_j^+\right\}$$

$$= P_\theta\{Z_j \leq -(c - \epsilon), j = 1,2,...,r\}$$
$$+ P_\theta\{Z_j \geq (c - \epsilon), j = 1,2,...,r\}$$

$$= \prod_{j=1}^{r} P_\theta\{Z_j \leq -(c - \epsilon)\}$$

$$+ \prod_{j=1}^{r} P_\theta\{Z_j \geq (c - \epsilon)\}$$

(since the Z_j are independent)

$$= P_\theta^r\{Z_1 \leq -(c - \epsilon)\} + P_\theta^r\{Z_1 \geq c - \epsilon\}$$

since the Z_j have the same distribution. Then from (13.20) it follows that

$$p = P_\theta\{|Z_1 + Z_2 + \cdots + Z_r| < c\} < 1.$$

For convenience, write $S_n = Z_1 + Z_2 + \cdots + Z_n$, $n = 1,2,...$. Note that $a < S_r < b$ and $a < S_{2r} < b$ imply that $|S_{2r} - S_r| < c$, where $c = |a| + |b|$. Similarly, for any integer m,

event $\{a < S_r < b, a < S_{2r} < b,...,a < S_{mr} < b\}$

$\subset$ event $\{|S_r| < c, |S_{2r} - S_r| < c,...,|S_{mr} - S_{(m-1)r}| < c\}$.

Also note that the sampling is done beyond the mrth state if an only if $a < S_j < b$ for $j = 1,2,...,mr$. Therefore, we have

$$P_\theta\{N > mr\} = P_\theta\{a < S_1 < b, a < S_2 < b,...,a < S_{mr} < b\}$$

$$\leq P_\theta\{a < S_r < b, b < S_{2r} < b,...,a < S_{mr} < b\}$$

$$\leq P_\theta\{|S_r| < c, |S_{2r} - S_r| < c,...,|S_{mr} - S_{(m-1)r}| < c\}$$

$$= P_\theta\left\{\left|\sum_{j=1}^{r} Z_j\right| < c, \left|\sum_{j=r+1}^{2r} Z_j\right| < c,..., \left|\sum_{j=(m-1)r+1}^{mr} Z_j\right| < c\right\}$$

$$= p^m$$

by (13.17), since the Z_j are independent and identically distributed. In view of what has been said earlier, this completes the proof of the theorem.   Q.E.D.

Theorem 13.3.2 If condition (13.15) holds, there exists a $t_0 > 0$ such that the moment-generating function $E_\theta e^{tN}$ of N exists for all $t < t_0$.

Proof. We have seen in the proof of Theorem 13.3.1 that under the condition (13.15) there exists an integer r and a $p < 1$ such that $P_\theta\{N > mr\} \le p^m$ for any integer m. For any fixed integer n, take $m = [n/r]$. Then $m = (n/r) - \delta$ for some δ, $0 \le \delta \le 1$. Thus

$$P_\theta\{N > n\} \le P_\theta\{N > mr\}$$

$$\le p^m$$

$$= (p)^{(n/r)-\delta}$$

$$\le (p)^{(n/r)-1} \qquad \text{(since } p < 1\text{)}$$

$$= K\delta^n, \tag{13.21}$$

where $K = 1/p$ and $\delta = p^{1/r}$ (if $p = 0$, we shall take the right-hand side to be 0). Now, by definition, for $t > 0$, we have

$$E_\theta e^{tN} = \sum_{n=1}^{\infty} e^{tn} P_\theta\{N = n\}$$

$$\le \sum_{n=1}^{\infty} e^{tn} P_\theta\{N \ge n\}$$

$$\le K \sum_{n=1}^{\infty} e^{tn} \delta^{n-1} \qquad \text{[by (13.21)]}$$

$$= \frac{K}{\delta} \sum_{n=1}^{\infty} (\delta e^t)^n,$$

which is finite for all $t < t_0$, where t_0 is such that $\delta e^{t_0} < 1$. This completes the proof of the theorem.   Q.E.D.

Corollary 13.3.2.1 If condition (13.15) holds, then $E_\theta N^s < \infty$ for all $s = 1, 2, \ldots$.

This follows immediately from Theorem 13.3.2.
It is well known that for any fixed integer n,

$$E_\theta \sum_{j=1}^{n} Z_j = \sum_{j=1}^{n} E_\theta Z_j = n E_\theta Z_1$$

since $E_\theta Z_1 = E_\theta Z_2 = \cdots = E_\theta Z_n$, the Z_j being identically distributed. The following theorem is a generalization of this fact.

Theorem 13.3.3: Wald's equation   If for any fixed $\theta \in \Omega$, $E_\theta|Z_1| < \infty$ and $E_\theta N < \infty$, then

$$E_\theta \sum_{j=1}^{N} Z_j = (E_\theta Z_1)(E_\theta N). \tag{13.22}$$

Proof.   Define the random variables $Y_1, Y_2, \ldots$, as follows:

$$Y_j = \begin{cases} 1 & \text{if } N \geq j \\ 0 & \text{otherwise} \end{cases}$$

for $j = 1, 2, \ldots$. Clearly, then, we have

$$E_\theta Y_j = P_\theta\{N \geq j\} = \sum_{n=j}^{\infty} P_\theta\{N = n\}.$$

Interchanging the order of summations, one easily gets

$$\sum_{j=1}^{\infty} E_\theta Y_j = \sum_{j=1}^{\infty} \sum_{n=j}^{\infty} P_\theta\{N = n\}$$

$$= \sum_{n=1}^{\infty} \sum_{j=1}^{n} P_\theta\{N = n\}$$

$$= \sum_{n=1}^{\infty} nP_\theta\{N = n\}$$

$$= E_\theta N. \tag{13.23}$$

Also note that the event $\{N \geq j\}$ depends only on $Z_1, Z_2, \ldots, Z_{j-1}$ (recall the definition of N). Therefore, Y_j and Z_j are independent, $j = 1, 2, \ldots$, and so

$$E_\theta Y_j Z_j = (E_\theta Y_j)(E_\theta Z_j). \tag{13.24}$$

Finally, it may be easily verified that

$$\sum_{j=1}^{N} Z_j = \sum_{j=1}^{\infty} Y_j Z_j.$$

Then we have

$$E_\theta \sum_{j=1}^{N} Z_j = E_\theta \sum_{j=1}^{\infty} Y_j Z_j = \sum_{j=1}^{\infty} E_\theta(Y_j Z_j)$$

$$= \sum_{j=1}^{\infty} (E_\theta Y_j)(E_\theta Z_j) \qquad \text{[by Eq. (13.24)]}$$

$$= (E_\theta Z_1) \sum_{j=1}^{\infty} E_\theta Y_j$$

$$= (E_\theta Z_1)(E_\theta N) \qquad \text{[by Eq. (13.23)]}$$

since the Z_j all have the same expectation. It was possible to interchange the order of summation and expectation because

$$\sum_{j=1}^{\infty} |E_\theta(Y_j Z_j)| = \sum_{j=1}^{\infty} |(E_\theta Y_j)(E_\theta Z_j)|$$

$$\le E_\theta |Z_1| \sum_{j=1}^{\infty} E_\theta Y_j$$

$$= E_\theta |Z_1|(E_\theta N)$$

$$< \infty$$

by the assumptions of our theorem. This completes the proof.   Q.E.D.

13.4   OPERATING CHARACTERISTIC FUNCTION

Consider a test T for testing H_0 against H_1. For $\theta \in \Omega$, write $\pi_T(\theta)$ to denote the probability that T leads to the acceptance of H_0 when θ is the true value of the parameter; $\pi_T(\theta)$ when considered as a function of θ is called the *operating characteristic* (OC) *function* of the test T. It gives an idea of the strength of the test in terms of its probabilistic capacity to distinguish between θ_0 and other possible values of θ. Clearly,

$$\pi_T(\theta_0) = 1 - \alpha, \qquad \pi_T(\theta_1) = \beta, \tag{13.25}$$

where α and β are the probabilities of the errors of the first and second kinds, respectively. It is difficult to compute the values of the OC function $\pi_S(\theta)$ of the SPRT $S(A, B; \theta_0, \theta_1)$; however, good approximations are available through the following lemma and theorem. The proofs are omitted here; they can be found in standard textbooks [see, e.g., Ghosh (1970), Wald (1947), or Zacks (1971)].

Lemma 13.4.1 Let Z be a random variable such that
  (a) $EZ \neq 0$.
  (b) $P\{Z > 0\} > 0$, $P\{Z < 0\} > 0$.
  (c)   Its moment-generating function $\phi(t) = Ee^{tZ} < \infty$ for all t.
Then there exists a $t_0 \neq 0$ such that $\phi(t_0) = 1$; t_0 is positive or negative according as EZ is negative or positive.

Theorem 13.4.1: Wald's fundamental identity Let $Z_1, Z_2, \ldots$ as defined in Eqs. (13.3) be independent and identically distributed and satisfy, for a given $\theta \in \Omega$, the conditions of Lemma 13.4.1, with $\phi_\theta(t) = E_\theta e^{tZ_1}$. Write $S_n = Z_1 + Z_2 + \cdots + Z_n$, $n = 1, 2, \ldots$, and let N be defined as in Eq. (13.5). Then

$$E_\theta\{e^{tS_N}[\phi_\theta(t)]^{-N}\} = 1. \tag{13.26}$$

Observe that the identity above is trivially true for any fixed-sample test.

To obtain an approximation for $\pi_S(\theta)$, assume that the conditions of Theorem 13.4.1 are satisfied by $Z_1, Z_2, \ldots$. Then by Lemma 13.4.1 there exists a $t_0 = t_0(\theta)$ such that

$$\phi_\theta(t_0(\theta)) = E_\theta e^{t_0(\theta)Z_1} = 1,$$

so that from Eq. (5.26) we have

$$E_\theta\{e^{t_0(\theta)S_N}\} = 1. \tag{13.27}$$

Now observe that $\pi_S(\theta)$ is the probability that $S_N \leq \log B$. Since, under the conditions of the theorem [because of condition (b) of Lemma 13.4.1], the SPRT $S(A, B)$ must terminate with probability 1 (see Theorem 13.3.1), $1 - \pi_S(\theta)$ is the probability of the event $S_N \geq \log B$. Therefore, from Eq. (13.27) we obtain

$$\pi_s(\theta)E_\theta\{e^{t_0(\theta)S_N}|S_N \leq \log A\} + [1 - \pi_s(\theta)]E_\theta\{e^{t_0(\theta)S_N}|S_N \geq \log B\} = 1,$$

so that

$$\pi_S(\theta) = \frac{1 - E_\theta\{e^{t_0(\theta)S_N}|S_N \geq \log B\}}{E_\theta\{e^{t_0(\theta)S_N}|S_N < \log A\} - E_\theta\{e^{t_0(\theta)S_N}|S_N \geq \log B\}}. \tag{13.28}$$

For all practical purposes, the overshoot of S_N beyond the boundaries $\log A$ and $\log B$ may be considered negligible; then

$$E_\theta\{e^{t_0(\theta)S_N}|S_n \leq \log A\} \simeq e^{t_0(\theta)\log A} = A^{t_0(\theta)}$$

$$E_\theta\{e^{t_0(\theta)S_N}|S_N \geq \log B\} \simeq e^{t_0(\theta)\log B} = B^{t_0(\theta)},$$

where "$\simeq$" means "approximately equal to." From these equations and Eq. (13.28) we now obtain the approximation

$$\pi_S(\theta) \simeq \frac{1 - B^{t_0(\theta)}}{B^{t_0(\theta)} - A^{t_0(\theta)}}, \tag{13.29}$$

where $t_0(\theta)$ is the solution of the equation

$$\phi_\theta(t) = E_\theta e^{tZ_1} = \int_{-\infty}^{+\infty} \left[\frac{f_X(x|\theta_1)}{f_X(x|\theta_0)} \right]^t f_X(x|\theta) \, dx = 1. \tag{13.30}$$

Example 13.4.1 Consider the problem described in Example 13.1.1. Here, for $-\infty < \mu < \infty$,

$$E_\mu(e^{tZ}) = \int_{-\infty}^{+\infty} \left\{ \frac{\exp[-(\frac{1}{2})(x - \mu_1)^2]}{\exp[-(\frac{1}{2})(x - \mu_0)^2]} \right\}^t \frac{1}{\sqrt{2\pi}} \exp[-(\frac{1}{2})(x - \mu)^2] \, dx$$

$$= \frac{1}{\sqrt{2\pi}} \int_\infty^\infty \exp\left[-\frac{t}{2} (2x\mu_0 - 2x\mu_1 + \mu_1^2 - \mu_0^2) \right]$$
$$\times \exp[-(\frac{1}{2})(x - \mu)^2] \, dx$$

$$= \frac{1}{\sqrt{2\pi}} \int_{-\infty}^{+\infty} \exp\{-(\frac{1}{2})[(x - \mu)^2 + 2xt(\mu_0 - \mu_1)$$
$$+ t(\mu_1^2 - \mu_0^2)]\} \, dx$$

$$= \left\{ \frac{1}{\sqrt{2\pi}} \int_{-\infty}^{+\infty} \exp\{-(\frac{1}{2})[(x - \mu)^2 + 2(x - \mu)t(\mu_0 - \mu_1)\right.$$

$$\left. + t^2(\mu_0 - \mu_1)^2]\} \, dx \right\} \exp\{-(\frac{1}{2})[t(\mu_1^2 - \mu_0^2)$$
$$- t^2(\mu_0 - \mu_1)^2 + 2\mu t(\mu_0 - \mu_1)]\}$$

$$= \exp(\frac{1}{2})[t^2(\mu_0 - \mu_1)^2 - t(\mu_1^2 - \mu_0^2) - 2\mu t(\mu_0 - \mu_1)]$$

since the first factor is equal to 1. Then the equation

$$E_\mu(e^{t_0 Z}) = 1 \qquad \text{for } t_0 = t_0(\mu)$$

implies that

$$t_0^2(\mu_0 - \mu_1)^2 - t_0(\mu_1^2 - \mu_0^2) - 2\mu t_0(\mu_0 - \mu_1) = 0,$$

that is,

$$t_0(\mu_1 - \mu_0) = (\mu_1 + \mu_0) - 2\mu,$$

so that

$$t_0(\mu) = \frac{\mu_1 + \mu_0 - 2\mu}{\mu_1 - \mu_0}. \tag{13.31}$$

Therefore, the OC $\pi_S(\mu)$ for the test described in Example 13.1.1 can be approximated by formula (13.29) with $t_0(\theta)$ replaced by $t_0(\mu)$, where the latter is given by Eq. (13.31).

Example 13.4.2 Consider the problem described in Example 13.1.2. Here, for $0 < p < 1$,

$$E_p(e^{tZ}) = (1 - p)\left(\frac{1 - p_1}{1 - p_0}\right)^t + p\left(\frac{p_1}{p_0}\right)^t,$$

which is rather difficult to solve for t theoretically. However, the relation

$$E_p(e^{t_0(p)Z}) = 1$$

yields the following functional relationship between p and $t_0(p)$:

$$p = \frac{1 - [(1 - p_1)/(1 - p_0)]^{t_0(p)}}{(p_1/p_0)^{t_0(p)} - [(1 - p_1)/(1 - p_0)]^{t_0(p)}}. \tag{13.32}$$

Formula (13.29) can now be used to compute approximations of $\pi_S(p)$ for values of p that are obtained via Eq. (13.32) for given values of $t_0(p)$.

13.5 AVERAGE SAMPLING NUMBER OF THE SPRT

As has been seen earlier, the number of observations N until a terminal decision is made is an integer-valued random variable. $E_\theta N$, when treated as function of $\theta \in \Omega$, is called the *average sampling number* (ASN) *function* of $S(A,B;\theta_0,\theta_1)$. The ASN function thus provides a rough idea about the sample size to be required and also a basis for comparison of $S(A,B;\theta_0,\theta_1)$ with other sequential procedures.

Although it may sometimes be difficult to compute the exact value of $E_\theta N$, it is possible to obtain an approximation of it with the help of Theorem 13.3.3. If $E_\theta Z_1 \neq 0$, then under all necessary conditions we have, from Eq. (13.22),

$$E_\theta N = \frac{E_\theta S_N}{E_\theta Z_1}, \tag{13.33}$$

where $S_N = Z_1 + Z_2 + \cdots + Z_N$. But, as before,

$$E_\theta S_N = \pi_S(\theta) E_\theta\{S_N | S_N \le \log A\}$$
$$+ (1 - \pi_S(\theta)) E_\theta\{S_N | S_N \ge \log B\}, \tag{13.34}$$

where $\pi_S(\theta)$ is the OC of $S(A, B; \theta_0, \theta_1)$. Again, for all practical purposes the overshoot of S_N beyond the boundaries $\log A$ and $\log B$ may be neglected. Thus

$$E_\theta\{S_N | S_N \le \log A\} \simeq \log A$$

and

$$E_\theta\{S_N | S_N \ge \log B\} \simeq \log B.$$

Therefore, from Eq. (13.34) we have

$$E_\theta N \simeq \frac{\pi_S(\theta) \log A + [1 - \pi_S(\theta)] \log B}{E_\theta Z_1}. \tag{13.35}$$

In practical applications, Eq. (13.35) with $\pi_S(\theta)$ replaced by its approximate value provides a satisfactory approximation for $E_\theta N$.

Example 13.5.1 Consider the problem described in Example 13.1.1. Here, for $-\infty < \mu < \infty$, we have

$$E_\mu(Z_1) = E_\mu\{(\mu_1 - \mu_0)X_1 + \tfrac{1}{2}(\mu_1^2 - \mu_0^2)\}$$
$$= (\mu_1 - \mu_0)\mu + \tfrac{1}{2}(\mu_1^2 - \mu_0^2).$$

This can now be plugged into the denominator of Eq. (13.35) to obtain an approximation for $E_\theta N$.

Example 13.5.2 Consider the problem described in Example 13.1.2. Here, for $0 < p < 1$, we have

$$E_p(Z_1) = p \log \frac{p_1}{p_0} + (1 - p) \log \frac{1 - p_1}{1 - p_0}.$$

EXERCISES

1. Assume that the left-hand side of Wald's fundamental identity permits differentiation with respect to t twice under the expectation in an open interval containing zero.
   (a) Show that the first derivative leads to Wald's equation.
   (b) What identity can you derive from the second derivative?

2. Show that if we choose $A = \beta/(1 - \alpha)$ and $B = (1 - \beta)/\alpha$, then the approximation (13.29) gives the correct values of $\pi_S(\theta_0)$ and $\pi_S(\theta_1)$. [*Hint*: Note that $t_0(\theta_0) = 1 = -t_0(\theta_1)$.]
3. Consider the testing problem described in Example 13.1.1. Take $\mu_0 = 0$, $\mu_1 = 1$, $A = \beta/(1 - \alpha)$, $B = (1 - \beta)/\alpha$ with $\alpha = 0.05$, $\beta = 0.10$. Calculate the approximate values of $E_{\mu_0}N$ and $E_{\mu_1}N$.
4. Let x_j, $j = 1,2,\ldots$ be a sequence of observations from a population having a pdf $f_X(x,\theta) = \theta e^{-\theta x}$, $\theta > 0$, $0 < x < \infty$. Take $\alpha = 0.05$, $\beta = 0.10$, $A = \beta/(1 - \alpha)$, $B = (1 - \beta)/\alpha$.
   (a) Determine the SPRT $S(A,B;\theta_0,\theta_1)$ with $\theta_1 > \theta_0$.
   (b) Calculate the approximate values of $E_\theta N$ for $\theta_0 = 1$, $\theta_1 = 2$.

BIBLIOGRAPHY

Ghosh, B. K. 91970). *Sequential Tests of Statistical Hypotheses*, Addison-Wesley, Reading, Mass.

Lehmann, E. L. (1959). *Testing Statistical Hypotheses*, Wiley, New York.

Wald, A. (1947). *Sequential Analysis*, Wiley, New York.

Wilks, S. S. (1962). *Mathematical Statistics*, Wiley, New York.

Zacks, S. (1971). *The Theory of Statistical Inference*, Wiley, New York.

14

Nonparametric Methods

In this chapter we describe and discuss some nonparametric methods of testing certain statistical hypotheses. As has been mentioned earlier, these methods, in contrast with the parametric methods described in Chapters 9 and 10, do not require specific knowledge of the functional forms of the underlying probability distributions of the random phenomenon.

14.1 ONE-SAMPLE METHODS

Of all the methods available, we describe only the sign test here.

14.1.1 Sign Test

Consider a random phenomenon specified by the continuous distribution function $F_X(x)$. Let there be a constant μ such that $F_X(\mu) = \frac{1}{2}$. In other words, μ is the median of the distribution $F_X(x)$. We want to test the null hypothesis H_0: $\mu = \mu_0$, where μ_0 is a given constant. Let $X_1, X_2, \ldots, X_n$ be a random sample of size n from this random phenomenon and let $x_1, x_2, \ldots, x_n$ be the corresponding sample of observations. Given $x_1, x_2, \ldots, x_n$, record only the signs of the differences $x_j - \mu_0$, $j = 1, 2, \ldots, n$, instead of their actual values. Suppose that n_1 of these n differences are negative, so that

$n - n_1$ of them are positive (observe that, X being a continuous random variable, $P\{X = \mu_0\} = 0$). Thus n_1 is an observed value of an integer-valued random variable N_1 assuming values between 0 and n.

The rationale of the method is as follows: If $\mu = \mu_0$, then the x_j are expected to be more or less evenly distributed about μ_0, so that the proportion of $+$ signs will be about the same as the proportion of $-$ signs. If $\mu > \mu_0$, more values are expected to be to the right of μ_0 than to its left, so that n_1 will tend to be small. A similar argument indicates that n_1 will tend to be large if $\mu < \mu_0$.

It is then clear that H_0: $\mu = \mu_0$ is equivalent to testing whether the corresponding binomial proportion of positive or negative signs is $\frac{1}{2}$. Under the null hypothesis, N_1 has a binomial distribution with parameters n and $p = \frac{1}{2}$. Therefore,

$$p_j = P\{N_1 = j\} = \begin{cases} \binom{n}{j} 2^{-n}, & j = 0,1,2,\ldots,n \\ 0, & \text{otherwise.} \end{cases} \tag{14.1}$$

For a given level of significance α, $0 < \alpha < 1$, let C_α be such that

$$\sum_{j=0}^{C_\alpha} p_j \leq \alpha < \sum_{j=0}^{C_\alpha+1} p_j. \tag{14.2}$$

If the alternative hypothesis is H_1: $\mu > \mu_0$, the critical region is specified by the inequality $n_1 \leq C_\alpha$. Similarly, if the alternative is H: $\mu < \mu_0$, the critical region is specified by $n_1 \geq C_\alpha$, while for the two-sided alternative H: $\mu \neq \mu_0$, one uses the critical region specified by

$$\{(x_1,\ldots,x_n) : n_1 \leq C_{\alpha/2}\} \cup \{(x_1,\ldots,x_n) : n_1 \geq n - C_{\alpha/2}\}.$$

If n is sufficiently large, one may alternatively use a test based on a contingency χ^2 with one degree of freedom (see Chapter 11).

It may be remarked here that the procedure above can easily be generalized to handle problems involving quantiles of any order of $F_X(x)$. Let ξ be the pth quantile of $F_X(x)$ [i.e., $F_X(\xi) = p, 0 < p < 1$]. The method of testing H_0: $\xi = \xi_0$ is exactly the same as that of testing H_0: $\mu = \mu_0$ in the case of the median, except that under H_0

$$p_j = P(N_1 = j) = \begin{cases} \binom{n}{j} p^j (1 - p)^{n-j}, & j = 0,1,2,\ldots,n \\ 0, & \text{otherwise.} \end{cases} \tag{14.3}$$

Example 14.1.1   The median height μ of a group of people is expected to be 65 inches. A random sample of size 10 yielded the following result:

    55, 63, 64, 70, 58, 72, 63, 71, 69, 66.

We shall use the sign test to determine if the data are consistent with the hypothesis $\mu = 65$.
    Note that the differences $x_j - \mu_0$ are

    $-10, -2, -1, 5, -7, 7, -2, 6, 4, 1,$

so that $n_1 = 5$. Also, from the binomial tables, for $n = 10$, $\alpha = 0.05$, we find that $C_{0.025} = 1$.
    Since the observed n_1 is greater than $C_{0.025} = 1$ but less than $n - C_{0.025} = 9$, we conclude that the data support the hypothesis that $\mu = 65$.

14.2   TWO-SAMPLE METHODS

Let X and Y be two random variables having continuous distribution functions F_1 and F_2, respectively. In what follows we shall describe a few methods of testing the null hypothesis H_0: $F_1(x) = F_2(x)$ for all x (which is equivalently expressed as $F_1 \equiv F_2$) against suitable alternatives.

14.2.1   Two-Sample Sign Test

Let the alternative hypothesis be $F_1(x) = F_2(x + a)$ for all x, where a may be positive, negative, or zero, depending on the situation.
    We take a random sample of size n from each population and pair them in a random fashion. Let the paired observations be $(x_1, y_1), (x_2, y_2), \ldots,$ (x_n, y_n), where the x_j are the observations on X and the y_j are the observations on Y. For each pair, we record the sign of the difference $(x_j - y_j)$, $j = 1, 2, \ldots, n$. (Note that since X and Y both have continuous pdf's, $P\{X = Y\} = 0$.) Observe that

$$P\{X - Y > 0\} = \int_{(x-y)>0} dF_1(x)\, dF_2(y)$$

$$= \int_{-\infty}^{\infty} dF_2(y) \int_{y}^{\infty} dF_1(x)$$

$$= \int_{-\infty}^{\infty} [1 - F_1(y)]\, dF_2(y).$$

Thus $P\{X - Y > 0\} = \frac{1}{2}$ if and only if $F_1 \equiv F_2$. Therefore, equivalently, we may test the hypothesis that the median μ of $Z = X - Y$ is zero, and for this we may apply the one-sample procedure that we have already described. The alternatives $F_1(x) = F_2(x + a)$ may be translated as $\mu < 0$, $\mu > 0$, $\mu \neq 0$ according as a is positive, negative, or just $a \neq 0$. (The reader is advised to prove this statement for practice.)

14.2.2 Median Test

If the data consist of unequal numbers of observations from the two populations, the sign test is not the natural one to apply, because one must then throw away the excess observations from the larger sample, and this would in turn mean the waste of a certain amount of readily available information. The median test may be used in such cases for testing the hypothesis described earlier. The procedure consists in combining the sample observations $x_1, x_2, \ldots, x_{n_1}$ on X and $y_1, y_2, \ldots, y_{n_2}$ on Y together and arranging them in order of magnitude. Let us denote this ordered set by

$$z_{(1)} < z_{(2)} < \cdots < z_{(n_1 + n_2)}.$$

Thus each $z_{(j)}$ is either an observation on X or on Y. Let r be an integer such that

$$r = \begin{cases} \dfrac{n_1 + n_2}{2} & \text{if } n_1 + n_2 \text{ is even} \\ \dfrac{n_1 + n_2 + 1}{2} & \text{if } n_1 + n_2 \text{ is odd.} \end{cases}$$

Then $z_{(r)}$ is the median of the combined sample, which we shall denote by z_{med}. Let m_1 be the number of x_i that exceed z_{med}, and m_2 be the number of y_j that exceed z_{med} (M_1 and M_2 will represent the corresponding random variables). The rationale of this method is as follows: If H_0 is true, the x_i and y_j should be well mixed, so that the ratio of the number of x_i to the number of y_j occurring to the right of z_{med} should be about the same as the ratio of the number of x_i to the number of y_j to the left of z_{med}. On the other hand, if $F_1(x) = F_2(x + a)$ and $a > 0$, then $F_1(x) \geq F_2(x)$ for all x, so that such a ratio should tend to be smaller to the right of z_{med} than to its left. Similar arguments hold for the cases $a < 0$ and $a \neq 0$.

The critical region for testing H_0: $F_1 \equiv F_2$ against the alternative H_1: $F_1(x) = F_2(x + a)$, $a > 0$, for all x, is specified by the inequality $m_1 \leq C_\alpha$, where C_α is such that the size of the test is α. The cases $a < 0$ and $a \neq 0$ can be treated similarly.

In order to determine C_α, one needs the distribution of M_1. Now recalling that $z_{(r)}$ may be either an x or y observation, and using the techniques of Chapter 8, the joint distribution of M_1, M_2, and $z_{(r)}$ under the null hypothesis is

$$
\begin{aligned}
dF(z_{(r)}, m_1, m_2) = \Bigg\{ & \frac{n_1!}{m_1! \, (n_1 - m_1 - 1)!} [F_1(z_{(r)})]^{n_1 - m_1 - 1} \\
& \times [1 - F_1(z_{(r)})]^{m_1} \, dF_1(z_{(r)}) \Bigg\} \\
& \times \left\{ \frac{n_2!}{m_2! \, (n_2 - m_2)!} [F_1(z_{(r)})]^{n_2 - m_2} [1 - F_1(z_{(r)})]^{m_2} \right\} \\
& + \left\{ \frac{n_1!}{m_1! \, (n_1 - m_1)!} [F_1(z_{(r)})]^{n_1 - m_1} [1 - F_1(z_{(r)})]^{m_1}] \right\} \\
& \times \left\{ \frac{n_2!}{m_2! \, (n_2 - m_2 - 1)!} [F_1(z_{(r)})]^{n_2 - m_2 - 1} \right. \\
& \times \left. [1 - F_1(z_{(r)})]^{m_2} \, dF_1(z_{(r)}) \right\}.
\end{aligned}
\tag{14.4}
$$

Integrating out $z_{(r)}$, one obtains

$$
P\{M_1 = m_1, M_2 = m_2\} = \frac{\binom{n_1}{m_1}\binom{n_2}{m_2}}{\binom{n_1 + n_2}{m_1 + m_2}}.
\tag{14.5}
$$

Since r is fixed, M_2 is determined once M_1 is given. Therefore,

$$
P\{M_1 = m_1\} = P\{M_1 = m_1, M_2 = m_2\} = \frac{\binom{n_1}{m_1}\binom{n_2}{m_2}}{\binom{n_1 + n_2}{m_1 + m_2}}.
\tag{14.6}
$$

When n_1 and n_2 are large, we can made use of a 2×2 contingency χ^2 test (Table 14.1). Under H_0, $EM_i = n_i/2$, $i = 1,2$. Now, using the formula given in Chapter 11, we construct the appropriate χ^2 statistic.

Table 14.1

	x observations	y observations	Total
Observations $> z_{\text{med}}$	m_1	m_2	$n_1 + n_2 - r$
Observations $\leq z_{\text{med}}$	$n_1 - m_1$	$n_2 - m_2$	r
Total	n_1	n_2	$n_1 + n_2$

14.2.3 Wald–Wolfowitz Run Test

As in the method of Section 14.2.2, we obtain $z_{(1)} < z_{(2)} \cdots < z_{(n_1 + n_2)}$. We then replace each $z_{(j)}$ by 0 or 1 according as it is an x or a y observation. We thus obtain a sequence of 0's and 1's. The run test consists of counting the total number of runs U of 0's and 1's. A run is a sequence of symobls (either 0 or 1) of the same kind preceded and followed by symbols of the other kind. Thus a run of the type, say,

$$0 \quad 0 \quad 1 \quad 0 \quad 0 \quad 0 \quad 1 \quad 1 \quad 1 \quad 0 \quad 1 \quad 0 \quad 0 \quad 0 \quad 1 \quad 0, \qquad (14.7)$$

has five runs of 0's and four runs of 1, so that $U = 5 + 4 = 9$. Observe that the minimum value of U is 2 (when the sequence of all the 0's is followed by the sequence of all the 1's, or conversely) and its maximum value is $2\min(n_1, n_2) + 1$ (when the sequence consist of 0's and 1's appearing alternately with all the extra 0's or 1's at the beginning or the end of it).

When the two distributions are completely separated, so that their ranges do not overlap, one of the random variables is always smaller than the other with probability 1, so that $U = 2$. U will be small even if F_1 and F_2 have the same location but different dispersions. For example, if F_1 has a larger dispersion than F_2, the 0's will tend to be at the beginning and the end of the sequence, and this means a reduction in the number of runs. On the other hand, when $F_1 \equiv F_2$, because of the fact that $P(X > Y) = P(X < Y) = \frac{1}{2}$ (as we have seen earlier), x's and y's will be well mixed, and U will be large. Summarizing the discussions above, we then find that the small values of U lead us to suspect the validity of H_0. Thus we shall reject H_0 whenever $U \leq U_\alpha$, where U_α is chosen to satisfy the size condition.

To determine U_α, we need the distribution of U under H_0. We shall need the following lemma.

Lemma 14.2.3.1 The number of distinguishable ways of dividing n identical objects into k groups is $\binom{n-1}{k-1}$.

Proof. Note that the required number is the same as the number of ways of distributing n balls in the $k - 1$ spaces between $k + 1$ bars: subject to the condition that no two bars are adjacent. Equivalently, this number is then the same as the number of ways of distributing $k - 1$ bars in $(n - 1)$ spaces between the n balls, that is, $\binom{n-1}{k-1}$.   Q.E.D.

Now, note that the sequence (14.7) can be written equivalently as

$$x_{(1)} < x_{(2)} < y_{(1)} < x_{(3)} < x_{(4)} < x_{(5)} < \cdots < x_{(10)}$$

and the probability of any such arrangement under H_0 can easily be found to be

$$\frac{1}{\binom{n_1 + n_2}{n_1}}$$

(For practice, the reader may verify this fact using techniques of Chapter 8.) Since there are altogether

$$\binom{n_1 + n_2}{n_1}$$

arrangements of n_1 x's and n_2 y's, this implies that all sequences such as (14.7) are equally likely to occur, under H_0. Let us now find $P\{U = 2k\}$ when H_0 is true. If there are $2k$ runs, there must be k runs of 0's and k runs of 1's. The number of different ways of arranging n_1 0's in k runs is, by the lemma above,

$$\binom{n_1 - 1}{k - 1}.$$

Similarly, the number of ways of forming k runs of 1's is

$$\binom{n_2 - 1}{k - 1}.$$

Since any sequence such as (14.7) starts either with a 0 or a 1, and since they are equally likely under H_0, we have

$$p_{2k} = P\{U = 2k\} = \frac{2\binom{n_1 - 1}{k - 1}\binom{n_2 - 1}{k - 1}}{\binom{n_1 + n_2}{n_1}}, \qquad k = 1,2,\ldots \qquad (14.8)$$

Similarly, under H_0,

$$p_{2k+1} = P\{U = 2k + 1\}$$

$$= \frac{\binom{n_1-1}{k-1}\binom{n_2-1}{k} + \binom{n_1-1}{k}\binom{n_2-1}{k-1}}{\binom{n_1+n_2}{n_1}}. \qquad (14.9)$$

U_α then is so chosen as to make

$$\sum_{j=0}^{U_\alpha} p_j = \alpha.$$

When both n_1 and n_2 are sufficiently large (say, when $n_1 \geq 10$, $n_2 \geq 10$), we may use instead an approximate value of U_α found from the asymptotic (i.e., as $n \to \infty$) distribution of U in the following way. Using the formula for p_j above, it can easily be seen that, under H_0,

$$E(U) = \frac{2n_1 n_2}{n_1 + n_2} + 1$$

$$\text{var}(U) = \frac{2n_1 n_2(2n_1 n_2 - n_1 - n_2)}{(n_1 + n_2)^2(n_1 + n_2 - 1)}. \qquad (14.10)$$

Write $n_1 + n_2 = n$, $n_1 = n\beta$, $n_2 = n\delta$. Then, for large n ($\simeq$ means approximately equal to),

$$E(U) \simeq 2n\beta\delta$$

$$\text{var}(U) \simeq 4n\beta^2\delta^2 \qquad (14.11)$$

and

$$\tau = \frac{U - 2n\beta\delta}{2\beta\delta\sqrt{n}}. \qquad (14.12)$$

τ can be shown to be asymptotically normally distributed with zero mean and unit variance. An approximation of U_α is then obtained by solving

$$-\tau_\alpha = \frac{U_\alpha - 2n\beta\delta}{2\beta\delta\sqrt{n}} \qquad (14.13)$$

for U_α, where τ_α is the upper $\alpha\%$ point of the unit normal distribution.

14.2.4 Wilcoxon Test (*T* Test) and Mann–Whitney Test (*U* Test)

As before, we combine the two samples $x_1, x_2, \ldots, x_{n_1}$ and $y_1, y_2, \ldots, y_{n_2}$ together and arrange the observations in order of magnitude. Let the new sequence be $z_{(1)} < \cdots < z_{(n_1 + n_2)}$. Wilcoxon's T statistic is defined as the sum of the ranks of y's in this arrangement. Mann and Whitney (1947) suggested an alternative statistic U, which is equal to the number of times a y precedes an x in the combined arrangement.

Let $r_{(i)}$ denote the rank of $y_{(i)}$ in the combined sample. Then we have

$$T = \sum_{i=1}^{n_2} r_i.$$

Also note that $y_{(i)}$ is followed by $(n_1 + n_2 - r_i)$ x and y observations, of which $n_2 - i$ are y's and hence $n_1 + i - r_i$ are x's. Therefore, we have

$$U = \sum_{i=1}^{n_2} (n_1 + i - r_i) = n_1 n_2 + \frac{n_2(n_2 + 1)}{2} - T. \tag{14.14}$$

Thus there exists a one-to-one relation between U and T. It is then sufficient to describe the test based on U only.

Observe that if the distributions of X and Y are the same except for their location, and if the distribution of X lies completely to the left of that of Y, then in the z arrangement all the y's should lie to the right of all the x's; in this case $U = 0$, since no y precedes an x, and the more the two distributions overlap, the higher will be the value of U.

To test H_0: $F_1(x) = F_2(x)$ for all x against the alternative H_1: $F_1(x) = F_2(x + a)$, $a > 0$, for all x, we determine U_α so that, under H_0

$$P\{U \le U_\alpha\} = \alpha, \tag{14.15}$$

where α is the level of significance; H_0 is rejected if $U \le U_\alpha$. Note that if the alternative is $F_1(x) = F_2(x - a)$, $a > 0$ for all x, then we may interchange the roles of X and Y and proceed as above.

Tables are available (Mann and Whitney, 1947) for the determination of U_α for small values of n_1 and n_2. However, for large values of n_1 and n_2 (say, when $n_1 \ge 8$, $n_2 \ge 8$) and under H_0, the distribution of $(U - E(U))/\sqrt{\mathrm{var}(U)}$ is close to that of a normal deviate, where

$$E(U) = \frac{n_1 n_2}{2} \quad \text{and} \quad \mathrm{var}(U) = \frac{n_1 n_2}{2}(n_1 + n_2 + 1). \tag{14.16}$$

Thus suitable normal approximations may be used for the determination of U_α in such cases.

Example 14.2.1   To compare the strengths of two preparations A and B of a drug, two batches of cats were used and one drug was administered to each. For each cat the fatal dose was determined; the tolerances on this basis are given below.

A	1.55	1.58	1.71	1.44	1.24	1.89	2.34	1.67	1.81	1.52	1.68	
B	2.42	1.85	2.00	2.27	1.70	1.47	2.20	2.21	2.02	2.16	1.98	2.09

Without assuming knowledge of underlying distributions, test whether drug B is not as strong as drug A (the higher the tolerance, the weaker the drug). We first combine the two samples and arrange them in order of magnitude, as in Table 14.2.

Median test.   The sample median in the combined sample is 1.85. Also observe that $m_1 = 2$, $m_2 = 9$. From the formula given in Section 6.2.2, $P\{M_1 \leq 2\} = 0.0027$. Therefore, at the 5% level of sginficance we reject the null hypotheses H_0 (median of A = median of B) in favor of the alternative H_1 (median of A < median of B). Thus on the basis of the data observed, we may conclude that drug B is weaker than drug A.

Wald–Wolfowitz run test.   Observe from Table 14.2 that the value of U is 10. Using the formula of Section 14.2.3, we have

$$E(U) = \frac{2 \cdot 11 \cdot 12}{23} + 1 = 12.4780$$

$$\text{var}(U) = 5.4629.$$

Table 14.2

Observations	Name of drug	Observations	Name of drug
1.24	A	1.89	A
1.44	A	1.98	B
1.47	B	2.00	B
1.52	A	2.02	B
1.55	A	2.09	B
1.58	A	2.16	B
1.67	A	2.20	B
1.68	A	2.21	B
1.70	B	2.27	B
1.71	A	2.34	A
1.81	A	2.42	B
1.85	B		

Under H_0 we have

$$\frac{U - E(U)}{\sqrt{\text{var}(U)}} = -1.06,$$

which is larger than -1.64, the lower 5% point of the normal distribution. Thus we are led to accept the null hypothesis that drug B is as strong as drug A.

Mann–Whitney test.   Here we have

$$T = 3 + 9 + 12 + 14 + 15 + 16 + 17 + 18 + 19 + 20 + 21$$
$$+ 23$$
$$= 187$$

$$U = 11 \times 12 + \frac{12 \times 13}{2} - 187$$

$$= 132 + 78 - 187$$

$$= 23$$

$$E(U) = 66 \qquad \text{var}(U) = 66 \times 24 = 1584,$$

so that

$$\frac{U - E(U)}{\sqrt{\text{var}(U)}} = -1.08.$$

Comparing this with the lower 5% point of the standard normal distribution, here we are also led to conclude that the drugs have equal strength.

14.3   RANK TEST FOR THE ONE-WAY CLASSIFICATION

Let

$$Y_{ij}, \qquad j = 1,\ldots,n_i, \quad i = 1,\ldots,p$$

denote a random sample of size n_i from the ith population with continuous unknown distribution functions

$$F_i(y) = F(y - \beta_i), \qquad i = 1,\ldots,p \tag{14.17}$$

for all y. We want to test the null hypothesis

$$H_0: \beta_1 = \cdots = \beta_p.$$

The alternative in this case is $H_1: \beta_i \neq \beta_j$ for some i, j ($i \neq j$). The case where $Y_{ij}, j = 1,\ldots,n_i$ is a random sample from a normal population with unknown mean β_i and unknown variance σ^2 is treated in Chapter 15. Here we treat the case where the form of F_i is unknown except that F_i satisfies (14.17).

Let $N = \sum_{i=1}^{p} n_i$. Let us rank N observations Y_{ij} in ascending order of magnitude, and let R_{ij} denote the rank of Y_{ij}. Define

$$R_{i\cdot} = \frac{1}{n_i} \sum_{j=1}^{n_i} R_{ij}, \qquad R_{\cdot\cdot} = \frac{1}{N} \sum_{i=1}^{p} \sum_{j=1}^{n_i} R_{ij}.$$

The Kruskal–Walis test of $H_0: \beta_1 = \cdots \beta_p$, based on R_{ij}, rejects H_0 whenever

$$Q = \frac{12}{N(N+1)} \sum_{i=1}^{p} n_i \left[R_{i\cdot} - \frac{1}{2}(N+1) \right]^2$$

$$= \frac{12}{N(N+1)} \sum_{i=1}^{p} n_i R_{i\cdot}^2 - 3(N+1)$$

$$\geq C,$$

where C is a constant such that

$$P(Q \geq C | H_0) = \alpha,$$

the level of significance of the test. When $p = 2$ the statistic Q reduces to the square of Wilcoxon's statistic T. For small N evaluation of the distribution of Q under H_0 is a laborious task. However, for large values of N, Q is approximately distributed as the central chi-square with $p - 1$ degrees of freedom.

EXERCISES

1. Prove Eq. (14.9).
2. A sample survey reveals the following information about the distribution of middle-class families by expenditure level per year in thousands of dollars for the years 1960 and 1965. Test if there is any change in the pattern of expenditure between the years.

					Expenditure				Above
Year	1–3	4–6	7–9	10–12	13–15	16–18	19–21	22–24	24
1960	152	227	202	181	110	94	93	73	63
1965	62	347	347	273	191	135	15	129	55

3. Four investigators, I_1, I_2, I_3, I_4, were asked to collect information regarding the proportion of school children in Calcutta, India, between the ages of 10 and 15 who take milk regularly. Four random samples of 250 school children each of age group 10–15 were prepared and each investigator was required to interview students included in his sample. The following results were obtained.

	I_1	I_2	I_3	I_4
Number taking milk	65	38	61	74
Number not taking milk	185	212	189	176

Should the difference in the proportions observed by different workers be ascribed to pure chance, or is there any chance to suspect bias or inaccuracy in the results submitted by any investigator?

4. It is suspected that observations on population A_1 will tend to be smaller than observations on population A_2. Two random samples are taken as follows:

A_1	2	7	15	20	28	35	52	
A_2	14	21	29	31	40	42	48	53

Find the level of significance for testing homogeneity, using:
(a) The median test.
(b) The U test.
(c) The Wilcoxon or Mann–Whitney test.

5. John and Richard decided to toss a coin 40 times, John or Richard winning according as a majority of heads or tails shows up. The game is played in two rounds of 20 tosses each. In the first round, John and Richard make the tosses alternately, beginning with John. In the second round John makes the first 10 tosses and Richard the next 10 tosses. The following results are obtained.

First round	T T H T H T H H T T H T H T H T
	H T H H
Second round	H H T H H H H T H H H T T T T H
	T T T H T

Do you support the suggestion that both players tried to manipulate the game in their favor (a) in the first round, and (b) in the second round?

6. Compute the distribution of the Mann–Whitney test statistic U under the null hypothesis H_0 in the following cases:
   (a) $n_1 = n_2 = 2$; (b) $n_1 = 1$, $n_2 = 3$.
7. Suppose that the null hypothesis H_0: $F_1(x) = F_2(x)$ for all x is not true, and let $p = P(X \le Y)$. Show that
   (a) $E(U/n_1 n_2) = p$.
   (b) $U/n_1 n_2$ converges in probability to p when $n_1 \to \infty$ and $n_2 \to \infty$.
8. The data in Table 14.3 relating to the lactation yield (10-lb unit) of daughters of five sires are taken from the records of the Kankrej herd maintained at the Institute of Agriculture, Anand, India. Analyze the data to test if the lactation yields for different sires are different.

Table 14.3 Lactation Yield (10-lb Unit) of Daughters

		Sire		
1	2	3	4	5
258	180	122	313	340
243	236	269	374	247
178	120	247	320	193
111	140	198	532	235
233	227	357	203	280
60	140	200	269	194
262	277	190	153	242
138	179	165	184	253

BIBLIOGRAPHY

Fraser, D. A. S. (1957). *Nonparametric Methods in Statistics*, Wiley, New York.

Hajek, J., and Sidak, Z. (1967). *Theory of Rank Tests*, Academic Press, New York.

Kruskal, W. H., and Wallis, W. A. (1952). Use of ranks in one-criterion variance analysis, *J. Am. Stat. Assoc.*, Vol. 47, pp. 583–612.

Lehmann, E. L. (1959). *Testing Statistical Hypotheses*, Wiley, New York.

Mood, A. M. (1950). *Introduction to the Theory of Statistics*, McGraw-Hill, New York.

Siegel, S. (1956). *Nonparametric Statistics for the Behavioral Sciences*, McGraw-Hill, New York.

Mann, H. B. and Whitney, W. R. (1947). On a test of whether one of two random variables is stochastically larger than the other, *Ann. Math. Stat.*, Vol. 18, pp. 50–60.

15
General Linear Hypothesis and Analysis of Variance

Suppose that $Y_1,\ldots,Y_n$ is a sequence of independent random variables with

$$E(Y_i) = \sum_{j=1}^{p} \beta_j x_{ij}$$

$$V(Y_i) = \sigma^2$$

(15.1)

for $i = 1,\ldots,n$, where $\beta_1,\ldots,\beta_p$, σ^2 are unknown constants (parameters) and x_{ij}, $i = 1,\ldots,n$, $j = 1,\ldots,p$ are known constants. Since $E(Y_i)$ are linear functions of the parameters $\beta_1,\ldots,\beta_p$, the equations given in (15.1) are called *general linear hypothesis models*. They are also called *linear models for the expectations with independent covariance structure*. Let

$$Y = \begin{pmatrix} Y_1 \\ \vdots \\ Y_n \end{pmatrix}, \qquad \beta = \begin{pmatrix} \beta_1 \\ \vdots \\ \beta_p \end{pmatrix}, \qquad X = \begin{pmatrix} x_{11} & \cdots & x_{1p} \\ \vdots & \cdots & \vdots \\ x_{n1} & \cdots & x_{np} \end{pmatrix}.$$

We can rewrite the equations given in (15.1) as

$$E(Y) = X\beta, \qquad E(Y - X\beta)(Y - X\beta)' = \sigma^2 I,$$

(15.2)

where I is the $n \times n$ identity matrix. Throughout this chapter we assume that $n \geq p$. If X has full rank, that is, if the rank of X is equal to p, these models are called *general linear hypotheses* or *linear models of full rank*.

In this chapter we are interested in making statistical inferences about the parameters $\beta_1,\dots,\beta_p,\sigma^2$. First we consider the problem of estimating these parameters, then we consider the problem of testing hypotheses about them. As an application of these models we discuss the problem of analysis of variance. The estimation of these parameters will be considered separately for the following different cases: (1) when the distributions of $Y_1,\dots,Y_n$ are not specified except for their means and variances, as given in (15.1); (2) when $Y_1,\dots,Y_n$ are independent with normal distribution and with means and variances given by (15.1). For the first case we use the least-squares method to find the estimates of the parameters. When the distributions of $Y_1,\dots,Y_n$ are normal we will find the maximum-likelihood estimates.

15.1 LEAST SQUARES ESTIMATES OF β

Let y_i be an observation on Y_i, $i = 1,\dots,n$. On the basis of observations $y_1,\dots,y_n$, we will find the least-squares estimates of $\beta_1,\dots,\beta_p$; that is, we will find the value $\overline{\beta}_i$ of β_i, $i = 1,\dots,p$, such that given $y_1,\dots,y_n$ and the matrix of constants X

$$
S = \sum_{i=1}^{n} \left(y_i - \sum_{j=1}^{p} \beta_j x_{ij} \right)^2
$$
$$
= (y - X\beta)'(y - X\beta) \tag{15.3}
$$

is minimum, where $y = (y_1,\dots,y_n)'$. Differentiating S with respect to β_k, we get

$$
\frac{\partial S}{\partial \beta_k} = -2 \sum_{i=1}^{n} x_{ik} \left(y_i - \sum_{j=1}^{p} \beta_j x_{ij} \right), \qquad k = 1,\dots,p \tag{15.4}
$$

The value $\overline{\beta}_k$ of β_k, $k = 1,\dots,p$, that minimizes S is given by

$$
\sum_{i=1}^{n} x_{ik} \left(y_i - \sum_{j=1}^{p} \overline{\beta}_j x_{ij} \right) = 0, \qquad k = 1,\dots,p. \tag{15.5}
$$

Rewriting Eq. (15.5) in matrix notation, we get

$$
(X'X)\overline{\beta} = X'y. \tag{15.6}
$$

The equation in β,

$$
(X'X)\beta = X'y, \tag{15.7}
$$

is called the *normal equation* for the general linear hypothesis model and is a nonhomogeneous system of linear equations in β. If X has full rank

p, then $X'X$ is of rank p and has a unique inverse $(X'X)^{-1}$. The least-squares estimate $\overline{\beta}$ of β, given by

$$\overline{\beta} = (X'X)^{-1}X'Y,$$

is unique. In the general case the reader is referred to Appendix A for the solution of the normal equation. The reader is referred to Rao (1965) for futher study. However, as we now show, the normal equation (15.7) always has a solution. Let $\overline{\beta}$ be any solution of (15.7). Then we have

$$\begin{aligned}
(y &- X\beta)'(y - X\beta) \\
&= [(y - X\overline{\beta} + X(\overline{\beta} - \beta)]'[y - X\overline{\beta} + X(\overline{\beta} - \beta)] \\
&= (y - X\overline{\beta})'(y - X\overline{\beta}) + (\overline{\beta} - \beta)'XX'(\overline{\beta} - \beta) \\
&\quad + 2(y - X\overline{\beta})'X(\overline{\beta} - \beta) \\
&= (y - X\overline{\beta})'(y - X\overline{\beta}) + (\overline{\beta} - \beta)'X'X(\overline{\beta} - \beta) \\
&= (y - X\overline{\beta})'(y - X\overline{\beta}) + [X(\overline{\beta} - \beta)]'[X(\overline{\beta} - \beta)] \\
&\geq (y - X\overline{\beta})'(y - X\overline{\beta})
\end{aligned}$$

since

$$\begin{aligned}
(y - X\overline{\beta})'X(\overline{\beta} - \beta) &= (y'X - \overline{\beta}'X'X)(\overline{\beta} - \beta) \\
&= (y'X - y'X)(\overline{\beta} - \beta) \\
&= 0
\end{aligned}$$

and

$$[X(\overline{\beta} - \beta)]'[X(\overline{\beta} - \beta)] \geq 0,$$

the minimum of $(y - X\beta)'(y - X\beta)$ is $(y - X\overline{\beta})'(y - X\overline{\beta})$ and is attained when $\beta = \overline{\beta}$ and is unique for all solutions $\overline{\beta}$ of (15.7).

Definition 15.1.1 A least-squares estimator of the linear parametric function $L'\beta$, where $L = (l_1,\ldots,l_p)'$ is a real p-vector, is defined to be $L'\overline{\beta}$, where $\overline{\beta}$ is a solution of (15.7). If X is of full rank, then $L'\overline{\beta}$ is unique.

Note: For notational convenience we shall not distinguish here between the least-squares estimator and the least-squares estimate of β. The estimator corresponding to any estimate is obtained by replacing y by Y in the expression for the estimate.

We now study the optimum character of the least-squares estimator of any linear parametric function. Since the distribution of $Y_1,\ldots,Y_n$ is not specified, it will not in general be possible to examine any optimum property of the least-squares estimator relative to all estimators. Thus we will consider only linear estimators (i.e., estimators that are linear functions of $Y_1,...,Y_n$ only). We show in Theorem 15.1.1 that in the case of general linear hypothesis models of full rank, the least-squares estimator of any linear parametric function $L'\beta$ is the best (minimum-variance) linear unbiased in the class of all linear unbiased estimators. As a special case of this theorem we prove the Gauss–Markoff theorem, which states that the least-squares estimator $\bar{\beta}_i$ of β_i is the best linear unbiased among all linear unbiased estimators. If X does not have full rank, the theorem also holds good provided that $L'\bar{\beta}$ is unique for all solutions of $X'X\bar{\beta} = X'Y$. The reader is referred to Rao (1965) for details.

Theorem 15.1.1   For the general linear hypothesis model of full rank, the least-squares estimator $L'\bar{\beta}$ of the linear parametric function $L'\beta$, where $\bar{\beta}$ is the unique solution of $X'X\beta = X'Y$, is the best linear unbiased.

Proof

$$L'\bar{\beta} = L'(X'X)^{-1}X'Y$$
$$E(L'\bar{\beta}) = L'(X'X)^{-1}X'X\beta = L'\beta$$

Obviously, $L'\bar{\beta}$ is a linear unbiased estimator of $L'\beta$. Let $K'Y$, $K = (k_1,...,k_n)'$ be any other linear unbiased estimator of $L'\beta$. Since $K'Y$ is unbiased, we get (for any β)

$$E(K'Y) = K'X\beta = L'\beta.$$

Hence $K'X = L'$. Now the variance of $K'Y$ is

$$
\begin{aligned}
V(K'Y) &= V(K'Y - L'\bar{\beta} + L'\bar{\beta}) \\
&= V(K'Y - L'\bar{\beta}) + V(L'\bar{\beta}) + 2\,\text{cov}(K'Y - L'\bar{\beta}, L'\bar{\beta}) \\
&= V(K'Y - L'\bar{\beta}) + V(L'\bar{\beta}) \\
&\geq V(L'\bar{\beta})
\end{aligned}
$$

since

$$
\begin{aligned}
\text{cov}(K'Y &- L'\bar{\beta}, L'\bar{\beta}) \\
&= \text{cov}(K'(Y - X\bar{\beta}), K'X\bar{\beta}) \\
&= \text{cov}(K'[I - X(X'X)^{-1}X']Y, K'X(X'X)^{-1}X'Y) \\
&= K'[I - X(X'X)^{-1}X']E((Y - X\beta)(Y - X\beta)')X(X'X)^{-1}X'K \\
&= \sigma^2 K'[I - X(X'X)^{-1}X']X(X'X)^{-1}X'K \\
&= \sigma^2 K'(X - X)(X'X)^{-1}X'K \\
&= 0.
\end{aligned}
$$

Thus $L'\overline{\beta}$ has minimum variance in the class of all linear unbiased estimators of L'.   Q.E.D.

Corollary 15.1.1.1   The least-squares estimator $\overline{\beta}_i$ of β_i, $i = 1,\ldots,p$, is best linear unbiased.

Proof.   Take L to be the unit vector whose ith component is 1, all other components being zero. Then $L'\beta = \beta_i$. The proof now follows from Theorem 15.1.1.   Q.E.D.

Minimization of the sum of squares S does not provide an estimate of σ^2. We show in the next section [Eq. (15.18)] that under the general linear hypothesis model of full rank, the unbiased estimator of σ^2 based on the least-squares estimator of β is given by

$$\frac{(Y - X\overline{\beta})(Y - X\overline{\beta})'}{n - p} = \frac{Y'[1 - X(X'X)^{-1}X']Y}{n - p}. \tag{15.8}$$

The matrix $I - X(X'X)^{-1}X'$ is symmetric and satisfies

$$[I - X(X'X)^{-1}X'][I - X(X'X)^{-1}X'] = I - X(X'X)^{-1}X'. \tag{15.9}$$

Now, a square-symmetric matrix A is called *idempotent* if $AA = A$. Thus $I - X(X'X)^{-1}X'$ is an idempotent matrix.

Some Properties of an Idempotent Matrix A (n × n)

1.   If A is idempotent and nonsingular, then $A = I$.

Proof. $AA = A$ implies that $A^{-1}AA = A^{-1}A$ or $A = I$.

2.   If $A = (a_{ij})$ is idempotent with $a_{ii} = 0$, then all elements of ith row and ith column are all identically zero.

Proof. $A = AA$ implies that

$$a_{ii} = \sum_{j=1}^{n} a_{ij}a_{ji} = \sum_{j=1}^{n} a_{ij}^2.$$

But if $a_{ii} = 0$, then $a_{ij} = 0$, $j = 1,\ldots,n$. But $A = A'$, so $a_{ji} = 0$, $j = 1,\ldots,n$.

3.   If A is idempotent and Θ is orthogonal, then $\Theta'A\Theta$ is idempotent.

Proof. $(\Theta'A\Theta)' = \Theta'A\Theta$ and $\Theta'A\Theta\Theta'A\Theta = \Theta'A\Theta$.

4.   The characteristic roots of an idempotent matrix A are either unity or zero.

Proof. Let λ be a nonzero characteristic root of A and let X $(\neq 0)$ be the corresponding characteristic vector; then

$$AX = \lambda X$$
$$AAX = \lambda AX = \lambda^2 X,$$

so that

$$\lambda X = \lambda^2 X \quad \text{or} \quad (\lambda^2 - \lambda)X = 0.$$

Since $X \neq 0$, $\lambda^2 = \lambda$; that is, $\lambda = 1$ or 0.

5. For any idempotent matrix A, rank of $A = \text{tr } A$.

Proof. Let k be the rank of A. By property (4) and Theorem A.2.6.5 in Appendix A, there exists an orthogonal matrix Θ such that $\Theta A \Theta' = E_k$, where E_k is the diagonal matrix with k diagonal elements equal to unity, the remaining ones being equal to zero. Now

$$\text{tr } A = \text{tr } \Theta A \Theta' = \text{tr } E_k = k = \text{rank of } A.$$

15.2 MAXIMUM-LIKELIHOOD ESTIMATES OF β and σ^2

Let us now assume that $Y_1,\ldots,Y_n$ are independently distributed normal random variables with means and variances given by the general linear hypothesis model of full rank. The likelihood of the observations $y_1,\ldots,y_n$ is given by

$$L(\beta,\sigma^2) = \frac{1}{(2\pi\sigma^2)^{n/2}} \exp\left[-\frac{1}{2\sigma^2} \sum_{i=1}^{n} \left(y_i - \sum_{j=1}^{p} \beta_j x_{ij} \right)^2 \right]$$

$$= \frac{1}{(2\pi\sigma^2)^{n/2}} \exp\left[-\frac{1}{2\sigma^2} (y - X\beta)'(y - X\beta) \right]. \tag{15.10}$$

Taking logarithms, we get

$$\log L(\beta,\sigma^2) = -\frac{n}{2} \log 2\pi - \frac{n}{2} \log \sigma^2 - \frac{1}{2\sigma^2} (y - X\beta)'(y - X\beta). \tag{15.11}$$

The maximum-likelihood estimate $\hat{\beta}_i$ of β_i, $i = 1,...,p$, and $\hat{\sigma}^2$ of σ^2 are obtained by solving the equations

$$\frac{\partial}{\partial \beta_k} \log L(\beta, \sigma^2)\bigg|_{\beta_i = \hat{\beta}_i, \sigma^2 = \hat{\sigma}^2} = 0$$

$$\frac{\partial}{\partial \sigma^2} \log L(\beta, \sigma^2)\bigg|_{\beta_i = \hat{\beta}_i, \sigma^2 = \hat{\sigma}^2} = 0$$

(15.12)

Differentiation $\log L(\beta, \sigma^2)$ with respect to β_k and σ^2, and using Eqs. (15.11) and (15.12), we obtain

$$\sum_{i=1}^{n} x_{ik}\left(y_i - \sum_{j=1}^{p} \hat{\beta}_j x_{ij}\right) = 0, \qquad k = 1,...,p$$

$$\hat{\sigma}^2 = \frac{(y - X\hat{\beta})'(y - X\hat{\beta})}{n},$$

(15.13)

where $\hat{\beta} = (\hat{\beta}_1,...,\hat{\beta}_p)$. Rewriting in matrix notation, we get

$$X'X\hat{\beta} = X'y$$

$$\hat{\sigma}^2 = \frac{(y - X\hat{\beta})'(y - X\hat{\beta})}{n}.$$

(15.14)

Hence the maximum-likelihood estimate of β (under the assumption of normal distribution) is the same as its least-squares estimate. Since X has full rank, $X'X$ is of rank p and has the unique inverse $(X'X)^{-1}$. From (15.14), the maximum-likelihood estimators of β and σ^2 are given by

$$\hat{\beta} = (X'X)^{-1}X'Y, \qquad \hat{\sigma}^2 = \frac{(Y - X\hat{\beta})'(Y - X\hat{\beta})}{n}.$$

(15.15)

For notational convenience we will be using the same symbol for the maximum-likelihood estimator and the corresponding estimate. Now

$$E(X'X)^{-1}X'Y = (X'X)^{-1}X'X\beta = \beta.$$

(15.16)

So the maximum-likelihood estimator $\hat{\beta}$ of β is unbiased. Obviously, the least-squares estimator $\bar{\beta} = \hat{\beta}$ of β is also unbiased. To examine $\hat{\sigma}^2$ for unbiasedness, we first write

$$n\hat{\sigma}^2 = Y'(I - X(X'X)^{-1}X')Y,$$

where $I - X(X'X)^{-1}X'$ is an idempotent matrix and the n-vector Y is normally distributed with $E(Y) = X\beta$ and covariance matrix $\sigma^2 I$. For the following computations it is not necessary to assume that Y is normally distributed. Obviously, they are also valid with normality assumptions

$$(Y - X\beta)'(I - X(X'X)^{-1}X')(Y - X\beta)$$
$$= Y'[I - X(X'X)^{-1}X']Y - \beta'X'[I - X(X'X)^{-1}X']Y$$
$$\quad - Y'[I - X(X'X)^{-1}X']X\beta + \beta'X'[I - X(X'X)^{-1}X']X\beta$$
$$= Y'[I - X(X'X)^{-1}X']Y.$$

$$(15.17)$$

Let

$$Z = Y - X\beta = (Z_1,\ldots,Z_n) \qquad A = (a_{ij}) = I - X(X'X)^{-1}X'.$$

Then

$$E(n\hat{\sigma}^2) = E(Z'AZ)$$

$$= E\left(\sum_{i=1}^{n} a_{ii}Z_i^2 + \sum_{\substack{i=1 \\ i \neq j}}^{n} \sum_{j=1}^{n} a_{ij}Z_iZ_j\right)$$

$$= \sum_{i=1}^{n} a_{ii}E(Z_i^2)$$

$$= \sigma^2 \sum_{i=1}^{n} a_{ii}$$

$$= \sigma^2 \, \mathrm{tr} \, A$$

$$= (n - p)\sigma^2 \tag{15.18}$$

since $\mathrm{tr}[I - X(X'X)^{-1}X'] = n - \mathrm{tr}[(X'X)(X'X)^{-1}] = n - p$ and $Z_1,\ldots,Z_n$ are independently distributed with mean zero and variance σ^2. Thus $n\hat{\sigma}^2/n - p$ is an unbiased estimator of σ^2.

15.3 PROPERTIES OF THE MAXIMUM-LIKELIHOOD ESTIMATORS $\hat{\beta}$ AND $\hat{\sigma}^2$

Theorem 15.3.1   Under the general linear hypothesis model of full rank, the maximum-likelihood estimator $(\hat{\beta},\hat{\sigma}^2)$ is jointly sufficient for (β,σ^2).

Proof. The joint probability density function of $Y_1, \ldots, Y_n$ is given by

$$\prod_{i=1}^{n} f_{Y_i}(y_i) = \frac{1}{(2\pi\sigma^2)^{n/2}} \exp\left[-\frac{1}{2\sigma^2} (y - X\beta)'(y - X\beta) \right]$$

$$= \frac{1}{(2\pi\sigma^2)^{n/2}} \exp\left[-\frac{1}{2\sigma^2} (y - X\hat{\beta})'(y - X\hat{\beta}) \right]$$

$$\times \exp\left\{ -\frac{1}{2\sigma^2} [(\hat{\beta} - \beta)'(X'X)(\hat{\beta} - \beta)] \right\}.$$

By Theorem 5.5.2, $(\hat{\beta}, \hat{\sigma}^2)$ is jointly sufficient for (β, σ^2).   Q.E.D.

Theorem 15.3.2 Under the general linear hypothesis model of full rank, the maximum-likelihood estimators $\hat{\beta}$, $\hat{\sigma}^2$ are independent.

Proof. The maximum-likelihood estimator

$$n\sigma^2 = Y'[I - X(X'X)^{-1}X']Y = Y'AY$$

is a quadratic form in Y and

$$\hat{\beta} = (X'X)^{-1}X'Y = BY$$

is a linear form in Y. (A and B are arbitrary designations.) Let Θ be an orthogonal matrix such that $\Theta'A\Theta$ is a diagonal matrix

$$D = \begin{pmatrix} D_1 & 0 \\ 0 & 0 \end{pmatrix},$$

where D_1 is a diagonal matrix of nonzero diagonal elements (see Appendix A), and let

$$E = B\Theta = \begin{pmatrix} E_{11} & E_{12} \\ E_{21} & E_{22} \end{pmatrix}.$$

Since

$$\begin{aligned} BA &= (X'X)^{-1}X'[I - X(X'X)^{-1}X'] \\ &= (X'X)^{-1}(X' - X') \\ &= 0, \end{aligned}$$

we get

$$\begin{aligned} 0 &= BA\Theta = B\Theta\Theta'A\Theta = ED \\ &= \begin{pmatrix} E_{11} & E_{12} \\ E_{21} & E_{22} \end{pmatrix} \begin{pmatrix} D_1 & 0 \\ 0 & 0 \end{pmatrix} \end{aligned}$$

so that $E_{11}D_1 = E_{21}D_1 = 0$. Thus $E_{11} = E_{21} = 0$, or, equivalently,

$$E = \begin{pmatrix} 0 & E_{12} \\ 0 & E_{22} \end{pmatrix} = (0 \quad E_2).$$

Let

$$\Theta'Y = Z = \begin{pmatrix} Z_{(1)} \\ Z_{(2)} \end{pmatrix}$$

such that

$$Y'AY = Z'\Theta'A\Theta Z = Z'_{(1)}D_1Z_{(1)} \tag{15.19}$$

and

$$BY = B\Theta Z = EZ = E_2Z_{(2)}. \tag{15.20}$$

Since Θ is orthogonal and the components of Y are independent, we know that (see Chapter 7) $Z_{(1)}$, $Z_{(2)}$ are independent subvectors. Hence from (15.19) and (15.20) we conclude that $n\hat{\sigma}^2$, $\hat{\beta}$ are independent.

Theorem 15.3.3 Under the general linear hypothesis model of full rank:
 (a) The p-vector $\hat{\beta} = (X'X)^{-1}X'Y$ is normally distributed with $E(\hat{\beta}) = \beta$ and covariance matrix $\sigma^2(X'X)^{-1}$.
 (b) $n\hat{\sigma}^2/\sigma^2$ has chi-square (central) distribution with $n - p$ degrees of freedom.

Proof of (a). Let $B = (X'X)^{-1}X'$. Then B is a $p \times n$ matrix of rank $p(p < n)$. By Theorem 7.3.2, the p-vector BY is normally distributed with

$$E(BY) = BE(Y) = (X'X)^{-1}X'X\beta = \beta$$

and covariance matrix

$$\begin{aligned}
E(BY - BE(Y))(BY - BE(Y))' &= BE(Y - E(Y))(Y - E(Y))'B' \\
&= (X'X)^{-1}X'(\sigma^2 I)X(X'X)^{-1} \\
&= \sigma^2(X'X)^{-1}.
\end{aligned}$$

Proof of (b)

$$\begin{aligned}
\frac{n\hat{\sigma}^2}{\sigma^2} &= \frac{Y'[I - X(X'X)^{-1}X']Y}{\sigma^2} \\
&= \frac{(Y - X\beta)'[I - X(X'X)^{-1}X'](Y - X\beta)}{\sigma^2} \\
&= Z'[I - X(X'X)^{-1}X']Z,
\end{aligned}$$

where $Z = (Y - X\beta)/\sigma = (Z_1,...,Z_n)'$ and the Z_i are independent and normally distributed with mean 0 and variance 1. Since $A = I - X(X'X)^{-1}X'$ is an (idempotent) matrix of rank $n - p$, from Theorem A.2.6.5 of Appendix A it follows that there exists an $n \times n$ orthogonal matrix Θ such that

$$\Theta'A\Theta = \begin{pmatrix} D_1 & 0 \\ 0 & 0 \end{pmatrix},$$

where D_1 is an $(n - p) \times (n - p)$ identity matrix. Let

$$U = (U_1,...,U_n)' = \Theta Z.$$

Then

$$\frac{n\hat{\sigma}^2}{\sigma^2} = (\Theta Z)'\Theta A \Theta'(\Theta Z) = \sum_{i=1}^{n-p} U_i^2.$$

By Theorem 7.3.3, $U_1,...,U_{n-p}$ are independently normally distributed with means zero and variance unity. Hence $n\hat{\sigma}^2/\sigma^2$ has chi-square distribution with $n - p$ degrees of freedom.   Q.E.D.

Remarks

1. Let $(X'X)^{-1} = C = (c_{ij})$. By Theorem 7.3.4. and the result of part (a), we get that $\hat{\beta}_i$ is normally distributed with mean β_i and variance $\sigma^2 c_{ii}$. Furthermore, $\text{cov}(\hat{\beta}_i,\hat{\beta}_j) = \sigma^2 c_{ij}$.
2. From the computation above, it follows that $E(\bar{\beta}) = \beta$, $\text{cov}(\bar{\beta}) = \sigma^2(X'X)^{-1}$.

Example 15.3.1.   Let $Y_1,...,Y_n$ be independently distributed random variables with

$$E(Y_i) = \beta_1 + \beta_2 x_i, \quad i = 1,...,n$$
$$V(Y_i) = \sigma^2, \tag{15.21}$$

where $x_1,...,x_n$ are known constants and β_1, β_2, σ_2 are unknown parameters. Here

$$X = \begin{pmatrix} 1 & x_1 \\ \vdots & \vdots \\ 1 & x_n \end{pmatrix}, \quad X'X = \begin{pmatrix} n & n\bar{x} \\ n\bar{x} & \sum_{i=1}^{n} x_i^2 \end{pmatrix}$$

$$(X'X)^{-1} = \frac{1}{n \sum_{i=1}^{n} (x_i - \bar{x})^2} \begin{pmatrix} \sum_{i=1}^{n} x_i^2 & -n\bar{x} \\ -n\bar{x} & n \end{pmatrix}, \quad \bar{x} = \frac{1}{n} \sum_{i=1}^{n} x_i.$$

The least-squares estimates of β_1, β_2 are

$$\bar{\beta}_1 = \bar{y} - \bar{\beta}_2\bar{x},$$

where

$$\bar{y} = \frac{1}{n}\sum_{i=1}^{n} y_i$$

$$\bar{\beta}_2 = \frac{\sum_{i=1}^{n}(x_i - \bar{x})(y_i - \bar{y})}{\sum_{i=1}^{n}(x_i - \bar{x})^2} = \frac{\sum_{i=1}^{n}(x_i - \bar{x})y_i}{\sum_{i=1}^{n}(x_i - \bar{x})^2}.$$

The corresponding estimators $\bar{\beta}_1$, $\bar{\beta}_2$ satisfy

$$E(\bar{\beta}_1) = \beta_1, \qquad E(\bar{\beta}_2) = \beta_2$$

$$V(\bar{\beta}_1) = \frac{\sigma^2 \sum_{i=1}^{n} x_i^2}{n\sum_{i=1}^{n}(x_i - \bar{x})^2}$$

$$V(\bar{\beta}_2) = \frac{\sigma^2}{\sum_{i=1}^{n}(x_i - \bar{x})^2}$$

$$\text{cov}(\bar{\beta}_1, \bar{\beta}_2) = \frac{-n\sigma^2\bar{x}}{n\sum_{i=1}^{n}(x_i - \bar{x})^2}.$$

The unbiased estimate of σ^2 based on the least-squares estimates of β_1, β_2 is given by

$$\frac{1}{n-2}(y - X\bar{\beta})'(y - X\bar{\beta})$$

$$= \frac{1}{n-2}\left[\sum_{i=1}^{n}(y_i - \bar{y})^2 - \frac{\sum_{i=1}^{n}(x_i - \bar{x})y_i}{\sum_{i=1}^{n}(x_i - \bar{x})^2}\right].$$

If $Y_1,\ldots,Y_n$ are independently normally distributed with means and variances as given in (15.21), the maximum-likelihood estimates of β_1, β_2, and σ^2 are

$$\hat{\beta}_1 = \bar{\beta}_1, \qquad \hat{\beta}_2 = \bar{\beta}_2, \qquad \hat{\sigma}^2 = \frac{1}{n}(y - X\bar{\beta})'(y - X\bar{\beta}),$$

where $\bar{\beta} = (\bar{\beta}_1, \bar{\beta}_2)'$.

In deriving the means, variances, and covariances of $\bar{\beta}_1$ and $\bar{\beta}_2$, we have used Theorem 15.3.3. Alternatively, they can be obtained by direct computation. Write

$$\bar{\beta}_1 = \sum_{i=1}^{n} d_i y_i, \qquad d_i = \frac{1}{n} - c_i \bar{x}, \qquad i = 1,\ldots,n$$

$$\bar{\beta}_2 = \sum_{i=1}^{n} c_i y_i, \qquad c_i = \frac{x_i - \bar{x}}{\sum_{i=1}^{n} (x_i - \bar{x})^2}, \qquad i = 1,\ldots,n.$$

Since

$$\sum_{i=1}^{n} c_i = 0, \qquad \sum_{i=1}^{n} c_i^2 = \frac{1}{\sum_{i=1}^{n} (x_i - \bar{x})^2},$$

$$\sum_{i=1}^{n} c_i x_i = 1, \qquad \sum_{i=1}^{n} d_i = 1, \qquad \sum_{i=1}^{n} d_i^2 = \frac{1}{n} + \frac{\bar{x}^2}{\sum_{i=1}^{n} (x_i - \bar{x})^2},$$

$$\sum_{i=1}^{n} d_i x_i = 0, \qquad \sum_{i=1}^{n} c_i d_i = \frac{-\bar{x}}{\sum_{i=1}^{n} (x_i - \bar{x})^2},$$

we get

$$E(\bar{\beta}_2) = \sum_{i=1}^{n} c_i(\beta_1 + \beta_2 x_i) = \beta_2, \qquad E(\bar{\beta}_1) = \sum_i d_i(\beta_1 + \beta_2 x_i) = \beta_1$$

$$\text{var}(\bar{\beta}_1) = \sigma^2 \sum_{i=1}^{n} d_i^2 = \sigma^2 \frac{\sum_{i=1}^{n} x_i^2}{\sum_{i=1}^{n} (x_i - \bar{x})^2}$$

$$\text{var}(\bar{\beta}_2) = \sigma^2 \sum_{i=1}^{n} c_i^2 = \frac{\sigma^2}{\sum_{i=1}^{n} (x_i - \bar{x})^2}$$

$$\text{cov}(\bar{\beta}_1,\bar{\beta}_2) = \sigma^2 \sum_{i=1}^{n} c_i d_i = \frac{-\sigma^2 \bar{x}}{\sum_{i=1}^{n} (x_i - \bar{x})^2}.$$

Example 15.3.2 In Example 15.3.1, let

$$E(Y_i) = \beta_1 + \beta_2(x_i - \bar{x}), \qquad \text{var}(Y_i) = \sigma^2, \qquad i = 1,\ldots,n$$

Here we get

$$\bar{\beta}_1 = \bar{y}, \qquad \bar{\beta}_2 = \sum_{i=1}^{n} c_i y_i.$$

Hence

$$E(\bar{\beta}_1) = \beta_1, \qquad E(\bar{\beta}_2) = \beta_2, \qquad \mathrm{var}(\bar{\beta}_1) = \frac{\sigma^2}{n}$$

$$\mathrm{var}(\bar{\beta}_2) = \frac{\sigma^2}{\sum_{i=1}^{n}(x_i - \bar{x})^2}, \qquad \mathrm{cov}(\bar{\beta}_1,\bar{\beta}_2) = 0.$$

15.4 TEST OF HYPOTHESES (ANALYSIS OF VARIANCE)

General linear hypotheses are hypotheses concerning the parameter β of the general linear hypothesis model. The technique of testing these hypotheses is called the *analysis of variance*. Let $Y_1,\dots,Y_n$ be independently and normally distributed, with means and variances given by the general hypothesis model of full rank. On the basis of observations $y_1,\dots,y_n$ we are interested in testing the following hypotheses:

A. H_0: $\beta_i = \beta_i^0$, $i = 1,\dots,p$, where β_i^0 are known constants and σ^2 is unknown.
B. H_0: $\beta_i = \beta_i^0$, $i = 1,\dots,q$, where $q < p$ and $\beta_{q+1},\dots,\beta_p,\sigma^2$ are unknown. We shall use the likelihood-ratio-test criterion to derive suitable tests in each case.
C. The null hypothesis that the linear function $l'\beta$ is equal to δ where $l = (l_1,\dots,l_p)'$ and δ are known constants.

Case A

The likelihood of $y_1,\dots,y_n$ is given by

$$L(y_1,\dots,y_n;\beta,\sigma^2) = \frac{1}{(2\pi\sigma^2)^{n/2}} \exp\left[-\frac{1}{2\sigma^2}(y - X\beta)'(y - X\beta)\right].$$

$$(15.22)$$

Given $y_1,\dots,y_n$, L is a function of β, σ^2 only; let it be $L(\beta,\sigma^2)$. The parametric space Ω is the $(p + 1)$-dimensional space

$$\Omega = \{(\beta_1,\dots,\beta_p,\sigma^2) : -\infty < \beta_i < \infty, \quad i = 1,\dots,p; \ \sigma^2 > 0\}.$$

Under the null hypothesis the parametric space is the one-dimensional space

$$\omega = \{(\beta_1^0,\dots,\beta_p^0,\sigma^2) : \sigma^2 > 0\}.$$

To find $L(\hat{\omega})$, the maximum of $L(\beta,\sigma^2)$ under H_0, we proceed as follows.

Observe that

$$\log L(\beta^0, \sigma^2) = -\frac{n}{2} \log 2\pi - \frac{n}{2} \log \sigma^2 - \frac{1}{2\sigma^2} (y - X\beta^0)'(y - X\beta^0),$$

(15.23)

where $\beta^0 = (\beta_1^0, \ldots, \beta_p^0)'$. Differentiating $\log L(\beta^0, \sigma^2)$ with respect to σ^2, it is easy to see that $L(\beta^0, \sigma^2)$ is maximum when

$$n\sigma^2 = (y - X\beta^0)'(y - X\beta^0).$$

Hence

$$L(\hat{\omega}) = \left[\frac{n}{2\pi(y - X\beta^0)'(y - X\beta^0)} \right]^{n/2} e^{-n/2}.$$

Similarly, from (15.14) we get

$$L(\hat{\Omega}) = \left[\frac{n}{2\pi(y - X\hat{\beta})'(y - X\hat{\beta})} \right]^{n/2} e^{-n/2}.$$

(15.24)

Thus the likelihood ratio is

$$\lambda = \frac{L(\hat{\omega})}{L(\hat{\Omega})} = \left[\frac{(y - X\hat{\beta})'(Y - X\hat{\beta})}{(y - X\beta^0)'(y - X\beta^0)} \right]^{n/2}.$$

(15.25)

Let

$$q_2 + q_1 = (y - X\beta^0)'(y - X\beta^0)$$
$$= [y - X\hat{\beta} + X(\hat{\beta} - \beta^0)]'[y - X\hat{\beta} + X(\hat{\beta} - \beta^0)]$$
$$= (y - X\hat{\beta})'(Y - X\hat{\beta}) + (\hat{\beta} - \beta^0)'X'X(\hat{\beta} - \beta^0).$$

Now

$$q_1 = [(X'X)^{-1}X'y - \beta^0]'X'X[(X'X)^{-1}X'y - \beta^0]$$
$$= [y'X(X'X)^{-1}X' - (\beta^0)'X'][X(X'X)^{-1}X'y - X\beta^0]$$
$$= (y - X\beta^0)'X(X'X)^{-1}X'(y - X\beta_0)$$
$$q_2 = y'[I - X(X'X)^{-1}X']y$$
$$= (y - X\beta^0)'[I - X(X'X)^{-1}X'](y - X\beta^0).$$

From the calculation above it can be easily verified that $q_1 + q_2$ is the minimum value of $(y - X\beta)'(y - X\beta)$ with respect to β when H_0: $\beta =$

β^0 is true, and that q_2 is the minimum value of $(y - X\beta)'(y - X\beta)$ without any restriction on β. Hence

$$q_1 = \min_{\omega}(y - X\beta)'(y - X\beta) - \min_{\Omega}(y - X\beta)'(y - X\beta).$$

The sum of squares $q_1 + q_2$ is called the *sum of squares due to the null hypothesis* and the sum of squares q_1 is called the *sum of squares due to deviation from the null hypothesis*, or the sum of squares due to β.

Let $Z = Y - X\beta^0 = (Z_1,\ldots,Z_n)$. Under H_0, $Z_1,\ldots,Z_n$ are independently and normally distributed with mean zero and variance σ^2. Let

$$Q_1 = Z'X(X'X)^{-1}X'Z, \qquad Q_2 = Z'[I - X(X'X)^{-1}X']Z.$$

Proceedings exactly in the same way as in part (b) of Theorem 15.3.3, we can conclude that under H_0, Q_1/σ^2 and Q_2/σ^2 have chi-square distribution with p and $n - p$ degrees of freedom, respectively. Furthermore, since

$$[I - X(X'X)^{-1}X']X(X'X)^{-1}X' = X(X'X)^{-1}X' - X(X'X)^{-1}X' = 0,$$

we conclude (see Exercises 11 of Chapter 7) that under H_0, Q_1 and Q_2 are independent. Thus, under H_0, $[(n - p)/p](Q_1/Q_2)$ has F distribution with p and $n - p$ degrees of freedom.

From Eq. (15.25), the likelihood-ratio test (at level α) of H_0 is equivalent to rejecting H_0 whenever

$$\frac{n - p}{p}\frac{q_1}{q_2} \geq c,$$

where c is a constant such that under H_0,

$$P\left(\frac{n - p}{p}\frac{Q_1}{Q_2} \geq c\right) = \alpha.$$

The computation of $[(n - p)/p](q_1/q_2)$ is generally done in the form of Table 15.1, which is called the *analysis-of-variance table*, or ANOVA table, for testing $H_0: \beta = \beta^0$.

Case B

Let X_1 be the $n \times q$ submatrix of X containing the first q columns of X, and let X_2 be the $n \times (p - q)$ submatrix of X containing the remaining $p - q$ columns of X. Writing

$$\beta_{(1)} = (\beta_1,\ldots,\beta_q)', \qquad \beta_{(2)} = (\beta_{q+1},\ldots,\beta_p)',$$

we get

$$X\beta = X_1\beta_{(1)} + X_2\beta_{(2)}.$$

Table 15.1 ANOVA for Testing H_0: $\beta = \beta^0$

Source of variation	Degrees of freedom	Sum of squares	Mean sum of squares	F
Due to β	p	q_1	$\dfrac{q_1}{p}$	$\dfrac{n-p}{p} \dfrac{q_1}{q_2}$
Error	$n-p$	q_2	$\dfrac{q_2}{n-p}$	
Total	n	$q_1 + q_2$		

We are interested here in testing the null hypothesis

$$H_0: \beta_{(1)} = \beta^0_{(1)} = (\beta^0_1, \ldots, \beta^0_q)'$$

when $\beta_{(2)}$, σ^2 are unknown. The parametric space Ω is the same as that of Case A. Under H_0, the parametric space is the $(p - q + 1)$-dimensional space

$$\omega = \{(\beta^0_{(1)}, \beta_{(2)}; \sigma^2) : -\infty < \beta_i < \infty, \quad i = q + 1, \ldots, p; \sigma^2 > 0\}.$$

Given $y_1, \ldots, y_n$, the maximum $L(\hat{\Omega})$ of the likelihood function $L(\beta, \sigma^2)$ is given by Eq. (15.24). To find $L(\hat{\omega})$ we observe that

$$(y - X\beta)'(y - X\beta) = (y - X_1\beta_{(1)} - X_2\beta_{(2)})'(y - X_1\beta_{(1)} - X_2\beta_{(2)}).$$

Hence, under H_0, the likelihood of $y_1, \ldots, y_n$ is

$$L(\beta^0_{(1)}, \beta_{(2)}, \sigma^2) = \frac{1}{(2\pi\sigma^2)^{n/2}} \exp\left[-\frac{1}{2\sigma^2} (u - X_2\beta_{(2)})'(u - X_2\beta_{(2)}) \right],$$

$$(15.26)$$

where $u = y - X_1\beta^0_{(1)}$. From Eq. (15.26) and the results of Section 15.2, it follows that the maximum-likelihood estimates $\hat{\beta}_{(2)}$ of $\beta_{(2)}$ and $\hat{\sigma}^2$ of σ^2 are given by

$$\hat{\beta}_{(2)} = (X'_2 X_2)^{-1} X'_2 u$$
$$n\hat{\sigma}^2 = (u - X_2\hat{\beta}_{(2)})'(u - X_2\hat{\beta}_{(2)}).$$
$$(15.27)$$

Note: $X'_2 X_2$, being a principal minor of the positive definite matrix $X'X$, is positive definite and hence has a unique inverse $(X'_2 X_2)^{-1}$.

From Eqs. (15.26) and (15.27) we get

$$L(\hat{\omega}) = \left[\frac{n}{2\pi(u - X\hat{\beta})'(u - X\hat{\beta})} \right]^{n/2} e^{-n/2}.$$
$$(15.28)$$

Now

$$(u - X_2\hat{\beta}_{(2)})'(u - X_2\hat{\beta}_{(2)})$$
$$= [u - X_2(X_2'X_2)^{-1}X_2'u]'[u - X_2(X_2'X_2)^{-1}X_2'u]$$
$$= u'[I - X_2(X_2'X_2)^{-1}X_2']u$$
$$= (y - X_1\beta_{(1)}^0)'[I - X_2(X_2'X_2)^{-1}X_2'](y - X_1\beta_{(1)}^0)$$
$$= (y - X_1\beta_{(1)}^0 - X_2\beta_{(2)})[I - X_2(X_2'X_2)^{-1}X_2](y - X_1\beta_{(1)}^0 - X_2\beta_{(2)})$$
$$(y - X\hat{\beta})'(y - X\hat{\beta})$$
$$= y'[I - X(X'X)^{-1}X']y$$
$$= (y - X_1\beta_{(1)}^0 - X_2\beta_{(2)})'[I - X(X'X)^{-1}X'](y - X_1\beta_{(1)}^0 - X_2\beta_{(2)})$$

as

$$X'[I - X(X'X)^{-1}X'] = X' - X' = 0$$

implies that

$$X_1'[I - X(X'X)^{-1}X'] = 0$$
$$X_2'[I - X(X'X)^{-1}X'] = 0. \tag{15.29}$$

Let

$$q_0 + q_2 = (u - X_2\hat{\beta}_{(2)})'(u - X_2\hat{\beta}_{(2)})$$
$$q_2 = (y - X\hat{\beta})'(y - X\hat{\beta}).$$

It is now obvious that $q_0 + q_2$ is the minimum value of $[(y - X\beta)'(y - X\beta)]$ with respect to β when H_0: $\beta_{(1)} = \beta_{(1)}^0$ is true, and it represents the sum of squares due to the null hypothesis H_0: $\beta_{(1)} = \beta_{(1)}^0$. Since q_2 is the minimum value of $(y - X\beta)'(y - X\beta)$ without any restriction on β, q_1 represents the sum of squares due to $\beta_{(1)}$ (adjusted for $\beta_{(2)}$) or the sum of squares due to deviation from the null hypothesis. Furthermore, $\hat{\beta}_{(2)}'X_2'u$ will be called the sum of squares due to $\beta_{(2)}$, ignoring $\beta_{(1)}$. From (15.25) and (15.28) the likelihood ratio is given by

$$\lambda = \frac{L(\hat{\omega})}{L(\hat{\Omega})} = \left(\frac{q_2}{q_0 + q_2}\right)^{n/2} = \left(1 + \frac{q_0}{q_2}\right)^{-n/2}. \tag{15.30}$$

Thus, to determine the likelihood-ratio test, we need the distribution of Q_0 and Q_2 under H_0, where

$$Q_0 + Q_2 = (Y - X_1\beta_{(1)}^0 - X_2\beta_{(2)})'[I - X_2(X_2'X_2)^{-1}X_2']$$
$$\times (Y - X_1\beta_{(1)}^0 - X_2\beta_{(2)})$$
$$Q_2 = (Y - X_1\beta_{(1)}^0 - X_2\beta_{(2)})'[I - X(X'X)^{-1}X']$$
$$\times (Y - X_1\beta_{(1)}^0 - X_2\beta_{(2)}).$$

Let

$$Z = (Y - X_1\beta_{(1)}^0 - X_2\beta_{(2)}) = (Z_1,...,Z_n)'.$$

Under H_0, $Z_1,...,Z_n$ are independently normally distributed with mean zero and common variance σ^2. Putting

$$B = X(X'X)^{-1}X' - X_2(X_2'X_2)^{-1}X_2'$$
$$A = I - X(X'X)^{-1}X',$$

we get $Q_2 = Z'AZ$, $Q_0 = Z'BZ$, where A, B are idempotent matrices of rank $n - p$, q, respectively. Exactly as in part (b) of Theorem 15.3.3, we conclude that under H_0, Q_2/σ^2 has chi-square distribution with $n - p$ degrees of freedom, and Q_0/σ^2 has chi-square distribution with q degrees of freedom. Furthermore, since, by (15.29),

$$BA = [X(X'X)^{-1}X' - X_2(X_2'X_2)^{-1}X'_2][I - X(X'X)^{-1}X']$$
$$= X(X'X)^{-1}X' - X(X'X)^{-1}X'$$
$$\quad - X_2(X_2'X_2)^{-1}X_2'[I - X(X'X)^{-1}X']$$
$$= 0,$$

by Exercise 13 of Chapter 7, Q_0, Q_2 are independent. Hence under H_0: $\beta_{(1)} = \beta_{(1)}^0$,

$$\frac{n - p}{q} \frac{Q_0}{Q_2}$$

has central F distribution with $(q, n - p)$ degrees of freedom. From Eq. (15.30), the likelihood-ratio test at level α for testing H_0: $\beta_{(1)} = \beta_{(1)}^0$ is thus equivalent to rejecting H_0 whenever

$$\frac{n - p}{q} \frac{q_0}{q_2} \geq c,$$

Table 15.2 ANOVA for Testing H_0: $\beta_{(1)} = \beta_{(1)}^0$

Source of variation	Degrees of freedom	Sum of squares	Mean sum of squares	F
Due to β	p	$\hat{\beta}'X'y$		
Due to $\beta_{(2)}$ (unadjusted)	$p - q$	$\hat{\beta}_{(2)}'X_2'u$		
Due to $\beta_{(1)}$ (adjusted)	q	q_0	$\dfrac{q_0}{q}$	$\dfrac{n-p}{q}\dfrac{q_0}{q_2}$
Error	$n - p$	q_2	$\dfrac{q_2}{n-p}$	
Total	n	$y'y$		

where c is a constant such that under H_0,

$$P\left(\frac{n-p}{q}\frac{Q_0}{Q_2} \geq c\right) = \alpha.$$

The ANOVA table for testing H_0: $\beta_{(1)} = \beta_{(1)}^0$ is given in Table 15.2.

Case C

Since $l'\hat{\beta}$ is normally distributed with mean $l'\beta$ and variance $\sigma^2 l'(X'X)^{-1}l$, it follows that under H_0: $l'\beta = \delta$,

$$t = \frac{l'\hat{\beta} - \delta}{\sqrt{\sigma^2[l'(X'X)^{-1}l]}} = \frac{(l'\hat{\beta} - \delta)/\sqrt{\sigma^2[l'(X'X)^{-1}l]}}{\sqrt{[n/(n-p)](\hat{\sigma}^2/\sigma^2)}}$$

has Student's t distribution with $n - p$ degrees of freedom. Hence an appropriate test of size α for testing H_0: $l'\beta = \delta$ against the alternative H_1: $l'\beta \neq \delta$ is given by

Reject H_0 whenever $|t| \geq t_{\alpha/2}$,

where $t_{\alpha/2}$ is a constant such that under H_0, $P(t \geq t_{\alpha/2}) = \alpha/2$.

Next we consider some specific simple cases in which the analysis-of-variance techniques can be applied.

One-Way Classification

One-way classification refers to the comparison of means of several univariate normal populations with the same variance σ^2. Let Y_{ij}, $j = 1,\dots,n_i$, $i = 1,\dots,p$, denote a random sample of size n_i from the ith population, and let the random samples from different populations be independent of each other. The general linear hypothesis model in this case is

$$E(Y_{ij}) = 0 \cdot \beta_1 + \cdots + 0 \cdot \beta_{i-1} + \beta_i + 0 \cdot \beta_{i+1} + \cdots + 0 \cdot \beta_p,$$

$$j = 1,\ldots,n_i, \quad i = 1,\ldots,p$$

$$V(Y_{ij}) = \sigma^2. \tag{15.31}$$

We shall also assume that the Y_{ij} are independent and normally distributed. An alternative representation of (15.31) is

$$Y_{ij} = \beta_i + e_{ij}, \quad j = 1,\ldots,n_i, \quad i = 1,\ldots,p,$$

where the e_{ij} are independent and normally distributed with mean 0 and variance σ^2.

Let $\sum_{i=1}^{p} n_i = n$ and let the n-vectors Y, y be given by

$$Y = (Y_{11},\ldots,Y_{1n_1},Y_{21},\ldots,Y_{pn_p})'$$

$$y = (y_{11},\ldots,y_{1n_1},y_{21},\ldots,y_{pn_p})'.$$

Denote

$$\beta = (\beta_1,\ldots,\beta_p)'.$$

Then the $n \times p$ matrix X has the following form: The first n_1 rows are all $\varepsilon_1' = (1,0,\ldots,0)$; the next n_2 rows are all $\varepsilon_2' = (0,1,0,\ldots,0),\ldots$, and the last n_p rows are all $\varepsilon_p' = (0,\ldots,0,1)$. Obviously, the rank of X is p. The model given in (15.31) can be written as

$$E(Y) = X\beta, \quad \text{cov}(Y) = \sigma^2 I.$$

We are interested in testing the null hypothesis

$$H_0: \beta_1 = \cdots = \beta_p \tag{15.32}$$

when σ^2 is unknown. The parametric space Ω is the $(p + 1)$-dimensional space

$$\Omega = \{(\beta_1,\ldots,\beta_p;\sigma^2) : -\infty < \beta_i < \infty, \quad i = 1,\ldots,p; \sigma^2 > 0\}.$$

Under H_0, the parametric space reduces to, for $\beta_1 = \cdots = \beta_p = \beta$,

$$\omega = \{(\beta,\sigma^2) : -\infty < \beta < \infty; \sigma^2 > 0\}.$$

Let $\hat{\beta}_i$, $\hat{\sigma}^2$ be the maximum-likelihood estimates of β_i and σ^2, respectively, under Ω, and let $\hat{\hat{\beta}}$, $\hat{\hat{\sigma}}^2$ be the maximum-likelihood estimates of β and σ^2, respectively, under ω. From the results of Sections 15.2 and 15.3, we get

$$\hat{\beta}_i = y_{i\cdot}, \qquad \hat{\sigma}^2 = \frac{1}{n} \sum_{i=1}^{p} \sum_{j=1}^{n_i} (y_{ij} - y_{i\cdot})^2$$

$$\hat{\beta} = y_{\cdot\cdot}, \qquad \hat{\sigma}^2 = \frac{1}{n} \sum_{i=1}^{p} \sum_{j=1}^{n_i} (y_{ij} - y_{\cdot\cdot})^2$$

$$L(\hat{\omega}) = \left(\frac{n}{2\pi\hat{\sigma}^2}\right)^{n/2} e^{-n/2}$$

$$L(\hat{\Omega}) = \left(\frac{n}{2\pi\hat{\sigma}^2}\right)^{n/2} e^{-n/2},$$

where

$$y_{\cdot\cdot} = \frac{1}{n} \sum_{i=1}^{p} \sum_{j=1}^{n_i} y_{ij}, \qquad y_{i\cdot} = \frac{1}{n_i} \sum_{j=1}^{n_i} y_{ij}.$$

Hence

$$\lambda = \frac{L(\hat{\omega})}{L(\Omega)} = \left[\frac{\sum_{i=1}^{p} \sum_{j=1}^{n_i} (y_{ij} - y_{i\cdot})^2}{\sum_{i=1}^{p} \sum_{j=1}^{n_i} (y_{ij} - y_{\cdot\cdot})^2}\right]^{n/2}$$

$$= \left[1 + \frac{\sum_{i=1}^{p} n_i(y_{i\cdot} - y_{\cdot\cdot})^2}{\sum_{i=1}^{p} \sum_{j=1}^{n_i} (y_{ij} - y_{i\cdot})^2}\right]^{-n/2} \tag{15.33}$$

since

$$\sum_{i=1}^{p} \sum_{j=1}^{n_i} (y_{ij} - y_{\cdot\cdot})^2 = \sum_{i=1}^{p} \sum_{j=1}^{n_i} (y_{ij} - y_{i\cdot})^2 + \sum_{i=1}^{p} n_i(y_{i\cdot} - y_{\cdot\cdot})^2$$

$$= q_2 + q_1,$$

where q_2 and q_1 are arbitrary designations. The sum of squares q_2 represents the variation within populations, and the sum of squares q_1 represents the variation between populations. From Eq. (15.33) it follows that to determine the likelihood-ratio test of H_0, we need to find the distribution of Q_1 and Q_2 under H_0, where

$$Q_2 = \sum_{i=1}^{p} \sum_{j=1}^{n_i} (Y_{ij} - Y_{i\cdot})^2 = \sum_{i=1}^{p} S_i^2$$

$$Q_1 = \sum_{i=1}^{p} n_i(Y_{i\cdot} - Y_{\cdot\cdot})^2$$

Table 15.3 ANOVA for One-Way Classification

Source of variation	Degrees of freedom	Sum of squares	Mean sum of squares	F
Between populations	$p - 1$	q_1	$\dfrac{q_1}{p - 1}$	$\dfrac{n - p}{p - 1}\dfrac{q_1}{q_2} \geq c$
Within populations	$n - p$	q_2	$\dfrac{q_2}{n - p}$	
Total	$n - 1$	$q_1 + q_2$		

$$Y_{i\cdot} = \frac{1}{n_i} \sum_{j=1}^{n_i} Y_{ij}, \qquad Y_{\cdot\cdot} = \frac{1}{n} \sum_{i=1}^{p} \sum_{j=1}^{n_i} Y_{ij}$$

$$S_i^2 = \sum_{j=1}^{n_i} (Y_{ij} - Y_{i\cdot})^2.$$

By Theorem 6.2.1.1 we get that S_i^2, $Y_{i\cdot}$ for each i are independent, S_i^2/σ^2 has chi-square distribution with $n_i - 1$ degrees of freedom, and $\sqrt{n_i}\, Y_{i\cdot}$ is normally distributed with mean $\sqrt{n_i}\,\beta$ (under H_0) and variance σ^2. Since random samples from different populations are independent of each other, we conclude that Q_2/σ^2 has chi-square distribution with $n - p$ degrees of freedom and is independent of $Y_{1\cdot},\dots,Y_{p\cdot}$. Furthermore, by Section 6.1.5, under H_0, Q_1/σ^2 has chi-square distribution with $p - 1$ degrees of freedom. Hence, under H_0, $[(n - p/p - 1](Q_1/Q_2)$ has F distribution with $(p - 1,$ $n - p)$ degrees of freedom. From Eq. (15.33) the likelihood-ratio test (at level α) for testing H_0 is equivalent to rejecting H_0:

$$\frac{n - p}{p - 1}\frac{q_1}{q_2} \geq c, \tag{15.34}$$

where the constant c is chosen in such a way that under H_0,

$$P\left(\frac{n - p}{p - 1}\frac{Q_1}{Q_2} \geq c\right) = \alpha.$$

The ANOVA table for the one-way classification is given in Table 15.3.

Two-Way Classification

We consider two-way classifications with one observation per cell only. Consider an experiment where I different varieties of rice are grown on I different plots of the same size in each location and the experiment is repeated in J such locations. Plots of all locations are equally treated with fertilizer. Such an experiment is called a *two-factor experiment*, the factors being location (A) and variety of rice (B). Different locations and different varieties of rice constitute different levels of factors A and B, respectively. In such a two-factor experiment the observations (yield per plot) can be arranged in a two-way classification or $I \times J$ table by letting the rows of the table correspond to the levels of the factor B and the columns correspond to the level of the factor A. The element of the ith row and jth column represents the yield of the ith variety of rice at jth location and is denoted by y_{ij}; the corresponding random variable is denoted by Y_{ij}. The general linear hypothesis model in this case is

$$E(Y_{ij}) = \alpha_i + \gamma_j, \quad i = 1,\dots,I, \quad j = 1,\dots,J$$
$$V(Y_{ij}) = \sigma^2, \quad\quad\quad\quad\quad\quad\quad\quad\quad\text{(15.35)}$$

where α_i is called the *ith-row effect* (*ith-variety effect*) and γ_j is called the *jth-column effect* (*jth-location effect*). When there is only one observation per cell, the combined effect of factors A and B cannot be evaluated and is taken to be zero. However, when there is more than one observation per cell, the combined effect of factors A and B (called the *interaction between factors A and B*) can be evaluated. The reader is referred to Kempthorne (1952) and Scheffé (1960) for the analysis of a two-way classification with interaction. The matrix X of the general linear hypothesis model (15.35) is given by the coefficients of $\alpha_1,\dots,\alpha_I$, $\gamma_1,\dots,\gamma_J$ in $E(Y_{ij})$. We shall assume that Y_{ij} are independently and normally distributed, with means and variances as given in (15.35). The hypotheses of chief interest are

$$H_A: \alpha_1 = \cdots = \alpha_I$$
$$H_B: \gamma_1 = \cdots = \gamma_J.$$

Let

$$n = IJ, \quad p = I + J$$

$$y_{i\cdot} = \frac{1}{J} \sum_{j=1}^{J} y_{ij}, \quad i = 1,\dots,I$$

$$y_{\cdot j} = \frac{1}{I} \sum_{i=1}^{I} y_{ij}, \quad j = 1,\dots,J$$

$$y_{\cdot\cdot} = \frac{1}{n} \sum_{i=1}^{I} \sum_{j=1}^{J} y_{ij}$$

$$Y_{i\cdot} = \frac{1}{J} \sum_{j=1}^{J} Y_{ij}, \quad i = 1,\dots,I$$

$$Y_{.j} = \frac{1}{I} \sum_{i=1}^{I} Y_{ij}, \qquad j = 1, \ldots, J$$

$$Y_{..} = \frac{1}{n} \sum_{i=1}^{I} \sum_{j=1}^{J} Y_{ij}$$

$$q_1 = \sum_{i=1}^{I} J(y_{i.} - y_{..})^2$$

$$q_2 = \sum_{j=1}^{J} I(y_{.j} - y_{..})^2$$

$$q_3 = \sum_{i=1}^{I} \sum_{j=1}^{J} (y_{ij} - y_{i.} - y_{.j} + y_{..})^2$$

$$Q_1 = \sum_{i=1}^{I} I(Y_{i.} - Y_{..})^2$$

$$Q_2 = \sum_{j=1}^{J} J(Y_{.j} - Y_{..})^2$$

$$Q_3 = \sum_{i=1}^{I} \sum_{j=1}^{J} (Y_{ij} - Y_{i.} - Y_{.j} + Y_{..})^2.$$

Proceedings exactly as for one-way classification, it may be shown that:

1. Q_3/σ^2 has chi-square distribution with $(I - 1)(J - 1)$ degrees of freedom.
2. Under H_A, Q_1/σ^2 has chi-square distribution with $I - 1$ degrees of freedom.
3. Under H_B, Q_2/σ^2 has chi-square distribution with $J - 1$ degrees of freedom.
4. Q_1 and Q_2 are independent.

Hence

5. Under H_A, $[(I - 1)(J - 1)/(I - 1)](Q_1/Q_3)$ has F distribution with $(I - 1, (I - 1)(J - 1))$ degrees of freedom.
6. Under H_B, $[(I - 1)(J - 1)/(J - 1)](Q_2/Q_3)$ has F distribution with $(J - 1, (I - 1)(J - 1))$ degrees of freedom.

Furthermore, it may be verified that:

7. The likelihood-ratio test (at level α) of H_A is equivalent to rejecting H_A if

Table 15.4　ANOVA for Two-Way Classification

Source of variation	Degrees of freedom	Sum of squares	Mean sum of squares	F
Between rows (varieties)	$I - 1$	q_1	$\dfrac{q_1}{I - 1}$	$(J - 1)\dfrac{q_1}{q_3}$
Between columns (locations)	$J - 1$	q_2	$\dfrac{q_2}{J - 1}$	$(I - 1)\dfrac{q_2}{q_3}$
Error	$(I - 1)(J - 1)$	q_3	$\dfrac{q_3}{(I - 1)(J - 1)}$	
Total	$IJ - 1$	$\sum_{i=1}^{I}\sum_{j=1}^{J}(y_{ij} - y_{..})^2$		

$$\frac{(I - 1)(J - 1)}{I - 1}\frac{q_1}{q_3} \geq c_1,$$

where the constant c_1 is chosen such that under H_A,

$$P\left(\frac{(I - 1)(J - 1)}{I - 1}\frac{Q_1}{Q_3} > c_1\right) = \alpha.$$

8. The likelihood-ratio test (at level α) of H_B is equivalent to rejecting H_B whenever

$$\frac{(I - 1)(J - 1)}{(J - 1)}\frac{q_2}{q_3} \geq c_2,$$

where the constant c_2 is chosen such that under H_B,

$$P\left(\frac{(I - 1)(J - 1)}{(J - 1)}\frac{Q_2}{Q_3} \geq c_2\right) = \alpha.$$

Furthermore, the maximum-likelihood estimates $\hat{\alpha}_i$, $\hat{\gamma}_j$ of α_i and γ_j, respectively, are given by $\hat{\gamma}_j = y_{\cdot j}$, $\hat{\alpha}_i = y_{i\cdot}$.
　The ANOVA table for the two-way classification is given in Table 15.4.

15.5　MULTIPLE COMPARISON: THE SCHEFFÉ METHOD

They are various multiple comparison procedures. We shall discuss here the Scheffé method, also known as the S-method. For others we refer to Giri (1986) or Scheffé (1960). The analysis-of-variance technique allows us to accept or reject the null hypothesis involving the equality of several parameters. For example, in the case of one-way classification, the null hypothesis H_0: $\beta_1 = \cdots = \beta_p$ involves the means of p different normal populations with the same unknown variance σ^2. The null hypothesis H_0 is rejected if

$$F = \frac{n - p}{p - 1} \frac{q_1}{q_2} \geq c,$$

where the constant c is such that under H_0,

$$P\left(\frac{n - p}{p - 1} \frac{Q_1}{Q_2} \geq c\right) = \alpha,$$

the level of significance of the test. Since under H_0,

$$\frac{n - p}{p - 1} \frac{Q_1}{Q_2}$$

is distributed as central $F_{p-1,n-p}$ with $p - 1$ and $n - p$ degrees of freedom, the value of c is

$$c = F_{1-\alpha,p-1,n-p},$$

which satisfies $P(F_{p-1,n-p} \leq c) = 1 - \alpha$.

If the hypothesis H_0 is rejected by the ANOVA technique, we conclude that the population means $\beta_1,...,\beta_p$ are not equal. Assume that the population means are ranked in the order of sample means, so that

$$y_{1.} \leq y_{2.} \leq \cdots \leq y_{p.}$$

We now want a statistical technique that will permit us to state, for example, that β_7 is greater than β_4 while there is not sufficient evidence that β_7 is greater than β_6 or β_5, and so on.

Let I_i, $i = 1,...,p$, be the confidence interval, derived from Example 10.7.1, for β_i in the one-way classification such that

$$P(\beta_i \in I_i) = 1 - \alpha \qquad \text{for each } i.$$

Hence

$$P(\beta_1 \in I_1, \beta_2 \in I_2, \ldots, \beta_p \in I_p) < (1 - \alpha).$$

In other words, the probability that we are right in our assertions for all β_i simultaneously is less than $1 - \alpha$. Hence the probability that at least one of our assertions about β_i is wrong may be quite large. Thus we need to find a simultaneous confidence interval for $\beta_1,...,\beta_p$.

Definition 15.5.1: Simultaneous confidence interval Let $\beta_1,\ldots,\beta_p$ be a set of parameters and $\{I_i, i = 1,\ldots,p\}$ be the set of confidence intervals for $\beta_1,\ldots,\beta_p$ satisfying

$$P((\beta_i \in I_i, i = 1,\ldots,p) = 1 - \alpha.$$

Then the I_i are called the $(1 - \alpha)\%$ *simultaneous confidence intervals* for $\beta_1,\ldots,\beta_p$.

The S-method gives the simultaneous confidence intervals of contrasts.

Definition 15.5.2: Contrast A linear function $\psi = \Sigma_{i=1}^p a_i\beta_i$, with known coefficients a_i, is called a *contrast* if $\Sigma_{i=1}^p a_i = 0$.

Examples of contrasts are:

(a) $\beta_1 - \beta_2$,
(b) $\beta_1 - 2\beta_2 + \beta_3$,
(c) $(1/q)(\beta_1 + \cdots + \beta_q) - \frac{1}{2}(\beta_{q+1} + \beta_{q+2})$.

Note: When the null hypothesis $H_0: \beta_1 = \cdots = \beta_p$ is rejected, the natural question arises if $\beta_i - \beta_j = 0$ for $i \neq j$.

For any contrast $\psi = \Sigma_{i=1}^p a_i\beta_i$, the unbiased estimate $\hat{\psi}$ of ψ is

$$\hat{\psi} = \sum_{i=1}^p a_i\hat{\beta}_i = \sum_{i=1}^p a_i y_i.$$

and its variance

$$\sigma_{\hat{\psi}}^2 = \sigma^2 \sum_{i=1}^p \frac{a_i^2}{n_i}.$$

It is estimated by

$$(\hat{\sigma}_{\hat{\psi}})^2 = \left(\sum_{i=1}^p \frac{a_i^2}{n_i}\right)\frac{q_2}{n - p}.$$

S-Method (for One-Way Classification)

Let $s^2 = (1 - p)F_{1-\alpha,p-1,n-p}$. The probability is $1 - \alpha$ that all contrasts ψ satisfy

$$\hat{\psi} - s\hat{\sigma}_{\hat{\psi}} \leq \psi \leq \hat{\psi} + s\hat{\sigma}_{\hat{\psi}}. \tag{15.36}$$

Hence the simultaneous confidence intervals of all contrasts ψ are given in (15.36). In particular, the simultaneous confidence intervals of all

$$\beta_i - \beta_j \qquad (i < j)$$

with confidence coefficient $1 - \alpha$ are given by

$$y_{i\cdot} - y_{j\cdot} - s\sqrt{\left(\frac{n_i + n_j}{n_i n_j}\right)\frac{q_2}{n - p}} \le \beta_i - \beta_j$$

$$\le y_{i\cdot} - y_{j\cdot} + s\sqrt{\left(\frac{n_i + n_j}{n_i n_j}\right)\frac{q_2}{n - p}}.$$

If this includes zero, we conclude that $\beta_i = \beta_j$.

To get the simultaneous confidence intervals for the two-way classification, we replace $n - p$ by $(I - 1)(J - 1)$, and take the error mean square for the estimate of σ^2. For H_A, p is equal to I, and for H_B, $p = J$.

EXERCISES

1. For the general linear hypothesis model

$$E(Y_i) = \beta_1 x_{1i} + \beta_2 x_{2i}, \qquad i = 1,\ldots,n$$

$$V(Y_i) = \sigma^2,$$

where $Y_1,\ldots,Y_n$ are independent and normally distributed, the following data obtain:

y_i	6.0	13.0	13.0	29.2	33.1	32.0	46.2
x_{1i}	1	2	3	4	5	6	8
x_{2i}	10	10	12	11	14	15	18

(a) Find the maximum-likelihood estimates of β_1, β_2, and σ^2.
(b) Test the null hypothesis H_0: $\beta_2 = 0$.

2. Let $Y_1,\ldots,Y_n$ be independently distributed normal random variables with means and variances given by the general linear hypothesis model (15.2) with full rank. Find the 95% confidence interval of (a) β_i, (b) σ^2, and (c) the linear parametric function $l'\beta$.

3. Let Y_{ij}, $i = 1,...,I$, $j = 1,...,J$, be independently distributed normal random variables with

$$E(Y_{ij}) = \mu + \alpha_i + \beta_j$$
$$V(Y_{ij}) = \sigma^2,$$

where

$$\sum_{i=1}^{I} \alpha_i = \sum_{k=1}^{J} \beta_j = 0.$$

Test the following null hypotheses:
(a) H_A: $\alpha_1 = \cdots = \alpha_I = 0.$
(b) H_B: $\beta_1 = \cdots = \beta_J = 0.$
(c) H_0: $\alpha_i - \alpha_{i'} = 0, i \neq i'.$
(d) H_0: $\beta_j - \beta_{j'} = 0, j \neq j'.$

4. For the two-way classification find the following:
(a) The 90% confidence interval of $\gamma_j - \gamma_{j'}, j \neq j'.$
(b) The 90% confidence interval of $\alpha_i - \alpha_{i'}, i \neq i'.$

5. For the one-way classification with general linear hypothesis model given by

$$E(Y_{ij}) = \mu + \alpha_i, \quad j = 1,...,n_i, \quad i = 1,...,p,$$
$$V(Y_{ij}) = \sigma^2$$

where $\sum_{i=1}^{p} \alpha_i = 0$, test the following:
(a) H_A: $\alpha_1 = \cdots = \alpha_p.$
(b) H_0: $\alpha_1 - 2\alpha_2 + \alpha_3 = 0.$

6. The following data give the birth weights of five litters of pigs, the litters being of different sizes. Analyze the data to find if the litter means are different, and also examine if the litter size is an assignable cause of variation in birth weights (in pounds).

Litter	Weights
1	3.5, 3.2, 3.8, 2.3, 1.6, 1.9, 2.0
2	3.2, 3.3, 3.2, 2.9, 2.5, 2.6, 2.8, 2.7
3	2.6, 2.6, 2.9, 2.0, 2.0, 2.1
4	3.1, 2.9, 3.1, 2.5
5	2.6, 2.2, 2.0, 2.5, 1.2, 1.2

7. Four objects, O_1, O_2, O_3, and O_4, whose individual weights are to be determined are weighed in a common balance according to the following design:

O_1	O_2	O_3	O_4	Weight
1	1	1	1	20.2
1	−1	1	−1	8.1
1	1	−1	−1	9.7
1	−1	−1	1	1.9
1	1	1	1	18.9
1	−1	1	−1	8.8
1	1	−1	−1	10.4
1	−1	−1	1	2.0

In the design above, 1 indicates that the object was placed in the right pan, and −1 indicates that the object was placed in the left pan of the balance. Assuming that errors of measurements are independent and normally distributed with zero mean and common unknown variance σ^2:

(a) Find the maximum-likelihood estimates of the four objects.
(b) Find the covariance matrix of these estimates.
(c) Find the estimate of σ^2.

8. Let $Y_1,...,Y_n$ be independently distributed normal random variables with

$$E(Y_i) = \beta_1 + \beta_2 x_i + \beta_3 z_i$$

$$\text{var}(Y_i) = \sigma^2$$

where β_1, β_2, β_3, and σ^2 are unknown parameters and x_i, z_i, $i = 1,...,n$ are known constants.

(a) Find the maximum-likelihood estimators $\hat{\beta}_1$, $\hat{\beta}_2$, $\hat{\beta}_3$, $\hat{\sigma}^2$ of β_1, β_2, β_3, σ^2, respectively.
(b) Find the distribution of $\hat{\beta}_i$, $i = 1,2,3$.
(c) Find the distribution of $\hat{\sigma}^2$.
(d) Do parts (a), (b) and (c) with $z_i = x_i^2$.

9. Let $Y_1,...,Y_6$ be independent with

$$E(Y_1) = \beta_1 + \beta_5 \qquad E(Y_4) = \beta_4 + \beta_6$$

$$E(Y_2) = \beta_2 + \beta_5 \qquad E(Y_5) = \beta_1 + \beta_3$$

$$E(Y_3) = \beta_3 + \beta_6 \qquad E(Y_6) = \beta_2 + \beta_4$$

and

$$\text{var}(Y_i) = \sigma^2, \qquad i = 1,\dots,6,$$

where $\beta_1,\dots,\beta_6$ and σ^2 are unknown.
(a) Find the best linear unbiased estimator of $\beta_2 - \beta_1$.
(b) Find at least three linear unbiased estimator of $\beta_2 - \beta_1$.
10. Do Problem 8 of Exercises 14 assuming that the distribution of lactation yields are normally distributed.

BIBLIOGRAPHY

Das, M. N., and Giri, N. (1986). *Design and Analysis of Experiment*, 2nd ed. Wiley, New York.

Giri, N. (1986). *Analysis of Variance*, South Asian Publishers, New Delhi, India.

Graybill, F. A. (1961). *An Introduction to Linear Statistical Models*, Vol. 1, McGraw-Hill, New York.

Kempthorne, O. (1952). *Design and Analysis of Experiments*. Wiley, New York.

Rao, C. R. (1952). *Advanced Statistical Methods in Biometrical Research*, Wiley, New York.

Rao, C. R. (1965). *Linear Statistical Inference and Its Applications*, Wiley, New York.

Scheffé, H. (1960). *The Analysis of Variance*, Wiley, New York.

16

Some Applications of Analysis of Variance

In this chapter we consider the randomized block designs and the Latin square designs as an application of ANOVA. Experimental designs are used in scientific experiments to know something unknown about a group of treatments and to prove or to disprove certain hypotheses about their effects. We use the term *treatment* in the general sense for a group, category, or class of something. The term *experimental unit* refers to the type of experimental materials that are being used to receive the application of treatments under investigation. When we are applying several types of fertilizers on potted plants, the potted plants are the experimental units and the different types of fertilizers are the treatments. In an agricultural field experiment with different varieties of wheat, the plots of land are the experimental units and the varieties of wheat are the treatments. The effects of the treatments on the experimental units, measured in certain units, are called the *yields* or *observations*.

Once we get the required number of experimental units needed to carry out the experiment efficiently, it is necessary to allocate the treatments to the experimental units in a random fashion, which is necessary to guard against any bias in the experimental results from the influence of some extraneous factors affecting the experiment. For example, if the experimental plots are not homogeneous with respect to soil fertility, we are bound to get different responses for different varieties of wheat, even if they are alike when they are allocated to experimental units with different soil fertilities. Random allocation also enables us to attach probabilistic statements to estimated treatment effects.

Design of experiment is the technique by which a group of treatments, to be compared among themselves, are placed at random on the experimental units and the different treatment responses are estimated with the utmost precision possible. Most experimental designs used in agricultural field experiment are block designs, where the available experimental plots are grouped into blocks of the same size having more or less identical characteristic properties. Different blocks, however, may have different properties. For example, in the case of an experiment with different fertilizers on a variety of wheat, the plots within a block need to be as homogeneous as possible with respect to soil fertility, but different blocks are permitted to be hetrogeneous in soil fertility. If the number of plots in a block is equal to the number of treatments in the experiment, we call it a complete block. Randomized block design and Latin square design are complete block designs.

16.1 RANDOMIZED BLOCK DESIGN

Suppose that there are v treatments to be compared in a randomized block design layout with b blocks each containing v plots of equal sizes. The experimental units are devided into b blocks of v plots each in such a way that the plots within a block are homogeneous with respect to some extraneous unknown factors affecting the experiment. The different blocks can be heterogeneous with respect to those extraneous factors. In a randomized block design the treatments are allocated randomly within each block. Since the block size is equal to the number of treatments, each treatment occurs once in each block.

Example 16.1.1 An experiment was conducted for the maximization of phosphate potential in the soils of Madras State, India, during 1951–1952. Treatments applied are (represented by numerals 1,2,...,10):

1. Control (no fertilizer).
2. Superphosphate 30 lb per acre.
3. Superphosphate 50 lb per acre.
4. Rockphosphate 30 lb per acre.
5. Rockphosphate 50 lb per acre.
6. Superphosphate 30 lb and 5 tons of farmyard manure per acre.
7. Superphosphate 50 lb and 5 tons of farmyard manure per acre.

Table 16.1 Yield of Ragi (lb per Plot)

Block	Plot	1	2	3	4	5	6	7	8	9	10
1	Treatment	5	1	9	2	4	6	7	8	10	3
	Yield	5.50	5.62	5.12	5.00	4.00	3.50	5.50	6.12	5.00	4.75
2	Treatment	10	5	8	7	9	3	10	6	2	4
	Yield	5.37	4.75	4.75	4.50	5.75	4.75	4.50	5.50	5.56	4.25
3	Treatment	2	7	1	4	8	9	3	5	10	6
	Yield	4.00	6.00	3.87	5.00	5.00	5.25	4.50	4.50	5.75	4.50
4	Treatment	1	4	9	8	10	2	6	3	7	5
	Yield	3.50	4.00	3.75	4.75	5.50	3.50	3.70	5.50	5.00	4.50

8. Rockphosphate 30 lb and 5 tons of farmyard manure per acre.
9. Rockphosphate 50 lb and 5 tons of farmyard manure per acre.
10. Farmyard manure 5 tons per acre.

Experimental layout: randomized block design with 4 blocks of 10 plots each. Crop sown: ragi.
Yield: see Table 16.1.

16.1.1 Statistical Analysis of Randomized Block Design

Let y_{ij}, $i = 1,\ldots,b$, $j = 1,\ldots,v$ denote the yield (or the measurement) of the effect of the jth treatment in the ith block. The corresponding random variables Y_{ij} are assumed to be independently distributed normals with means and variances given by

$$E(Y_{ij}) = \mu + \beta_i + \alpha_j, \qquad i = 1,\ldots,b, \quad j = 1,\ldots,v$$
$$\text{var}(Y_{ij}) = \sigma^2, \tag{16.1}$$

where μ is the overall mean, α_j is the jth treatment effect satisfying $\Sigma_{j=1}^{v} \alpha_j = 0$, and β_i is the ith block effect satisfying $\Sigma_{i=1}^{b} \beta_i = 0$.

As discussed in Chapter 15, the model given (16.1) is a linear model of a two-way classification with one observation per cell. The hypotheses of interest are

$$H_T: \alpha_1 = \cdots = \alpha_v = 0$$

and $\tag{16.2}$

$$H_B: \beta_1 = \cdots = \beta_b = 0.$$

In the context of randomized block design H_T is the most meaningful one. However, if H_B is accepted, one may argue that blocking may not be necessary for future experimentation. The treatments could be allocated randomly over the bv plots with the restriction that each treatment is repeated b times in the experiment. From the analysis of the two-way classification (Chapter 15), the ANOVA table for testing H_B and H_T is given in Table 16.2.

Under H_B, $(v - 1)(Q_1/Q_3)$ is distributed as the central $F_{b-1,(b-1)(v-1)}$ with $(b - 1)$, $(b - 1)(v - 1)$ degrees of freedom. Hence the hypothesis H_B of the equality of block effects is rejected if

$$(v - 1) \frac{q_1}{q_3} \geq F_{1-\alpha,b-1,(b-1)(v-1)} \tag{16.3}$$

where $P(F_{b-1,(b-1)(v-1)} \leq F_{1-\alpha,b-1,(b-1)(v-1)}) = 1 - \alpha$. Similarly, the hypothesis H_T of the equality of treatment effects is rejected if

$$(b - 1) \frac{q_2}{q_3} \geq F_{1-\alpha,v-1,(b-1)(v-1)}. \tag{16.4}$$

Example 16.1.2 Consider Example 16.1.1. The ANOVA table for testing

$$H_T: \alpha_1 = \cdots = \alpha_{10} = 0$$

and

$$H_B: \beta_1 = \cdots = \beta_4 = 0$$

is given in Table 16.3.

Since

$$F_{1-\alpha,9,27} = \begin{cases} 3.65 & \text{if } \alpha = 0.025 \\ 2.89 & \text{if } \alpha = 0.05 \\ 2.17 & \text{if } \alpha = 0.10, \end{cases}$$

we accept the hypothesis H_T of the equality of treatment effects at least for $\alpha \leq 0.10$.

16.2 LATIN SQUARE DESIGN

In the randomized block design we devide the experimental units into blocks of homogeneous units. For a Latin square design the experimental units are divided into v rows and v columns, each containing v experimental units (equal to the number of treatments to be compared). The treatments are allocated to v rows and v columns in such a way that each treatment occurs once and only once in each row and in each column. Thus in a Latin square design the treatments are grouped into replicates in two ways, once

Table 16.2 ANOVA for Randomized Block Design

Source of variation	d.f.	Sum of squares (S.S.)	Mean sum of squares (M.S.S.)	F
Blocks	$b - 1$	$q_1 = \sum_i v(y_{i\cdot} - y_{\cdot\cdot})^2$	$\dfrac{q_1}{b - 1}$	$(v - 1)\dfrac{q_1}{q_3}$
Treatments	$v - 1$	$q_2 = \sum_j b(y_{\cdot j} - y_{\cdot\cdot})^2$	$\dfrac{q_2}{v - 1}$	$(b - 1)\dfrac{q_2}{q_3}$
Error	$(b - 1)(v - 1)$	$q_3 = \sum_i \sum_j (y_{ij} - y_{i\cdot} - y_{\cdot j} + y_{\cdot\cdot})^2$	$\dfrac{q_3}{(b - 1)(v - 1)}$	
Total	$bv - 1$	$\sum_i \sum_j (y_{ij} - y_{\cdot\cdot})^2$		

Table 16.3 ANOVA for Data in Table 16.1

Source of variation	d.f.	S.S.	M.S.S.	F
Blocks	3	3.0131	1.0044	
Treatments	9	5.9305	0.6589	1.62
Error	27	10.9553	0.4057	
Total	39	19.8989		

in rows and once in columns. In randomized block design the treatments are grouped into replicates in only one way, namely, the blocks. As a result of this, the estimate of the error variance (σ^2) in Latin square designs can be considerably reduced.

Definition 16.2.1: Latin square A *Latin square* of order v (number of treatments) is an arrangement of v symbols (representing treatments) in v^2 cells arranged in v rows and v columns such that each symbol occurs once and only once in each row and each column.

There are at most $v - 1$ different Latin squares of order v. A Latin square design is a three-way classification in which each of the three factors, row, column, and treatment, is at v levels and observations on only v^2 of v^3 possible combinations, containing each level of each factor, are taken.

Example 16.2.1 Two different Latin squares of order 4, with Latin letters A, B, C, D as treatments, are given below.

Latin Square 1			
A	B	C	D
B	C	D	A
C	D	A	B
D	A	B	C

Latin Square 2			
A	B	C	D
B	A	D	C
C	D	A	B
D	C	B	A

16.2.1 Analysis of Latin Square Design

Suppose that we have selected a Latin square design of order v. Let

$$y_{ijk} \quad i = 1,\ldots,v, \quad j = 1,\ldots,v, \quad k = 1,\ldots,v$$

denote the observation (yield) on the kth treatment in the ith row and the jth column. The triplet (i,j,k) takes only v^2 values. Let

$$S = \{(i,j,k)\}$$

denote the set of all v^2 values of the triplet (i,j,k) indicated by the particular Latin square selected for the experiment. Let the corresponding random variables Y_{ijk}, with values y_{ijk}, be independently and normally distributed with

$$E(Y_{ijk}) = \mu + \alpha_i + \beta_j + \tau_k$$

$$\text{var}(Y_{ijk}) = \sigma^2,$$

(16.5)

where μ is the overall mean, α_i is the ith row effect satisfying $\Sigma_{i=1}^{v} \alpha_i = 0$, β_j is the jth column effect satisfying $\Sigma_{j=1}^{v} \beta_j = 0$, and τ_k is the kth treatment effect satisfying $\Sigma_{k=1}^{v} \tau_k = 0$. In a Latin square layout we can test the following hypothesis,

$$H_T: \alpha_1 = \cdots = \alpha_v = 0$$

$$H_C: \beta_1 = \cdots = \beta_v = 0$$

$$H_T: \gamma_1 = \cdots = \gamma_v = 0.$$

But H_T is the most meaningful one in this context. For detailed calculations we refer to Scheffé (1967) or Giri (1986) or Das and Giri (1986).

Let

$y_{...}$ denote the average of v^2 observations y_{ijk}

$y_{i..}$ denote the average of v observations in the ith row

$y_{.j.}$ denote the average of v observations in the jth column

$y_{..k}$ denote the average of v observations of the kth treatment.

The ANOVA table of the $v \times v$ Latin square design is given in Table 16.4. Here

$$q_4 = \sum_{(i,j,k)\in S} (y_{ijk} - y_{i..} - y_{.j.} - y_{..k} + 2y_{...})^2.$$

The hypothesis H_T is rejected at the level of significance α if [using the notation of (16.3)]

$$(v - 2) \frac{q_3}{q_4} \geq F_{1-\alpha, v-1, (v-1)(v-2)}.$$

Table 16.4 ANOVA for $v \times v$ Latin Square

Source of variation	d.f.	S.S.	M.S.S.	F
Rows	$v - 1$	$q_1 = \sum_i v(y_{i..} - y_{...})^2$	$\dfrac{q_1}{v - 1}$	$(v - 2)\dfrac{q_1}{q_4}$
Columns	$v - 1$	$q_2 = \sum_j v(y_{.j.} - y_{...})^2$	$\dfrac{q_2}{v - 1}$	$(v - 2)\dfrac{q_2}{q_4}$
Treatments	$v - 1$	$q_3 = \sum_k v(y_{..k} - y_{...})^2$	$\dfrac{q_3}{v - 1}$	$(v - 2)\dfrac{q_3}{q_4}$
Error	$v^2 - 3v + 2$	q_4	$\dfrac{q_4}{v^2 - 3v + 2}$	
Total	$v^2 - 1$	$\sum_{(i,j,k) \in S} (y_{ijk} - y_{...})^2$		

The hypothesis H_R is rejected at the level of significance α if

$$(v - 1)\frac{q_1}{q_4} \geq F_{1-\alpha, v-1, (v-1)(v-2)}.$$

The hypothesis H_C is rejected at the level of significance α if

$$(v - 1)\frac{q_2}{q_4} \geq F_{1-\alpha, v-1, (v-1)(v-2)}.$$

Example 16.2.2 An experiment was conducted at the Sugarcane Research Station, Rayaquda, Orissa, India with a 5×5 Latin square design to study the effects of varying doses of superphosphate on the yield of sugarcane crop under unirrigated condition.

Table 16.5 Layout of 5×5 Latin Square (Yields in Parentheses)

2	4	5	3	1
(269)	(262)	(322)	(320)	(202)
4	2	3	1	5
(164)	(248)	(200)	(258)	(229)
5	1	4	2	3
(210)	(159)	(185)	(224)	(225)
1	3	2	5	4
(217)	(257)	(210)	(158)	(151)
3	5	1	4	2
(189)	(230)	(220)	(218)	(205)

Table 16.6 ANOVA of Sugarcane Data

Source of variation	d.f.	S.S.	F
Rows	4	19,544.64	
Columns	4	4,145.84	
Treatments	4	5,989.84	1.28
Error	12	17,476.72	
Total	24	47,157.04	

The five treatments were:

1. Control 0 lb of P_2O_2 as superphosphate.
2. 40 lb of P_2O_2 per acre as superphosphate.
3. 80 lb of P_2O_2 per acre as superphosphate.
4. 120 lb of P_2O_2 per acre as superphosphate.
5. 160 lb of P_2O_2 per acre as superphosphate.

Plot size: 39 ft $\times$ 39 ft.
Treatments: 1, 2, 3, 4, 5.
The layout and the yield in kilograms of sugarcane per plot are given in Table 16.5. The ANOVA table for the sugarcane data is given in Table 16.6. Since $F_{1-\alpha,4,12} = 2.48$, for $\alpha = 0.10$ we accept the hypothesis H_T of the equality of treatment effects with high probability.

EXERCISES

1. An agricultural field experiment was conducted on cotton to find out a suitable level of farmyard manure for the maximum yield. The treatments were:
   (1) No farmyard manure.
   (2) 15 tons of farmyard manure per hectare.
   (3) 30 tons of farmyard manure per hectare.
   (4) 75 tons of farmyard manure per hectare.
   (5) 150 tons of farmyard manure per hectare.
   (6) 300 tons of farmyard manure per hectare.
   The layout was a randomized block design with four blocks. The layout with the yield of cotton per plot in kilograms is given below. The numbers in parentheses indicate the treatment number.

Block		Yield/treatment number				
1	5.40	8.90	5.85	3.10	5.05	2.25
	(4)	(6)	(2)	(1)	(3)	(2)
2	8.40	4.50	3.90	6.40	3.60	6.70
	(4)	(1)	(6)	(5)	(2)	(3)
3	3.20	3.40	4.45	8.05	6.70	5.40
	(3)	(2)	(1)	(6)	(5)	(4)
4	5.50	5.80	3.70	4.50	5.56	4.30
	(5)	(6)	(1)	(3)	(4)	(2)

Analyze the experimental data.

2. An experiment was conducted at the Indian Agricultural Research Institute, New Delhi, India during 1970–1971 under irrigation conditions to determine the optimum dose of ammonium sulfate for the Kalyan Sona variety of wheat. The layout was a 6×6 Latin square design with plot size 8 m $\times$ 0.60 m with the following treatments:
(1) 0 kg of ammonium sulfate per hectare.
(2) 40 kg of ammonium sulfate per hectare.
(3) 80 kg of ammonium sulfate per hectare.
(4) 120 kg of ammonium sulfate per hectare.
(5) 160 kg of ammonium sulfate per hectare.
(6) 200 kg of ammonium sulfate per hectare.
The layout with wheat yield per plot in kilograms is given below. The numbers in parentheses indicate the treatment number.

2.19	2.50	2.27	1.62	1.82	0.91
(6)	(5)	(4)	(3)	(2)	(1)
2.27	1.41	0.91	1.91	2.13	1.95
(5)	(3)	(1)	(4)	(6)	(2)
2.04	0.91	2.25	2.29	2.50	2.07
(2)	(1)	(6)	(5)	(4)	(3)
0.77	2.04	2.40	1.99	1.82	2.50
(1)	(2)	(5)	(6)	(3)	(4)
2.50	2.31	2.09	2.04	0.91	2.27
(4)	(6)	(3)	(2)	(1)	(5)
1.52	1.86	1.91	0.77	2.30	1.98
(3)	(4)	(2)	(1)	(5)	(6)

Analyze the data.

BIBLIOGRAPHY

Das, M. N., and Giri, N. (1986). *Design and Analysis of Experiments*, 2nd ed., Wiley, New York.

Giri, N. (1986). *Analysis of Variance*, South Asian Publisher, New Delhi, India.

Scheffé, H. (1967). *The Analysis of Variance*, Wiley, New York.

Appendix A
Vectors and Matrices

A.1 VECTORS

We shall represent by E^p the p-dimensional Euclidean space.

A.1.1 Definition of and Operations on Vectors

A vector x is an ordered p-tuple of real numbers $x_1,\ldots,x_p$ and is written as

$$x = \begin{pmatrix} x_1 \\ \cdot \\ \cdot \\ \cdot \\ x_p \end{pmatrix}.$$

By *ordered*, we mean, for example, that the vector $\begin{pmatrix} 1 \\ 2 \\ 3 \end{pmatrix}$ is different from the vector $\begin{pmatrix} 2 \\ 1 \\ 3 \end{pmatrix}$, and by *p-tuple*, we mean that the vector x has p components.

Actually, such a vector is called a *p-dimensional column vector*, but for convenience we shall call it simply a *p-vector* or a *vector*. Geometrically, a *p*-vector

$$x = \begin{pmatrix} x_1 \\ \cdot \\ \cdot \\ \cdot \\ x_p \end{pmatrix}$$

represents the point $A = (x_1,\ldots,x_p)$ in E^p or the directed line $\overline{OA}$, joining the origin O of E^p with the point A.

For a given p, the set of all p-vectors will be denoted by V^p; obviously, $V^p = E^p$. The transpose of a vector x is given by $x' = (x_1,\ldots,x_p)$ and is a row vector. Obviously, $(x')' = x$. If all components of a vector are zero, it is called a *zero* or a *null vector*.

We now define several operations on vectors.

1. *Vector Addition.* For any two p-vectors $x = (x_1,\ldots,x_p)$, $y = (y_1,\ldots,y_p)'$ we define the sum

$$x + y = \begin{pmatrix} x_1 + y_1 \\ \cdot \\ \cdot \\ \cdot \\ x_p + y_p \end{pmatrix}. \tag{A.1}$$

It is obvious that

$$x + y = y + x \tag{A.2}$$

and for $z = (z_1,\ldots,z_p)'$,

$$(x + y) + z = x + (y + z). \tag{A.3}$$

In other words, vector addition is an associative and commutative operation.

2. *Scalar Multiplication.* For any scalar constant a and for any $x = (x_1,\ldots,x_p)'$, we define

$$ax = \begin{pmatrix} ax_1 \\ \cdot \\ \cdot \\ \cdot \\ ax_p \end{pmatrix}. \tag{A.4}$$

Obviously, for scalars a, b and for vectors x, y in V^p, we have

$$(a + b)x = ax + bx$$

$$a(x + y) = ax + ay \qquad \text{(A.5)}$$

$$a(bx) = b(ax) = abx.$$

The first equality in (A.5) tells us that scalar multiplication is a distributive operation.

Definition A.1.1.1: Linear combination of vectors   The vector z is a *linear combination of vectors* $\alpha_i = (a_{i1},\dots,a_{ip})'$, $i = 1,\dots,n$, if z can be expressed as

$$z = \sum_{i=1}^{n} c_i \alpha_i,$$

where $c_1,\dots,c_n$ are scalar constants.

Obviously, z is a p-vector and hence $\alpha_1,\dots,\alpha_n$, z all belong to V^p. In particular, the linear combinations

$$\alpha_1 + \cdots + \alpha_n, \qquad \sum_{1}^{n} \frac{\alpha_i}{n}$$

are, respectively, called the *sum* and the *mean* of $\alpha_1,\dots,\alpha_n$.

Definition A.1.1.2: Scalar product of vectors   The *scalar product* (or the inner product) of two vectors $x = (x_1,\dots,x_p)'$, $y = (y_1,\dots,y_p)'$ is given by the scalar quantity $\sum_{i=1}^{p} x_i y_i$ and is written $x'y$.

The notation $x'y$ is used in conformity with the notation of matrix multiplication to be defined in Section A.2. Sometimes the scalar product $x'y$ is written $x \cdot y$ or (x,y). The scalar product is commutative:

$$x'y = \sum_{i=1}^{p} x_i y_i = \sum_{i=1}^{p} y_i x_i = y'x.$$

The scalar product of a vector $x = (x_1,\dots,x_p)'$ with itself is denoted by

$$\|x\|^2 = x'x = \sum_{i=1}^{p} x_i^2.$$

$\|x\|$ is called the *norm* of the vector x. Some important geometrical applications of the scalar product follow.

1. $\|x\|^2$ is equal to the square of the distance of the point x from the origin in E^p.
2. The square of the distance between two points $x = (x_1,...,x_p)'$, $y = (y_1,...,y_p)'$ in E^p is given by

$$(x - y)'(x - y) = \sum_{i=1}^{p} (x_i - y_i)^2. \tag{A.6}$$

3. The angle θ between two vectors x, y is given by

$$\cos \theta = \left(\frac{x}{\|x\|}\right)'\left(\frac{y}{\|y\|}\right). \tag{A.7}$$

A.1.2 Orthogonal, Orthonormal, and Linearly Independent Vectors, Rank of a Vector Space

Definition A.1.2.1: Orthogonal vectors Two vectors x, y in V^p are said to be *orthogonal* if and only if $x'y = y'x = 0$. Geometrically, the condition $x'y = 0$ implies, in view of Eq. (A.7), that the angle between the vectors x and y is $\pi/2$. A set of vectors is orthogonal if they are pairwise orthogonal.

Example A.1.2.1 $x' = (1,2,-1,1)$, $y' = (1,1,2,-1)$ are two orthogonal vectors.

Definition A.1.2.2: Projection of a vector on another vector For x, y in V^p and $y \neq 0$, the projection of x on y is given by $\|y\|^{-2}(x'y)y$.

It can easily be verified that if $x = \overline{OA}$, $y = \overline{OB}$ and P is the foot of the perpendicular from A to OB, then $\|y\|^{-2}(x'y)y = \overline{OP}$, where O is the origin of E^p. Thus two vectors x, y are orthogonal if and only if the projection of x on y is zero.

Definition A.1.2.3: Normalized vector and orthonormal vectors A vector x is said to be *normalized* if $x'x = 1$. A set of p p-vectors in V^p is *orthonormal* if each one is a normal vector and if they are pairwise orthogonal.

Example A.1.2.2 In V^p the vectors

$$\epsilon_1 = (1,0,0,...,0)', \epsilon_2 = (0,1,0,...,0),...,\epsilon_p = (0,0,...,0.1)'$$

are orthonormal; $\epsilon_1,\epsilon_2,...,\epsilon_p$ are called unit vectors of V^p.

Definition A.1.2.4: Linearly independent vectors A set of vectors $\alpha_1,\dots,\alpha_k$ in V^p is said to be *linearly independent* if none of them can be expressed as a linear combination of the others; in other words, there does not exist a set of scalar constants $(c_1,\dots,c_k)$ not all zero such that $c_1\alpha_1 + \cdots + c_k\alpha_k = 0$.

Example A.1.2.3 The vectors $\epsilon_1,\dots,\epsilon_p$ defined in Example A.1.2.2 are linearly independent.

Lemma A.1.2.1 Orthogonal nonzero vectors $\alpha_1,\dots,\alpha_k$, x in V^p are necessarily linearly independent.

Proof. Suppose that they are not linearly independent and let $x = c_1\alpha_1 + \cdots + c_k\alpha_k$, where $c_1,\dots,c_k$ are scalar constants, not all zero. Then $x'x = c_1\alpha_1'x + \cdots + c_k\alpha_k'x = 0$, which implies that x is a null vector, which is a contradiction. Hence the lemma.   Q.E.D.

Definition A.1.2.5: Vector space spanned by a set of vectors Let $\alpha_1,\dots,\alpha_k$ be a set of k vectors in V^p. Then the vector space V spanned by $\alpha_1,\dots,\alpha_k$ is the set of all vectors which can be expressed as linear combinations of $\alpha_1,\dots,\alpha_k$ and the null vector.

From Definition A.1.2.5 it follows that if x and y belong to the vector space V, then for scalar constants a and b, $ax + by$ and ax also belong to V. Furthermore, since $\alpha_1,\dots,\alpha_k$ belong to V^p and the elements of V are linear combinations of these vectors, any element of V is an element of V^p (i.e., $V \subset V^p$). Because of these facts we call V a linear subspace of V^p.

Definition A.1.2.6: Basis of a vector space A basis of a vector space V is a set of linearly independent vectors that span V.

Example A.1.2.4 In V^p the unit vectors $\epsilon_1 = (1,0,0,\dots,0)'$, $\epsilon_2 = (0,1,0,\dots,0)',\dots,\epsilon_p = (0,\dots,0,1)'$ form a basis. Obviously, $\epsilon_1,\dots,\epsilon_p$ are linearly independent and any vector in V^p, say $x' = (x_1,\dots,x_p)$, can be expressed as a linear combination of these vectors:

$$x = \sum_{i=1}^{p} x_i\epsilon_i.$$

We now prove some important properties of the basis of a vector space in the following theorems.

Theorem A.1.2.1   If $\alpha_1,\ldots,\alpha_k$ span a vector space V consisting of at least one nonzero vector, then a subset of $(\alpha_1,\ldots,\alpha_k)$ forms a basis of V.

Proof.   If $\alpha_1,\ldots,\alpha_k$ are linearly independent, then $\alpha_1,\ldots,\alpha_k$ themselves form a basis of V. Otherwise, for scalar constants $c_1,\ldots,c_k$, not all zero we have $\Sigma_1^k c_i\alpha_i = 0$. Without any loss of generality let us assume that $c_1 \neq 0$. Then

$$\alpha_1 = -\sum_{i=2}^{k} \frac{c_i}{c_1}\,\alpha_i.$$

Thus $\alpha_2,\ldots,\alpha_k$ span V. Now if $\alpha_2,\ldots,\alpha_k$ are not linearly independent, we repeat the argument above until we arrive at a subset of $\alpha_2,\ldots,\alpha_k$ the elements of which are linearly independent. Obviously, this subset of linearly independent vectors is not empty, nor does it consist of all zero vectors, because in either case the space V will consist of just the zero vector, contrary to assumption.   Q.E.D.

From Definition A.1.2.5 and Theorem A.1.2.1 we get the following theorem.

Theorem A.1.2.2   Every vector space V has a basis.

Theorem A.1.2.3   Any two bases of a vector space V have the same number of elements.

Proof.   Let $(\alpha_1,\ldots,\alpha_m)$, $(\beta_1,\ldots,\beta_n)$ be two bases of V and let $m < n$. Since $\alpha_1,\ldots,\alpha_m$ span V and β_1 is in V, $\alpha_1,\ldots,\alpha_m,\beta_1$ also span V, and we can express β_1 as

$$\beta_1 = \sum_{j=1}^{m} a_{1j}\alpha_j,$$

where $a_{11},\ldots,a_{1m}$ are scalar constants and not all a_{1j} are zero. Without any loss of generality, let us assume that $a_{11} \neq 0$. Then

$$\alpha_1 = \beta_1 - \frac{a_{12}}{a_{11}}\alpha_2 - \cdots - \frac{a_{1m}}{a_{11}}\alpha_m.$$

In other words, α_1 can be expressed as a linear combination of $\alpha_2,\ldots,\alpha_m$ and β_1, and hence $\alpha_2\ldots,\alpha_m$, β_1 span V. Since β_2 is in V, we can write

$$\beta_2 = a_{22}\alpha_2 + \cdots + a_{2m}\alpha_m + a_{21}\beta_1,$$

where $a_{21},\ldots,a_{2m}$ are scalar constants and not all of them are zero. Without any loss of generality, let us assume that $a_{22} \neq 0$. Then α_2 can be expressed as a linear combination of $\alpha_3,\ldots,\alpha_m$, β_1, β_2 and hence $\alpha_3,\ldots,\alpha_m$, β_1, β_2 span V. Repeating the argument above, we can assert that $\beta_1,\ldots,\beta_m$ span V. Since β_{m+1} is in V, β_{m+1} can be expressed as a linear combination of

$\beta_1,\ldots,\beta_m$, which contradicts the assumption that $\beta_1,\ldots,\beta_n$ is a basis of V. Thus $m \nless n$. Similarly, we can establish that $n \nless m$. Thus $m = n$. Q.E.D.

Theorem A.1.2.3 asserts that although the basis of a vector space can be chosen in more than one way, the number of elements in any basis is constant and depends on the nature of the vector space.

Theorem A.1.2.4 If a vector space V is spanned by the vectors $\alpha_1,\ldots,\alpha_k$, any element x in V is uniquely expressed as

$$x = \sum_{i=1}^{k} a_i\alpha_i,$$

where $a_1,\ldots,a_k$ are scalar constants not all zero, if and only if $\alpha_1,\ldots,\alpha_k$ is a basis of V.

Proof. Assume the contrary, and let $\{b_i\}$ be a set of scalar constants such that

$$x = \sum_{i=1}^{k} b_i\alpha_i.$$

Then

$$0 = \sum_{i=1}^{k} (b_i - a_i)\alpha_i. \tag{A.8}$$

If $\alpha_1,\ldots,\alpha_k$ are linearly independent, Eq. (A.8) implies that $b_i - a_i = 0$ for all i, and hence that $a_1,\ldots,a_k$ are unique. Conversely, if $\alpha_1,\ldots,\alpha_k$ are not linearly independent, Eq. (A.8) implies that not all $(b_i - a_i)$ are zero. In other words, $a_1,\ldots,a_k$ are not unique. Q.E.D.

Definition A.1.2.7: Coordinates of a vector with respect to a basis If $(\alpha_1,\ldots,\alpha_k)$ is a basis of a vector space V and if x in V is expressed as $x = \sum_{i=1}^{k} a_i\alpha_i$, which is a unique representation by Theorem A.1.2.4, the coefficient a_i of the vector α_i is called the ith *coordinate* of x with respect to the basis $(\alpha_1,\ldots,\alpha_k)$.

Example A.1.2.5 The following equation is clearly valid:

$$\begin{pmatrix} 1 \\ 2 \\ 3 \end{pmatrix} = 1\begin{pmatrix} 1 \\ 0 \\ 0 \end{pmatrix} + 2\begin{pmatrix} 0 \\ 1 \\ 0 \end{pmatrix} + 3\begin{pmatrix} 0 \\ 0 \\ 1 \end{pmatrix}.$$

Thus 1, 2, and 3 are the coordinates of the vector $(1,2,3)'$ with respect to the basis $((1,0,0)', (0,1,0)', (0,0,1)')$.

Definition A.1.2.8: Orthonormal basis of a vector space   A basis $(\alpha_1,\ldots,\alpha_k)$ of a vector space V is orthonormal if the vectors $\alpha_1,\ldots,\alpha_k$ are normalized and are pairwise orthogonal.

Definition A.1.2.9: Rank of a vector space   The number of vectors in any basis of a vector space V is called the *rank* (or *dimension*) of V.

From Theorem A.1.2.3 it follows that the rank of a vector space V is a unique number.

Example A.1.2.6   The vectors $\alpha_1,\ldots,\alpha_p$, defined in Example A.1.2.4 form an orthonormal basis of the vector space V^p of rank p.

Given a basis of a vector space V it is always possible to orthonormalize it by using only rational operations. This method is called the Gram–Schmidt orthogonalization process and is given in the following theorem.

Theorem A.1.2.5: Gram–Schmidt orthogonalization method   For any basis $(\alpha_1,\ldots,\alpha_k)$ of a vector space V_k of rank k there exists an orthonormal basis $(\gamma_1,\ldots,\gamma_k)$ of V_k.

Proof.   Let the vectors $\beta_1,\ldots,\beta_k$ be defined by

$$\beta_i = \alpha_i - \sum_{j=1}^{i-1} c_{ij}\beta_j, \qquad i = 1,\ldots,k \tag{A.9}$$

where the constants c_{ij} are determined in such a way that β_i is orthogonal to each of $\beta_1,\ldots,\beta_{i-1}$; in other words, we choose

$$c_{ij} = \frac{\beta_j'\alpha_i}{\sqrt{\beta_j'\beta_j}}.$$

Using (A.9), we can then write

$$\beta_i = \alpha_i - \sum_{j=1}^{i-1} d_{ij}\alpha_j$$

where d_{ij} are scalar constants and $\beta_i = 0$ implies that $\alpha_1,\ldots,\alpha_i$ are linearly dependent. Thus the vectors $\beta_1,\ldots,\beta_k$ are nonzero and orthogonal, and by Lemma A.1.2.1 they are linearly independent. Hence $\beta_1,\ldots,\beta_k$ form an orthogonal basis of V_k. Defining $\gamma_i = \beta_i/\|\beta_i\|$, we get $\gamma_1,\ldots,\gamma_k$ as the required orthonormal basis of V_k.   Q.E.D.

EXERCISES

1. Find a basis of the vector space V^3 of rank 3 which contains the vector $(1,2,3)'$.
2. Let $\alpha_1 = (1,1,-2,1)'$, $\alpha_2 = (3,0,4,-1)'$, $\alpha_3 = (-1,2,5,2)'$. Which of the three vectors $(-1,1,0,1)'$, $(4,-5,8,-7)'$, $(3,2,-4,4)'$ are in the space spanned by α_1, α_2, α_3?
3. If $\alpha_1,\ldots,\alpha_k$ form a basis of a vector space V, show that no set of $k + 1$ vectors are linearly independent.
4. Let $\alpha_1,\ldots,\alpha_m$ be a set of vectors in V^p. Show that the set of all linear combinations $\sum_{i=1}^{m} c_i\alpha_i$, where c_i are scalar constants, form a linear subspace of V^p.
5. Find the coordinates of the vectors $(1,0,0,0)'$, $(0,1,0,0)'$, $(0,0,1,0)'$, $(0,0,0,1)'$ with respect to the basis $((1,1,0,0)'$, $(0,0,1,1)'$, $(1,0,0,4)'$, $(0,0,0,2)')$.
6. Find the orthogonal projection of the vector $(1,2,3)'$ on the vector $(1,0,1)'$.
7. If $(\gamma_1,\ldots,\gamma_k)$ is an orthonormal basis of the subspace V^k of a vector space V^p, show that it can be extended to an orthonormal basis $(\gamma_1,\ldots,\gamma_k, \gamma_{k+1},\ldots,\gamma_p)$ of V^p.
8. Show that for any three vectors x, y, z in V^p, the function d defined by

$$d(x,y) = \max_{1 \le i < p} |x_i - y_i|$$

satisfies the following relations:

(a) $d(x,y) = d(y,x) \ge 0$ (symmetry).
(b) $d(x,z) \le d(x,y) + d(y,z)$ (triangular inequality), and thus $d(x,y)$ can be taken as another definition of distance in E^p in place of Eq. (A.6).

9. Let W be a vector subspace of a vector space V. Show that rank of $W \le$ rank of V.
10. (Cauchy–Schwartz inequality) For any two vectors x, y in V^p, show that $|x'y| \le \|x\| \|y\|$.
11. (Triangle inequality) For any two vectors x, y in V^p, show that $\|x + y\| \le \|x\| + \|y\|$. [*Hint:*

$$\|x + y\|^2 = (x + y)'(x + y)$$
$$= \|x\|^2 + \|y\|^2 + 2x'y$$
$$\le \|x\|^2 + \|y\|^2 + 2\|x\| \|y\|.]$$

A.2 MATRICES

A.2.1 Definition of and Operations on Matrices

Definition A.2.1.1: Matrix   A real *matrix A* is an ordered rectangular array of elements a_{ij} (real numbers)

$$
A = \begin{pmatrix}
a_{11} & \cdots & a_{1q} \\
\cdot & \cdots & \cdot \\
\cdot & \cdots & \cdot \\
\cdot & \cdots & \cdot \\
a_{p1} & \cdots & a_{pq}
\end{pmatrix}
\tag{A.10}
$$

and is written as $A_{p \times q} = (a_{ij})$.

A matrix with p rows and q columns is called a matrix of dimension $p \times q$ (read "p by q"), the number of rows always being listed first. If $p = q$, we call it a square matrix. A p-dimensional column vector is a matrix of dimension $p \times 1$. We will normally use capital letters to denote matrices, the elements of which are the corresponding lowercase letters with appropriate subscripts. Two matrices of the same dimension $A_{p \times q} = (a_{ij})$ and $B_{p \times q} = (b_{ij})$ are said to be equal (written as $A = B$) if $a_{ij} = b_{ij}$ for $i = 1,...,p; j = 1,...,q$. If all $a_{ij} = 0$, then A is called a *null matrix* and is denoted by the symbol 0.

Definition A.2.1.2: Transpose of a matrix   The *transpose* of a $p \times q$ matrix $A_{p \times q} = (a_{ij})$ is a $q \times p$ matrix A', given by

$$
A' = \begin{pmatrix}
a_{11} & \cdots & a_{p1} \\
\cdot & \cdots & \cdot \\
\cdot & \cdots & \cdot \\
\cdot & \cdots & \cdot \\
a_{1q} & \cdots & a_{pq}
\end{pmatrix}.
$$

In other words, $A'_{q \times p} = (a_{ji})$, and is obtained by interchanging the rows and columns of A. It is easy to see that $(A')' = A$.

Definition A.2.1.3: Symmetric matrix   A square matrix A is said to be *symmetric* if $A = A'$; in other words, $A_{p \times p} = (a_{ij})$ is symmetric if $a_{ij} = a_{ji}$ for all $i,j = 1,...,p$.

Definition A.2.1.4: Skew-symmetric matrix   A square matrix A is said to be *skew-symmetric* if $A = -A'$; in other words, $A_{p \times p} = (a_{ij})$ is skew-symmetric if $a_{ij} = -a_{ji}$ for all $i,j = 1,...,p$.

For example,

$$A = \begin{pmatrix} 1 & 2 & 3 \\ 2 & 2 & 4 \\ 3 & 4 & 5 \end{pmatrix}$$

is a symmetric matrix and

$$A = \begin{pmatrix} 0 & 2 & 3 \\ -2 & 0 & 4 \\ -3 & -4 & 0 \end{pmatrix}$$

is a skew-symmetric matrix.

We now define several operations on matrices.

1. *Matrix Addition.* For any two matrices

$$A = \begin{pmatrix} a_{11} & \cdots & a_{1q} \\ \cdot & & \cdot \\ \cdot & \cdots & \cdot \\ \cdot & & \cdot \\ a_{p1} & \cdots & a_{pq} \end{pmatrix}, \qquad B = \begin{pmatrix} b_{11} & \cdots & b_{1q} \\ \cdot & & \cdot \\ \cdot & \cdots & \cdot \\ \cdot & & \cdot \\ b_{p1} & \cdots & b_{pq} \end{pmatrix}$$

of the same dimension $p \times q$ we define the sum $A + B$ as a matrix of dimensions $p \times q$ given by

$$A + B \begin{pmatrix} a_{11} + b_{11} & \cdots & a_{1q} + b_{1q} \\ \cdot & & \cdot \\ \cdot & \cdots & \cdot \\ \cdot & & \cdot \\ a_{p1} + b_{p1} & \cdots & a_{pq} + b_{pq} \end{pmatrix}.$$

The matrix $A - B$ is to be understood in the same sense as $A + B$, where $+$ is replaced by $-$. For example,

$$\begin{pmatrix} 1 & 2 \\ 2 & 1 \end{pmatrix} + \begin{pmatrix} 3 & 2 \\ 2 & 3 \end{pmatrix} = \begin{pmatrix} 4 & 4 \\ 4 & 4 \end{pmatrix}.$$

Clearly, the operation of matrix addition is commutative and associative; that is, for any three matrices A, B, C of the same dimensions:

(a) $A + B = B + A$ (commutative)
(b) $(A + B) + C = A + (B + C)$ (associative)
Obviously, $(A + B)' = A' + B'$.

2. *Scalar Multiplication of a Matrix.* For any matrix A of dimension $p \times q$ and a scalar constant c,

$$Ac = cA = \begin{pmatrix} ca_{11} & \cdots & ca_{1q} \\ \cdot & & \cdot \\ \cdot & \cdots & \cdot \\ \cdot & & \cdot \\ ca_{p1} & \cdots & ca_{pq} \end{pmatrix}.$$

Obviously, $(cA)' = cA'$, and for scalar constants a, b, $(a + b)A = aA + bA$. Thus scalar multiplication is a distributive operation. For example,

$$2 \cdot \begin{pmatrix} 1 & 2 \\ 2 & 1 \end{pmatrix} = \begin{pmatrix} 2 & 4 \\ 4 & 2 \end{pmatrix}.$$

3. *Matrix Product.* The product of two matrices $A_{p \times q} = (a_{ij})$ and $B_{q \times r} = (b_{ij})$, where A is of dimension $p \times q$ and B is of dimension $q \times r$, is a matrix $C_{p \times r} = AB = (c_{ij})$ of dimension $p \times r$ with elements

$$c_{ij} = \sum_{k=1}^{q} a_{ik}b_{kj}, \qquad i = 1,\ldots,p, \quad j = 1,\ldots,r.$$

It is to be remembered that the matrix product AB is defined if the number of columns of A is equal to the number of rows of B, and in general, $AB \neq BA$. Furthermore, $(AB)' = B'A'$ and is defined if AB is defined. It may be verified that matrix multiplication is distributive and associative; that is, for any three matrices A, B, C:

(a) $A(B + C) = AB + AC$ (distributive)
(b) $(AB)C = A(BC)$ (associative)

provided the products are defined. As an example, we have

$$\begin{pmatrix} 1 & 2 & 2 \\ 2 & 1 & 0 \end{pmatrix} \times \begin{pmatrix} 2 & 1 & 2 \\ 0 & 1 & 3 \\ 2 & 2 & 1 \end{pmatrix}$$

$$= \begin{pmatrix} 1(2) + 2(0) + 2(2) & 1(1) + 2(1) + 2(2) & 1(2) + 2(3) + 2(1) \\ 2(2) + 1(0) + 0(2) & 2(1) + 1(1) + 0(2) & 2(2) + 1(3) + 0(1) \end{pmatrix}$$

$$= \begin{pmatrix} 6 & 7 & 10 \\ 4 & 3 & 7 \end{pmatrix}.$$

A.2.2 Diagonal and Triangular Matrices

Definition A.2.2.1: Diagonal matrix A square matrix A is said to be a *diagonal matrix* if all off-diagonal elements of A are 0. For example,

$$\begin{pmatrix} \frac{1}{2} & 0 & 0 \\ 0 & \frac{1}{3} & 0 \\ 0 & 0 & \frac{1}{4} \end{pmatrix}$$

is a diagonal matrix.

Definition A.2.2.2: Identity matrix A diagonal matrix whose diagonal elements are 1 is called an *identity matrix* and is denoted by I.

For any square matrix A, $AI = IA = A$.

Definition A.2.2.3: Triangular matrix A square matrix A whose elements a_{ij} satisfies the conditions $a_{ij} = 0$ if $j < i$ is called an *upper triangular matrix*. If $a_{ij} = 0$ for $j > i$, then A is called a *lower triangular matrix*.

Definition A.2.2.4: Orthogonal matrix A square matrix A is said to be *orthogonal* if $AA' = A'A = I$.

A.2.3 Determinant, Minor, Cofactor; Nonsingular and Inverse Matrices

Associated with any square matrix A of dimension $p \times p$ is a unique scalar number $|A|$ [also written as det (A)] called its *determinant* and defined by

$$|A| = \sum_{\pi} \delta(\pi) a_{1\pi(1)} a_{2\pi(2)} \cdots a_{p\pi(p)}$$

where π runs over all $p!$ permutations of column subscripts $(1,2,\ldots,p)$ and $\delta(\pi) = 1$ if the number of inversions $\pi(1),\ldots,\pi(p)$ from the standard order $1,\ldots,p$ is even, and $\delta(\pi) = -1$ if the number of such inversions is odd.

(The number of inversions in a particular permutation is the total number of times in which an element is followed by numbers that would ordinarily precede it in the standard order $1,\ldots,p$.)

For example,

$$\begin{vmatrix} a_{11}, & a_{12} \\ a_{21}, & a_{22} \end{vmatrix} = a_{11}a_{22} - a_{12}a_{21}$$

$$\begin{vmatrix} a_{11}, & a_{12}, & a_{13} \\ a_{21}, & a_{22}, & a_{23} \\ a_{31}, & a_{32}, & a_{33} \end{vmatrix} = \begin{matrix} a_{11}a_{22}a_{33} - a_{12}a_{21}a_{33} + a_{13}a_{21}a_{32} \\ - a_{11}a_{23}a_{32} + a_{12}a_{23}a_{31} - a_{13}a_{22}a_{31}. \end{matrix}$$

Definition A.2.3.1: Minor and Cofactor For any square matrix A of dimension $p \times p$, the minor of the element a_{ij} of A is the determinant of the matrix formed by deleting the ith row and the jth column of A. The *cofactor* of a_{ij}, written as A_{ij}, is defined as the *minor* of a_{ij} multiplied by $(-1)^{i+j}$.

Definition A.2.3.2: Principal minor The determinant of a submatrix of A of dimension $i \times i$ whose diagonal elements are diagonal elements of A is called a *principal minor* of order i. The set of leading principal minors is a set of p principal minors of orders $1,\ldots,p$ such that the matrix of the principal minor of order i is a submatrix of the matrix of the principal minor of order $i + 1$.

The reader may find it instructive to verify the following properties of the determinant.

P.1. For any scalar c and a matrix A of dimensions $p \times p$, $|cA| = c^p|A|$.

P.2. $|A|$ changes sign if any two columns (or rows) are interchanged.

P.3. If two columns (or rows) of a matrix A are proportional or equal, then $|A| = 0$.

P.4. The determinant of a diagonal matrix $A_{p \times p} = (a_{ij})$ is

$$|A| = \prod_{i=1}^{p} a_{ii}.$$

P.5. The determinant of a triangular matrix $A_{p \times p} = (a_{ij})$ (upper or lower) is

$$|A| = \prod_{i=1}^{p} a_{ii}.$$

P.6. For any two square matrices A, B of the same dimension,

$$|AB| = |A|\,|B|, \qquad |A'| = |A|.$$

P.7. If A is symmetric, then $A_{ij} = A_{ji}$ for all i, j.

P.8. $|A| = a_{i1}A_{i1} + \cdots \, a_{ip}A_{ip}$

$\qquad = a_{1j}A_{1j} + \cdots + a_{pj}A_{pj}.$

P.9. For $i \neq j$,

$$a_{1j}A_{1i} + \cdots + a_{pj}A_{pi} = 0$$

$$a_{j1}A_{i1} + \cdots + a_{jp}A_{ip} = 0.$$

Definition A.2.3.3: Singular and nonsingular matrices   A square matrix A is called *singular* if $|A| = 0$. It is called *nonsingular* if $|A| \neq 0$.

Clearly, all rows and columns of a nonsingular matrix are linearly independent. The product of two nonsingular matrices is a nonsingular matrix. However, the sum of two nonsingular matrices is not necessarily a nonsingular matrix: A trivial example is $A = -B$, where both A and B are nonsingular matrices, but $A + B$ is the null matrix 0.

Definition A.2.3.4: Inverse matrix   The *inverse* of a nonsingular matrix A of dimensions $p \times p$ is the unique matrix A^{-1} such that $AA^{-1} = A^{-1}A = I$.

Let A_{ij} be the cofactor of the element a_{ij} of A and let

$$
C = \begin{pmatrix} \dfrac{A_{11}}{|A|} & \cdots & \dfrac{A_{1p}}{|A|} \\ \cdot & & \cdot \\ \cdot & \cdots & \cdot \\ \cdot & & \cdot \\ \dfrac{A_{p1}}{|A|} & \cdots & \dfrac{A_{pp}}{|A|} \end{pmatrix}.
$$

From properties P.8 and P.9 of the determinant we get

$$AC' = I.$$

Hence by Definition A.2.3.4, we get

$$
A^{-1} = C' = \begin{pmatrix} \dfrac{A_{11}}{|A|} & \cdots & \dfrac{A_{p1}}{|A|} \\ \cdot & & \cdot \\ \cdot & & \cdot \\ \cdot & & \cdot \\ \dfrac{A_{1p}}{|A|} & \cdots & \dfrac{A_{pp}}{|A|} \end{pmatrix}. \tag{A.11}
$$

From above it is obvious that the inverse of a singular matrix does not exist. If A is symmetric, then A^{-1} is also symmetric. This follows from the fact that for any symmetric matrix A, $A_{ij} = A_{ji}$.

Theorem A.2.3.1   If A, B are nonsingular matrices of the same dimension, then (a) $|A^{-1}| = 1/|A|$, (b) $(A^{-1})' = (A')^{-1}$, and (c) $(AB)^{-1} = B^{-1}A^{-1}$.

Proof of (a).   Since $AA^{-1} = I$, $|AA^{-1}| = |A|\,|A^{-1}| = 1$. Hence $|A^{-1}| = 1/|A|$.

Proof of (b).   Since A is nonsingular, $|A'| = |A| \neq 0$, and hence A' is nonsingular. Thus $A'(A')^{-1} = I$. Taking the transpose we get

$$((A')^{-1})'A = I. \tag{A.12}$$

But $A^{-1}A = I$. Thus from the uniqueness of the inverse and from Eq. (A.12) we get

$$A^{-1} = ((A')^{-1})'$$

or, equivalently,

$$(A^{-1})' = (A')^{-1}.$$

Proof of (c).   Since A, B are nonsingular $|AB| = |A|\,|B| \neq 0$, and hence AB is nonsingular. Now $B^{-1}A^{-1}AB = B^{-1}IB = B^{-1}B = I$. But $(AB)^{-1}AB = I$. Hence $(AB)^{-1} = B^{-1}A^{-1}$.   Q.E.D.

A.2.4   Rank and Trace of a Matrix

Let us consider a matrix of dimension $p \times q$ as a set of q p-dimensional column vectors or p q-dimensional row vectors written in some particular order. Let $R(A)$ be the vector space spanned by the row vectors of A, and let $C(A)$ be the vector space spanned by the column vectors of A.

Definition A.2.4.1: Row rank and column rank   The space $R(A)$ is called the *row space* of A and the rank of $R(A)$, denoted by $r(A)$, is called the *row rank* of A. The space $C(A)$ is called the *column space* of A and the rank of $C(A)$, denoted by $c(A)$, is called the *column rank* of A.

In what follows we write $A_{(i)}$ for the ith row vector of A and $A^{(j)}$ for the jth column vector of A.

Theorem A.2.4.1   For any matrix A of dimension $p \times q$ the row rank is equal to the column rank.

Proof.   Let $r(A) = r \leq p$ and let the row vectors of A, numbered $\omega_1,\ldots,\omega_r$, form a basis of $R(A)$. Consider the matrix B composed of these rows only.

Obviously, $r(B) = r$. Since $A_{(\omega_1)},\ldots,A_{(\omega_r)}$ form a basis of $R(A)$, any row $A_{(i)}$ of A can be written as

$$A_{(i)} = \sum_{k=1}^{r} c_{ik}A_{(\omega_k)}, \qquad i = 1,\ldots,p, \tag{A.13}$$

where c_{ik} are scalar constants not all zero. From Eq. (A.13) it follows that the elements a_{ij} of A can be expressed as

$$a_{ij} = \sum_{k=1}^{r} c_{ik}a_{\omega_k j}, \qquad i = 1,\ldots,p, \quad j = 1,\ldots,q. \tag{A.14}$$

We will now establish that if the column vectors of B are linearly dependent, the column vectors of A are also linearly dependent and $c(B) \geq c(A)$. To do this, let us first assume that the column vectors of B are linearly dependent. In other words, let us assume that there exist constants $d_1,\ldots,d_q$, not all zero, such that

$$\sum_{j=1}^{q} d_j B^{(j)} = 0. \tag{A.15}$$

This implies that

$$\sum_{j=1}^{q} d_j a_{\omega_k j} = 0, \qquad k = 1,\ldots,r. \tag{A.16}$$

From Eqs. (A.16) and (A.14) it follows that for $i = 1,\ldots,q$,

$$\sum_{j=1}^{q} d_j a_{ij} = \sum_{j=1}^{q} d_j \left(\sum_{k=1}^{r} c_{ik}a_{\omega_k j} \right)$$
$$= \sum_{k=1}^{r} c_{ik} \left(\sum_{j=1}^{q} d_j a_{\omega_k j} \right).$$
$$= 0. \tag{A.17}$$

Hence

$$\sum_{j=1}^{q} d_j A^{(j)} = 0. \tag{A.18}$$

Further, let us assume that $A^{(i_1)},\ldots,A^{(i_m)}$, $m \leq q$ form a basis of $C(A)$. If $B^{(i_1)},\ldots,B^{(i_m)}$ are linearly dependent, then for scalar constants $c_1,\ldots,c_m$ not all zero we will have

$$\sum_{k=1}^{m} c_k B^{(i_k)} = 0, \tag{A.19a}$$

or, equivalently,

$$\sum_{i=1}^{q} c_i B^{(i)} = 0, \tag{A.19b}$$

where $c_i = 0$ for $i \neq i_1,\dots,i_m$.

From Eq. (A.19) it then follows that

$$\sum_{i=1}^{q} c_i A^{(i)} = 0 \tag{A.20}$$

and hence $\sum_{k=1}^{m} c_k A^{(i_k)} = 0$.

This is impossible, as it contradicts the fact that $A^{(i_1)},\dots,A^{(i_m)}$ are linearly independent. Thus $c(B) \geq c(A)$.

Since $c(B)$ is the rank of the vector space spanned by q r-dimensional column vectors of the matrix B, by Problem 9 in Exercises, Appendix A, we get

$$r \geq c(B) \geq c(A)$$

or, equivalently,

$$r(A) \geq c(A). \tag{A.21}$$

Repeating the arguments above for the transpose matrix A', we get from Eq. (A.21),

$$r(A') \geq c(A').$$

But clearly, $r(A') = c(A)$ and $c(A') = r(A)$. Therefore,

$$c(A) \geq r(A). \tag{A.22}$$

Thus from Eqs. (A.21) and (A.22) we get $c(A) = r(A)$.   Q.E.D.

Definition A.2.4.2: Rank of a matrix   The common value of the row rank and the column rank of the matrix A is called the *rank* of the matrix A and is denoted by $\rho(A)$.

For any matrix of dimension $p \times q$ with $p < q$, $\rho(A)$ may vary from 0 to p. If $\rho(A) = p$, we call A a matrix of full rank. The rank of a null matrix is 0.

Theorem A.2.4.2   For any two matrices A, B whose product AB is defined, $\rho(AB) \leq \min[\rho(A),\rho(B)]$.

Proof.   The columns of AB are linear combinations of the columns of A. Thus the number of linearly independent columns of AB can not exceed the number of linearly independent columns of A. Hence $\rho(AB) \leq \rho(A)$.

Arguing similarly for the rows of AB, we get $\rho(AB) \leq \rho(B)$. Hence $\rho(AB) \leq \min[\rho(A),\rho(B)]$.   Q.E.D.

Theorem A.2.4.3 If A is a matrix of dimension $p \times q$ and B, C are nonsingular matrices of dimension $p \times p$, $q \times q$, respectively, then $\rho(A) = \rho(AC) = \rho(BA) = \rho(BAC)$.

Proof.   By Theorem A.2.4.2, we have

$$\rho(BA) \leq \rho(A) = \rho(B^{-1}(BA)) \leq \rho(BA). \tag{A.23}$$

Therefore, $\rho(BA) = \rho(A)$. Similarly,

$$\rho(AC) \leq \rho(A) = \rho((AC)C^{-1}) \leq \rho(AC),$$

which implies that $\rho(AC) = \mathrm{p}(A)$. Finally, by Eq. (A.23),

$$\rho(BAC) = \rho(B(AC)) = \rho(AC) = \rho(A).   \text{Q.E.D.}$$

Definition A.2.4.3: Trace of a matrix The *trace* of a square matrix $A_{p \times p} = (a_{ij})$ of dimension $p \times p$ is defined by the sum of its diagonal elements and is written as

$$\mathrm{tr}\ A = \sum_{i=1}^{p} a_{ii}.$$

Obviously, $\mathrm{tr}\ A = \mathrm{tr}\ A'$, and for any two matrices A, B, $\mathrm{tr}(A + B) = \mathrm{tr}\ A + \mathrm{tr}\ B$.

Theorem A.2.4.4   For any two matrices A, B, $\mathrm{tr}\ AB = \mathrm{tr}\ BA$, provided that both AB and BA are defined.

Proof.   By definition,

$$\mathrm{tr}\ AB = \sum_{i=1}^{p} \sum_{j=1}^{p} a_{ij}b_{ji}, \qquad \mathrm{tr}\ BA = \sum_{i=1}^{p} \sum_{k=1}^{p} b_{ik}a_{ki}.$$

But it is clear that

$$\sum_{i=1}^{p} \sum_{j=1}^{p} a_{ij}b_{ji} = \sum_{i=1}^{p} \sum_{k=1}^{p} b_{ik}a_{ki}.$$

Hence $\mathrm{tr}\ AB = \mathrm{tr}\ BA$.   Q.E.D.

From Theorem A.2.4.4 it thus follows that for any three matrices A, B, C, $\mathrm{tr}(ABC) = \mathrm{tr}(BCA) = \mathrm{tr}(CAB)$. In particular, for any orthogonal matrix O, $\mathrm{tr}(O'AO) = \mathrm{tr}(AOO') = \mathrm{tr}\ A$.

A.2.5   Quadratic Forms and Positive Definite Matrices

A quadratic form in the variables (real) $x_1,\ldots,x_p$ is an expression of the form

$$Q = \sum_{i,j=1}^{p} a_{ij}x_ix_j, \tag{A.24}$$

where a_{ij} are real constants.

Writing x as a column p-vector $x = (x_1,\ldots,x_p)'$ and A as a $p \times p$ matrix with elements a_{ij}, the quadratic form can be written as $Q = x'Ax$. Without any loss of generality the matrix A in the quadratic form Q can be taken to be symmetric, since, Q being a scalar quantity, we can write

$$Q = x'Ax = Q' = x'A'x$$

and then we can write, say,

$$Q = \frac{1}{2}(Q + Q') = x'\left(\frac{A + A'}{2}\right)x = x'Bx,$$

where $B = (A + A')/2$ and is thus symmetric. Thus if A is not symmetric, we can replace it by a symmetric matrix $\frac{1}{2}(A + A')$ without changing the value of the quadratic form. Henceforth, whenever we refer to a quadratic form $Q = x'Ax$, we shall assume A to be symmetric.

Definition A.2.5.1: Positive definite matrix   A square matrix A and the associated quadratic form $x'Ax$ are called *positive definite* if $x'Ax > 0$ for every nonnull vector $x = (x_1,\ldots,x_p)'$, and are called *positive semidefinite* if $x'Ax \geq 0$ for all x.

Definition A.2.5.2: Negative definite matrix   A square matrix A and the associated quadratic form $x'Ax$ are called *negative definite* or *negative semidefinite* if $-x'Ax$ is positive definite or positive semidefinite, respectively.

For example, let

$$Q = (x_1,x_2)\begin{pmatrix} 2 & 1 \\ 1 & 2 \end{pmatrix}(x_1,x_2)'$$

$$= 2(x_1^2 + x_2^2 + x_1x_2)$$

$$= 2[(\tfrac{1}{2}x_1 + x_2)^2 + \tfrac{3}{4}x_2^2]$$

$$> 0$$

for all $(x_1,x_2)' \neq 0$. Hence the matrix $\begin{pmatrix} 2 & 1 \\ 1 & 2 \end{pmatrix}$ and the associated quadratic form are positive definite.

A.2.6 Characteristic Roots and Vectors

Definition A.2.6.1: Characteristic roots of a matrix The characteristic roots of a square matrix A of dimension $p \times p$ are defined as the roots of the characteristic equation

$$|A - \lambda I| = 0, \tag{A.25}$$

where λ is real.

Since $|A - \lambda I|$ is a polynomial in λ of degree p, the equation $|A - \lambda I| = 0$ has exactly p roots. For example, let

$$A = \begin{pmatrix} a_{11} & a_{12} \\ a_{21} & a_{22} \end{pmatrix}.$$

Then

$$\begin{aligned} |A - \lambda I| &= \begin{vmatrix} a_{11} - \lambda & a_{12} \\ a_{21} & a_{22} - \lambda \end{vmatrix} \\ &= (a_{11} - \lambda)(a_{22} - \lambda) - a_{12}a_{21} \\ &= \lambda^2 - \lambda(a_{11} + a_{22}) + (a_{11}a_{22} - a_{12}a_{21}) \\ &= \lambda^2 - \lambda \operatorname{tr} A + |A|. \end{aligned}$$

In general, if A is a square matrix of dimension $p \times p$, then

$$\begin{aligned} |A - \lambda I| &= (-\lambda)^p + (-\lambda)^{p-1}S_1 + (-\lambda)^{p-2}S_2 \\ &\quad + \cdots + (-\lambda)S_{p-1} + |A| \end{aligned} \tag{A.26}$$

where S_i is the sum of the $i \times i$ principal minors of A. In particular, $S_1 = \operatorname{tr} A$. Thus the product of the characteristic roots of A is equal to $|A|$ and the sum of the characteristic roots of A is equal to $\operatorname{tr} A$. Obviously the characteristic roots of a diagonal matrix are the diagonal elements themselves.

Definition A.2.6.2: Characteristic vector If λ is a characteristic root of a square matrix A of dimension $p \times p$, then a vector $x = (x_1, \ldots, x_p)'$, not identically zero, satisfying

$$(A - \lambda I)x = 0 \tag{A.27}$$

is called a *characteristic vector* of the matrix A corresponding to the characteristic root λ.

From the definition it follows that if x is a characteristic vector of A corresponding to λ, then any scalar multiple of x is also a characteristic vector of A corresponding to λ.

Theorem A.2.6.1 If P is an orthogonal matrix, then the matrix PAP' has the same characteristic roots as A.

Proof. We have

$$\begin{aligned}
|PAP' - \lambda I| &= |PAP' - \lambda PP'| \\
&= |P(A - \lambda I)P'| \\
&= |P|\,|A - \lambda I|\,|P'| \\
&= |PP'|\,|A - \lambda I| \\
&= |A - \lambda I|.
\end{aligned}$$

Thus the characteristic equations

$$|PAP' - \lambda I| = 0, \qquad |A - \lambda I| = 0$$

have the same roots.   Q.E.D.

Theorem A.2.6.2 If A is a symmetric matrix of dimension $p \times p$, then all its characteristic roots are real.

Proof. Assume the contrary. Let λ be a complex characteristic root of A and let $x + iy$, where $x = (x_1, \ldots, x_p)'$, $y = (y_1, \ldots, y_p)'$, and $i = \sqrt{-1}$, be the characteristic vector corresponding to λ. Then

$$A(x + iy) = \lambda(x + iy)$$

and hence

$$\begin{aligned}
(x - iy)'A(x + iy) &= (x' - iy')A(x + iy) \\
&= x'Ax - iy'Ax + ix'Ay + y'Ay \\
&= \lambda x'x + \lambda y'y.
\end{aligned}$$

Since $x'Ax + y'Ay$ is real, λ must be real.   Q.E.D.

It may be noted that the characteristic vector z corresponding to a complex characteristic root must be complex. Otherwise, $Az = \lambda z$ will imply that a real vector is equal to a complex vector.

Theorem A.2.6.3 If λ_i, λ_j are two distinct characteristic roots of a symmetric square matrix $A_{p \times p}$, then their associated characteristic vectors $x_i = (x_{i1}, \ldots, x_{ip})'$, $x_j = (x_{j1}, \ldots, x_{jp})'$ are orthogonal.

Proof. By hypothesis,

$$Ax_i = \lambda_i x_i, \; Ax_j = \lambda_j x_j.$$

Thus

$$x_j' A x_i = \lambda_i x_j' x_i, \quad x_i' A x_j = \lambda_j x_i' x_j.$$

Since $x_i' A x_j$ is a scalar quantity and A is symmetric, we get

$$x_i' A x_j = x_j' A x_i.$$

Hence

$$\lambda_i x_j' x_i = \lambda_j x_i' x_i.$$

Since $\lambda_i \neq \lambda_j$, we conclude that $x_j' x_i = 0$. In other words, x_i, x_j are orthogonal.   Q.E.D.

Theorem A.2.6.4   If A is a positive definite (symmetric) matrix, then all its characteristic roots are positive.

Proof.   Let λ be a characteristic root of A and let x be its corresponding characteristic vector. Then

$$Ax = \lambda x,$$

or, equivalently,

$$x' A x = \lambda x' x.$$

Since $x' A x > 0$, $x' x > 0$, we get $\lambda > 0$.   Q.E.D.

Corollary A.2.6.4.1   If A is positive definite, then $|A| > 0$.

Proof.   From Eq. (A.26), $|A|$ is equal to the product of the characteristic roots of A. Since each characteristic root of A is positive, it follows that $|A| > 0$.   Q.E.D.

Theorem A.2.6.5   For every real symmetric matrix A of dimension $p \times p$ there exists an orthogonal matrix P such that PAP' is a diagonal matrix of the roots of A.

Proof.   Let $\lambda_1 \geq \lambda_2 \geq \cdots \geq \lambda_p$ denote the characteristic roots of A, including multiplicities, and let $x_1, \ldots, x_p$ be the corresponding characteristic vectors of A. Let $y_i = x_i / \|x_i\|$, $i = 1, \ldots, p$. Then $y_1, \ldots, y_p$ are the normalized characteristic vectors of A.

Suppose that there exist $s(\leq p)$ orthonormal vectors $y_1, \ldots, y_s$ such that [cf. Eq. (A.27)]

$$(A - \lambda_i I) y_i = 0, \quad i = 1, \ldots, s.$$

Writing A^r as the matrix that is the product of r matrices each equal to A, we get

$$A^r y_i = \lambda_i A^{r-1} y_i = \cdots = \lambda_i^r y_i, \qquad i = 1,\dots,s. \tag{A.28}$$

Now let us choose a vector x orthogonal to the vector space spanned by $y_1,\dots,y_s$. Then we have

$$(A^r x)' y_i = x' A^r y_i = \lambda_i^r x' y_i = 0 \tag{A.29}$$

for all values of r including zero and for $i = 1,\dots,s$. Hence any vector belonging to the vector space spanned by the vectors x, Ax, $A^2 x,\dots$ is orthogonal to any vector belonging to the vector space spanned by $y_1,\dots,y_s$. Obviously, the vectors x, Ax, $A^2 x,\dots$ can not all be linearly independent. Let k be the smallest value of r such that for real constants $b_1,\dots,b_k$,

$$A^k x + b_1 A^{k-1} x + \cdots + b_k x = 0. \tag{A.30}$$

Factoring the left-hand side of Eq. (A.30), we see that for constants $\mu_1,\dots,\mu_k$,

$$\prod_{i=1}^{k} (A - \mu_i I) x = 0.$$

Now choose the vector y_{s+1} such that

$$y_{s+1} = \prod_{i=2}^{k} (A - \mu_i I) x.$$

Then

$$(A - \mu_i I) y_{s+1} = 0.$$

In other words, there exists a vector y_{s+1} (normalized) in the space spanned by $(x, Ax, A^2 x,\dots)$ which is a characteristic vector of A corresponding to its characteristic root $\mu_1(=\lambda_{s+1}$, say) and which is orthogonal to $y_1,\dots,y_s$. Since y_1 can be chosen corresponding to any characteristic vector to start with, we have proved the existence of p orthonormal characteristic vectors $y_1,\dots,y_p$:

$$A y_i = \lambda_i y_i, \qquad i = 1,\dots,p. \tag{A.31}$$

In other words, writing P as an orthogonal matrix of dimension $p \times p$ with y_i as its rows, and $\triangle$ as a diagonal matrix with λ_i as its diagonal elements, we can write Eq. (A.31) as

$$AP' = P'\triangle, \qquad \text{where } P'P = I,$$

or, equivalently,

$$PAP' = \triangle. \qquad \text{Q.E.D.}$$

Corollary A.2.6.5.1 Any positive definitive quadratic form $x'Ax$ can be transformed to a diagonal form $\sum_{i=1}^{p} \lambda_i y_i^2$, where $y = (y_1,...,y_p)' = Px$, such that PAP' is a diagonal matrix with diagonal elements $\lambda_1,...,\lambda_p$ (characteristic roots of A); P is the orthogonal matrix of Theorem A.2.6.5.

Proof

$$\sum_{i=1}^{p} \lambda_i y_i^2 = y'(PAP')y$$

$$= (Px)'PAP'(Px)$$
$$= x'P'PAP'Px$$
$$= x'Ax. \qquad \text{Q.E.D.}$$

Theorem A.2.6.6 If A is a positive definite symmetric matrix of dimension $p \times p$, then there exists a nonsingular matrix C of the same dimension as A such that $A = C'C$.

Proof. By Theorem A.2.6.5 there exists an orthogonal matrix P such that

$$PAP' = D = \begin{pmatrix} \lambda_1 & 0 & \cdots & 0 \\ 0 & \lambda_2 & \cdots & \cdot \\ \cdot & \cdot & & \cdot \\ \cdot & \cdot & & \cdot \\ \cdot & \cdot & \cdots & \cdot \\ 0 & \cdots & 0 & \lambda_p \end{pmatrix}.$$

Let

$$D^{1/2} = \begin{pmatrix} \lambda_1^{1/2} & 0 & \cdots & 0 \\ 0 & \lambda_2^{1/2} & \cdots & 0 \\ & & \cdots & \\ 0 & \cdots & 0 & \lambda_p^{1/2} \end{pmatrix}$$

and $PD^{1/2} = C$. Then

$$A = P'PAP'P = P'DP = P'D^{1/2}D^{1/2}P = C'C$$

and $|C| = |D^{1/2}| \neq 0.$ \qquad Q.E.D.

Corollary A.2.6.6.1 Any positive definite quadratic form $x'Ax$ can be transformed to a diagonal form $y'y = \sum_{i=1}^{p} y_i^2$ where $y = Cx$, $C = PD^{1/2}$ as in Theorem A.2.6.6.

Proof. By Theorem A.2.6.6 there exists a nonsingular matrix C such that $A = C'C$. Hence $x'Ax = x'C'Cx = (Cx)'(Cx) = y'y.$ \qquad Q.E.D.

Corollary A.2.6.6.2 If A is a positive definite matrix, then there exists a nonsingular matrix B such that $BAB' = I$.

Proof. Let $B = D^{-1/2}P$ such that $PAP' = D = D^{1/2}D^{1/2}$, where P and D are as defined in Theorem A.2.6.6. Then

$$BAB' = D^{-1/2}PAP'D^{-1/2} = D^{-1/2}DD^{-1/2} = D^{-1/2}D^{1/2}D^{1/2}D^{-1/2} = I.$$

Q.E.D.

Corollary A.2.6.6.3 If A is positive definite, then A^{-1} is also positive definite.

Proof. Let $A = C'C$, where C is nonsingular. Then $A^{-1} = C^{-1}(C^{-1})'$. For any nonnull x let $y = C^{-1}x$; then y is a nonnull vector and

$$x'A^{-1}x = (C^{-1}x)'(C^{-1}x) = y'y > 0.$$

Thus A^{-1} is positive definite. Q.E.D.

Theorem A.2.6.7 If A is a positive definite matrix of dimension $p \times p$ and if B of dimension $p \times q$, $q < p$, is of rank q, then $B'AB$ is positive definite.

Proof. Given any nonnull vector x, let $y = Bx$. Since B is of rank q, y is also a nonnull vector, and hence

$$x'(B'AB)x = (Bx)'A(Bx) = y'Ay > 0.$$

Thus $B'AB$ is positive definite. Q.E.D.

In Theorem A.2.6.7, if B is nonsingular, then obviously $B'AB$ is positive definite.

Theorem A.2.6.8 If A is positive definite and B is positive semidefinite, then there exists a nonsingular matrix E such that $EAE' = I$ and EBE' is a diagonal matrix with diagonal elements $\lambda_1,...,\lambda_p$, the characteristic roots of the equation $|B - \lambda A| = 0$.

Proof. The equations $|B - \lambda A| = 0$ and $|C'BC - \lambda C'AC| = 0$, where C is any nonsingular matrix, have the same roots. This is because

$$|C'BC - \lambda C'AC| = |C|^2 |B - \lambda A|.$$

By Corollary A.2.6.6.2 there exists a nonsingular matrix D such that $DAD' = I$. Let $DBD' = B^*$; then B^* is real and symmetric. By Theorem A.2.6.5 there exists an orthogonal matrix P such that PB^*P' is a diagonal matrix. Defining $PD = E$ (nonsingular) we get $EAE' = I$, and EBE' is a diagonal matrix whose diagonal elements are the characteristic roots of B^*, which in turn are the characteristic roots of $|B - \lambda A| = 0$. Q.E.D.

A.2.7 Partitioned Matrices

Definition A.2.7.1: Partitioned Matrix A matrix A of dimension $p \times q$ is said to be *partitioned* into submatrices A_{ij}, $i,j = 1,2$ if A can be written as

$$A = \begin{pmatrix} A_{11} & A_{12} \\ A_{21} & A_{22} \end{pmatrix}, \tag{A.32}$$

where

$$A_{11} = (a_{ij}), \quad i = 1,\ldots,m, \quad\quad j = 1,\ldots,n$$
$$A_{12} = (a_{ij}), \quad i = 1,\ldots,m, \quad\quad j = n + 1,\ldots,q$$
$$A_{21} = (a_{ij}), \quad i = m + 1,\ldots,p, \quad j = 1,\ldots,n$$
$$A_{22} = (a_{ij}), \quad i = m + 1,\ldots,p, \quad j = n + 1,\ldots,q.$$

Since a p-dimensional column vector is a matrix of dimension $p \times 1$, the definition of partitioned vector follows from the definition of partitioned matrix. If A, B are matrices of the same dimension and are similarly partitioned, then

$$A + B = \begin{pmatrix} A_{11} + B_{11} & A_{12} + B_{12} \\ A_{21} + B_{21} & A_{22} + B_{22} \end{pmatrix}$$

$$A' = \begin{pmatrix} A_{11}' & A_{21}' \\ A_{12}' & A_{22}' \end{pmatrix}.$$

Let

$$A = \begin{pmatrix} A_{11} & A_{12} \\ A_{21} & A_{22} \end{pmatrix}$$

be partitioned as in Eq. (A.32), and let C be a matrix of dimension $q \times r$ and be partitioned into submatrices C_{ij}, $i,j = 1,2$, where C_{11}, C_{12} have n rows. Then it may be verified that

$$AC = \begin{pmatrix} A_{11}C_{11} + A_{12}C_{21} & A_{11}C_{12} + A_{12}C_{22} \\ A_{21}C_{11} + A_{22}C_{21} & A_{21}C_{12} + A_{22}C_{22} \end{pmatrix}.$$

Theorem A.2.7.1 If A is a positive definite or nonsingular matrix and if

$$A = \begin{pmatrix} A_{11} & 0 \\ 0 & A_{22} \end{pmatrix},$$

where A_{11} is a square submatrix of A, then $|A| = |A_{11}||A_{22}|$, and

$$A^{-1} = \begin{pmatrix} A_{11}^{-1} & 0 \\ 0 & A_{22}^{-1} \end{pmatrix}.$$

Proof

$$\begin{vmatrix} A_{11} & 0 \\ 0 & A_{22} \end{vmatrix} = \left| \begin{pmatrix} A_{11} & 0 \\ 0 & I \end{pmatrix} \begin{pmatrix} I & 0 \\ 0 & A_{22} \end{pmatrix} \right|$$

$$= \begin{vmatrix} A_{11} & 0 \\ 0 & I \end{vmatrix} \begin{vmatrix} I & 0 \\ 0 & A_{22} \end{vmatrix}$$

$$= |A_{11}||A_{22}|.$$

The evaluation of the determinants, for example,

$$\begin{vmatrix} I & 0 \\ 0 & A_{22} \end{vmatrix}$$

is made by expanding according to minors of the first row, the only nonzero element in the sum being the first, which is 1 times a determinant of the same form with I or order one less. This is continued until $|A_{22}|$ is the minor. Since $|A|$ is nonzero, A_{11} and A_{22} are both nonsingular, and thus A_{11}^{-1}, A_{22}^{-1} exist. Hence

$$\begin{pmatrix} A_{11} & 0 \\ 0 & A_{22} \end{pmatrix} \begin{pmatrix} A_{11}^{-1} & 0 \\ 0 & A_{22}^{-1} \end{pmatrix} = \begin{pmatrix} I & 0 \\ 0 & I \end{pmatrix}.$$

Thus, from the uniqueness of the inverse matrix,

$$A^{-1} = \begin{pmatrix} A_{11}^{-1} & 0 \\ 0 & A_{22}^{-1} \end{pmatrix}. \qquad \text{Q.E.D.}$$

In a similar fashion one can show that

$$\begin{vmatrix} A_{11} & A_{12} \\ 0 & A_{22} \end{vmatrix} = \begin{vmatrix} I & 0 \\ 0 & A_{22} \end{vmatrix} \begin{vmatrix} A_{11} & A_{12} \\ 0 & I \end{vmatrix} = |A_{11}||A_{22}|.$$

Theorem A.2.7.2   For any nonsingular square matrix

$$A = \begin{pmatrix} A_{11} & A_{12} \\ A_{21} & A_{22} \end{pmatrix},$$

where A_{11} is a square submatrix of A, $|A| = |A_{22}||A_{11} - A_{12}A_{22}^{-1}A_{21}|$.

Proof.   Since

$$\begin{vmatrix} I & 0 \\ -A_{22}A_{21} & I \end{vmatrix} = 1,$$

$$\begin{vmatrix} A_{11} & A_{12} \\ A_{21} & A_{22} \end{vmatrix} = \left| \begin{pmatrix} A_{11} & A_{12} \\ A_{21} & A_{22} \end{pmatrix} \begin{pmatrix} I & 0 \\ -A_{22}^{-1}A_{21} & I \end{pmatrix} \right|$$

$$= \begin{vmatrix} A_{11} - A_{12}A_{22}^{-1}A_{21} & A_{12} \\ 0 & A_{22} \end{vmatrix}$$

$$= |A_{11} - A_{12}A_{22}^{-1}A_{21}||A_{22}|. \qquad \text{Q.E.D.}$$

In a similar way one can show that

$$|A| = |A_{11}||A_{22} - A_{21}A_{11}^{-1}A_{12}|.$$

From these results it follows that if A is nonsingular, then the submatrices A_{22}, $A_{11} - A_{12}A_{22}^{-1}A_{21}$ (or, equivalently, A_{11}, $A_{22} - A_{21}A_{11}^{-1}A_{12}$) are non-singular.

Theorem A.2.7.3 If

$$A = \begin{pmatrix} A_{11} & A_{12} \\ A_{21} & A_{22} \end{pmatrix}$$

is positive definite or nonsingular and A_{11} is a square submatrix of A, and if $A^{-1} = B$ is similarly partitioned as

$$B = \begin{pmatrix} B_{11} & B_{12} \\ B_{21} & B_{22} \end{pmatrix},$$

then

$$A_{11}^{-1} = B_{11} - B_{12}B_{22}^{-1}B_{21}, \qquad A_{22}^{-1} = B_{22} - B_{21}B_{11}^{-1}B_{12}.$$

Proof. Since $A^{-1}B = I$, we get

$$\begin{pmatrix} A_{11} & A_{12} \\ A_{21} & A_{22} \end{pmatrix}\begin{pmatrix} B_{11} & B_{12} \\ B_{21} & B_{22} \end{pmatrix} = I.$$

Thus

$$A_{11}B_{11} + A_{12}B_{21} = I, \qquad A_{11}B_{12} + A_{12}B_{22} = 0,$$
$$A_{21}B_{11} + A_{22}B_{21} = 0, \qquad A_{21}B_{12} + A_{22}B_{22} = I.$$

Solving the matrix equations above, we get

$$A_{11}B_{11} - A_{11}B_{12}B_{22}^{-1}B_{21} = I$$
$$A_{22}B_{22} - A_{22}B_{21}B_{11}^{-1}B_{12} = I,$$

or, equivalently,

$$A_{11}^{-1} = (B_{11} - B_{12}B_{22}^{-1}B_{21}), \qquad A_{22}^{-1} = (B_{22} - B_{21}B_{11}^{-1}B_{12}). \qquad \text{Q.E.D.}$$

Theorem A.2.7.4 A square matrix (of dimension $p \times p$)

$$A = \begin{pmatrix} A_{11} & A_{12} \\ A_{21} & A_{22} \end{pmatrix},$$

where A_{11} is a square submatrix of dimension $q \times q$, is positive definite if and only if A_{11}, $A_{22} - A_{21}A_{11}^{-1}A_{12}$ are positive definite.

Proof. Let

$$x = (x_1,\ldots,x_p)' = \begin{pmatrix} x_{(1)} \\ x_{(2)} \end{pmatrix},$$

where $x_{(1)} = (x_1,\ldots,x_q)'$, $x_{(2)} = (x_{q+1},\ldots,x_p)'$. Now

$$x'Ax = x'_{(1)}A_{11}x_{(1)} + x'_{(1)}A_{12}x_{(2)} + x'_{(2)}A_{21}x_{(1)} + x'_{(2)}A_{22}x_{(2)}$$
$$= (x_{(1)} + A_{11}^{-1}A_{12}x_{(2)})'A_{11}(x_{(1)} + A_{11}^{-1}A_{12}x_{(2)})$$
$$+ x'_{(2)}(A_{22} - A_{21}A_{11}^{-1}A_{12})x_{(2)}.$$

Thus from above we get that if A is positive definite, then A_{11} is positive definite. Now if $A_{11}, A_{22} - A_{21}A_{11}^{-1}A_{12}$ are positive definite, then A is positive definite. Conversely, if A and thus A_{11} are positive definite, then by taking $x \neq 0$ such that $x_{(1)} + A_{11}^{-1}A_{12}x_{(2)} = 0$, we get that $A_{22} - A_{21}A_{11}^{-1}A_{12}$ is positive definite. Hence the theorem is proved.   Q.E.D.

Writing

$$x'Ax = (x_{(2)} + A_{22}^{-1}A_{21}x_{(1)})'A_{22}(x_{(2)} + A_{22}^{-1}A_{21}x_{(1)})$$
$$+ x'_{(1)}(A_{11} - A_{12}A_{22}^{-1}A_{21})x_{(1)}$$

we can assert that A is positive definite if and only if A_{22} and $A_{11} - A_{12}A_{22}^{-1}A_{21}$ are positive definite. We now state a theorem which is useful in our context.

Theorem A.2.7.5 If A is a $p \times p$ positive definite matrix, there exists a $p \times p$ lower triangular matrix T such that $TAT = I$.

A.2.8 Linear Equations, Method of Sweep-out, and Generalized Inverse

Let us consider a system of p linear equations in q unknowns $x_1,\ldots,x_q$ (reals).

$$a_{11}x_1 + \cdots + a_{1q}x_q = b_1$$
$$\vdots \tag{A.33}$$
$$a_{p1}x_1 + \cdots + q_{pq}x_q = b_p,$$

where a_{ij}, $i = 1,\ldots,p$, $j = 1,\ldots,q$, and $b_1,\ldots,b_p$ are real known constants. In terms of the matrix $A = (a_{ij})$ of dimension $p \times q$ and vectors $x = (x_1,\ldots,x_q)'$, $b = (b_1,\ldots,b_p)'$ the equations above can be written as

$$Ax = b. \tag{A.34}$$

If $b = 0$, the system (A.33) is called a *homogeneous system of linear equations*; if, on the other hand, $b \neq 0$, it is called a *nonhomogeneous system of linear equations*.

Solution of a Homogeneous System of Linear Equations

A homogeneous system of linear equations $Ax = 0$ is always solvable, $x = 0$ being a trivial solution. The system has a nontrivial solution $x \neq 0$ if and only if the q column vectors of A are linearly dependent. This follows from the fact that $Ax = 0$ can be rewritten as

$$x_1\beta^1 + \cdots + x_q\beta^q = 0,$$

where $\beta^j = (a_{1j},\ldots,a_{pj})'$ is the jth column vector of A, $j = 1,\ldots,q$. We shall now show that the solutions of a homogeneous system of linear equations form a vector space. For this we use the following result: A collection V of vectors forms a vector space if for any two vectors α and β in V we have $\alpha + \beta \in V$ and $c\alpha \in V$, where c is any real scalar. The proof of this result is left to the reader as an exercise.

Lemma A.2.8.1 The solutions of a homogeneous system of linear equations $Ax = 0$ form a vector space.

Proof. Let $x^1 = (x_1^1,\ldots,x_q^1)'$ and $x^2 = (x_1^2,\ldots,x_q^2)$ be two solutions of the system $Ax = 0$. Then

$$Ax^1 = 0 \quad \text{and} \quad Ax^2 = 0.$$

Thus we have

$$A(x^1 + x^2) = 0 \quad \text{and} \quad A(cx^1) = cAx^1 = 0,$$

where c is a real scalar. Hence $x^1 + x^2$ and cx^1 are also solutions of the system $Ax = 0$. Hence the result.   Q.E.D.

We shall represent by X the vector space of all solutions of the homogeneous system of linear equations $Ax = 0$. We shall now find the rank of X. For this we note that if $\alpha^i = (a_{i1},\ldots,a_{iq})'$, $i = 1,\ldots,p$, is the ith row vector of A, then every solution $x = (x_1,\ldots,x_q)'$ of $Ax = 0$ is orthogonal to $\alpha^1,\ldots,\alpha^p$, the p row vectors of A, and conversely, any vector orthogonal to $\alpha^1,\ldots,\alpha^p$ is a solution of $Ax = 0$. This implies that for any two vectors $x \in X$ and $\alpha \in R(A)$, the row vector space of A, x, is orthogonal to α. Thus X is orthogonal to $R(A)$. To find the rank and a basis of X, we introduce the method of sweep-out of a matrix.

Method of Sweep-out of a Matrix

Consider a vector space V spanned by p q-vectors $\alpha^i = (a_{i1},\ldots,a_{iq})'$, $i = 1,\ldots,p$ (or equivalently, the row vectors of a matrix $A = (a_{ij})$ of dimension $p \times q$)

$$A = \begin{pmatrix} a_{11} & \cdots & a_{1q} \\ \cdot & & \cdot \\ \cdot & \cdots & \cdot \\ \cdot & & \cdot \\ a_{p1} & \cdots & a_{pq} \end{pmatrix}.$$

A convenient method of finding the rank and a basis of V [or those of $R(A)$] is given by the method of sweep-out of the matrix A, which consists of the following operations:

1. Choose a row vector of A whose first component is not zero and divide all its components by the first component. Let the reduced form of the row vector be

    $$(1,c_2,\ldots,c_q),$$

    which will be called the first pivotal row of A.

2. Subtract from every row vector of A other than the first pivotal row $k(1,c_2,\ldots,c_q)$, where k is the first component of the row vector. Obviously, all the resulting row vectors, except the first pivotal row, will thereby have zero as their first component, and thus the first column in the reduced form of A will have all its components equal to zero except only the one corresponding to the pivotal row, which will be unity. This is expressed by saying that the first column of A has been *swept out*. Note that as a result of sweeping out the first column of A it may happen that all components in one or more columns of A may be zero.

3. Repeat operations (1) and (2) on the reduced row vectors of A until a single nonzero row vector (called the last pivotal row vector) is left, while all the remaining row vectors are equal to the 0 vector. The final reduced form of the matrix A so obtained is called its *swept-out form*.

If r ($\leq p$) is the total number of pivotal rows in the swept-out form of matrix A, by a suitable interchange of the rows we can throw the swept-out form of A in the form

$$H = \begin{pmatrix} 1 & 0 & \cdots & 0 & c_{1,r+1} & \cdots & c_{1q} \\ 0 & 1 & \cdots & 0 & c_{2,r+1} & \cdots & c_{2q} \\ \cdots\cdots\cdots\cdots\cdots\cdots\cdots\cdots\cdots \\ 0 & 0 & \cdots & 1 & c_{r,r+1} & \cdots & c_{rq} \\ 0 & 0 & \cdots & 0 & 0 & & 0 \\ \cdots\cdots\cdots\cdots\cdots\cdots\cdots\cdots\cdots \\ 0 & 0 & \cdots & 0 & 0 & & 0 \end{pmatrix} = \begin{pmatrix} I & C \\ 0 & 0 \end{pmatrix}, \tag{A.35}$$

where the first r rows of H correspond to the r pivotal rows of A; c_{ij}'s, $i = 1,...,r, j = r + 1,...,q$, are real numbers some of which may be zero; I is the $r \times r$ unit matrix; and C is the $r \times (q - r)$ matrix (c_{ij}). The first 0 matrix in the second row is of dimension $(p - r) \times r$, and the second 0 matrix is of dimension $(p - r) \times (q - r)$.

Since operations 1 to 3 on the row vectors of A and the interchange of rows of A do not affect the dependence or independence of these vectors, it is quite obvious that the rank of V [or of $R(A)$ and thus of A] is r, which gives the total number of pivotal rows in the swept-out form of A, and a basis of V [or of $R(A)$] is defined by the set of nonzero row vectors

$$\mu^i = (0,0,...,1,0,...,0,c_{i,r+1},...,c_{iq}), \quad i = 1,...,r$$

of H. That the vectors $\mu^1, \mu^2,...,\mu^r$ are independent is evident.

To demonstrate the method of sweep-out, take the matrix

$$A = \begin{pmatrix} 1 & 2 & 3 \\ 2 & 1 & 3 \\ 3 & 1 & 2 \end{pmatrix}.$$

The results of operations 1 to 3 on the row vectors of A allowing interchange of rows is given below.

	Rows	Columns			Operations
		(1)	(2)	(3)	
	(1)	1	2	3	First pivotal row = row (1)
	(2)	2	1	3	Row (5) = row (2) − 2 · row (1)
	(3)	3	1	2	Row (6) = row (3) − 3 · row (1)
	(4)	1	2	3	Row (7) = row (4) − 2 · row (8)
Step I	(5)	0	−3	−3	Second pivotal row = row (5)/(−3)
	(6)	0	−5	−7	Row (9) = row (6) − (−5) · row (8)
	(7)	1	0	1	Row (10) = row (7) − row (12)
Step II	(8)	0	1	1	Row (11) = row (8) − row (12)
	(9)	0	0	−2	Third pivotal row = row (9)/(−2)
	(10)	1	0	0	
Step III	(11)	0	1	0	
	(12)	0	0	1	

Thus the swept-out form of A is

$$\begin{pmatrix} 1 & 0 & 0 \\ 0 & 1 & 0 \\ 0 & 0 & 1 \end{pmatrix}.$$

Steps I, II, and III sweep out the first, second, and third columns of A, respectively. The rank of A is 3 and a basis of $R(A)$ is the set of unit vectors $(1,0,0)$, $(0,1,0)$, and $(0,0,1)$.

We observed earlier that $x = (x_1,\ldots,x_q)'$ is a solution of a homogeneous system $Ax = 0$, where A is of dimension $p \times q$, if and only if x is orthogonal to every vector in $R(A)$. If the rank of A is r and H at (A.35) is the swept-out form of A after suitable interchange of rows so that the first r row vectors of H form a basis of $R(A)$, this implies that any x such that

$$Hx = 0 \quad \text{or} \quad x_{(1)} = -Cx_{(2)}, \quad \text{i.e., } x_i = -\sum_{j=r+1}^{q} c_{ij}x_j, \quad i = 1,\ldots,r$$

(A.36)

where $x_{(1)} = (x_1,\ldots,x_r)'$ and $x_{(2)} = (x_{r+1},\ldots,x_q)'$ is a solution of the system $Ax = 0$. Thus the solution vector space X for the system $Ax = 0$ consists of all x conforming to the relation (A.36). Note that the relation (A.36) is such that we can give any values to the variables $x_{r+1},\ldots,x_q$ and then the values of the variables $x_1,\ldots,x_r$ are thereby fixed through the relation (A.36). The number of free variables in a solution is thus $q - r$, namely $x_{r+1},\ldots,x_q$. Giving the following $q - r$ independent sets of values to these free variables [the $q - r$ unit $(q - r)$-vectors]:

x_{r+1}	x_{r+2} $\cdots$ x_q
1	0 0
0	1 0
....
0	0 1

we obtain a basis of X given by the vectors

$$\lambda^1 = (-c_{1,r+1}, -c_{2,r+1},\ldots, -c_{r,r+1},1,0,\ldots,0)$$

$$\vdots$$

(A.37)

$$\lambda^{q-r} = (-c_{1q}, -c_{2q},\ldots, -c_{rq},0,0,\ldots,1).$$

Thus the rank of X is $q - r = q - $ rank of A. We thus have the following theorem.

Theorem A.2.8.1 If X is the solution vector space for a homogeneous system of linear equations $Ax = 0$ in the components of $x = (x_1,\ldots,x_q)'$, where A is of dimension $p \times q$, then

$$\rho(X) + \rho(A) = q = \text{number of variables in the system}$$

where ρ represents rank.

The form (A.35) of a matrix obtained by suitably interchanging the pivotal rows in the final swept-out form of a matrix inspires the following definitions.

Definition A.2.8.1: Echelon form of a matrix A matrix A is said to be in an *echelon form* if and only if:

(a)   Each row of A consists of zero elements only, or a unity preceded by zeros and not necessarily succeded by zeros.

(b)   Any column of A containing the first unity of a row has the rest of its elements zeros.

Definition A.2.8.2: Hermite canonical form of a matrix A matrix A is said to be in a *Hermite canonical form* if its principal diagonal consists of only zeros and ones such that when a diagonal element is zero, all the elements in the corresponding row are zeros, and when a diagonal element is unity, all the remaining elements in the corresponding column are zeros.

It is quite obvious from the definitions above that the swept-out form of a matrix is a matrix in an echelon form, and that a matrix in a Hermite canonical form is obtained from a matrix in an echelon form by an interchange of rows. In particular, the matrix H in (A.35) is in a Hermite canonical form.

In the above, the sweeping-out of a matrix A has been done through operations 1 to 3 on the row vectors of A, thereby sweeping out the columns of A, leading to the matrix H by allowing interchange of rows. In a similar fashion, the matrix A could have been swept out by performing the corresponding operations 1 to 3 on the columns of A, thereby sweeping the rows of A, leading to the same matrix H by allowing interchange of columns.

It is now evident that any matrix A can be swept out by successive applications of the following three operations, which summarize operations 1 to 3 detailed earlier on rows (columns) of A, and the operation of interchange of rows (columns) of A:

O_1:   Multiply a row (column) by a scalar c.

O_2:   Replace the rth row (column) by the rth row (column) + k times the sth row (column), where k is a scalar constant.

O_3:   Interchange any two rows (columns).

It is easy to see that the operations above are rank preserving. We will now show that these operations are equivalent to multiplication by non-singular elementary matrices as follows:

1. The operation O_1 of multiplying the rth row of a matrix A of dimension $p \times q$ by a scalar c is equivalent to premultiplying A by the elementary matrix $E_r(c) = (e_{ij})$ of dimension $p \times p$, where

    $$e_{rr} = c; \quad e_{ii} = 1 \quad \text{for } i \neq r; \quad e_{ij} = 0 \quad \text{for } i \neq j.$$

2. The operation O_2 on the rth row of A is equivalent to premultiplying A by the elementary matrix $E_{rs}(k) = (e_{ij})$ of dimension $p \times p$, where

    $$e_{rs} = k, \quad e_{ii} = 1 \quad \text{for all } i \quad \text{and} \quad e_{ij} = 0, \quad \text{otherwise.}$$

3. The operation O_3 for interchanging say the rth and the sth rows of A is equivalent to premultiplying A successively by the elementary matrices of dimension $p \times p$:

    $$E_{rs}(-1), \quad E_{rs}(1), \quad E_{sr}(-1), \quad E_r(-1).$$

For similar operations on the columns, we need to postmultiply by elementary matrices of dimension $q \times q$ of the type $E_r(c)$ and $E_{rs}(k)$.

By repeated operations of types O_1 and O_2 on a matrix A, it can be reduced to a matrix in echelon form. By application of O_3 on the echelon form, it can be reduced to the Hermite canonical form. Symbolically, the entire process can be represented as follows: Let B be the product of all elementary matrices such that BA is in echelon form. It needs only an interchange of rows of BA to reduce it to a Hermite canonical form. As we have seen above, premultiplication with elementary matrices does this. If the product of elementary matrices to reduce BA to the Hermite canonical form is G, then we have $GBA = H$, where H is a matrix in Hermite canonical form.

Lemma A.2.8.2   Let A be any matrix of dimension $p \times q$. Then there exists a nonsingular matrix D of dimension $p \times p$ (or a nonsingular matrix E of dimension $q \times q$) such that

$$DA = H \qquad (\text{or} \quad AE = H),$$

where H is a matrix in the Hermite canonical form. If, in particular, A is a square matrix of order $p \times p$, then H is also a square matrix of order $p \times p$ and we have $H^2 = H \cdot H = H$. In other words, H is an idempotent matrix.

Proof.   The proof of the first part follows by putting $GB = D$, where $GBA = H$ in the discussion above. The second part of the lemma follows from the fact if a square matrix is in Hermite canonical form, it is necessarily idempotent.   Q.E.D.

Solution of a Nonhomogeneous System of Linear Equations

We consider the nonhomogeneous system of linear equations (A.33) or (A.34) for which $b \neq 0$. We observe that the system may not have any solution. This will be the case when the system of equations (A.33) are not inherently consistent. An example of an inconsistent system of equations with no solutions is as follows:

$$2 x_1 + 3 x_2 - 5 x_3 = 2$$

$$4 x_1 + 6 x_2 - 10 x_3 = 3.$$

We can rewrite system (A.33) as follows:

$$x_1\alpha^1 + x_2\alpha^2 + \cdots + x_q\alpha^q = b, \tag{A.38}$$

where α^i is the ith column vector of A, $i = 1,\dots,q$. It is quite evident from (A.38) that the system of equations $Ax = b$ admits of a solution if and only if the vector b is linearly dependent on the column vectors of A, or in other words, if b is contained in $C(A)$, the column vector space of A. This, on the other hand, implies that the rank of A is the same as the rank of (Ab), which is the matrix A augmented by the addition of the column b. We thus have the following theorem.

Theorem A.2.8.2 A necessary and sufficient condition that the nonhomogeneous system of linear equations $Ax = b$ has a solution is that rank $A = \text{rank}(Ab)$.

If $x^1 = (x_1^1,\dots,x_q^1)'$ and $x^2 = (x_1^2,\dots,x_q^2)'$ are any two solutions of the nonhomogeneous system $Ax = b$, then $x^1 - x^2$ is a solution of the homogeneous system $Ax = 0$; on the other hand, if x^1 is any particular solution of $Ax = b$ and x^2 is any solution of $Ax = 0$, then $x^1 + x^2$ is a solution of $Ax = b$. We thus have the following lemma.

Lemma A.2.8.3 All solutions, if any, of a nonhomogeneous system of linear equations $Ax = b$ are obtained by adding to a particular solution of $Ax = b$ all solutions of the homogeneous system $Ax = 0$.

Lemma A.2.8.3 resolves the problem of solving a nonhomogeneous system $Ax = b$ to that of finding a particular solution of $Ax = b$; all the solutions of $Ax = b$ can then be obtained by adding to this particular solution of $Ax = b$ all the solutions of the homogeneous system $Ax = 0$.

For finding a particular solution of $Ax = b$ consider the associated homogeneous system

$$Ax - bz = 0 \qquad (A.39)$$

by introducing an additional dummy variable z. It can be easily verified that if the system $Ax = b$ admits of a solution; that is, if rank $A = \text{rank}(Ab)$, then the homogeneous system (A.39) admits of a solution $(x_1, x_2, \ldots, x_q, z)$ with $z \neq 0$. Such a solution can be found out by sweeping out the matrix $B = (Ab)$ and finding all solutions of (A.39) in the form (A.36) in relation to the matrix B and picking therefrom one with $z \neq 0$. If we now write

$$x_i^0 = \frac{x_i}{z}, \qquad i = 1, \ldots, q,$$

it follows from (A.39) that $x^0 = (x_1^0, \ldots, x_q^0)'$ is a particular solution of the system $Ax = b$.

If $q = p$ (i.e., A is a square matrix of dimension $p \times p$) and A is nonsingular, the system $Ax = b$ has a unique solution

$$x = A^{-1}b.$$

The case where A is a singular matrix is of particular interest to us. In this case the inverse A^{-1} does not exist, but it will be of advantage to us if even in this case, or in the general case where A is not a square matrix, we can express a particular solution of $Ax = b$ in terms of a matrix D as $x = Db$. It is in this context we define an operator, similar to the inverse A^{-1}, which will be called a generalized inverse of A.

Definition A.2.8.3: Generalized inverse of a matrix Let A be a matrix of dimension $p \times q$, and let the matrix B of dimension $q \times p$ be such that for every $y = (y_1, \ldots, y_p)'$ for which the system of equations $Ax = y$ in the elements of $x = (x_1, \ldots, x_q)'$ are solvable, $x = By$ is a particular solution. Then B is called the *generalized inverse* (or *conditional inverse* or *pseudo-inverse*) of A, and is represented as A^-.

If $p = q$ in Definition A.2.8.3, so that A is a square matrix, and A is nonsingular, the generalized inverse A^- of A is unique and is equal to the ordinary inverse A^{-1}. Otherwise, however, there would be more than one generalized inverse of a matrix.

It is evident that the concept of a generalized inverse of a matrix is closely connected with that of swept-out form of the matrix (echelon form) and the Hermite canonical form of the matrix. Lemma A.2.8.2 asserts that if H (of dimension $p \times q$) is the Hermite canonical form of a matrix A of dimension $p \times q$, there exists a nonsingular matrix D of dimension $p \times p$ such that $DA = H$. Now let the rank of A be r, so that H can be expressed as in (A.35):

$$DA = H = \begin{pmatrix} I & C \\ 0 & 0 \end{pmatrix},$$

where I is the unit matrix of dimension $r \times r$ and the dimension of C is $r \times (q - r)$. Let the matrix D be partitioned as

$$D = \begin{pmatrix} D_{(1)} \\ D_{(2)} \end{pmatrix},$$

where $D_{(1)}$ is made up of the first r rows of D and $D_{(2)}$ is made up of the remaining $(p - r)$ rows of D. It is easy to see that if the system $Ax = b$ is solvable, we must have

$$D_{(2)}b = 0$$

and that the Hermite canonical form of the matrix $(A, -b)$ of the homogeneous system (A.39) of equations is G:

$$G = \begin{pmatrix} I & C & -D_{(1)}b \\ 0 & 0 & 0 \end{pmatrix},$$

and thus the solutions of the system (A.39) in accordance with (A.36) are given by

$$x_{(1)} = -Cx_{(2)} + (D_{(1)}b)z, \tag{A.40}$$

where $x_{(2)} = (x_{r+1},...,x_q)'$ and z are the free variables and the values of $x_{(1)} = (x_1,...,x_r)'$ are thereby fixed through the relation (A.40). For the values $x_{(2)} = 0$ and $z = 1$, a particular solution of the system $Ax = b$ is obtained from (A.40) as

$$x^0 = \begin{pmatrix} D_{(1)}b \\ 0 \end{pmatrix} = D^*b, \tag{A.41}$$

where D^* is a matrix of dimension $q \times p$ such that the first r rows of D^* give the matrix $D_{(1)}$ and the remaining $q - r$ rows of D^* give a matrix $D_{(2)}^*$ such that $D_{(2)}^*b = 0$; that is, if we partition D^* as

$$D^* = \begin{pmatrix} D_{(1)}^* \\ D_{(2)}^* \end{pmatrix},$$

we must have $D_{(1)}^* = D_{(1)}$ and $D_{(2)}^*b = 0$. Then D^* is a generalized inverse of A.

Note that in the definition of a generalized inverse of A by D^*, although the submatrix $D^*_{(1)}$ is uniquely defined being equal to $D_{(1)}$, the definition of the submatrix $D^*_{(2)}$ is arbitrary in the sense that $D^*_{(2)}$ may be any matrix of dimension $(q - r) \times p$ such that $D^*_{(2)}b = 0$. There may be thus more than one generalized inverse of a matrix. Moreover, if $q = p$ (i.e., if A is a square matrix of dimension $p \times p$), we can take, in view of $D_{(2)}b = 0$, $D^*_{(2)} = D_{(2)}$, and thus in this case D is a generalized inverse of A and a particular solution of the system $Ax = b$ is

$$x^0 = Db. \tag{A.43}$$

We give below an example to illustrate the evaluation of a generalized inverse of a singular matrix.

Example A.2.8.1   Consider the matrix

$$A = \begin{pmatrix} 1 & -2 & -3 \\ -2 & 5 & -3 \\ 1 & -3 & 6 \end{pmatrix}.$$

A is a singular matrix (last row $= -[$first row $+$ second row$]$), so that the ordinary inverse is not defined. Consider the equations

$$\begin{pmatrix} 1 & -2 & -3 \\ -2 & 5 & -3 \\ 1 & -3 & 6 \end{pmatrix}\begin{pmatrix} x_1 \\ x_2 \\ x_3 \end{pmatrix} = \begin{pmatrix} y_1 \\ y_2 \\ y_3 \end{pmatrix}, \qquad \text{i.e.,} \quad \begin{aligned} x_1 - 2x_2 - 3x_3 &= y_1 \\ 2x_1 + 5x_2 - 3x_3 &= y_2 \\ x_1 - 3x_2 + 6x_3 &= y_3. \end{aligned}$$

Sweeping out the first and second columns of the matrix on the left, we get

$$\begin{pmatrix} 1 & 0 & -21 \\ 0 & 1 & -9 \\ 0 & 0 & 0 \end{pmatrix}\begin{pmatrix} x_1 \\ x_2 \\ x_3 \end{pmatrix} = \begin{pmatrix} 5y_1 + 2y_2 \\ 2y_1 + y_2 \\ y_1 + y_2 + y_3 \end{pmatrix}.$$

Thus the equations are solvable if and only if

$$y_1 + y_2 + y_3 = 0.$$

For any y_1, y_2, y_3 for which the equations are solvable, we get a particular solution putting $x_3 = 0$, and this solution is

$$x_1 = 5y_1 + 2y_2$$

$$x_2 = 2y_1 + y_2$$

$$x_3 = 0,$$

which can be expressed as

$$\begin{pmatrix} x_1 \\ x_2 \\ x_3 \end{pmatrix} = \begin{pmatrix} 5 & 2 & 0 \\ 2 & 1 & 0 \\ 0 & 0 & 0 \end{pmatrix} \begin{pmatrix} y_1 \\ y_2 \\ y_3 \end{pmatrix},$$

so that the matrix

$$\begin{pmatrix} 5 & 2 & 0 \\ 2 & 1 & 0 \\ 0 & 0 & 0 \end{pmatrix}$$

is a generalized inverse of the given matrix A. Again as $y_1 + y_2 + y_3 = 0$, we could write the particular solution as

$$\begin{pmatrix} x_1 \\ x_2 \\ x_3 \end{pmatrix} = \begin{pmatrix} 5 & 2 & 0 \\ 2 & 1 & 0 \\ 1 & 1 & 1 \end{pmatrix} \begin{pmatrix} y_1 \\ y_2 \\ y_3 \end{pmatrix},$$

so that the matrix on the right is another generalized inverse of A. Alternatively, we could put $x_3 = y_3$ to get a different particular solution

$$\begin{pmatrix} x_1 \\ x_2 \\ x_3 \end{pmatrix} = \begin{pmatrix} 5 & 2 & 21 \\ 2 & 1 & 9 \\ 0 & 0 & 1 \end{pmatrix} \begin{pmatrix} y_1 \\ y_2 \\ y_3 \end{pmatrix},$$

which gives the matrix on the right as another generalized inverse of A.

EXERCISES

12. Show that for any two matrices A, B, $(A + B)(A - B) = A^2 - B^2$ if and only if $AB = BA$.
13. Show that for any matrix A, $\rho(AA') = \rho(A'A) = \rho(A)$.
14. Show that for any three matrices A, B, C,

$$\rho(AB) + \rho(BC) \le \rho(A) + \rho(ABC)$$

provided that the products AB, BC, ABC are defined.
15. (Vandermonde's matrix) Let

$$V = \begin{pmatrix} 1 & x_1 & x_1^2 & \cdots & x_1^{p-1} \\ 1 & x_2 & x_2^2 & \cdots & x_2^{p-1} \\ 1 & x_p & x_p^2 & \cdots & x_p^{p-1} \end{pmatrix}.$$

Show that

$$|v| = [(x_2 - x_1)][(x_3 - x_2)(x_3 - x_1)] \cdots [(x_p - x_{p-1})(x_p - x_{p-2}) \cdots (x_p - x_1)].$$

16. If (x_1y_1), (x_2y_2) are two different points in the real plane, show that

$$\begin{vmatrix} x & y & 1 \\ x_1 & y_1 & 1 \\ x_2 & y_2 & 1 \end{vmatrix} = 0$$

represents a straight line through these two points.

17. Find the rank and the inverse of the matrix

$$\begin{pmatrix} 1 & 2 & 1 \\ 3 & 4 & 8 \\ 6 & 2 & 5 \end{pmatrix}.$$

18. (Skew matrix) A matrix A is skew if $A' = -A$. Show that:
    (a) For matrix A, A^2 is symmetric if A is skew.
    (b) A skew matrix with an odd number of rows is singular.
    (c) The determinant of a skew-symmetric matrix is nonnegative.

19. (Idempotent matrix) A square matrix A is an idempotent matrix if $AA = A$. Show that:
    (a) The characteristic roots of an idempotent matrix are either unity or zero.
    (b) If A is idempotent and nonsingular, then $A = I$.
    (c) All idempotent matrices not of full rank are positive semidefinite.
    (d) If A is idempotent and $A + B = I$, then B is idempotent and $AB = BA = 0$.
    (e) If A is idempotent of rank r, then tr $A = r$.

20. (Jacobian of a nonsingular transformation) Let $y = Ax$, where $y = (y_1,\ldots,y_p)'$, $x = (x_1,\ldots,x_p)'$; and A is a nonsingular matrix of dimension $p \times p$. show that

$$J = \left| \left| \begin{pmatrix} \dfrac{\partial x_1}{\partial y_1} & \cdots & \dfrac{\partial x_1}{\partial y_p} \\ \vdots & \cdots & \vdots \\ \dfrac{\partial x_p}{\partial y_1} & \cdots & \dfrac{\partial x_p}{\partial y_p} \end{pmatrix} \right| \right| = |A^{-1}|.$$

21. Let $F(x)$ be a differentiable function of $x_1,\ldots,x_p$, and let

$$x = (x_1,\ldots,x_p)', \qquad \frac{\partial f}{\partial x} = \left(\frac{\partial f}{\partial x_1},\ldots,\frac{\partial f}{\partial x_p} \right)'.$$

Show that:
    (a) $\partial x'x/\partial x = 2x$.
    (b) $\partial x'Ax/\partial x = 2Ax$.

22. Show that a generalized inverse exists for every matrix.

23. Show that a matrix B (of dimension $q \times p$) is generalized inverse of a matrix A (of dimension $p \times q$) if and only if $ABA = A$.
24. Let A be a $p \times p$ positive definite matrix and B be a $p \times p$ nonsingular matrix. Show that BAB' is a positive definite matrix.
25. Solve the following systems of equations:
    (a) $2x_1 - 3x_2 + 4x_4 - x_4 = 0$
    $\quad x_1 + 2x_2 - 5x_3 + x_4 = 0$
    $\quad\quad\quad 3x_2 - 4x_3 + 2x_4 = 0$
    (b) $5x_1 + 2x_2 - 3x_3 + 4x_4 = 3$
    $\quad x_1 - 2x_2 + 4x_3 + x_4 = 2$
    $\quad 2x_1 + 3x_2 - 5x_3 - x_4 = 0$
    $\quad x_1 + 3x_2 - 6x_3 + 3x_4 = -1$

    and evaluate generalized inverse of the matrices of systems (a) and (b).

BIBLIOGRAPHY

Birkoff, G., and S. Maclane (1953). *A Survey of Modern Algebra*, Macmillan, New York.

Levi, F. W. (1900). *Algebra*, Calcutta University Publications, Calcutta, India.

Marcus, M., and H. Mink (1967). *Introduction to Linear Algebra*, Macmillan, New York.

Perlis, S. (1952). *Theory of Matrices*, Addison-Wesley, Reading, Mass.

Appendix B
Statistical Tables

Table 1B Binomial Distribution Function*

$$F_X(j) = \sum_{x=0}^{j} \binom{n}{x} p^x (1 - p)^{n-x}, \quad j = 0,1,...,n$$

n	j	0.10	0.20	p 0.30	0.40	0.50
2	0	.8100	.6400	.4900	.3600	.2500
	1	.9900	.9600	.9100	.8400	.7500
3	0	.7290	.5120	.3430	.2160	.1250
	1	.9720	.8960	.7840	.6480	.5000
	2	.9990	.9920	.9730	.9360	.8750
4	0	.6561	.4096	.2401	.1296	.0625
	1	.9477	.8192	.6517	.4752	.3125
	2	.9963	.9728	.9163	.8208	.6875
	3	.9999	.9984	.9919	.9744	.9375
5	0	.5905	.3277	.1681	.0778	.0312
	1	.9185	.7373	.5282	.3370	.1875
	2	.9914	.9421	.8369	.6826	.5000
	3	.9995	.9933	.9692	.9130	.8125
	4	1.0000	.9997	.9976	.9898	.9688
6	0	.5314	.2621	.1176	.0467	.0156
	1	.8857	.6554	.4202	.2333	.1094
	2	.9842	.9011	.7443	.5443	.3428
	3	.9987	.9830	.9295	.8208	.6562
	4	.9999	.9984	.9891	.9590	.8906
	5	1.0000	.9999	.9993	.9959	.9844

*Reproduced from *Introduction to Probability Theory and Statistical Inference* by H. Larson, with kind permission of the publisher, John Wiley & Sons, Inc., New York.

Table 1B *(continued)*

n	j	0.10	0.20	p 0.30	0.40	0.50
7	0	.4783	.2097	.0824	.0280	.0078
	1	.8503	.5767	.3294	.1586	.0625
	2	.9743	.8520	.6471	.4199	.2266
	3	.9973	.9667	.8740	.7102	.5000
	4	.9998	.9953	.9712	.9037	.7734
	5	1.0000	.9996	.9962	.9812	.9375
	6	1.0000	1.0000	.9998	.9984	.9922
8	0	.4305	.1678	.0576	.0168	.0039
	1	.8131	.5033	.2553	.1064	.0352
	2	.9619	.7969	.5518	.3154	.1445
	3	.9950	.9437	.8059	.5941	.3633
	4	.9996	.9896	.9420	.8263	.6367
	5	1.0000	.9988	.9887	.9502	.8555
	6	1.0000	.9999	.9987	.9915	.9648
	7	1.0000	1.0000	.9999	.9993	.9961
9	0	.3874	.1342	.0404	.0101	.0020
	1	.7748	.4362	.1960	.0705	.0195
	2	.9470	.7382	.4628	.2318	.0898
	3	.9917	.9144	.7297	.4826	.2539
	4	.9991	.9804	.9012	.7334	.5000
	5	.9999	.9969	.9747	.9006	.7461
	6	1.0000	.9997	.9957	.9750	.9102
	7	1.0000	1.0000	.9996	.9962	.9805
	8	1.0000	1.0000	1.0000	.9997	.9980
10	0	.3487	.1074	.0282	.0060	.0010
	1	.7361	.3758	.1493	.0464	.0107
	2	.9298	.6778	.3828	.1673	.0547
	3	.9872	.8791	.6496	.3823	.1719
	4	.9984	.9672	.8497	.6331	.3770
	5	.9999	.9936	.9527	.8338	.6230
	6	1.0000	.9991	.9894	.9452	.8281
	7	1.0000	.9999	.9984	.9877	.9453
	8	1.0000	1.0000	.9999	.9983	.9893
	9	1.0000	1.0000	1.0000	.9999	.9990
11	0	.3138	.0859	.0198	.0036	.0005
	1	.6974	.3221	.1130	.0302	.0059
	2	.9104	.6174	.3127	.1189	.0327
	3	.9815	.8389	.5696	.2963	.1133
	4	.9972	.9496	.7897	.5328	.2744

Table 1B *(continued)*

n	j	0.10	0.20	P 0.30	0.40	0.50
11	5	.9997	.9883	.9218	.7535	.5000
	6	1.0000	.9980	.9784	.9006	.7256
	7	1.0000	.9998	.9957	.9707	.8867
	8	1.0000	1.0000	.9994	.9941	.9673
	9	1.0000	1.0000	1.0000	.9993	.9941
	10	1.0000	1.0000	1.0000	1.0000	.9995
12	0	.2824	.0687	.0138	.0022	.0002
	1	.6590	.2749	.0850	.0196	.0032
	2	.8891	.5583	.2528	.0834	.0193
	3	.9744	.7946	.4925	.2253	.0730
	4	.9957	.9274	.7237	.4382	.1938
	5	.9995	.9806	.8822	.6652	.3872
	6	.9999	.9961	.9614	.8418	.6128
	7	1.0000	.9994	.9905	.9427	.8062
	8	1.0000	.9999	.9983	.9847	.9270
	9	1.0000	1.0000	.9998	.9972	.9807
	10	1.0000	1.0000	1.0000	.9997	.9968
	11	1.0000	1.0000	1.0000	1.0000	.9998
13	0	.2542	.0550	.0097	.0013	.0001
	1	.6213	.2336	.0637	.0126	.0017
	2	.8661	.5017	.2025	.0579	.0112
	3	.9658	.7473	.4206	.1686	.0461
	4	.9935	.9009	.6543	.3530	.1334
	5	.9991	.9700	.8346	.5744	.2905
	6	.9999	.9930	.9376	.7712	.5000
	7	1.0000	.9988	.9818	.9023	.7095
	8	1.0000	.9998	.9960	.9679	.8666
	9	1.0000	1.0000	.9993	.9922	.9539
	10	1.0000	1.0000	.9999	.9987	.9888
	11	1.0000	1.0000	1.0000	.9999	.9983
	12	1.0000	1.0000	1.0000	1.0000	.9999
14	0	.2288	.0440	.0068	.0008	.0001
	1	.5846	.1979	.0475	.0081	.0009
	2	.8416	.4481	.1608	.0398	.0065
	3	.9559	.6982	.3552	.1243	.0287
	4	.9908	.8702	.5842	.2793	.0898
	5	.9985	.9561	.7805	.4859	.2120
	6	.9998	.9884	.9067	.6925	.3953
	7	1.0000	.9976	.9685	.8499	.6047
	8	1.0000	.9996	.9917	.9417	.7880
	9	1.0000	1.0000	.9983	.9825	.9102

Table 1B *(continued)*

n	j	0.10	0.20	p 0.30	0.40	0.50
14	10	1.0000	1.0000	.9998	.9961	.9713
	11	1.0000	1.0000	1.0000	.9994	.9935
	12	1.0000	1.0000	1.0000	.9999	.9991
	13	1.0000	1.0000	1.0000	1.0000	.9999
15	0	.2059	.0352	.0047	.0005	.0000
	1	.5490	.1671	.0353	.0052	.0005
	2	.8159	.3980	.1268	.0271	.0037
	3	.9444	.6482	.2969	.0905	.0176
	4	.9873	.8358	.5155	.2173	.0592
	5	.9978	.9389	.7216	.4032	.1509
	6	.9997	.9819	.8689	.6098	.3036
	7	1.0000	.9958	.9500	.7869	.5000
	8	1.0000	.9992	.9848	.9050	.6964
	9	1.0000	.9999	.9963	.9662	.8491
	10	1.0000	1.0000	.9993	.9907	.9408
	11	1.0000	1.0000	.9999	.9981	.9824
	12	1.0000	1.0000	1.0000	.9997	.9963
	13	1.0000	1.0000	1.0000	1.0000	.9995
	14	1.0000	1.0000	1.0000	1.0000	1.0000
16	0	.1853	.0281	.0033	.0003	.0000
	1	.5147	.1407	.0261	.0033	.0003
	2	.7892	.3518	.0994	.0183	.0021
	3	.9316	.5981	.2459	.0651	.0106
	4	.9830	.7982	.4499	.1666	.0384
	5	.9967	.9183	.6598	.3288	.1051
	6	.9995	.9733	.8247	.5272	.2272
	7	.9999	.9930	.9256	.7161	.4018
	8	1.0000	.9985	.9743	.8577	.5982
	9	1.0000	.9998	.9929	.9417	.7728
	10	1.0000	1.0000	.9984	.9809	.8949
	11	1.0000	1.0000	.9997	.9951	.9616
	12	1.0000	1.0000	1.0000	.9991	.9894
	13	1.0000	1.0000	1.0000	.9999	.9979
	14	1.0000	1.0000	1.0000	1.0000	.9997
	15	1.0000	1.0000	1.0000	1.0000	1.0000
17	0	.1668	.0225	.0023	.0002	.0000
	1	.4818	.1182	.0193	.0021	.0001
	2	.7618	.3096	.0774	.0123	.0012
	3	.9174	.5489	.2019	.0464	.0064
	4	.9779	.7582	.3887	.1260	.0245

Table 1B *(continued)*

n	j	0.10	0.20	p 0.30	0.40	0.50
17	5	.9953	.8943	.5968	.2639	.0717
	6	.9992	.9623	.7752	.4478	.1662
	7	.9999	.9891	.8954	.6405	.3145
	8	1.0000	.9974	.9597	.8011	.5000
	9	1.0000	.9995	.9873	.9081	.6855
	10	1.0000	.9999	.9968	.9652	.8338
	11	1.0000	1.0000	.9993	.9894	.9283
	12	1.0000	1.0000	.9999	.9975	.9755
	13	1.0000	1.0000	1.0000	.9995	.9936
	14	1.0000	1.0000	1.0000	.9999	.9988
	15	1.0000	1.0000	1.0000	1.0000	.9999
	16	1.0000	1.0000	1.0000	1.0000	1.0000
18	0	.1501	.0180	.0016	.0001	.0000
	1	.4503	.0991	.0142	.0013	.0001
	2	.7338	.2713	.0600	.0082	.0007
	3	.9018	.5010	.1646	.0328	.0038
	4	.9718	.7164	.3327	.0942	.0154
	5	.9936	.8671	.5344	.2088	.0481
	6	.9988	.9487	.7217	.3743	.1189
	7	.9998	.9837	.8593	.5634	.2403
	8	1.0000	.9957	.9404	.7368	.4073
	9	1.0000	.9991	.9790	.8653	.5927
	10	1.0000	.9998	.9939	.9424	.7597
	11	1.0000	1.0000	.9986	.9797	.8811
	12	1.0000	1.0000	.9997	.9942	.9519
	13	1.0000	1.0000	1.0000	.9987	.9846
	14	1.0000	1.0000	1.0000	.9998	.9962
	15	1.0000	1.0000	1.0000	1.0000	.9993
	16	1.0000	1.0000	1.0000	1.0000	.9999
	17	1.0000	1.0000	1.0000	1.0000	1.0000
19	0	.1351	.0144	.0011	.0001	.0000
	1	.4203	.0829	.0104	.0008	.0000
	2	.7054	.2369	.0462	.0055	.0004
	3	.8850	.4551	.1332	.0230	.0022
	4	.9648	.6733	.2822	.0696	.0096
	5	.9914	.8369	.4739	.1629	.0318
	6	.9983	.9324	.6655	.3081	.0835
	7	.9997	.9767	.8180	.4878	.1796
	8	1.0000	.9933	.9161	.6675	.3238
	9	1.0000	.9984	.9674	.8139	.5000

Table 1B *(continued)*

n	j	0.10	0.20	p 0.30	0.40	0.50
19	10	1.0000	.9997	.9895	.9115	.6762
	11	1.0000	1.0000	.9972	.9648	.8204
	12	1.0000	1.0000	.9994	.9884	.9165
	13	1.0000	1.0000	.9999	.9969	.9682
	14	1.0000	1.0000	1.0000	.9994	.9904
	15	1.0000	1.0000	1.0000	.9999	.9978
	16	1.0000	1.0000	1.0000	1.0000	.9996
	17	1.0000	1.0000	1.0000	1.0000	1.0000
20	0	.1216	.0115	.0008	.0000	.0000
	1	.3917	.0692	.0076	.0005	.0000
	2	.6769	.2061	.0355	.0036	.0002
	3	.8670	.4114	.1071	.0160	.0013
	4	.9568	.6296	.2375	.0510	.0059
	5	.9887	.8042	.4164	.1256	.0207
	6	.9976	.9133	.6080	.2500	.0577
	7	.9996	.9679	.7723	.4159	.1316
	8	.9999	.9900	.8868	.5956	.2517
	9	1.0000	.9974	.9520	.7533	.4119
	10	1.0000	.9994	.9829	.8725	.5881
	11	1.0000	.9999	.9949	.9435	.7483
	12	1.0000	1.0000	.9987	.9790	.8684
	13	1.0000	1.0000	.9997	.9935	.9423
	14	1.0000	1.0000	1.0000	.9984	.9793
	15	1.0000	1.0000	1.0000	.9997	.9941
	16	1.0000	1.0000	1.0000	1.0000	.9987
	17	1.0000	1.0000	1.0000	1.0000	.9998
	18	1.0000	1.0000	1.0000	1.0000	1.0000

Table 2B Poisson Distribution Function

$$F_X(x) = \sum_{j=0}^{x} \frac{e^{-\lambda}\lambda^j}{j!}$$

x	λ 0.5	1.0	1.5	2.0	3.0	4.0
0	0.6065	0.3679	0.2231	0.1353	0.0498	0.0183
1	0.9098	0.7358	0.5578	0.4060	0.1991	0.0916
2	0.9856	0.9197	0.8088	0.6747	0.4232	0.2381
3	0.9982	0.9810	0.9343	0.8571	0.6472	0.4335
4	0.9998	0.9963	0.9814	0.9473	0.8153	0.6288
5	1.0000	0.9994	0.9955	0.9834	0.9161	0.7851
6		0.9999	0.9990	0.9955	0.9665	0.8893
7		1.0000	0.9998	0.9989	0.9881	0.9480
8			1.0000	0.9998	0.9962	0.9786
9				1.0000	0.9989	0.9919
10					0.9997	0.9972
11					0.9999	0.9991
12					1.0000	0.9997
13						0.9999
14						1.0000

Table 2B *(continued)*

x	5.0	6.0	7.0	8.0	9.0	10.0
0	0.0067	0.0025	0.0009	0.0003	0.0001	0.0000
1	0.0404	0.0174	0.0073	0.0030	0.0012	0.0005
2	0.1246	0.0620	0.0296	0.0138	0.0062	0.0028
3	0.2650	0.1512	0.0818	0.0424	0.0212	0.0103
4	0.4405	0.2851	0.1730	0.0996	0.0550	0.0293
5	0.6160	0.4457	0.3007	0.1912	0.1157	0.0671
6	0.7622	0.6063	0.4497	0.3134	0.2068	0.1301
7	0.8666	0.7440	0.5987	0.4530	0.3239	0.2202
8	0.9319	0.8472	0.7291	0.5925	0.4557	0.3328
9	0.9682	0.9161	0.8305	0.7166	0.5874	0.4579
10	0.9863	0.9574	0.9015	0.8159	0.7060	0.5830
11	0.9945	0.9799	0.9467	0.8881	0.8030	0.6968
12	0.9980	0.9912	0.9730	0.9362	0.8758	0.7916
13	0.9993	0.9964	0.9872	0.9658	0.9261	0.8645
14	0.9998	0.9986	0.9943	0.9827	0.9585	0.9165
15	0.9999	0.9995	0.9976	0.9918	0.9780	0.9513
16	1.0000	0.9998	0.9990	0.9963	0.9889	0.9730
17		0.9999	0.9996	0.9984	0.9947	0.9857
18		1.0000	0.9999	0.9994	0.9976	0.9928
19			0.9999	0.9997	0.9989	0.9965
20			1.0000	0.9999	0.9996	0.9984
21				1.0000	0.9998	0.9993
22					0.9999	0.9997
23					1.0000	0.9999
24						1.0000

Table 3B Unit Normal Distribution*

$$P_X(x) = \int_{-\infty}^{x} \frac{1}{\sqrt{2\pi}} e^{-(1/2)y^2} dy, \qquad x \geq 0$$

x	$P_X(x)$	x	$P_X(x)$	x	$P_X(x)$
0.00	0.5000	1.10	0.8643	2.10	0.9821
0.05	0.5199	1.15	0.8749	2.15	0.9842
0.10	0.5398	1.20	0.8849	2.20	0.9861
0.15	0.5596	1.25	0.8943	2.25	0.9878
0.20	0.5793	1.30	0.9032	2.30	0.9893
0.25	0.5897	1.35	0.9115	2.326	0.9900
0.30	0.6179	1.40	0.9192	2.35	0.9906
0.35	0.6368	1.45	0.9265	2.40	0.9918
0.40	0.6554	1.50	0.9332	2.45	0.9929
0.45	0.6736	1.55	0.9394	2.50	0.9938
0.50	0.6915	1.60	0.9452	2.55	0.9946
0.55	0.7088	1.645	0.9500	2.576	0.9950
0.60	0.7257	1.65	0.9505	2.60	0.9953
0.65	0.7421	1.70	0.9554	2.65	0.9960
0.70	0.7580	1.75	0.9599	2.70	0.9965
0.75	0.7734	1.80	0.9641	2.75	0.9970
0.80	0.7881	1.85	0.9678	2.80	0.9974
0.85	0.8023	1.90	0.9713	2.85	0.9978
0.90	0.8159	1.95	0.9744	2.90	0.9981
0.95	0.8289	1.96	0.9750	2.95	0.9984
1.00	0.8413	2.00	0.9772	3.00	0.9986
1.05	0.8531	2.05	0.9798		

*$P_X(-x) = 1 - P_X(x)$.

Table 4B χ^2 Distribution*

Values of χ_α^2 for some values of α, where the probability that a random variable χ^2 having the χ^2 distribution with n degrees of freedom is not smaller than χ_α^2 is equal to α.

$$P(\chi^2 \geq \chi_\alpha^2) = \frac{1}{2^{n/2}\Gamma(n/2)} \int_{\chi_\alpha^2}^{\infty} e^{-u/2} u^{n/2-1} \, du = \alpha$$

n	α								
	0.80	0.70	0.50	0.30	0.20	0.10	0.05	0.02	0.01
1	0.064	0.148	0.455	1.074	1.642	2.706	3.841	5.412	6.635
2	0.446	0.713	1.386	2.408	3.219	4.605	5.991	7.824	9.210
3	1.005	1.424	2.366	3.665	4.642	6.251	7.815	9.837	11.345
4	1.649	2.195	3.357	4.878	5.989	7.779	9.488	11.668	13.277
5	2.343	3.000	4.351	6.064	7.289	9.236	11.070	13.388	15.086
6	3.070	3.828	5.348	7.231	8.558	10.645	12.592	15.033	16.812
7	3.822	4.671	6.346	8.383	9.803	12.017	14.067	16.622	18.475
8	4.594	5.527	7.344	9.524	11.030	13.362	15.507	18.168	20.090
9	5.380	6.393	8.343	10.656	12.242	14.684	16.919	19.679	21.666
10	6.179	7.267	9.342	11.781	13.442	15.987	18.307	21.161	23.209
11	6.989	8.148	10.341	12.899	14.631	17.275	19.675	22.618	24.725
12	7.807	9.034	11.340	14.011	15.812	18.549	21.026	24.054	26.217
13	8.634	9.926	12.340	15.119	16.985	19.812	22.362	25.472	27.688
14	9.467	10.821	13.339	16.222	18.151	21.064	23.685	26.873	29.141
15	10.307	11.721	14.339	17.322	19.311	22.307	24.996	28.259	30.578
16	11.152	12.624	15.338	18.418	20.465	23.542	26.296	29.633	32.000
17	12.002	13.531	16.338	19.511	21.615	24.769	27.587	30.995	33.409
18	12.857	14.440	17.338	20.601	22.760	25.989	28.869	32.346	34.805
19	13.716	15.352	18.338	21.689	23.900	27.204	30.144	33.687	36.191
20	14.578	16.266	19.337	22.775	25.038	28.412	31.410	35.020	37.566
21	15.445	17.182	20.337	23.858	26.171	29.615	32.671	36.343	38.932
22	16.314	18.101	21.337	24.939	27.301	30.813	33.924	37.659	40.289
23	17.187	19.021	22.337	26.018	28.429	32.007	35.172	38.968	41.638
24	18.062	19.943	23.337	27.096	29.553	33.196	36.415	40.270	42.980
25	18.940	20.867	24.337	28.172	30.675	34.382	37.652	41.566	44.314
26	19.820	21.792	25.336	29.246	31.795	35.563	38.885	42.856	45.642
27	20.703	22.719	26.336	30.319	32.912	36.741	40.113	44.140	46.963
28	21.588	23.647	27.336	31.391	34.027	37.916	41.337	45.419	48.278
29	22.475	24.577	28.336	32.461	35.139	39.087	42.557	46.693	49.588
30	23.364	25.508	29.336	33.530	36.250	40.256	43.773	47.962	50.892

*Reproduced from *Probability Theory and Mathematical Statistics* by M. Fisz, with kind permission of the publisher, John Wiley & Sons, Inc., New York.

Table 5B Student's t Distribution*

Values of t_α for some values of α, where the probability that a random variable t having Student's t distribution with n degrees of freedom is not smaller than t_α in absolute value is equal to α.

$$P(|t| \geq t_\alpha) = \frac{2}{\sqrt{n}\ B(\frac{1}{2},n/2)} \int_{t_\alpha}^{\infty} \frac{1}{(1 + t^2/n)^{(n+1)2}}\ dt = \alpha$$

n	α							
	0.80	0.60	0.40	0.20	0.10	0.05	0.02	0.01
1	0.325	0.727	1.376	3.078	6.314	12.706	31.821	63.657
2	0.289	0.617	1.061	1.886	2.920	4.303	6.965	9.925
3	0.277	0.584	0.978	1.638	2.353	3.182	4.541	5.841
4	0.271	0.569	0.941	1.533	2.132	2.776	3.747	4.604
5	0.267	0.559	0.920	1.476	2.015	2.571	3.365	4.032
6	0.265	0.553	0.906	1.440	1.943	2.447	3.143	3.707
7	0.263	0.549	0.896	1.415	1.895	2.365	2.998	3.499
8	0.262	0.546	0.889	1.397	1.860	2.306	2.896	3.355
9	0.261	0.543	0.883	1.383	1.833	2.262	2.821	3.250
10	0.260	0.542	0.879	1.372	1.812	2.228	2.764	3.169
11	0.260	0.540	0.876	1.363	1.796	2.201	2.718	3.106
12	0.259	0.539	0.873	1.356	1.782	2.179	2.681	3.055
13	0.259	0.538	0.870	1.350	1.771	2.160	2.650	3.012
14	0.258	0.537	0.868	1.245	1.761	2.145	2.624	2.977
15	0.258	0.536	0.866	1.341	1.753	2.131	2.602	2.947
16	0.258	0.535	0.865	1.337	1.746	2.120	2.583	2.921
17	0.257	0.534	0.863	1.333	1.740	2.110	2.567	2.898
18	0.257	0.534	0.862	1.330	1.734	2.101	2.552	2.878
19	0.257	0.533	0.861	1.328	1.729	2.093	2.539	2.861
20	0.257	0.533	0.860	1.325	1.725	2.086	2.528	2.845
21	0.257	0.532	0.859	1.323	1.721	2.080	2.518	2.831
22	0.256	0.532	0.858	1.321	1.717	2.074	2.508	2.819
23	0.256	0.532	0.858	1.319	1.714	2.069	2.500	2.807
24	0.256	0.531	0.857	1.318	1.711	2.064	2.492	2.797
25	0.256	0.531	0.856	1.316	1.708	2.060	2.485	2.787
26	0.256	0.531	0.856	1.315	1.706	2.056	2.479	2.779
27	0.256	0.531	0.855	1.314	1.703	2.052	2.473	2.771
28	0.256	0.530	0.855	1.313	1.701	2.048	2.467	2.763
29	0.256	0.530	0.854	1.311	1.699	2.045	2.462	2.756
30	0.256	0.530	0.854	1.310	1.697	2.042	2.457	2.750
40	0.255	0.529	0.851	1.303	1.684	2.021	2.423	2.704
60	0.254	0.527	0.848	1.296	1.671	2.000	2.390	2.660
120	0.254	0.526	0.845	1.289	1.658	1.980	2.358	2.617
∞	0.253	0.524	0.842	1.282	1.645	1.960	2.326	2.576

*Reproduced from *Probability Theory and Mathematical Statistics* by M. Fisz, with kind permission of the publisher, John Wiley & Sons, Inc., New York.

Table 6B F Distribution*

$$\Pr(F \le f) = \int_0^f \frac{\Gamma[(r_1 + r_2)/2](r_1/r_2)^{r_1/2}w^{r_1/2-1}}{\Gamma(r_1/2)\Gamma(r_2/2)((1 + r_1w/r_2)^{(r_1+r_2)/2}} \, dw$$

Pr(F≤f)	r_2	1	2	3	4	5	6 r_1	7	8	9	10	12	15
0.95	1	161	200	216	225	230	234	237	239	241	242	244	246
0.975		648	800	864	900	922	937	948	957	963	969	977	985
0.99		4052	4999	5403	5625	5764	5859	5928	5982	6023	6056	6106	6157
0.95	2	18.5	19.0	19.2	19.2	19.3	19.3	19.4	19.4	19.4	19.4	19.4	19.4
0.975		38.5	39.0	39.2	39.2	39.3	39.3	39.4	39.4	39.4	39.4	39.4	39.4
0.99		98.5	99.0	99.2	99.2	99.3	99.3	99.4	99.4	99.4	99.4	99.4	99.4
0.95	3	10.1	9.55	9.28	9.12	9.01	8.94	8.89	8.85	8.81	8.79	8.74	8.70
0.975		17.4	16.0	15.4	15.1	14.9	14.7	14.6	14.5	14.5	14.4	14.3	14.3
0.99		34.1	30.8	29.5	28.7	28.2	27.9	27.7	27.5	27.3	27.2	27.1	26.9
0.95	4	7.71	6.94	6.59	6.39	6.26	6.16	6.09	6.04	6.00	5.96	5.91	5.86
0.975		12.2	10.6	9.98	9.60	9.36	9.20	9.07	8.98	8.90	8.84	8.75	8.66
0.99		21.2	18.0	16.7	16.0	15.5	15.2	15.0	14.8	14.7	14.5	14.4	14.2
0.95	5	6.61	5.79	5.41	5.19	5.05	4.95	4.88	4.82	4.77	4.74	4.68	4.62
0.975		10.0	8.43	7.76	7.39	7.15	6.98	6.85	6.76	6.68	6.62	6.52	6.43
0.99		16.3	13.3	12.1	11.4	11.0	10.7	10.5	10.3	10.2	10.1	9.89	9.72
0.95	6	5.99	5.14	4.76	4.53	4.39	4.28	4.21	4.15	4.10	4.06	4.00	3.94
0.975		8.81	7.26	6.60	6.23	5.99	5.82	5.70	5.60	5.52	5.46	5.37	5.27
0.99		13.7	10.9	9.78	9.15	8.75	8.47	8.26	8.10	7.98	7.87	7.72	7.56
0.95	7	5.59	4.74	4.35	4.12	3.97	3.87	3.79	3.73	3.68	3.64	3.57	3.51
0.975		8.07	6.54	5.89	5.52	5.29	5.12	4.99	4.90	4.82	4.76	4.67	4.57
0.99		12.2	9.55	8.45	7.85	7.46	7.19	6.99	6.84	6.72	6.62	6.47	6.31
0.95	8	5.32	4.46	4.07	3.84	3.69	3.58	3.50	3.44	3.39	3.35	3.28	3.22
0.975		7.57	6.06	5.42	5.05	4.82	4.65	4.53	4.43	4.36	4.30	4.20	4.10
0.99		11.3	8.65	7.59	7.01	6.63	6.37	6.18	6.03	5.91	5.81	5.67	5.52
0.95	9	5.12	4.26	3.86	3.63	3.48	3.37	3.29	3.23	3.18	3.14	3.07	3.01
0.975		7.21	5.71	5.08	4.72	4.48	4.32	4.20	4.10	4.03	3.96	3.87	3.77
0.99		10.6	8.02	6.99	6.42	6.06	5.80	5.61	5.47	5.35	5.26	5.11	4.96
0.95	10	4.96	4.10	3.71	3.48	3.33	3.22	3.14	3.07	3.02	2.98	2.91	2.85
0.975		6.94	5.46	4.83	4.47	4.24	4.07	3.95	3.85	3.78	3.72	3.62	3.52
0.99		10.0	7.56	6.55	5.99	5.64	5.39	5.20	5.06	4.94	4.85	4.71	4.56
0.95	12	4.75	3.89	3.49	3.26	3.11	3.00	2.91	2.85	2.80	2.75	2.69	2.62
0.975		6.55	5.10	4.47	4.12	3.89	3.73	3.61	3.51	3.44	3.37	3.28	3.18
0.99		9.33	6.93	5.95	5.41	5.06	4.82	4.64	4.50	4.39	4.30	4.16	4.01
0.95	15	4.54	3.68	3.29	3.06	2.90	2.79	2.71	2.64	2.59	2.54	2.48	2.40
0.975		6.20	4.77	4.15	3.80	3.58	3.41	3.29	3.20	3.12	3.06	2.96	2.86
0.99		8.68	6.36	5.42	4.89	4.56	4.32	4.14	4.00	3.89	3.80	3.67	3.52

*Abridged and adapted from "Tables of Percentage Points of the Inverted Beta Distribution," *Biometrika*, 33 (1943). It is reproduced with the kind permission of the Editors of *Biometrika*, on behalf of the authors, Maxine Merrington and Catherine M. Thompson.

Index

Additive set function, 48
Additive system of sets, 47
Admissibility of Bayes' rules, 396
Admissible decision rules, 389
Algebra of sets, 35
Alternative hypothesis, 314
Analysis of variance table, 447
A posteriori probabilities, 52
Apostol, T., 135, 167
Average sampling number, 415
Axiomatic definition of
    probability, 34, 50

Bahadur, R. R., 183, 191
Basis of a vector space, 479
Bayes decision rules, 384
Bayes theorem, 52
Berger, J. O., 400
Bernoulli distribution, 59
Bernoulli, J. A., 54

Bernoulli random variable, 59,
    321
Best critical region, 317
Best unbiased estimator, 268
Beta distribution, 67
Beta integral, 67
Beta random variable, 67
Bickel, P. J., 210
Bienayme–Chebychev inequality,
    133
Binomial distribution, 188, 192,
    287
Binomial random variable, 59,
    100
Birkoff, G., 517
Bivariate normal, 73, 74, 215
Blackwell, D., 266, 312
Bochner, S., 135
Borel–Cantelli lemma, 150
Borel, E., 167

Borel function, 87, 115
Born, M., 8

Cauchy distribution, 113
Central limit theorems, 155
Central moments, 106
Characteristic functions, 108
Characteristic roots, 495
Characteristic vectors, 495
Chebychev inequality, 131
Chi distribution, 198
Chi-square distribution
    central, 192
    noncentral 198
Chi-square test, 351
Chu, Kai-Ching, 229
Chung, K. L., 54, 167
Class of sets, 46
Classical definition of probability,
    9, 15
Closed class of sets, 46
Cochran's theorem, 239
Cochran, W. G., 175, 191
Cofactor, 488
Complement of an event, 40
Complement of a set, 37
Complete family of distributions,
    269
Completely additive class of sets,
    47
Completely additive set function,
    47, 49
Composite hypothesis, 331
Compound events, 20
Compound normal, 231
Conditional distribution, 78
Conditional mass function, 79
Conditional mathematical
    expectation, 129
Conditional probability, 26
Confidence interval, 358, 362
Consistent estimator, 275, 282

Contaminated normal, 230
Contingency table, 377
Contingent event, 11
Convergence
    in distribution, 155
    in probability, 138
    with probability one, 139
Coordinates of a vector, 481
Covariance matrix, 221
Cramér, H., 135, 167, 312
Critical region, 317

Das, M. N., 463, 470, 474
David, F. N., 368
DeGroot, M. H., 400
DeMoivre–Laplace limit
    theorem, 161
Design of experiments, 174
Determinant, 487
Diagonal matrix, 487
Difference of sets, 37
Dirichlet distribution, 245
Discrete univariate random
    variable, 58
Distribution of
    bivariate continuous random
        variable, 73
    bivariate discrete random
        variable, 72
    univariate random variable, 57
Doksum, K. A. 210
Domain of a function, 45
Duality in set relation, 40

Echelon form of a matrix, 509
Edgeworth approximation, 373
Element of a set, 35
Elementary events, 9, 10
Elliptically symmetric
    distribution, 229
Equal sets, 36
Equalizer decision rule, 392

Equally likely events, 14
Equiprobability, 14
Error probabilities, 405
Estimate, 259
Estimation problem, 172
Estimator, 259
Exhaustive events, 14
Experimental units, 464
Exponential family of
    distributions, 185, 208, 271
Exponential probability
    distribution, 283
Extended Bayes decisions, 393

F-random variable, 97
F-test, 352
Favorable cases, 15
Feller, W., 54, 164, 167
Ferguson, T. S., 400
Finitely additive class of sets, 47
Finitely additive set function, 48,
    49
Fisher–Neyman factorization
    theorem, 181, 185
Fisher, R. A., 179, 191, 275, 312,
    368
Fisz, M., 135, 164, 167, 369
Fixed sample-size design, 176
Fraser, D. A. S., 431
Function of random variables,
    87, 91

Galton, F., 225, 245
Gamma random variable, 63
    distribution of, 64
    probability density function of,
    64
General linear hypothesis, 432
General linear hypothesis model,
    432
Generalized inverse, 512

Geometric random variable, 60,
    192
Ghosh, B. K., 412
Giri, N., 167, 242, 245, 312, 457,
    463, 470, 474
Glivenko–Cantelli theorem, 257
Gredenko, B. V., 164
Gram–Schmidt orthogonalization
    method, 482
Graybill, F. A., 463

Hajek, J., 431
Hajek–Renyi inequality, 147
Hausdorff, F., 54
Helly's lemma, 156
Hermite canonical form of a
    matrix, 509
History of statistical
    methodology, 178
Homogeneous system of linear
    equations, 505
Hotelling, H., 369
Hotelling's T^2-distribution, 242
Hotelling's T^2-test, 360
Hypergeometric distribution, 192,
    195
Hypothesis
    alternative, 314
    composite, 314
    null, 314
    simple, 314

Indempotent matrix, 436
Independence of
    finite sequence of random
    variables, 84
    pair of random variables, 82
Independent events, 27
Inductive inference, 169
Infinitely divisible distribution,
    208

Intersectional sets, 38
Invariance property, 304

Jacobian, 91
Johnson, R. A., 245
Joint moments
  central, 116
  raw, 116

Kempthorne, O., 455
Kolmogorov, A. N., 135
Kolmogorov–Smirnov theorem,
  258
Kruskal, W. H., 431

Laplace distribution, 192
Laplace, P. S., 8
Latin squares, 469
Latin square designs, 467
Laws of large numbers
  Bernoulli, 140
  Borel's strong law, 144
  Chebychev, 141
  Khinchin's theorem, 142
  Kolmogorov's law, 147
  Kolmogorov's theorem, 150
  Markov, 144
  Poisson, 142
  strong law, 140
  weak law, 140
Least-square estimate, 434
Least-square estimator, 435
Least-square method, 434
Lebesgue's dominated
  convergence theorem, 157
Lehmann, E. L., 183, 191, 312,
  336, 365, 369, 417, 431
Level of significance, 316
Levi, F. W., 517
Lévy–Cramér theorem, 156
Liapounov, A., 164
Likelihood ratio test, 342

Limit of a sequence of sets, 42
Lindeberg, J. W., 164
Lindeberg–Feller theorem, 164
Lindeberg–Levy theorem, 163
Lindley, D. V., 130
Linear equations, 504
Linear model, 432
Linear model of full rank, 432
Linear regression, 130
Linearly independent vectors,
  479
Loève, M., 54
Log normal, 192

Maclane, S., 517
Mann, H. B., 426, 431
Mann–Whitney test, 426
Marcus, M., 517
Matrix, 475
Matrix of regression coefficients,
  305
Matrix product, 486
Matrix of sample regression
  coefficients, 305
Mathematical expectation
  function of continuous random
    variables, 100
  function of discrete random
    variables, 99
  function of random matrix, 214
Maximum likelihood estimate,
  286
McGraw, D. K., 229, 245
Measure, 45
Measurable sets, 49
Median, 245
Median test, 421
Method of moments, 285
Method of sweep-out of a
  matrix, 504
Minimax decision rules, 340
Minor, 488

Moment-generating function, 106
Monotone likelihood ratio, 331
Monotone sequence of
  decreasing sets, 42
  increasing sets, 42
Mood, A. M., 431
Most efficient estimator, 275
Most powerful tests, 317
Multinomial distribution
  function, 235
Multiple classification problem,
  173
Multiple comparison, 457
Multiple correlation coefficient,
  226
Multiplication of probabilities, 28
Multivariate
  normal distribution, 219
  t-distribution, 231
Mutually exclusive events, 14

Negative binomial, 133
Negative definite matrix, 494
Neyman, J., 179, 369
Neyman–Pearson lemma, 317
Nonparametric problems, 173
Nonsingular matrix, 489
Null hypothesis, 314

One-way classification, 451
Operating characteristic function,
  412
Order statistic, 245
Ordered parameters, 246
Orthogonal basis, 482
Orthogonal matrix, 487
Orthogonal vectors, 478

Paired t-test, 349
Pairwise independent events, 30
Parametric problems, 173
Partial correlation coefficient, 225

Partitioned matrices, 501
Parzen, E., 54, 135
Pearson, E. S., 179, 191, 369
Pearson, K., 369, 376, 380
Pearson probability distribution,
  208
Perlis, S., 517
Point estimation, 259
Poisson
  approximation, 61
  distribution function, 172, 273
Polya distribution, 195
Positive definite matrix, 494
Posterior risk, 385
Power of a test, 316
Prior probabilities, 52
Prior probability density
  function, 384
Probability generating function,
  114
Probability mass function, 58
Probability space, 50
Projection of a vector, 478

Quadratic forms, 237
Quantiles, 246

Random events, 16
Random experiment, 9
Random variable, 55
Randomized block design, 465
Randomized test, 319
Range, 247
Range of a function, 45
Rank
  column, 490
  of a matrix, 492
  row, 490
Rank test, 428
Rao, C. R., 210, 245, 266, 312,
  380, 463
Rao–Blackwell theorem, 264

Rao–Cramér inequality, 284
Rao–Cramér lower bound, 278
Rayleigh distribution, 193, 197,
202
Rectangular
distribution, 67
random variable, 65
Regression curve, 130
Regression surface, 225
Regularity conditions, 184
Relative frequency, 6
Rice, S. O., 210
Rieman–Stieltjes
integral, 102
sum, 102

S-method, 459
Sample correlation matrix, 305
Sample covariance matrix, 305
Sample multiple correlation
coefficient, 306
Sample space, 10
Sampling distribution,
multivariate
of covariance matrix, 240
of mean, 240
Sampling distribution, univariate
of mean, 204
of variance, 204
Scheffé, H., 455, 457, 463, 470,
474
Scheffé's solution, 350
Schwarz's inequality, 132
Semi-interquantile range, 246
Sequence of sets, 41
Sequential experimental design,
176
Sequential probability ratio test,
408
Set, 35
Set function, 45
Sidak, Z., 431

Siegel, S., 431
Sign test, 418
Simple random sampling, 175
Simultaneous confidence interval,
459
Singular matrix, 489
Size of a test, 316
Skew-symmetric matrix, 484
Spherically symmetric
distributions, 230
Square error loss, 382
Statistic, 137
Statistical inference problem, 174
Stein estimator, 310
Stirling's approximation, 255
Stopping rule, 407
Stopping variable, 403
Stratified random sampling, 175
Stratified sampling theorem, 189
Strong consistency, 282
Strong convergence, 139
Student's t-test, 344
Sufficient statistic, 181
Symmetric matrix, 484
Symmetric normal dsitribution,
243

t-distribution
central, 193
noncentral, 193
Test of equality of two means,
346
Test of goodness of fit, 375
Test of zero correlation, 355
Theorem of compound
probability, 18
Theorem of total probability, 18
Trace of a matrix, 493
Transpose of a matrix, 484
Triangular matrix, 487
Tukey, J. R., 184, 191
Two-sample methods, 420

Two-sample sign test, 420
Two-way classification, 454

Unbiased estimator, 260
Unbiased test, 335
Uniform distribution, 67
Uniformly most powerful test, 328
Union of sets, 37
Unit normal distribution, 63
Universal set, 37
Uspensky, J. V., 54

Vandermonde's matrix, 295
Variance stabilizing transformations, 374
Vector, 476

Venn diagram, 38
Von Mises, R., 8

Wagner, J. F., 229, 245
Wald, A., 180, 191, 402, 412, 417
Wald's fundamental identity, 413
Wald–Wolfowitz's run test, 423
Weak consistency, 282
Weak convergence, 138
Weibull random variable, 193
Whitney, D. R., 426, 431
Wichern, D. W., 245
Wilcoxon test, 426
Wilks, S. S. 417
Wishart distribution, 241

Zacks, S., 312, 412, 417
Zellner, A., 229, 245